Operational Calculus and Related Topics

ANALYTICAL METHODS AND SPECIAL FUNCTIONS

An International Series of Monographs in Mathematics

FOUNDING EDITOR: A.P. Prudnikov (Russia)
SERIES EDITORS: C.F. Dunkl (USA), H.-J. Glaeske (Germany) and M. Saigo (Japan)

Volume 1
Series of Faber Polynomials, *P.K. Suetin*

Volume 2
Inverse Spectral Problems for Linear Differential Operators
and Their Applications, *V.A. Yurko*

Volume 3
Orthogonal Polynomials in Two Variables, *P.K. Suetin*

Volume 4
Fourier Transforms and Approximations, *A.M. Sedletskii*

Volume 5
Hypersingular Integrals and Their Applications, *S. Samko*

Volume 6
Methods of the Theory of Generalized Functions, *V.S. Vladimirov*

Volume 7
Distributions, Integral Transforms and Applications, *W. Kierat
and U. Sztaba*

Volume 8
Bessel Functions and Their Applications, *B. Korenev*

Volume 9
H-Transforms: Theory and Applications, *A.A. Kilbas and M. Saigo*

Volume 10
Operational Calculus and Related Topics, *H.-J. Glaeske,
A.P. Prudnikov, and K.A. Skórnik*

Operational Calculus and Related Topics

H.-J. Glaeske

*Friedrich-Schiller University
Jena, Germany*

A.P. Prudnikov

(Deceased)

K.A. Skòrnik

*Institute of Mathematics
Polish Academy of Sciences
Katowice, Poland*

CRC Press
Taylor & Francis Group
Boca Raton London New York

CRC Press is an imprint of the
Taylor & Francis Group, an **informa** business

A CHAPMAN & HALL BOOK

CRC Press
Taylor & Francis Group
6000 Broken Sound Parkway NW, Suite 300
Boca Raton, FL 33487-2742

First issued in paperback 2019

© 2006 by Taylor & Francis Group, LLC
CRC Press is an imprint of Taylor & Francis Group, an Informa business

No claim to original U.S. Government works

ISBN-13: 978-1-58488-649-5 (hbk)
ISBN-13: 978-0-367-39049-5 (pbk)

Library of Congress Card Number 2006045622

Library of Congress Cataloging-in-Publication Data

Glaeske, Hans-Jürgen.
    Operational calculus and related topics / Hans-Jürgen Glaeske, Anatoly P. Prudnikov, Krystyna A. Skórnik.
        p. cm. -- (Analytical methods and special functions ; 10)
    ISBN 1-58488-649-8 (alk. paper)
    1. Calculus, Operational. 2. Transformations (Mathematics) 3. Theory of distributions (Functional analysis) I. Prudnikov, A. P. (Anatolii Platonovich) II. Skórnik, Krystyna. III. Title. IV. Series.

QA432.G56 2006
515'.72--dc22   2006045622

Visit the Taylor & Francis Web site at
http://www.taylorandfrancis.com

and the CRC Press Web site at
http://www.crcpress.com

In memory of

Λ. P. Prudnikov

January 14, 1927 — January 10, 1999

Contents

Preface

The aim of this book is to provide an introduction to operational calculus and related topics: integral transforms of functions and generalized functions. This book is a cross between a textbook for students of mathematics, physics and engineering and a monograph on this subject. It is well known that integral transforms, operational calculus and generalized functions are the backbone of many branches of pure and applied mathematics. Although centuries old, these subjects are still under intensive development because they are useful in various problems of mathematics and other disciplines. This stimulates continuous interest in research in this field.

Chapter 1 deals with integral transforms (of functions), historically the first method to justify Oliver Heaviside's (algebraic) operational calculus in the first quarter of the twentieth century. Methods connected with the use of integral transforms have gained wide acceptance in mathematical analysis. They have been sucessfully applied to the solution of differential and integral equations, the study of special functions, the evaluation of integrals and the summation of series.

The sections deal with conditions for the existence of the integral transforms in consideration, inversion formulas, operational rules, as for example, differentiation rule, integration rules and especially the definition of a convolution $f * g$ of two functions f and g, such that for the transform $\mathfrak{T}$ it holds that

$$\mathfrak{T}[f * g] = \mathfrak{T}[f] \cdot \mathfrak{T}[g].$$

Sometimes applications are given. Because of the special nature of this book some extensive proofs are only sketched. The reader interested in more detail is referred for example to the textbooks of R.V. Churchill [CH.2], I.W. Sneddon [Sn.2], and A.H. Zemanian [Ze.1]. Short versions of many integral transforms can be found in A.I. Zayed's handbook *Function and Generalized Function Transformations* [Za]. For tables of integral transforms we refer to [EMOT], [O.1]-[O.3], [OB], [OH], and [PBM], vol. IV, V.

In this book we deal only with integral transforms for $\mathbb{R}^1$ - functions. The reader interested in the multidimensional case is referred to [BGPV].

In Chapter 2 (algebraic) operational calculus is considered. This complete return to the original operator point of view of Heaviside's operational calculus was done by Jan Mikusiński; see [Mi.7]. He provided a strict operator basis without any references to the theory of the Laplace transform. His theory of convolution quotients provides a clear and simple basis for an operational calculus. In contrast to the definition of the multiplication of functions f and g, continuous on $[0, \infty)$ given by J. Mikusiński,

$$(1) \qquad (f * g)(t) = \int_0^t f(x)\, g(t - x)\, dx,$$

in Chapter 2 functions with a continuous derivative on $[0, \infty)$ are considered and the multiplication is defined by means of

$$(2) \qquad (f * g)(t) = \frac{d}{dt} \int_0^t f(x)\, g(t - x)\, dx.$$

Both definitions have advantages and disadvantages. Some formulas are simpler in the one case, otherwise in the case of definition (2). In the case of definition (2) the $*$-product of two functions constant on $[0, \infty)$,

$$f(x) = a, \quad g(x) = b, \quad x \in [0, \infty)$$

equals a function h with $h(x) = ab, \quad x \in [0, \infty)$, such that the $*$-product of two numbers equals their usual product. In the case of definition (1) this product equals abt. In both cases the field of operators generated by the original space of functions is the same; the field of Mikusiński operators. For our version of the starting point we refer to L. Berg, [Be.1] and [DP]. After an introduction a proof of Titchmarsh's theorem is given. Then the operator calculus is derived and the basis of the analysis of operators is developed. Finally, applications to the solution of ordinary and partial differential equations are given.

Chapter 3 consists of the theory of generalized functions. Various investigations have been put forward in the middle of the last century. The mathematical problems encountered are twofold: first, to find an analytical interpretation for the operations performed and to justify these operations in terms of the interpretation and, second, to provide an adequate theory of Dirac's, δ-function, which is frequently used in physics. This "function" is often defined by means of

$$\delta(x) = 0, \quad x \neq 0, \quad \int_{-\infty}^{+\infty} \delta(x)\, \varphi(x)\, dx = \varphi(0),$$

for an arbitrary continuous function φ. It was introduced by the English physicist Paul Dirac in his quantum mechanics in 1927; see [Dir]. It was soon pointed out that from the purely mathematical point of view this definition is meaningless. It was of course clear to Dirac himself that the δ-function is not a function in the classical meaning and, what is important, that it operates as an operator (more precisely as a functional), that related to each continuous function φ its value at the point zero, $\varphi(0)$; see Laurent Schwartz [S.2]. Similar to the case of operational calculus of Chapter 2, J. Mikusiński together with R. Sikorski developed an elementary approach to generalized functions, a so-called sequential approach; see [MiS.1] and [AMS]. They did not use results of functional analysis, but only basic results of algebra and analysis. In Chapter 3 we follow this same line.

Because this book is not a monograph, the reference list at the end of the book is not complete.

We assume that the reader is familiar with the elements of the theory of algebra and analysis. We also assume a knowledge of the standard theorems on the interchange of limit processes. Some knowledge of Lebesgue integration, such Fubini's theorem, is necessary because integrals are understood as Lebesgue integrals. Finally, the reader should be familiar with the basic subject matter of a one-semester course in the theory of functions of a complex variable, including the theory of residues. Formulas for special functions are taken from textbooks on special functions, such as [E.1], [PBM] vols. I-III, [NU], and [Le].

The advantage of this book is that both the analytical and algebraic aspects of operational calculus are considered equally valuable. We hope that the most important topics of this book may be of interest to mathematicians and physicists interested in application-relevant questions; scientists and engineers working outside of the field of mathematics who apply mathematical methods in other disciplines, such as electrical engineering; and undergraduate- and graduate-level students researching a wide range of applications in diverse areas of science and technology.

The idea for this book began in December 1994 during A.P. Prudnikov's visit to the Mathematical Institute of the Friedrich Schiller University in Jena, Germany. The work was envisioned as the culmination of a lengthy collaboration. Unfortunately, Dr. Prudnikov passed away on January 10, 1999. After some consideration, we decided to finish our joint work in his memory. This was somewhat difficult because Dr. Prudnikov's work is very extensive and is only available in Russian. We were forced to be selective. We hope that our efforts accurately reflect and respect the memory of our colleague.

Hans-Juergen Glaeske and Krystyna A. Skórnik

List of Symbols

xvi

Chapter 1

Integral Transforms

1.1 Introduction to Operational Calculus

In the nineteenth century mathematicians developed a "symbolic calculus," a system of rules for calculation with the operator of differentiation $D := \frac{d}{dt}$.

The papers of Oliver Heaviside (1850–1925) were instrumental in promoting operational calculus methods. Heaviside applied his calculus in the solution of differential equations, especially in the theory of electricity. He had a brilliant feel for operator calculus, but because he did not consider the conditions for the validity of his calculations his results were sometimes wrong. Heaviside published his results in some papers about operators in mathematical physics in 1892–1894 and also in his books *Electrical Papers* (1892) and *Electromagnetic Theory* (1899); see [H.1]–[H.3].

Heaviside used the operator D and calculated with D in an algebraic manner, defining

$$D^0 := I, \quad D^k := \frac{d^k}{dt^k}, \quad k \in \mathbb{N},$$

where I is the identity. This method seems to be clear because of the following rules of calculus

$$D(cf)(t) = cDf(t) \tag{1.1.1}$$

$$D(f + g)(t) = Df(t) + Dg(t), \tag{1.1.2}$$

$$D^k(D^l f)(t) = D^{k+l} f(t), \quad k, l \in \mathbb{N}_o. \tag{1.1.3}$$

If one replaces derivatives in differential equations by means of the operator D, then there certain functions of D appear. Because of (1.1.1) through (1.1.3) it is easy to understand the meaning of $P_n(D)$, where P_n is a polynomial of degree $n \in \mathbb{N}_o$. Because of

$$f(t + h) = \sum_{n=0}^{\infty} \frac{h^n}{n!} f^{(n)}(t) = \sum_{n=0}^{\infty} \frac{(hD)^n}{n!} f(t) =: e^{hD} f(t)$$

one can interpret the operator e^{hD} as the translation operator:

$$e^{hD} f(t) = f(t + h).$$

According to the rules of algebra one has to define $D^{-1} := \frac{1}{D}$ as

$$D^{-1}f(t) = \int_0^t f(\tau)d\tau.$$

Example 1.1.1 *Look for a solution of*

$$y'(t) + y(t) = t^2. \tag{1.1.4}$$

The solution of this first-order linear differential equation is well known as

$$y(t) = c\,e^{-t} + (t^2 - 2t + 2), \tag{1.1.5}$$

with some arbitrary constant c. By means of the operator D, equation (1.1.4) can be rewritten as

$$(1 + D)y = t^2$$

and therefore,

$$y(t) = \frac{1}{1 + D}\, t^2. \tag{1.1.6}$$

There are various interpretations of $\frac{1}{1+D}$. Quite formally one has, for example,

$$\frac{1}{1 + D} = 1 - D + D^2 \mp \cdots,$$

and applying the right-hand side to (1.1.6) one has

$$y(t) = t^2 - 2t + 2;$$

this is the solution (1.1.5) of (1.1.4) with $y(0) = 2$. On the other hand one can write

$$\frac{1}{1 + D} = \frac{1}{D\left(1 + \frac{1}{D}\right)} = \frac{1}{D} - \frac{1}{D^2} \pm \cdots.$$

Interpreting

$$\frac{1}{D^k}f = \frac{1}{D}\left(\frac{1}{D^{k-1}}f\right), \qquad k \in \mathbb{N}$$

as k-time integration of f from 0 to t we obtain from (1.1.6)

$$
\begin{aligned}
y(t) &= \frac{t^3}{3} - \frac{t^4}{3 \cdot 4} + \frac{t^5}{3 \cdot 4 \cdot 5} \mp \cdots \\
&= -2\left[\frac{t^0}{0!} - \frac{t^1}{1!} + \frac{t^2}{2!} - \frac{t^3}{3!} \pm \cdots\right] + t^2 - 2t + 2 \\
&= -2e^{-t} + t^2 - 2t + 2,
\end{aligned}
$$

and this is the solution (1.1.5) of (1.1.4) with $y(0) = 0$. So depending on the interpretation of the expression $\frac{1}{1+D}$, one obtains different solutions of (1.1.4). In this manner one can develop an elementary legitimate calculus for the solution of linear ordinary differential equations with constant coefficients. Let us look for a solution of

$$L_n[y](t) := y^{(n)}(t) + a_1 y^{(n-1)}(t) + \cdots + a_n y(t) = h(t)$$

$$a_k \in \mathbb{R}, \qquad k = 0, 1, \cdots, n, \tag{1.1.7}$$

with the initial value conditions

$$y(0) = y'(0) = \cdots = y^{(n-1)}(0) = 0. \tag{1.1.8}$$

Setting $L_n(D) = D^n + a_1 D^{n-1} + \cdots + a_n$ one has

$$L_n(D)y(t) = h(t)$$

or

$$y(t) = \frac{1}{L_n(D)} h(t).$$

For the interpretation of $\frac{1}{L_n(D)}$ we start with $L_1(D) = D$. Then we have the equation

$$Dy = h$$

and

$$y(t) = \frac{1}{D} h(t) = \int_0^t h(\tau) d\tau. \tag{1.1.9}$$

In the case of $L_1(D) = D - \lambda$ we have $(D - \lambda)y = h$ or

$$e^{\lambda t} D \left(e^{-\lambda t} y \right) = (D - \lambda) y = h.$$

Therefore,

$$D \left(e^{-\lambda t} y(t) \right) = e^{-\lambda t} h(t),$$

and according to (1.1.9),

$$y(t) = e^{\lambda t} \int_0^t e^{-\lambda \tau} h(\tau) d\tau =: \frac{1}{D - \lambda} h(t) \tag{1.1.10}$$

Similarly one can perform the case of a polynomial $L_n(\lambda)$ with a degree $n > 1$ and n zeros λ_j, $j = 1, 2, \cdots, n$, where $\lambda_i \neq \lambda_j$ if $i \neq j$. Then

$$L_n(\lambda) = (\lambda - \lambda_1)(\lambda - \lambda_2) \cdots (\lambda - \lambda_n),$$

and

$$\frac{1}{L_n(\lambda)} = \frac{A_1}{\lambda - \lambda_1} + \frac{A_2}{\lambda - \lambda_2} + \cdots + \frac{A_n}{\lambda - \lambda_n}$$

with

$$A_k = \frac{1}{L_n'(\lambda_k)}, \qquad k = 1, 2, \cdots, n.$$

4 *Integral Transforms*

From (1.1.7) we obtain, applying (1.1.10)

$$y(t) = \sum_{k=1}^{n} \frac{A_k}{D - \lambda_k} h(t) = \sum_{k=1}^{n} A_k e^{\lambda_k t} \int_0^t e^{-\lambda_k \tau} h(\tau) d\tau. \tag{1.1.11}$$

One can easily verify that (1.1.11) is the solution of (1.1.7) with vanishing initial values (1.1.8).

This method can also be extended to polynomials L_n with multiple zeros.

Problems arose applying this method to partial differential equations. Then one has to "translate" for example functions of the type $D^{n-1/2}$, $n \in \mathbb{N}$ or $e^{-x\sqrt{D}}$. Heaviside gave a translation rule for such functions in the so-called "Expansion Theorem." The solutions often took the form of asymptotic series, often better suited for applications than convergent series. Sometimes incorrect results appeared because conditions for the validity were missing. In Heaviside's opinion:

"It is better to learn the nature of and the application of the expansion theorem by actual experience and practice."

There were various attempts to justify Heaviside's quite formal operational methods. At the beginning of the twentieth century mathematicians such as Wagner (1916), Bromwich (1916), Carson (1922), and Doetsch used a combination of algebraic and analytic methods. They used two different spaces: A space of originals f and a space of images F, connected with the so-called Laplace transform (see section 1.4)

$$F(p) = \mathcal{L}[f](p) = \int_0^\infty e^{-pt} f(t) dt, \qquad p \in \mathbb{C}, \tag{1.1.12}$$

provided that the integral exists. Integrating by part one has

$$\mathcal{L}[f'](p) = pF(p) - f(0). \tag{1.1.13}$$

This formula can be extended to higher derivatives. So Heaviside's "mystique" multiplication with the operator D is replaced by the multiplication of the image F with the complex variable p. From (1.1.13) we see that nonvannishing initial values also can be taken into consideration. In the space of images the methods of the theory of functions of a complex variable can be used. Of course one needs a formula for the transform of the images into the space of originals. This is explained in section 1.4. The disadvantage of this method is that it is a mixture of analysis and algebra. Because of the convergence of the integral (1.1.12) quite unnatural restrictions appear. So, for example, $\mathcal{L}[e^{t^2}]$ does not exist, and Dirac's δ also cannot be included in this theory. Nevertheless, the Laplace transform was used and is still used today in many applications in electrotechnics, physics, and engineering. In the following, similar to the Laplace transform, many other integral transforms are investigated and constructed for the solution of linear differential equations with respect to special differential operators of first or second order.

A radical return to the algebraic methods was given by J. Mikusiński. His theory is free of the convergence–restrictions of integral transforms, and Dirac's δ appears as a special operator of the field of Mikusiński operators. This is explained in Chapter 2.

Chapter 3 introduces spaces of generalized functions. Their elements have derivatives of arbitrary order and infinite series can be differentiated term-wise. Moreover, they include subspaces of "ordinary" functions. They are linear spaces in which a multiplication of its elements, called convolution "$*$" is defined, such that, for example,

$$(D^n \delta) * f = D^n f$$

is valid. So one again has an operational calculus for the solution of linear differential equations with constant coefficients.

1.2 Integral Transforms – Introductory Remarks

In Chapter 1 we deal with (one-dimensional) linear integral transforms. These are mappings of the form

$$F(x) = \mathfrak{T}[f](x) = \int_a^b f(t) K(x,t) dt. \qquad (1.2.1)$$

Here K is some given kernel, $f : \mathbb{R} \to \mathbb{C}$ is the original function and F is the image of f under the transform $\mathfrak{T}$. Sometimes x belongs to an interval on the real line, sometimes it belongs to a domain in the complex plane $\mathbb{C}$. In these cases the transform $\mathfrak{T}$ is called a continuous transform; see sections 1.3 through 1.9. If the domain of definition of the images F is a subset of the set of integers $\mathbb{Z}$ the transform $\mathfrak{T}$ is sometimes called discrete, sometimes finite; see section 1.10. We prefer the latter. Sometimes the variable of the images appears in the kernel as an index of a special function. Yakubovich [Ya] called these transforms index transforms. Index transforms can be continuous transforms (see sections 1.8.3 and 1.9) or finite transforms (see section 1.10).

In the following chapters we deal with transforms of interest for applications in mathematical physics, engineering, and mathematics. The kernels $K(x,t)$ "fall down from heaven," since otherwise the sections would become too voluminous. The kernels can be determined by means of the differential operators in which one is interested. For example, to find a kernel for the operator D with $Df = f'$ on $\mathbb{R}_+$ one has

$$\mathcal{F}[Df](x) = \int_0^\infty f'(t) K(x,t) dt = [f(t) K(x,t)]_0^\infty - \int_0^\infty f(t) \frac{\partial}{\partial t} K(x,t) dt.$$

To obtain a kernel K such that the operation of differentiation is transformed into multiplication with the variables of images one can choose

$$\frac{\partial}{\partial t}K(x,t) = -xK(x,t)$$

and

$$\lim_{t\to 0+} K(x,t) = 1, \qquad \lim_{t\to +\infty} K(x,t) = 0.$$

A special solution is

$$K(x,t) = e^{-xt}, \quad x,t \in \mathbb{R}_+,$$

and so we derived the kernel of the Laplace transform (1.1.12), with the differentiation rule (1.1.13). This transform is considered in detail in section 1.4.

Another problem is as follows: Let $u(x,y)$ be a solution of the Laplace equation on the upper half plane

$$\triangle_2 u(x,y) = u_{xx}(x,y) + u_{yy}(x,y) = 0, \qquad x \in \mathbb{R}, y \in \mathbb{R}_+ \qquad (1.2.2)$$

with the boundary conditions

$$u(x,0) = e^{i\xi x}, \qquad \xi, x \in \mathbb{R}, \qquad (1.2.3)$$

$$\lim_{|x|,y\to +\infty} u(x,y) = 0. \qquad (1.2.4)$$

One can easily verify that

$$u(x,y) = e^{i\xi x - |\xi| y}$$

is a solution of the problem. To solve the problem under a more general condition than (1.2.3)

$$u(x,0) = f(x) \qquad (1.2.5)$$

one can choose the superposition principle since $\triangle_2$ is a linear differential operator. This leads to the attempt to set

$$u(x,y) = \int_{-\infty}^{\infty} F(\xi)e^{i\xi x - |\xi| y} d\xi \qquad (1.2.6)$$

for some function F. Condition (1.2.5) yields

$$f(x) = \int_{-\infty}^{\infty} F(\xi)e^{i\xi x} d\xi, \quad x \in \mathbb{R}. \qquad (1.2.7)$$

This is an integral equation for the function F and the solution leads to an integral transform

$$F = \mathfrak{T}[f]. \qquad (1.2.8)$$

The formulas (1.2.8) and (1.2.7) are a pair consisting of an integral transform and its inversion. In this special example we have the Fourier transform, investigated in section 1.3.

Readers interested in the derivation of the kernel of an integral transform are referred to Sneddon [Sn.2], Churchill [Ch.2], and especially to [AKV].

The sections that follow start with the definition of a transform, conditions of the existence, inversion formulas and operational rules for the application of the transforms, such as differentiation rules considered in the examples above. A convolution theorem plays an important part. Here a relation $f, g \to f * g$ has to be defined such that

$$\mathfrak{T}[f * g] = \mathfrak{T}[f] \cdot \mathfrak{T}[g].$$

All these operational rules are derived under relatively simple conditions, since in applications one has to use the rules in the sense of Heaviside; see section 1.1. One applies the rule, not taking note of the conditions of their validity (pure formally), and afterward one has to verify the result and state the conditions under which the formally derived solution solves this problem. Here often the conditions are much less restrictive than the set of conditions for the validity of the operational rules that have been used for the calculation of the solution.

Remark 1.2.1 *For every transformation there is a special definition of the convolution. Because there are few unique signs "*" sometimes the same sign is used for different transforms and therefore for different convolutions. In this case this sign is valid for the transform discussed in the section under consideration. If in such a section the convolution of another transform is used, then we will make additional remarks.*

Notations. In the following, $\mathbb{N}$ is the set of natural numbers $\mathbb{N} = \{1, 2, 3, \cdots\}$, $\mathbb{N}_o = \mathbb{N} \cup \{0\}$, $\mathbb{Z}$ is the set of integers, $\mathbb{Q}$ the field of rational numbers, $\mathbb{R}$ the field of real numbers, $\mathbb{R}_+$ the set of positive real numbers, $\bar{\mathbb{R}}_+ = \mathbb{R}_+ \cup \{0\}$, and $\mathbb{C}$ is the set of complex numbers. All other notations are defined, when they first appear; see also the "List of Symbols" at the beginning of this volume.

1.3   The Fourier Transform

1.3.1   Definition and Basic Properties

Definition 1.3.1 *The Fourier transform (FT) of a function $f : \mathbb{R} \to \mathbb{C}$ is the function $f^\wedge$ defined by*

$$f^\wedge(\tau) = \mathcal{F}[f](\tau) = \int\limits_{-\infty}^{\infty} f(t)e^{-i\tau t}dt, \qquad \tau \in \mathbb{R}, \tag{1.3.1}$$

provided that the integral exists.

Remark 1.3.2 *Instead of the kernel $e^{-i\tau t}$ sometimes $e^{i\tau t}$, $e^{-2\pi i \tau t}$, $(2\pi)^{-1/2}e^{\pm i\tau t}$ are chosen and in certain instances these kernels are more convenient.*

Remark 1.3.3 *The convergence of the integral (1.3.1) can be considered in a different manner: As pointwise convergence, as uniformely convergence, in the sense of the principal value of Cauchy, in the sense of L_p–spaces or others.*

We consider the Fourier transform in the space

$$L_1(\mathbb{R}) = L_1 = \{f : f \text{ measurable on } \mathbb{R}, \qquad \|f\|_1 = \int\limits_{-\infty}^{\infty} |f(t)|dt < \infty\}.$$

The space L_1 is obviously suited as the space of originals for the Fourier transform. The Fourier transforms of L_1–functions are proved to belong to the space

$$\mathcal{C}(\mathbb{R}) = \mathcal{C} = \{f : f \text{ continuous on } \mathbb{R}, \qquad \|f\| = \sup_{t \in \mathbb{R}} |f(t)| < \infty\}.$$

Theorem 1.3.1 *Let $f \in L_1$, then $f^\wedge = \mathcal{F}[f] \in \mathcal{C}$. The FT is a continuous linear transformation, i.e.,*

$$\mathcal{F}[\alpha f + \beta g] = \alpha f^\wedge + \beta g^\wedge, \quad \alpha, \beta \in \mathbb{C}, \quad f, g \in L_1 \tag{1.3.2}$$

and if a sequence $(f_n)_{n \in \mathbb{N}}$ is convergent with the limit f in L_1 then the sequence $(f_n^\wedge)_{n \in \mathbb{N}}$ of their Fourier transforms is convergent with the limit $f^\wedge$ in $\mathcal{C}$.

Proof.   We have

$$|f^\wedge(\tau)| \leq \int\limits_{-\infty}^{\infty} |f(t)|dt = \|f\|_1$$

and, therefore, there exists $\|f^\wedge\| = \sup_{\tau \in \mathbb{R}} |f^\wedge(\tau)|$. If $h \in \mathbb{R}$ and $T > 0$ then

$$|f^\wedge(\tau + h) - f^\wedge(\tau)| \le \int_{-\infty}^{\infty} |e^{-iht} - 1| \, |f(t)| dt$$

$$\le \int_{-T}^{T} |e^{-iht} - 1| \, |f(t)| dt + 2 \left(\int_{-\infty}^{-T} |f(t)| dt + \int_{T}^{\infty} |f(t)| dt \right) < \varepsilon,$$

since $|e^{-iht} - 1|$ becomes arbitrarily small if $|h|$ is sufficient small and the last two integrals become arbitrary small if T is sufficiently large. As such we have $f^\wedge \in \mathcal{C}$. The *FT* is obviously linear. From

$$\|f^\wedge - f_n^\wedge\| = \sup_{\tau \in \mathbb{R}} |(f - f_n)^\wedge(\tau)| \le \|f - f_n\|_1$$

we obtain the continuity of the *FT*. ⬜

Example 1.3.2 *If $f \in L_1$ then $f^\wedge \in \mathcal{C}$, but the image $f^\wedge$ must not belong to L_1. Let*

$$f(t) = \begin{cases} 1, & |t| \le 1 \\ 0, & |t| > 1. \end{cases}$$

Then $f \in L_1$, but

$$f^\wedge(\tau) = \int_{-1}^{1} e^{-i\tau t} dt = 2\tau^{-1} \sin \tau$$

does not belong to L_1. But there holds

Theorem 1.3.2 *Let $f \in L_1$. Then $f^\wedge(\tau) = \mathcal{F}[f](\tau)$ tends to zero as τ tends to $\pm\infty$.*

Proof.

Step **1.** Let f be the characteristic function of an interval $[a,b]$, $-\infty < a < b < \infty$, i.e.,

$$f(t) = \chi_{[a,b]}(t) = \begin{cases} 1, & t \in [a,b] \\ 0, & t \in \mathbb{R} \setminus [a,b]. \end{cases}$$

Then we have

$$f^\wedge(\tau) = \int_{a}^{b} e^{-i\tau t} dt = i \frac{e^{-ib\tau} - e^{-ia\tau}}{\tau}$$

and $f^\wedge(\tau)$ tends to zero as $\tau \to \pm\infty$.

Step **2.** Let f be a "simple function," i.e.,

$$f(t) = \sum_{j=1}^{n} \alpha_j \chi_{[a_j,b_j]}, \qquad \alpha_j \in \mathbb{C}, \quad j = 1, 2, \ldots, n,$$

where the intervals $[a_j, b_j]$ are disjointed. Then

$$f^{\wedge}(\tau) = \sum_{j=1}^{n} \frac{i}{\tau} \alpha_j (e^{-ib_j\tau} - e^{-ia_j\tau})$$

and $f^{\wedge}(\tau)$ tends also to zero as $\tau \to \pm\infty$.

Step **3**. The set of simple functions are dense in L_1. Therefore, for every $\varepsilon > 0$ there exists a simple function f_o such that

$$\|f - f_0\|_1 < \varepsilon/2,$$

and there exists a number $T > 0$ such that

$$|f_o^{\wedge}(\tau)| < \varepsilon/2, \qquad |\tau| > T$$

according to step 2.

Therefore,

$$|f^{\wedge}(\tau)| = |(f - f_o)^{\wedge}(\tau) + f_o^{\wedge}(\tau)| \leq |(f - f_o)^{\wedge}(\tau)| + |f_o^{\wedge}(\tau)|$$
$$\leq \|f - f_0\|_1 + |f_o^{\wedge}(\tau)| < \frac{\varepsilon}{2} + \frac{\varepsilon}{2} = \varepsilon$$

if $|\tau| > T$. $\square$

Remark 1.3.4 *Not every function g, continuous on $\mathbb{R}$, uniformly bounded with $g(\tau) \to 0$ as $\tau \to \pm\infty$ is an image of an L_1–function f under the FT. One can prove that the function g which is defined by means of*

$$g(\tau) = \begin{cases} 1/\log \tau, & \tau > e \\ \tau/e, & 0 \leq \tau \leq e \end{cases}$$

and $g(-\tau) = -g(\tau)$, is not a FT of a function $f \in L_1$ (see [Ob.], pp. 22–24).

Remark 1.3.5 *Let $\mathcal{C}_o$ be the Banach space*

$$\mathcal{C}_o(\mathbb{R}) = \mathcal{C}_o = \{f : f \in \mathcal{C} : \lim_{\tau \to \pm\infty} f(\tau) = 0\}.$$

Then because of Remark 1.3.4 *we have:*

Theorem 1.3.3 *The FT is a continuous linear mapping of L_1 into $\mathcal{C}_o$.*

Finally we obtain by straightforward calculation:

Proposition 1.3.1 *If f is even (respectively odd), then $f^{\wedge}$ is even (respectively odd) and we have*

$$f^{\wedge}(-\tau) = f^{\wedge}(\tau) = 2 \int_0^{\infty} f(t) \cos \tau t \, dt \qquad (1.3.3)$$

and

$$f^{\wedge}(-\tau) = -f^{\wedge}(\tau) = -2i \int_0^{\infty} f(t) \sin \tau t \, dt. \tag{1.3.4}$$

The integrals in (1.3.3) respectively (1.3.4) are called the Fourier–cosine respectively Fourier–sine transform:

$$\mathcal{F}_c[f](\tau) = \int_0^{\infty} f(t) \cos \tau t \, dt, \qquad \tau > 0 \tag{1.3.5}$$

$$\mathcal{F}_s[f](\tau) = \int_0^{\infty} f(t) \sin \tau t \, dt, \qquad \tau > 0. \tag{1.3.6}$$

1.3.2 Examples

Example 1.3.3 *Let 1_+ be Heaviside's step function:*

$$1_+(t) = \begin{cases} 1 & \text{if } t > 0 \\ 0 & \text{if } l < 0. \end{cases} \tag{1.3.7}$$

Then we obtain

$$\mathcal{F}[1_+(T - |t|)](\tau) = 2\frac{\sin T\tau}{\tau}, \qquad T > 0. \tag{1.3.8}$$

Example 1.3.4 *Let $\alpha > 0$. Then we have*

$$\int_{-\infty}^{\infty} e^{-[\alpha|t| + i\tau t]} dt = \int_0^{\infty} [e^{-(\alpha - i\tau)t} + e^{-(\alpha + i\tau)t}] dt = \frac{1}{\alpha - i\tau} + \frac{1}{\alpha + i\tau} = \frac{2\alpha}{\tau^2 + \alpha^2},$$

i.e.,

$$\mathcal{F}[e^{-\alpha|t|}](\tau) = \frac{2\alpha}{\tau^2 + \alpha^2}, \qquad \alpha > 0. \tag{1.3.9}$$

Example 1.3.5 *Using the Fresnel integral*

$$\int_{-\infty}^{\infty} e^{ix^2} dx = \sqrt{\pi} e^{\pi i/4} \tag{1.3.10}$$

(see [PBM], vol. I, 2.3.15, 2) we obtain

$$\int_{-\infty}^{\infty} e^{it^2 - i\tau t} dt = \lim_{a,b \to \infty} \int_{-a}^{b} e^{i(t - \frac{\tau}{2})^2 - \frac{i}{4}\tau^2} dt = e^{-\frac{i}{4}\tau^2} \int_{-\infty}^{\infty} e^{ix^2} dx = \sqrt{\pi} e^{-\frac{i}{4}(\tau^2 - \pi)},$$

i.e.,

$$\mathcal{F}[e^{it^2}](\tau) = \sqrt{\pi} e^{-\frac{i}{4}(\tau^2 - \pi)}. \tag{1.3.11}$$

Remark 1.3.6 *The original* $\exp(it^2)$ *does not belong to* L_1. *So one should not wonder that the right-hand side of* (1.3.11) *tends not to zero as* τ *tends to* $\pm\infty$.

Example 1.3.6 *Let* $\alpha > 0$. *Then we have*

$$\mathcal{F}[e^{-\alpha^2 t^2}]\tau = \alpha^{-1} \int_{-\infty}^{\infty} e^{-(x^2 + \frac{i\tau}{\alpha}x)}\,dx = \alpha^{-1} e^{-\tau^2/4\alpha^2} \int_{-\infty}^{\infty} e^{-(x + \frac{i\tau}{2\alpha})^2}\,dx$$

$$= \alpha^{-1} e^{-\tau^2/4\alpha^2} \int_{-\infty + i\tau/2\alpha}^{i\tau/2\alpha + \infty} e^{-z^2}\,dz.$$

By means of the theory of residues one can easily prove that the integral on the right-hand side is equal to

$$\int_{-\infty}^{\infty} e^{-x^2}\,dx = \sqrt{\pi}.$$

Therefore we have

$$\mathcal{F}[e^{-\alpha^2 t^2}](\tau) = \frac{\sqrt{\pi}}{\alpha} e^{-\tau^2/4\alpha^2}, \qquad \alpha > 0. \tag{1.3.12}$$

Example 1.3.7 *Now we are going to prove that*

$$\mathcal{F}[|t|^{p-1}](\tau) = 2\cos\pi p/2\,\Gamma(p)|\tau|^{-p}, \qquad 0 < p < 1. \tag{1.3.13}$$

For the proof of formula (1.3.13) *we consider the function*

$$f : z \to z^{p-1}e^{-\alpha z}, \qquad 0 < p < 1,\ \alpha > 0,\ 0 \le arg(z) \le \pi/2.$$

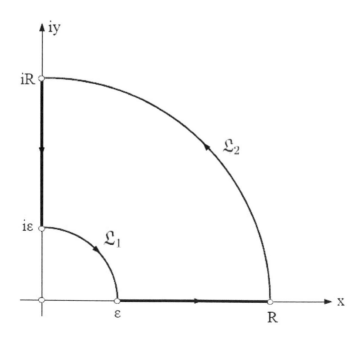

Figure 1

If $\mathfrak{L}$ is the closed contour of Figure 1 by means of the theorem of residues we have

$$\oint_{\mathfrak{L}} f(z)\,z = 0 = \int_{\mathfrak{L}_1} f(z)dz + \int_{\varepsilon}^{R} f(x)dx + \int_{\mathfrak{L}_2} f(z)\,dz - i\int_{\varepsilon}^{R} f(iy)dy.$$

The integrals on $\mathfrak{L}_1$ respectively $\mathfrak{L}_2$ tend to zero as $\varepsilon \to +0$ respectively $R \to +\infty$. Therefore, we have

$$e^{\pi ip/2}\int_{0}^{\infty} y^{p-1}e^{-i\alpha y}dy = \int_{0}^{\infty} x^{p-1}e^{-\alpha x}dx = \alpha^{-p}\Gamma(p)$$

by means of the integral representation of the Gamma function. Furthermore, we obtain

$$e^{-\pi ip/2}\int_{-\infty}^{0} |y|^{p-1}e^{-i\alpha y}dy = e^{-\pi ip/2}\int_{0}^{\infty} y^{p-1}e^{i\alpha y}\,dy = \alpha^{-p}\Gamma(p)$$

because this is the conjugate complex value of the upper integral and the result is real-valued. Adding the last two formulas and substituting $y \to t$, $\alpha \to \tau$ leads to the result (1.3.13).

Analogously taking the difference of the last two formulas we obtain by means of 1.3.1, Proposition 1.3.1:

$$\mathcal{F}[|t|^{p-1}\mathrm{sgn}\ t](\tau) = -2i\sin\pi p/2\ \Gamma(p)|\tau|^{-p}\mathrm{sgn}\ \tau, \qquad 0 < p < 1. \tag{1.3.14}$$

For many examples of Fourier transforms we refer to the tables [O.1], [EMOT], vol. I.

1.3.3   Operational Properties

For the application of the *FT* we need certain operational properties. By straightforward calculation we obtain:

Proposition 1.3.2 *Let $f \in L_1$, $a, b \in \mathbb{R}$, $b \neq 0$. Then*

$$\mathcal{F}[f(t-a)](\tau) = e^{-ia\tau}f^{\wedge}(\tau) \tag{1.3.15}$$

$$\mathcal{F}[e^{iat}f(t)](\tau) = f^{\wedge}(\tau - a) \tag{1.3.16}$$

$$\mathcal{F}[f(bt)](\tau) = |b|^{-1}f^{\wedge}(\tau/b). \tag{1.3.17}$$

For the application on differential equations the *FT* of derivatives is of interest. Let as usual

$$C^k = \{f : f \quad k - \text{times continuous differentiable on} \quad \mathbb{R}\}, \quad k \in \mathbb{N}.$$

Then the following holds:

Proposition 1.3.3 *Let* $f \in L_1 \cap C^1$, $f' \in L_1$. *Then we have*

$$\mathcal{F}[f'](\tau) = i\tau f^{\wedge}(\tau). \tag{1.3.18}$$

Proof. From $f \in C^1$ we know

$$f(t) = f(0) + \int_0^t f'(x)\, dx.$$

Since $f' \in L_1$ there exists the limit of the right-hand side as t tends to $\pm\infty$. Therefore, $f(t)$ tends to zero as t tends to $\pm\infty$. So we obtain

$$\mathcal{F}[f'](\tau) = \int_{-\infty}^{\infty} f'(t)e^{-i\tau t}dt = f(t)\, e^{-it\tau}\, |_{t=-\infty}^{\infty} + i\tau f^{\wedge}(\tau).$$

From the consideration above we conclude that the first expression on the right-hand side is zero.

By complete induction we obtain   ☐

Corollary 1.3.1 *Let* $f \in C^k$, $D^j f \in L_1$, $j = 0, 1, \dots, k$, $k \in \mathbb{N}$. *Then*

$$\mathcal{F}[D^k f](\tau) = (i\tau)^k f^{\wedge}(\tau) \tag{1.3.19}$$

and

$$f^{\wedge}(\tau) = o(|\tau|^{-k}), \qquad \tau \to \pm\infty. \tag{1.3.20}$$

Remark 1.3.7 *Analogous to formula (1.3.18) one can derive differentation rules for the Fourier cosine and Fourier sine transform, defined by 1.3.1., (1.3.18), and (1.3.19),*

$$\mathcal{F}_c[f'](\tau) = \tau \mathcal{F}_s[f] - f(0)$$
$$\mathcal{F}_s[f'](\tau) = -\tau \mathcal{F}_c[f](\tau).$$

For the proof see [Ch.2], 140, (2), and (1).

The differentiation rule is the basis for operational calculus with respect to the operator $-iD$ (respectively $D = d/dt$). It can be used for the solution of linear differential equations with respect to this operator; see 1.3.5, Example 1.3.5. A differentiation rule in the space of images is the following:

Proposition 1.3.4 *Let* $f, tf \in L_1$. *Then there exists* $Df^{\wedge}$ *and*

$$Df^{\wedge}(\tau) = \mathcal{F}[-itf(t)](\tau). \tag{1.3.21}$$

Proof. From $f, tf \in L_1$ we see that the *FT* of f and of tf converge absolutely and uniformly with respect to τ and therefore one can interchange differentiation and integration of the *FT* of f.   ☐

Corollary 1.3.2 *Let* $t^j f \in L_1$, $j = 0, 1, \ldots, k$, $k \in \mathbb{N}$. *Then there exist* $D^k f^\wedge$ *and*

$$D^k f^\wedge(\tau) = \mathcal{F}[(-it)^k f(t)](\tau). \tag{1.3.22}$$

Remark 1.3.8 *Analogously one can derive formulas for the Fourier cosine and the Fourier sine transform; see [Ch. 2], 140, (1.3.19), and (1.3.18), namely,*

$$D\mathcal{F}_c[f](\tau) = \mathcal{F}_s[-tf(t)](\tau), \qquad D = d/d\tau,$$
$$D\mathcal{F}_s[f](\tau) = \mathcal{F}_c[tf(t)](\tau).$$

Conversely, one can easily prove rules for the *FT* of integrals in the domain of originals and of images.

Proposition 1.3.5 *Let* $f \in L_1 \cap C$, $\int_o^t f(x)dx \in L_1$. *Then we have*

$$\mathcal{F}\left[\int_0^t f(x)dx\right](\tau) = (i\tau)^{-1} f^\wedge(\tau). \tag{1.3.23}$$

Proof. Let $\varphi : t \to \int_o^t f(x)\,dx$. Then φ fulfills the requirements of Proposition 1.3.3 and therefore from (1.3.18) we get

$$\mathcal{F}[\varphi'](\tau) = f^\wedge(\tau) = i\tau\varphi^\wedge(\tau).$$

By means of Fubini's theorem we obtain a rule for the integration in the domain of images. ☐

Proposition 1.3.6 *Let* $f \in L_1$. *Then we have*

$$\int_0^t f^\wedge(x)\,dx = i \int_{-\infty}^\infty \frac{e^{-ity} - 1}{y} f(y)dy \tag{1.3.24}$$

or

$$f^\wedge(t) = iD \int_{-\infty}^\infty \frac{e^{-ity} - 1}{y} f(y)dy. \tag{1.3.25}$$

Finally we would like to derive some product formulas. From $f, g \in L_1$ we get $f(t)g(\tau) \in L_1(\mathbb{R} \times \mathbb{R})$. Using

$$|f(t)g(\tau)e^{-it\tau}| = |f(t)g(\tau)|$$

and Fubini's theorem we have:

Proposition 1.3.7 *If $f, g \in L_1$, then $f^\wedge g, f g^\wedge \in L_1$ and*

$$\int_{-\infty}^{\infty} f^\wedge(\tau) g(\tau) d\tau = \int_{-\infty}^{\infty} f(\tau) g^\wedge(\tau) d\tau. \tag{1.3.26}$$

Now we are going to define a "product" of two originals such that its *FT* yields the product of the Fourier transforms of the two originals.

Definition 1.3.2 *Let $f, g : \mathbb{R} \to \mathbb{C}$. Then as the (Fourier) convolution $f * g$ of f and g we define*

$$(f * g)(t) = \int_{-\infty}^{\infty} f(x) g(t - x) dx, \tag{1.3.27}$$

provided that the integral exists (in some sense).

Theorem 1.3.4 (Convolution Theorem) *Let $f, g \in L_1$. Then $f * g \in L_1$. The convolution is commutative, associative, and*

$$\|f * g\|_1 \le \|f\|_1 \|g\|_1. \tag{1.3.28}$$

For the FT it holds that

$$(f * g)^\wedge = f^\wedge \cdot g^\wedge. \tag{1.3.29}$$

Proof. For every $x \in \mathbb{R}$ we have $\|g\|_1 = \int_{-\infty}^{\infty} |g(t - x)| dt$. Therefore,

$$\|f\|_1 \|g\|_1 = \int_{-\infty}^{\infty} |f(x)| \int_{-\infty}^{\infty} |g(t - x)| dt dx = \int_{-\infty}^{\infty} \int_{-\infty}^{\infty} |f(x) g(t - x)| dt dx$$

$$= \int_{-\infty}^{\infty} \int_{-\infty}^{\infty} |f(x) g(t - x)| dx dt \ge \int_{-\infty}^{\infty} |(f * g)(t)| dt = \|f * g\|_1$$

by means of Fubini's theorem and this yields the first part of Theorem 1.3.4. Furthermore, again by means of Fubini's theorem,

$$(f * g)^\wedge(\tau) = \int_{-\infty}^{\infty} \left(\int_{-\infty}^{\infty} f(x) g(t - x) dx \right) e^{-i\tau t} dt = \int_{-\infty}^{\infty} f(x) \int_{-\infty}^{\infty} g(t - x) e^{-i\tau t} dt dx$$

$$= \int_{-\infty}^{\infty} f(x) \int_{-\infty}^{\infty} g(y) e^{-i(x + y)\tau} dy dx = \int_{-\infty}^{\infty} f(x) e^{-i\tau x} dx \int_{-\infty}^{\infty} g(y) e^{-i\tau y} dy$$

$$= f^\wedge(\tau) \cdot g^\wedge(\tau).$$

The commutativity and the associativity of the convolution can be proved by straightforward calculation. Another proof can be performed by use of the *FT*, but then Theorem 1.3.7 is necessary. □

1.3.4   The Inversion Formula

An inversion formula can easily be proved for piecewise smooth L_1-functions.

Theorem 1.3.5 *Let $f \in L_1$, piecewise smooth in each interval $[a,b] \subset \mathbb{R}$. Then we have for every $t_0 \in \mathbb{R}$*

$$(2\pi)^{-1} \int_{-\infty}^{\infty} e^{it_0\tau} f^{\wedge}(\tau)\, d\tau = \begin{cases} f(t_0), & \text{if } f \text{ continuous at } t_0 \\ (Af)(t_0), & \text{if } f \text{ discontinuous at } t_0, \end{cases}$$

where

$$(Af)(t_0) = \frac{1}{2}[f(t_0 + 0) - f(t_0 - 0)] \tag{1.3.30}$$

is the arithmetical mean value of the right-sided and of the left-sided limit of f at t_0 and the integral has to be understood in the sense of Cauchy's principal value (PV) in the second case.

Proof.   We follow the lines of a proof of J. Koekock, [Koe]. We choose $t_0 = 0$. If $t_0 \neq 0$ we choose instead of f the function

$$f_0(t) = f(t + t_0).$$

Using Proposition 1.3.2, we have $f_o^{\wedge}(\tau) = e^{it_o\tau} f^{\wedge}(\tau)$ and therefore

$$(2\pi)^{-1} \int_{-\infty}^{\infty} f^{\wedge}(\tau)e^{it_o\tau}d\tau = (2\pi)^{-1} \int_{-\infty}^{\infty} f_o^{\wedge}(\tau)d\tau = (2\pi)^{-1} \int_{-\infty}^{\infty} f_o^{\wedge}(\tau)e^{io\tau}d\tau.$$

Case 1: f continuous at $t = 0$, $f(0) = 0$.

By the mean value theorem we have as well in a right-sided neighborhood as in a left-sided neighborhood of the origin

$$f(t) = tf'(\vartheta t), \qquad 0 < \vartheta < 1.$$

Therefore, the function $g(t) = f(t)/t$ belongs to L_1. Setting

$$I_a^b f^{\wedge}(t) = (2\pi)^{-1} \int_{-a}^{b} f^{\wedge}(\tau)e^{it\tau}d\tau, \qquad a,b > 0,$$

we obtain by means of Fubini's theorem

$$2\pi I_a^b f^{\wedge}(0) = \int_{-a}^{b} f^{\wedge}(\tau)d\tau = \int_{-a}^{b}\int_{-\infty}^{\infty} f(t)e^{-i\tau t}dt\, d\tau = \int_{-\infty}^{\infty} f(t) \int_{-a}^{b} e^{-i\tau t}d\tau dt$$

$$= i \int_{-\infty}^{\infty} g(t)\big(e^{-ibt} - e^{iat}\big)dt = i[g^{\wedge}(b) - g^{\wedge}(-a)].$$

Since $g \in L_1$ the right-hand side tends to zero as a, b tend to infinity and this is the statement of the theorem in the case of $t_0 = 0$, $f(0) = 0$, f continuous at the origin.

Case 2: $f(0-) = p$, $f(0+) = q$ ($p = q$ is allowed!).

We consider the function h defined by

$$h(t) = \begin{cases} pe^t, & -\infty < t < 0 \\ qe^{-t}, & 0 < t < \infty. \end{cases}$$

The function φ with $\varphi = f - h$ fulfills the assumptions of case 1 and therefore we have

$$\lim_{a,b \to \infty} I_a^b f^\wedge(0) = \lim_{a,b \to \infty} I_a^b h^\wedge(0). \tag{$*$}$$

But

$$h^\wedge(\tau) = q \int_0^\infty e^{-t(1+i\tau)} dt + p \int_{-\infty}^0 e^{t(1-i\tau)} dt = \frac{1}{1+\tau^2}[p + q + i(p-q)\tau].$$

It follows that

$$I_a^b h^\wedge(0) = (2\pi)^{-1} \int_{-a}^b h^\wedge(\tau) d\tau = \frac{p+q}{2\pi}(\operatorname{arctg}(a) + (\operatorname{arctg}(b)) + \frac{q-p}{2\pi} \log \left(\frac{1+b^2}{1+a^2} \right)^{\frac{1}{2}}.$$

If $q = p$ we obtain

$$\lim_{a,b \to \infty} I_a^b h^\wedge(0) = p.$$

If $q \neq p$ we have to choose $b = a$ (principal value) and get

$$\lim_{a \to \infty} I_a^a h^\wedge(0) = \frac{p+q}{2}.$$

This completes the proof of Theorem 1.3.5 because of $(*)$. $\square$

Corollary 1.3.3 *Let $f \in L_1 \cap C$, piecewise smooth in each interval $[a,b] \subset \mathbb{R}$. Then it holds that*

$$f(t) = (2\pi)^{-1} \int_{-\infty}^\infty f^\wedge(\tau) e^{it\tau} d\tau = \mathcal{F}^{-1}[f^\wedge](t). \tag{1.3.31}$$

Remark 1.3.9 *The integral in (1.3.31) is called the inverse FT of $f^\wedge$.*

Remark 1.3.10 *If (1.3.31) is fulfilled then we have*

$$\mathcal{F}^{-1}[f^\wedge](t) = (2\pi)^{-1} \mathcal{F}[f^\wedge](-t) \tag{1.3.32}$$

$$\mathcal{F}[f(-t)](\tau) = (2\pi)\mathcal{F}^{-1}[f](\tau) \tag{1.3.33}$$

$$\mathcal{F}[f](\tau) = 2\pi \mathcal{F}^{-1}[f](-\tau). \tag{1.3.34}$$

Because of the last property the tables of Fourier transforms contain only the Fourier transforms $f^\wedge$ of originals f and not conversely the originals of given Fourier transforms $f^\wedge$. So from section 1.3.2, Example 1.3.3, we obtain

$$\mathcal{F}\left[\frac{\sin at}{at}\right](\tau) = \frac{\pi}{a}1_+(a-|\tau|), \quad a > 0. \tag{1.3.35}$$

Remark 1.3.11 *One can prove that the inversion formula (1.3.31) holds if $f \in L_1 \cap C$, $f^\wedge \in L_1$ and it holds a.e. if $f, f^\wedge \in L_1$. Of course under these conditions the proof is not so simple. In [SW] it is proved that from $f \in L_1$, f continuous at zero and $f^\wedge \geq 0$ it follows that $f^\wedge \in L_1$ and, therefore, (1.3.31) holds a.e.*

Remark 1.3.12 *Combining the convolution theorem, Theorem 1.3.4, with the inversion formula (1.3.31) and Remark 1.3.11, we obtain*

$$f * g = \mathcal{F}^{-1}[f^\wedge g^\wedge]$$

provided that $f, g, f^\wedge \in L_1$.

Remark 1.3.13 *If $f, g \in L_1$ such that $f^\wedge, g^\wedge \in L_1$ we have*

$$(f^\wedge * g^\wedge)^\wedge = 2\pi\mathcal{F}^{-1}[f^\wedge * g^\wedge] = (f^\wedge)^\wedge(g^\wedge)^\wedge = (2\pi\mathcal{F}^{-1}[f^\wedge])(2\pi\mathcal{F}^{-1}[g^\wedge])$$

and applying the FT on both sides we get

$$f^\wedge * g^\wedge = 2\pi(fg)^\wedge.$$

Remark 1.3.14 *Analogously one can derive inversion formulas for the Fourier–cosine and for the Fourier–sine transforms (1.3.33) and (1.3.34):*

$$f(x) = \mathcal{F}_c^{-1}[\mathcal{F}_c[f]](t) = \frac{2}{\pi}\int_0^\infty \mathcal{F}_c[f](\tau)\cos t\tau d\tau = \frac{2}{\pi}\mathcal{F}_c[\mathcal{F}_c[f]](t)$$

and

$$f(x) = \mathcal{F}_s^{-1}[\mathcal{F}_s[f]](t) = \frac{2}{\pi}\int_0^\infty \mathcal{F}_s[f](\tau)\sin t\tau d\tau = \frac{2}{\pi}\mathcal{F}_s[\mathcal{F}_s[f]](t),$$

see [Ch. 2] 139, (3) and 138, (3).

Another inversion formula for L_1-functions is given in the following

Theorem 1.3.6 *Let $f \in L_1$. Then a.e. on $\mathbb{R}$ it holds that*

$$f(t) = \frac{1}{2\pi}D\int_{-\infty}^\infty f^\wedge(\tau)\frac{e^{it\tau}-1}{i\tau}d\tau, \quad D = d/dt, \tag{1.3.36}$$

the integral has to be understood in the sense of the Cauchy principal value.

Proof.   For every $t \in \mathbb{R}$ the function $\tau \to (i\tau)^{-1}(e^{i\tau} - 1)$ belongs to $\mathcal{C}_0$. Since the Fourier integral (1.3.30) is uniformly convergent by means of Fubini's theorem we have

$$\frac{1}{2\pi} \int_{-T}^{T} f^{\wedge}(\tau) \frac{e^{it\tau} - 1}{i\tau} d\tau = \frac{1}{2\pi} \int_{-\infty}^{\infty} f(x) \int_{-T}^{T} \frac{e^{it\tau} - 1}{i\tau} c^{-i\tau x} d\tau dx \qquad (1.3.37)$$

$$= \int_{-\infty}^{\infty} f(x) K_T(t, x) d\tau dx.$$

The kernel $K_T(t, x)$ is continuous and bounded on $\mathbb{R}^2$ for every $T > 0$, since

$$K_T(t, x) = \frac{1}{2\pi} \int_{0}^{T} \left(\frac{e^{i\tau(t-x)} - e^{-i\tau x}}{i\tau} - \frac{e^{-i\tau(t-x)} - e^{i\tau x}}{i\tau} \right) d\tau = \frac{1}{\pi} \int_{0}^{T} \frac{\sin(t - x)\tau + \sin \tau x}{\tau} d\tau$$

and

$$\lim_{T \to +\infty} K_T(t, x) = \begin{cases} 1, & t > x > 0 \\ 0, & 0 < t < x \quad \text{and} \quad x < t < 0 \\ -1, & t < x < 0. \end{cases} \qquad (1.3.38)$$

Therefore, we have

$$|f(x) K_T(t, x)| \le M |f(x)|, \qquad T > 0, (t, x) \in \mathbb{R}^2.$$

By means of the Lebesgue theorem of dominated convergence the limit $T \to +\infty$ may be performed under the x-integral:

$$\lim_{T \to +\infty} \int_{-\infty}^{\infty} f(x) K_T(t, x) dx = \int_{-\infty}^{\infty} f(x) \lim_{T \to +\infty} K_T(t, x) dx = \int_{0}^{t} f(x) dx, \qquad (1.3.39)$$

where formula (1.3.38) was used. From $f \in L_1$ we know that the right-hand side of formula (1.3.39) has a.e. a derivative and this derivative is $f(t)$. This is formula (1.3.36). $\quad\Box$

From (1.3.39) we conclude a uniqueness property.

Theorem 1.3.7 *Let $f \in L_1$. Then from $f^{\wedge} = 0$ it follows that $f = 0$ a.e.*

Corollary 1.3.4 *Let $f, g \in L_1$. Then from $f^{\wedge} = g^{\wedge}$ it follows that $f = g$ a.e.*

Remark 1.3.15 *If $f \in L_1 \cap C$ then the supplement "a.e." in Theorem 1.3.7 and in Corollary 1.3.4 can be omitted.*

1.3.5 Applications

Now we consider some applications of the FT to the theory of special functions, integral equations, and partial differential equations. The method is the following: The problems are formulated in some space of originals. Then the problems are transformed by means of the integral transformation under consideration, here the FT, into the space of images. The transformed problem is solved if it is easier to handle than in the original space. Finally the solution is transformed into the original space by means of the inversion formula, tables or other rules of operational calculus of the FT. In every case the rules of the operational calculus are used quite formally. It is not proved if the conditions of the validity of the rules are fulfilled. At the end one has to consider if the result is really a solution of the original problem and under which conditions it is a solution. These conditions are often proved to be less strong than the conditions for the validity of the rules of operational calculus.

Example 1.3.8 *Let h_n be the Hermite functions*

$$h_n(t) = e^{-t^2/2} H_n(t), \qquad n \in \mathbb{N}_0 \tag{1.3.40}$$

where H_n are the Hermite polynomials

$$H_n(t) = (-1)^n e^{t^2} D^n e^{-t^2}, \qquad n \in \mathbb{N}_0. \tag{1.3.41}$$

One can easily prove that

$$Dh_n(t) = -th_n(t) + 2nh_{n-1}(t) \tag{1.3.42}$$

and

$$h_{n+1}(t) = 2th_n(t) - 2nh_{n-1}(t). \tag{1.3.43}$$

By means of complete induction one can easily prove

$$H_n(iD)e^{-t^2/2} = (-i)^n h_n(t).$$

Using the differentiation rule of the FT in the space of images, see 1.3.3, Corollary 1.3.2 in the form

$$\mathcal{F}[t^n f(t)](\tau) = (iD)^n f^\wedge(\tau)$$

we obtain with $f(t) = e^{-t^2/2}$

$$\mathcal{F}[h_n](\tau) = H_n(iD)\mathcal{F}[e^{-t^2/2}](\tau),$$

and by use of 1.3.2, Example 1.3.6, and (1.3.11) we have

$$\mathcal{F}[h_n](\tau) = \sqrt{2\pi} H_n(i\,D)e^{-\tau^2/2} = \sqrt{2\pi}(-i)^n h_n(\tau),$$

i.e.,

$$\mathcal{F}[h_n](\tau) = \sqrt{2\pi}(-i)^n h_n(\tau). \tag{1.3.44}$$

Result. The Hermite functions h_n are the eigenfunctions of the operator $\mathcal{F}$ of the *FT* with respect to the eigenvalues $\lambda_n = \sqrt{2\pi}\,(-i)^n$.

Example 1.3.9 *By means of a simple example we would like to illustrate the evaluation of definite integrals using FT-technique. Let us consider the integral*

$$I(\alpha, \beta) = \int\limits_{-\infty}^{\infty} \frac{dt}{(t^2 + \alpha^2)(t^2 + \beta^2)}, \qquad \alpha, \beta > 0.$$

By means of formula (1.3.9) and $f(t) = e^{-\alpha|t|}$, $g(t) = e^{-\beta|t|}$ we obtain $f^\wedge(\tau) = \frac{2\alpha}{\tau^2+\alpha^2}$, $g^\wedge(\tau) = \frac{2\beta}{\tau^2+\beta^2}$ and the convolution theorem in the form 1.3.4, Remark 1.3.12, specialized for the case $t = 0$, i.e.,

$$\int\limits_{-\infty}^{\infty} f^\wedge(\tau)g^\wedge(\tau)d\tau = 2\pi \int\limits_{-\infty}^{\infty} f(x)g(-x)dx$$

leads to

$$I(a,b) = (2\alpha\beta)^{-1}\pi \int\limits_{-\infty}^{\infty} e^{-(\alpha+\beta)|x|}dx = (\alpha\beta)^{-1}\pi \int\limits_{0}^{\infty} e^{-(\alpha+\beta)x}dx = \frac{\pi}{\alpha\beta(\alpha+\beta)}.$$

Example 1.3.10 *Now we consider linear differential equations with constant coefficients:*

$$Px(t) = h(t), \tag{1.3.45}$$

where

$$P = P(D) = a_n D^n + a_{n-1}D^{n-1} + \cdots + a_1 D + a_0,$$

with $a_j \in \mathbb{C}$, $j = 0, 1, \ldots, n$ and $D = d/dt$. Applying the FT to (1.3.45) by means of the differentiation rule (1.3.18) we have

$$P(i\tau)x^\wedge(\tau) = h^\wedge(\tau).$$

The image $x^\wedge(\tau)$ of the solution $x(t)$ we are looking for is

$$x^\wedge(\tau) = \frac{h^\wedge(\tau)}{P(i\tau)} = Q(\tau)h^\wedge(\tau),$$

where $Q(\tau) = 1/P(i\tau)$. If there exists $q = \mathcal{F}^{-1}[Q]$ then we obtain a solution of (1.3.45) in the form

$$x(t) = (q * h)(t).$$

If, for example,

$$P(D) = -D^2 + a^2, \qquad a > 0,$$

then

$$Q(\tau) = \frac{1}{\tau^2 + a^2}$$

and by means of formula (1.3.8) we have

$$q(t) = \frac{1}{2a} e^{-a|t|},$$

such that a solution of (1.3.45) in this special case is given by

$$u(t) = \frac{1}{2a} \int_{-\infty}^{\infty} h(x) e^{-a|t-x|} dx,$$

provided that $h \in \mathcal{C}$ is bounded.

Example 1.3.11 *We are going to solve Volterra integral equations of the first kind and of convolutional type*

$$\int_{-\infty}^{\infty} k(t-x)u(x)dx = f(t) \qquad x \in \mathbb{R}, \tag{1.3.46}$$

where k and f are given functions. Quite formally by means of the convolution theorem, Theorem 1.3.4, we obtain

$$k^\wedge u^\wedge = f^\wedge$$

or

$$u^\wedge = Rf^\wedge, \qquad R = 1/k^\wedge. \tag{1.3.47}$$

If R is the FT of a function r, $R = r^\wedge$, resp. $r = \mathcal{F}^{-1}[1/k^\wedge]$ we get the solution u of (1.3.46) by means of the convolution theorem:

$$u = r * f. \tag{1.3.48}$$

Often r does not exist, because, for example, if $k \in L_1$ then $k^\wedge(\tau)$ tends to zero as $\tau \to \pm\infty$ and therefore $1/k^\wedge(\tau)$ is not bounded. But if it happens that there exists some $n \in \mathbb{N}$ such that there exists

$$m = \mathcal{F}^{-1}\left[\frac{(i\tau)^{-n}}{k^\wedge(\tau)}\right]$$

or

$$m^\wedge(\tau) = \frac{(i\tau)^{-n}}{k^\wedge(\tau)}$$

then from (1.3.47) it follows that

$$u^\wedge(\tau) = [(i\tau)^n f^\wedge(\tau)]m^\wedge(\tau).$$

From the differentiation rule 1.3.3, Corollary 1.3.1 we have

$$u = (D^n f) * m. \tag{1.3.49}$$

Let us consider for example $k(t) = |t|^{-1/2}$, *i.e.,*

$$\int_{-\infty}^{\infty} |t - x|^{-1/2} u(x) dx = f(t). \tag{1.3.50}$$

From 1.3.2, Example 1.3.7 we have

$$k^\wedge(\tau) = \sqrt{2\pi} |\tau|^{-1/2}.$$

Therefore $\mathcal{F}^{-1}[1/k^\wedge]$ *does not exist. But by means of* (1.3.14) *and* (1.3.34) *we get*

$$m(t) = \mathcal{F}^{-1}\left[\frac{1}{i\tau k^\wedge(\tau)}\right](t) = -\frac{i}{\sqrt{2\pi}} \mathcal{F}^{-1}\left[|\tau|^{-1/2} \operatorname{sgn}\tau\right](t) = \frac{-1}{2\pi} |t|^{-1/2} \operatorname{sgn}t$$

and by means of (1.3.49) *we obtain*

$$u(t) = \frac{1}{2\pi} \int_{-\infty}^{t} \frac{f'(x)}{\sqrt{t - x}} dx - \frac{1}{2\pi} \int_{t}^{\infty} \frac{f'(x)}{\sqrt{x - t}} dx. \tag{1.3.51}$$

One easily can prove that (1.3.51) *is a solution of equation* (1.3.50) *provided that* $f \in L_1 \cap C^1$, $f' \in L_1$.

Example 1.3.12 *Now we are going to solve the Cauchy problem for the wave equation on the real line, i.e., we are looking for the solution* $u(x,t)$ *of the wave equation*

$$a^2 u_{xx}(x,t) - u_{tt}(x,t) = 0, \qquad x, t \in \mathbb{R}, \; x > 0$$

with the initial value conditions

$$u(x,0) = u_0(x), \qquad u_t(x,0) = u_1(x),$$

with given functions u_0, u_1. *By means of the FT with respect to* x

$$u^\wedge(\xi, t) = \mathcal{F}[u(\cdot, t)](\xi)$$

we have with 1.3.3, Corollary 1.3.1

$$u_{tt}^\wedge(\xi, t) + (a\xi)^2 u^\wedge(\xi, t) = 0.$$

The solution of this ordinary differential equation is well known:

$$u^\wedge(\xi, t) = A(\xi) \cos a\xi t + B(\xi) \sin a\xi t.$$

The transformation of the initial value conditions leads to

$$u^\wedge(\xi,0) = u_0^\wedge(\xi) = A(\xi)$$

$$u_t^\wedge(\xi,0) = u_1^\wedge(\xi) = a\xi B(\xi).$$

In all three steps we assume that the Fourier transform and the limit processes (differentiation with respect to t, t → 0) can be interchanged. So we obtain for the FT of the solution under consideration

$$u^\wedge(\xi,t) = u_0^\wedge(\xi)\cos a\xi t + \frac{u_1^\wedge(\xi)}{a\xi}\sin a\xi t.$$

Expressing the trigonometrical functions by the exponential function and using the inversion formula (1.3.31) we have

$$u(x,t) = \frac{1}{2}\left\{\frac{1}{2\pi}\int_{-\infty}^{\infty} u_0^\wedge(\xi)\left[e^{i\xi(x+at)} + e^{i\xi(x-at)}\right]d\xi\right\}$$

$$+ \frac{1}{2a}\left\{\frac{1}{2\pi}\int_{-\infty}^{\infty}\frac{u_1^\wedge(\xi)}{i\xi}\left[e^{i\xi(x+at)} - e^{i\xi(x-at)}\right]d\xi\right\}.$$

From the inversion formula (1.3.31), and the integration rule, Proposition 1.3.5, we have

$$u(x,t) = \frac{1}{2}\left[u_0(x+at) + u_0(x-at)\right] + (2a)^{-1}\int_{x-at}^{x+at} u_1(\xi)d\xi. \qquad (1.3.52)$$

This is the well-known d'Alembert solution of the Cauchy problem.

If $u_0 \in C^2$, $u_1 \in C^1$ the formula (1.3.52) is the classical solution of the Cauchy problem. The existence of the FT and all other conditions for the operational rules used above are not necessary in the final form (1.3.52) of the solution!

Example 1.3.13 *Now we are going to solve the Dirichlet problem of the Laplace equation for the upper half-plane, i.e., the solution of*

$$u_{xx}(x,y) + u_{yy}(x,y) = 0, \qquad x,y \in \mathbb{R}, \ y > 0$$

under the conditions

$$u(x,0) = u_o(x), \qquad x \in \mathbb{R}$$

$$u(x,y) \to 0 \quad if \quad x^2 + y^2 \to +\infty, \ y > 0.$$

Denoting the FT with respect to x by

$$u^\wedge(\xi,y) = \mathcal{F}[u(\cdot,y)](\xi)$$

again with 1.3.3, Corollary 1.3.1 our problem is transformed into the ordinary differential equation

$$u_{yy}^\wedge(\xi,y) - \xi^2 u(\xi,y) = 0$$

and the conditions

$$u^\wedge(\xi, 0) = u_0(\xi)$$

$$u^\wedge(\xi, y) \to 0 \quad if \quad y \to +\infty.$$

The solution of the ordinary differential equation is

$$u^\wedge(\xi, y) = A(\xi)e^{|\xi|y} + B(\xi)e^{-|\xi|y}.$$

Because of the behavior as $y \to +\infty$ *we conclude that* $A(\zeta) = 0$. *From the boundary condition on* $y = 0$ *we have* $B(\xi) = u_0^\wedge(\xi)$. *Therefore,*

$$u^\wedge(\xi, y) = u_o^\wedge(\xi)e^{-|\xi|y}.$$

By means of the convolution theorem, section 1.3.3, Theorem 1.3.4 *we have*

$$u(x, y) = u_o(x) * \mathcal{F}^{-1}[e^{-y|\xi|}](x).$$

Making use of 1.3.2, Example 1.3.4 *this yields*

$$u(x, y) = u_o(x) * \frac{1}{\pi}\frac{y}{x^2 + y^2},$$

i.e.,

$$u(x, y) = \frac{y}{\pi}\int_{-\infty}^{\infty}\frac{u_0(t)}{(x - t)^2 + y^2}dt. \tag{1.3.53}$$

One can verify that (1.3.53) is the classical solution of the Dirichlet problem, provided that $u_0 \in L_1 \cap C$ *is bounded.*

Example 1.3.14 *Now we consider the Cauchy problem for the heat conduction on the real line. We look for the solution of the heat equation:*

$$u_t(x, t) - a^2 u_{xx}(x, t) = 0, \qquad x, t \in \mathbb{R}, \ t > 0$$

with the initial condition

$$u(x, 0) = u_o(x), \qquad x \in \mathbb{R}.$$

As usual we consider the FT with respect to x. *By means of the differentiation rule 1.3.3,* Corollary 1.3.1 *we obtain*

$$u_t^\wedge(\xi, t) + (a\xi)^2 u^\wedge(\xi, t) = 0$$

and

$$u^\wedge(\xi, 0) = u_o^\wedge(\xi).$$

The solution of this initial value problem is

$$u^\wedge(\xi, t) = u_o^\wedge(\xi)e^{-(a\xi)^2 t},$$

and the convolution theorem, section 1.3.3, Theorem 1.3.4, leads to

$$u(x,t) = (u_0 * \mathcal{F}^{-1}[e^{-(a\xi)^2 t}])(x).$$

By means of 1.3.2, Example 1.3.6 we obtain

$$u(x,t) = (4\pi a^2 t)^{-1/2} \int_{-\infty}^{\infty} u_o(y) e^{-(x-y)^2/4a^2 t} dy. \tag{1.3.54}$$

It can be proved that (1.3.54) is the classical solution of the Cauchy problem if $u_o \in L_1 \cap \mathcal{C}$ or if u_0 is continuous and bounded on $\mathbb{R}$.

1.4 The Laplace Transform

1.4.1 Definition and Basic Properties

The application of the *FT* is restricted to a relatively poor class of functions. Polynomials, exponential functions, for example, cannot be transformed. One possibility is to extend the domain of originals to distributions (see Chapter 3).

Another possibility is to change the kernel of the transformation in such a manner that the integral converges for a larger class of functions. For example, one can consider the integral

$$\int_{-\infty}^{\infty} (f(t)e^{-\sigma t}) e^{-i\tau t} dt, \qquad \sigma \in \mathbb{R}. \tag{1.4.1}$$

This leads to the two-sided or bilateral Laplace transform (see section 1.4.7). In applications there often appear functions $f(t)$, which vanish for $t < 0$. Then we obtain the one-sided or unilateral Laplace transform. Putting $p = \sigma + i\tau \in \mathbb{C}$ we obtain the following definition.

Definition 1.4.3 *The (one-sided) Laplace transform (LT) of a function $f : \mathbb{R}_+ \to \mathbb{C}$ is the function F defined by*

$$F(p) = \mathcal{L}[f](p) = \int_0^{\infty} f(t)e^{-pt} dt, \tag{1.4.2}$$

provided that the integral exists.

Definition 1.4.4 *As the space of originals of the LT we consider the space E_a of functions $f : \mathbb{R} \to \mathbb{C}$, $f \in L_1^{loc}(\mathbb{R})$, and that there exists a number $a \in \mathbb{R}$ such that $f \in L_1(\mathbb{R}_+; e^{-at})$ and $f(t)$ vanishes if $t < 0$, equipped with the norm*

$$\|f\|_{E_a} = \int_0^{\infty} e^{-at} |f(t)| dt. \tag{1.4.3}$$

Remark 1.4.16 *Originals are sometimes written by means of the Heaviside function as* $1_+(t)f(t)$. *We usually omit the factor* $1_+(t)$ *and we assume that originals* f *have the property* $f(t) = 0$ *for* $t < 0$. *So in concrete cases we give only the formula for* $f(t)$ *if* $t \geq 0$.

Remark 1.4.17 *Functions* $f \in L_1^{loc}(\mathbb{R}_+)$ *with the property*

$$|f(t)| \leq Me^{at}, \qquad t \geq T > 0$$

(functions of exponential growth) belong to $E_{a+\varepsilon}$, $\varepsilon > 0$, *arbitrary.*

Remark 1.4.18 *Sometimes instead of the LT* (1.4.2) *the Laplace–Carson transform*

$$\mathcal{LC}[f](p) = pF(p)$$

is considered (see, for example, [DP]). In this notation some formulas become more simple.

Remark 1.4.19 *The advantage of the LT is that the images are functions of a complex variable and so the method of the theory of functions can be used in the space of images. The Laplace transforms appear to be analytic functions in some half-plane.*

Theorem 1.4.8 *Let* $f \in E_a$. *Then the Laplace integral* (1.4.2) *is absolutely and uniformly convergent on* $\bar{H}_a = \{p : p \in \mathbb{C}, \ Re(p) \geq a\}$. *The LT* F *is bounded on* $\bar{H}_a$ *and it is an analytic function on* $H_a = \{p : p \in \mathbb{C}, \ Re(p) > a\}$ *and it holds that*

$$D^k F(p) = (-1)^k \mathcal{L}[t^k f(t)](p), \qquad k \in \mathbb{N}. \tag{1.4.4}$$

Furthermore, it is a linear transformation, i.e.,

$$\mathcal{L}[\alpha f + \beta g] = \alpha F + \beta G, \qquad \alpha, \beta \in \mathbb{C}, \ f, g \in E_a.$$

Proof.   Let $\sigma = Re(p) \geq a$. Then for every $v \in \mathbb{R}_+$ we have

$$\int_0^v |e^{-pt} f(t)| dt \leq \int_0^\infty e^{-at} |f(t)| dt < \infty,$$

because of Definition 1.4.4. Therefore, the Laplace integral (1.4.2) is absolutely and uniformly convergent on $\bar{H}_a$. From

$$|F(p)| \leq \int_0^\infty e^{-\sigma t} |f(t)| dt \leq \int_0^\infty e^{-at} |f(t)| dt < \infty$$

we see that $F(p)$ is bounded on $\bar{H}_a$. Since

$$\int_0^v e^{-pt} f(t) \, dt$$

is an entire function the Laplace transform

$$F(p) = \lim_{v \to +\infty} \int_0^v e^{-pt} f(t) dt$$

is analytical on the interior of the domain of convergence, i.e., on H_a.

Now let $\varepsilon \in \mathbb{R}_+$, arbitrarily, and $p \in H_{a+\varepsilon}$. Because of

$$|tf(t)e^{-pt}| \le e^{-at}|f(t)| t e^{-\varepsilon t} \le \varepsilon^{-1} e^{-at} |f(t)|$$

one can differentiate (1.4.2) under the integral sign and we obtain

$$DF(p) = \int_0^\infty f(t) \frac{d}{dp} e^{-pt} dt = -\mathcal{L}[tf(t)](p).$$

By induction we have formula (1.4.4). The linearity of the LT is obviously true. ▯

If $f \in E_a$ then $e^{-at} f(t) \in L_1$ and together with Theorem 1.4.8 we have the following connection between the FT and the LT.

Corollary 1.4.5 *Let $f \in E_a$, then for $\sigma \ge a$ we have*

$$\mathcal{L}[f](p) = \mathcal{F}[e^{-\sigma t} f(t)](\tau). \tag{1.4.5}$$

By means of this connection and section 1.3.4, Corollary 1.3.4 we obtain immediately

Theorem 1.4.9 *Let $f \in E_a$, $g \in E_b$ and $c = \max(a,b)$. If*

$$F(p) = G(p), \qquad p \in H_c,$$

then $f(t) = g(t)$ a.e.

The number a in Definition 1.4.4 is not uniquely determined. Therefore, one can define

$$\sigma_{ac} = \inf\{\sigma : \int_0^\infty e^{-\sigma t} |f(t)| dt < \infty\}. \tag{1.4.6}$$

σ_{ac} is called the abscissa of absolute convergence.

Proposition 1.4.8 $\mathcal{L}[f](p)$ *is (absolutely) convergent on $H_{\sigma_{ac}}$, it is not (absolutely) convergent on $\mathbb{R} \setminus \bar{H}_{\sigma_{ac}}$.*

Proof. Let $\sigma > \sigma_{ac}$. Then there exists a number $\sigma' \in (\sigma_{ac}, \sigma)$ such that

$$\int_0^\infty e^{-\sigma' t} |f(t)| dt < \infty.$$

From Theorem 1.4.8 we conclude that $\mathcal{L}[f](p)$ is absolutely convergent on $H_{\sigma_{ac}}$.

It $\sigma < \sigma_{ac}$ and $\sigma'' \in (\sigma, \sigma_{ac})$ and $\mathcal{L}[f](p)$ absolutely convergent then from Theorem 1.4.8 we have that $\mathcal{L}[f](p'')$ is also absolutely convergent and this is a contradiction to the definition (1.4.6) of σ_{ac}. □

Remark 1.4.20 *The domain of absolute convergence of a Laplace integral (1.4.2) is $H_{\sigma_{ac}}$ or $\bar{H}_{\sigma_{ac}}$.*

Example 1.4.15 *Let $f(t) = \frac{1}{1+t^2}$. Then $\sigma_{ac} = 0$ and the half-plane of absolute convergence is $\bar{H}_0$.*

Example 1.4.16 *Let $f(t) = 1$. Then $\sigma_{ac} = 0$ but the half-plane of absolute convergence is H_0 since $\int\limits_0^\infty |e^{-it\tau}| dt$ is divergent.*

Remark 1.4.21 *The Laplace transform $F(p)$ of a function f is defined only in a half-plane. By analytical continuation one can sometimes obtain a larger domain of definition for the image F. Relations in the image domain proved for some half-plane are then also true in this larger domain.*

Remark 1.4.22 *Instead of the investigation of the absolute convergence of the Laplace integral (1.4.2) one can consider its ordinary convergence. Similar to the case of the absolute convergence one can prove: If the Laplace integral (1.4.2) is convergent for $p = p_1$, then it converges in the half-plane H_{σ_1}, One defines an abscissa of convergence σ_c by means of*

$$\sigma_c = \inf\{\sigma : \int\limits_0^\infty e^{-\sigma t} f(t) dt \text{ finite}\}. \tag{1.4.7}$$

The domain of convergence is the half-plane H_{σ_c}. Obviously $\sigma_c \leq \sigma_{ac}$ and the case $\sigma_c < \sigma_{ac}$ is possible (see, for example, [Doe. 3], p. 29. Analogous to the asymptotic behavior of the FT $f^\wedge(\tau)$ of L_1-functions as τ tends to $\pm\infty$ (see 1.3.1, Theorem 1.3.2) one can prove:

Theorem 1.4.10 *Let $f \in E_a$. Then $F(p) = \mathcal{L}[f](p)$ tends to zero as p tends to ∞ in the half-plane $\bar{H}_a$.*

Proof. Let $0 < u < v$. Then

$$F(p) = \left(\int\limits_o^u + \int\limits_u^v + \int\limits_v^\infty \right) e^{-pt} f(t) dt.$$

Let $\varepsilon \in \mathbb{R}_+$ arbitrarily. Then the absolute value of the first integral becomes less than $\varepsilon/3$ as u is sufficiently small ($\sigma \in \bar{\mathbb{R}}_+$). The absolute value of the third integral becomes less than $\varepsilon/3$ as v is sufficiently large (since the LT is absolutely and uniformly convergent on $\bar{H}_a$). Now let $\sigma \geq \sigma_0 > \max(a, 0)$. Then for the second integral we have

$$\left| \int_u^v e^{-pt} f(t) dt \right| \leq e^{-\sigma_0 u} \int_u^v |f(t)| dt < \varepsilon/3$$

and therefore $|F(p)| < \varepsilon$ in the half-plane $\bar{H}_{\sigma_0}$.

Remark 1.4.23 *The functions* $\sin p$ *and* p^α, $\alpha \in \bar{\mathbb{R}}_+$, *for example, cannot be Laplace transforms of originals of some space* E_a. *The function* e^{-p} *also cannot be a Laplace transform of a function of* E_a, *although it tends to zero as* σ *tends to* $\pm\infty$, *since it does not tend to zero as* τ *tends to* $\pm\infty$, $\sigma \geq a$, *fixed.*

From Theorem 1.4.10 we obtain immediately:

Corollary 1.4.6 *Let* $f \in E_a$ *and* $F(p) = \mathcal{L}[f](p)$ *analytical at* $p = \infty$. *Then* $F(\infty) = 0$.

1.4.2 Examples

In the following examples let λ, ν be complex parameters.

Example 1.4.17 *From*

$$\int_0^\infty e^{-(p-\lambda)t} dt = \frac{1}{p - \lambda}$$

we obtain

$$\mathcal{L}[e^{\lambda t}](p) = \frac{1}{p - \lambda}, \qquad \sigma > Re(\lambda) = \sigma_{ac}. \tag{1.4.8}$$

Example 1.4.18 *By means of the linearity of the LT we obtain by means of*

$$\cos \lambda t = \frac{1}{2}\left(e^{i\lambda t} + e^{-i\lambda t}\right), \qquad \sin \lambda t = \frac{1}{2i}\left(e^{i\lambda t} - e^{-i\lambda t}\right)$$

and formula (1.4.8) by straightforward calculation

$$\mathcal{L}[\cos \lambda t](p) = \frac{p}{p^2 + \lambda^2}, \qquad \sigma > |Im(\lambda)| = \sigma_{ac} \tag{1.4.9}$$

and

$$\mathcal{L}[\sin \lambda t](p) = \frac{\lambda}{p^2 + \lambda^2}, \qquad \sigma > |Im(\lambda)| = \sigma_{ac}. \tag{1.4.10}$$

Example 1.4.19 *Now we are going to calculate the LT of the function* $f(t) = t^\lambda$, $Re(\lambda) > -1$. *For* $p > \sigma$ *and substituting* $st = x$ *we obtain*

$$\int_0^\infty e^{-pt} t^\lambda dt = p^{-\lambda-1} \int_0^\infty e^{-x} x^\lambda dx = \Gamma(\lambda + 1) p^{-\lambda-1},$$

where the Gamma function is defined by

$$\Gamma(z) = \int_0^\infty e^{-t} t^{z-1} dt, \qquad Re(z) > 0. \tag{1.4.11}$$

By means of analytical continuation we have

$$\mathcal{L}[t^\lambda](p) = \frac{\Gamma(\lambda+1)}{p^{\lambda+1}} \qquad \sigma > 0 = \sigma_{ac}, Re(\lambda) > -1. \tag{1.4.12}$$

Especially for $\lambda = n \in \mathbb{N}_0$ we have

$$\mathcal{L}[t^n](p) = \frac{n!}{p^{n+1}}, \qquad \sigma > 0 = \sigma_{ac}. \tag{1.4.13}$$

Example 1.4.20 *Using formula (1.3.12) we obtain, substituting $t = u^2$,*

$$\mathcal{L}\left[\frac{1}{\pi\sqrt{t}}\cos x\sqrt{t}\right](p) = \frac{2}{\pi}\int_0^\infty e^{-pu^2}\cos xu\, du = \frac{1}{\pi}\int_{-\infty}^\infty e^{-pu^2} e^{-ixu} du$$

$$= \frac{1}{\pi}\mathcal{F}[e^{-pu^2}](x) = \frac{1}{\sqrt{\pi p}} e^{-x^2/4p}, \qquad p \in \mathbb{R}_+.$$

By means of analytical continuation with respect to p we get

$$\mathcal{L}\left[\frac{1}{\pi\sqrt{t}}\cos x\sqrt{t}\right](p) = \frac{1}{\sqrt{\pi p}} e^{-x^2/4p}, \qquad p \in H_0. \tag{1.4.14}$$

Example 1.4.21 *Now we are going to evaluate the LT of the image (1.4.14). Preparing the calculation we prove first that*

$$\int_0^\infty v^{-2} e^{\left(\frac{\alpha}{v}-v\right)^2} dv = \frac{\sqrt{\pi}}{2}, \qquad \alpha \in \mathbb{R}_+. \tag{1.4.15}$$

The proof runs as follows. Substituting $\alpha/u = v$ we obtain

$$\int_0^\infty e^{-\left(\frac{\alpha}{u}-u\right)^2} du = \alpha \int_0^\infty v^{-2} e^{-\left(v-\frac{\alpha}{v}\right)^2} dv.$$

We then obtain

$$2\int_0^\infty e^{-\left(\frac{\alpha}{u}-u\right)^2} du = \int_0^\infty \left(1 + \frac{\alpha}{v^2}\right) e^{-\left(\frac{\alpha}{v}-v\right)^2} dv = \int_{-\infty}^\infty e^{-w^2} dw = \sqrt{\pi},$$

where we substituted $\frac{\alpha}{v} - v = w$. This is formula (1.4.15). Now let $x, p > 0$. Substituting $x/2\sqrt{t} = u$ we have

$$\mathcal{L}\left[\frac{1}{\sqrt{\pi t}} e^{-x^2/4t}\right](p) = e^{-x\sqrt{p}} \int_0^\infty e^{-(\sqrt{pt}-x/2\sqrt{t})^2} \frac{dt}{\sqrt{\pi t}}$$

$$= \frac{x}{\sqrt{\pi}} e^{-x\sqrt{p}} \int_0^\infty u^{-2} e^{-\left(\frac{x\sqrt{p}}{2u}-u\right)^2} du = e^{-x\sqrt{p}}/\sqrt{p},$$

where formula (1.4.15) with $\alpha = \sqrt{p}\,x/2$ was used. If $x < 0$ then we have to set $|x|$ instead of x. So we have proved

$$\mathcal{L}\left[\frac{1}{\sqrt{\pi t}}e^{-x^2/4t}\right](p) = \frac{1}{\sqrt{p}}e^{-|x|\sqrt{p}}, \qquad x \in \mathbb{R},\ p \in H_0, \qquad (1.4.16)$$

where analytical continuation with respect to p was performed. For a list of Laplace transforms and inverse Laplace numbers we refer to the tables [PBM], vol. IV, V, [EMOT], vol. I. and [OB].

1.4.3   Operational Properties

In the following we assume that the originals belong to E_a such that their Laplace transform converges absolutely in the half-plane $\bar{H}_a$. So we obtain the following three rules by straightforward calculation.

Proposition 1.4.9 (Shifting Rule) *Let $t_0 \geq 0$. Then we have*

$$\mathcal{L}[f(t - t_0)](p) = e^{-t_0 p}F(p), \qquad \sigma \geq a. \qquad (1.4.17)$$

Proposition 1.4.10 (Similarity Rule) *Let $c > 0$. Then it holds that*

$$\mathcal{L}[f(ct)](p) = \frac{1}{c}F\left(\frac{p}{c}\right) \qquad \sigma \geq ca. \qquad (1.4.18)$$

Proposition 1.4.11 (Damping Rule) *Let $\mu \in \mathbb{C}$. Then we have*

$$\mathcal{L}[e^{-\mu t}f(t)](p) = F(p + \mu), \qquad \sigma \geq a - Re(\mu). \qquad (1.4.19)$$

Example 1.4.22 *From 1.4.2, Example 1.4.19 and using the Damping Rule (1.4.19) we obtain immediately*

$$\mathcal{L}[e^{\mu t}t^\lambda](p) = \frac{\Gamma(\lambda + 1)}{(p - \mu)^{\lambda+1}}, \qquad Re(\lambda) > -1,\ \sigma > Re(\mu). \qquad (1.4.20)$$

Proposition 1.4.12 (Multiplication Rule) *Let $n \in \mathbb{N}_0$. Then we have*

$$\mathcal{L}[(-t)^n f(t)](p) = D^n F(p), \qquad \sigma > a. \qquad (1.4.21)$$

For the proof we remark that this rule is only part of 1.4.1, Theorem 1.4.8.

Proposition 1.4.13 (Division Rule) *Let $f(t)/t \in E_a$. Then it holds that*

$$\mathcal{L}[t^{-1}f(t)](p) = \int_p^\infty F(u)du \qquad \sigma > a. \qquad (1.4.22)$$

Proof. Putting $\mathcal{L}[t^{-1}f(t)](p) = \phi(p)$ from Proposition 1.4.12 we obtain

$$\mathcal{L}[f] = -D\phi = F$$

and therefore

$$\phi(p) = \int_p^{p_0} F(u)du + \phi(p_0).$$

Let $p_0 \to \infty$. Then from Theorem 1.4.10 we know that $\phi(p_0) \to 0$. Therefore, we arrive at formula (1.4.22).   □

Example 1.4.23 *If the integrals in* (1.4.22) *converge if* $p = 0$, *then we obtain*

$$\int_o^\infty \frac{f(t)}{t}dt = \int_o^\infty F(p)dp.$$

So, for example,

$$\int_o^\infty \frac{\sin t}{t}dt = \int_o^\infty \mathcal{L}[\sin t](p)dp = \int_o^\infty \frac{dp}{1+p^2} = (\text{arctg } p)\ |_o^\infty = \frac{\pi}{2},$$

i.e.,

$$\int_o^\infty \frac{\sin t}{t}dt = \frac{\pi}{2}. \tag{1.4.23}$$

Example 1.4.24 *From*

$$\int_o^\infty u^{-1/2}e^{-x\sqrt{u}}du = -\frac{2}{x}\int_p^\infty \frac{\partial}{\partial u}\left(e^{-x\sqrt{u}}\right)du = \frac{2}{x}e^{-x\sqrt{p}}, \qquad x > 0$$

and formula 1.4.2, (1.4.16) we have

$$\mathcal{L}\left[\frac{1}{\sqrt{\pi}}t^{-3/2}e^{-x^2/4t}\right](p) = \frac{2}{x}e^{-x\sqrt{p}}, \qquad x \in \mathbb{R}_+, \ p \in H_0. \tag{1.4.24}$$

Now we are going to prove two product formulas. By means of the definition of the Laplace transform and of Fubini's theorem the proof is straightforward.

Proposition 1.4.14 *Let* $g \in E_0$ *and* $F = \mathcal{L}[f]$ *be defined and bounded on* H_0. *It then holds that*

$$\int_o^\infty f(t)G(t)dt = \int_o^\infty F(t)g(t)dt.$$

Now we are going to formulate the convolution theorem for the *LT*. Since functions of E_a vanish on the negative real axis the convolution (1.3.27) has the form

$$(f * g)(t) = \int_0^t f(x)g(t-x)dx. \tag{1.4.25}$$

Theorem 1.4.11 (Convolution Theorem) *Let $f \in E_a$, $g \in E_b$. Then $f * g \in E_c$, $c = \max(a, b)$. Furthermore, we have*

$$\|f * g\|_{E_c} \le \|f\|_{E_a} \|g\|_{E_b} \qquad (1.4.26)$$

and the convolution is commutative and associative and it holds that

$$\mathcal{L}[f * g](p) = F(p)G(p), \qquad p \in \bar{H}_c. \qquad (1.4.27)$$

Proof. Making use of Fubini's theorem and of the substitution $y = t - x$ we have

$$\|f\|_{E_a}\|g\|_{E_b} = \int_0^\infty e^{-ax}|f(x)|dx \int_0^\infty e^{-by}|g(y)|dy \ge \int_0^\infty |f(x)| \int_0^\infty e^{-c(x+y)}|g(y)|dydx$$

$$= \int_0^\infty |f(x)| \int_0^\infty e^{-ct}|g(t-x)|dtdx = \int_0^\infty e^{-ct} \int_0^t |f(x)g(t-x)|dxdt$$

$$\ge \int_0^\infty e^{-ct}|(f * g)(t)|dt = \|f * g\|_{E_c}$$

where the double integral was considered as a two-dimensional integral on the second octant $\{0 \le x < \infty, \, x \le t < \infty\} = \{0 \le x \le t, \, 0 \le t < \infty\}$ of the (x,t)-plane. Therefore, (1.4.26) holds.

For the LT we obtain again by virtue of Fubini's theorem

$$\mathcal{L}[f * g](p) = \int_0^\infty f(x) \int_0^\infty e^{-pt}g(t-x)dtdx = \int_0^\infty e^{-px}f(x)G(p) = F(p)G(p).$$

Here in the first integral $(0, \infty)$ was used as the interval of integration because $g(t - x)$ vanishes if $x > t$ and then the shifting rule (1.4.17) was applied. So formula (1.4.27) is proved. Taking the LT of $f*g$ resp. $(f*g)*h$ we conclude by means of 1.4.1, Theorem 1.4.9 that the convolution is commutative and also associative. ∎

Next we consider two applications.

Example 1.4.25 *Let*

$$I_n(t) = \int_0^t dx_n \int_0^{x_n} dx_{n-1} \cdots \int_0^{x_3} dx_2 \int_0^{x_2} f(x_1)dx_1. \qquad (1.4.28)$$

Then this can be written as

$$I_n(t) = f * \overbrace{1 * 1 * \cdots * 1}^{n \ times}.$$

The convolution theorem leads to

$$\mathcal{L}[I_n](p) = F(p)(\mathcal{L}[1](p))^n = F(p)p^{-n} = F(p)\mathcal{L}\left[\frac{t^{n-1}}{(n-1)!}\right](p)$$

$$= \mathcal{L}\left[f(t) * \frac{t^{n-1}}{(n-1)!}\right](p).$$

Therefore, we obtain

$$I_n(t) = \frac{1}{(n-1)!} \int\limits_0^t f(x)(t-x)^{n-1}dx. \tag{1.4.29}$$

Example 1.4.26 *Let $B(u,v)$ be Euler's first integral or Beta function, defined by*

$$
\begin{aligned}
B(u,v) &= \int\limits_0^1 x^{u-1}(1-x)^{v-1}dx, \qquad Re(u), Re(v) > 0 \\
&= (t^{u-1} * t^{v-1})_{t=1}.
\end{aligned}
\tag{1.4.30}
$$

By means of 1.4.2, formula (1.4.12) we obtain

$$\mathcal{L}[(x^{u-1} * x^{v-1})(t)](p) = \frac{\Gamma(u)\Gamma(v)}{p^{u+v}},$$

where Γ is Euler's Gamma function; see equation (1.4.12). Again making use of formula (1.4.12) (in the opposite direction) we have

$$(x^{u-1} * x^{v-1})(t) = \frac{\Gamma(u)\Gamma(v)}{\Gamma(u+v)}t^{u+v-1},$$

and for $t=1$ we have

$$B(u,v) = \frac{\Gamma(u)\Gamma(v)}{\Gamma(u+v)}. \tag{1.4.31}$$

Now we will derive rules for the *LT* of the primitive and of the derivative of a function.

Proposition 1.4.15 (Integration Rule) *Let $f \in E_a$ and $\varphi(t) = \int_0^t f(x)dx$. Then $\varphi \in E_c$, where $c = a$ if $a \in \mathbb{R}_+$ and $c = \varepsilon \in \mathbb{R}_+$, arbitrarily if $a \leq 0$ and it holds that*

$$\mathcal{L}\left[\int\limits_0^t f(x)dx\right](p) = p^{-1}F(p), \qquad \sigma > \max(0,a). \tag{1.4.32}$$

Proof. From $\varphi = 1 * f$ and applying the convolution theorem we obtain formula (1.4.32). Since $1 \in E_\varepsilon$ for arbitrary $\varepsilon \in \mathbb{R}_+$ from Remark 1.4.16 we obtain the rest.   □

Proposition 1.4.16 (Differentiation Rule) *Let $f \in E_a$ and let there exist $Df \in E_a$ on $[0,\infty)$. Then it holds that*

$$\mathcal{L}[Df](p) = pF(p) - f(+0), \qquad \sigma > \max(0,a). \tag{1.4.33}$$

Proof. Because of $f(t) - f(0+) = \int\limits_0^t f'(u)du$ and Proposition 1.4.15 we get

$$\mathcal{L}[f(t) - f(+0)](p) = \frac{1}{p}\mathcal{L}[f'](p) = F(p) - p^{-1}f(0+),$$

and this is formula (1.4.33).   □

Corollary 1.4.7 *If there exists $f^{(k)} \in E_a$ on $[0, \infty)$, $k = 0, 1, \ldots, n$, then it holds for $\sigma > \max(0, a)$, that*

$$\mathcal{L}[D^n f](p) = p^n F(p) - p^{n-1} f(0+) - p^{n-2} f'(+0) \cdots - f^{(n-1)}(+0). \tag{1.4.34}$$

Remark 1.4.24 *One can prove that if $f(t)$ is n-time differentiable on $\mathbb{R}_+$ and $\mathcal{L}[f^{(n)}](p)$ converges in a real point $p = \sigma_0 > 0$, then $\mathcal{L}[f]$ also converges in the point σ_o, the limits $f^{(k)}(0+)$, $k = 0, 1, \ldots, n-1$ exist and (1.4.34) holds for $\sigma > \sigma_0$ and $\mathcal{L}[f^k](p)$, $k = 0, 1, \ldots, n-1$ converge absolutely on H_{σ_0} (see [Doe.3], Th. 9.3).*

1.4.4 The Complex Inversion Formula

The connection between the *FT* and the *LT*, see 1.4.1 Corollary 1.4.5, allows us very easily to derive an inversion formula for the *LT*. If $f \in E_a$ is a original of the *LT* which is piecewise smooth in every interval $[a, b] \subset [0, \infty)$, then $e^{-\sigma t} f(t) \in L_1$ for every $\sigma \geq a$ and one can use formula (1.4.5) and the inversion theorem, Theorem 1.3.5, of the *FT*. We obtain

$$e^{-\sigma t} f(t) = \frac{1}{2\pi} \int_{-\infty}^{\infty} F(\sigma + i\tau) e^{it\tau} d\tau$$

if f is continuous in the point t. Otherwise we obtain $e^{-\sigma t} A_f(t)$, where the integral has to be chosen as the Cauchy *PV*. Substituting $p = \sigma + i\tau$ leads to the complex inversion formula

$$f(t) = \frac{1}{2\pi i} \int_{(c)} F(p) e^{tp} dp = \mathcal{L}^{-1}[F](t), \qquad c > a \tag{1.4.35}$$

in points of continuity, where the path (c) of integration is the vertical line from $c - i\infty$ to $c + i\infty$. $\mathcal{L}^{-1}$ is called the inverse *LT*. So we have:

Theorem 1.4.12 *Let $f \in E_a$ be smooth on every interval $(a, b) \subset \mathbb{R}_+$. Then in points t of continuity the complex inversion formula (1.4.35) holds. In points of discontinuity the integral in (1.4.35) represents the arithmetical mean of the left-sided and right-sided limit of f in the point t, and the integral in this case has to be taken as Cauchy's PV.*

Now we are going to derive the inversion formula (1.4.35) under conditions on the image F. Not every function F analytical on a half-plane H_a is an image of a function $f \in E_a$ (see 1.4.1, Remark 1.4.23). But we have

Theorem 1.4.13 *Let $F(p)$ be analytic on a half-plane $\bar{H}_a$ and let $F(p) \to 0$, $|p| \to \infty$, $p \in H_a$, uniformly with respect to $\arg(p)$. Furthermore, let*

$$\int_{(c)} |F(p)| dp < \infty \tag{1.4.36}$$

for every $c \geq a$. Then F is the Laplace transform of a function $f \in E_{a+\varepsilon}$, $\varepsilon > 0$, arbitrarily, and f is continuous on $\bar{\mathbb{R}}_+$.

Proof.

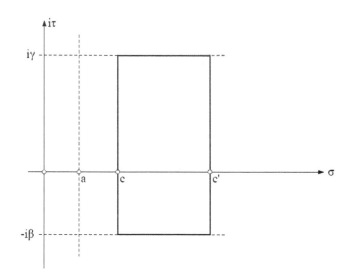

Figure 2

Step 1. We have

$$\left| \int_{c-i\beta}^{c+i\gamma} e^{pt} F(p)\, dp \right| = e^{ct} \left| \int_{-\beta}^{\gamma} e^{it\tau} F(c+i\tau)\, d\tau \right| = e^{ct} \int_{-\beta}^{\gamma} |F(c+i\tau)|\, d\tau.$$

As $\beta, \gamma \to +\infty$ the latter integral is convergent and therefore the integral in formula (1.4.35) is absolutely convergent and defines a function f. Since the integral in (1.4.35) because of (1.4.36) is uniformly convergent with respect to t, if one separates the factor e^{ct}, the function f is continuous. The value $f(t)$ does not depend on the number $c \geq a$. For the proof from Figure 2 we obtain

$$0 = \oint_{\mathfrak{R}} e^{pt} f(t)\, dt = \int_{c-i\beta}^{c'-i\beta} \cdots + \int_{c'-i\beta}^{c'+i\gamma} \cdots + \int_{c'+i\gamma}^{c+i\gamma} \cdots + \int_{c+i\gamma}^{c-i\beta} \cdots.$$

As $\beta, \gamma \to +\infty$ the integrals on the horizontal segments vanish according to our assumptions and therefore

$$\int_{(c)} e^{pt} F(p)dp = \int_{(c')} e^{pt} F(p)dp. \tag{1.4.37}$$

If $t < 0$ then the integral on the right-hand side of formula (1.4.37) tends to zero as $c' \to +\infty$ and therefore we conclude that $f(t) = 0, t < 0$.

Step 2. From step 1 we have for each $c \geq a$

$$|f(t)| \leq (2\pi)^{-1}e^{ct} \int_{-\infty}^{\infty} |F(c+i\tau)|\, d\tau = Me^{ct},$$

i.e., for every $\varepsilon \in \mathbb{R}_+$ we have

$$\int_0^{\infty} e^{-(a+\varepsilon)t}|f(t)|dt \leq M \int_0^{\infty} e^{-(a+\varepsilon-c)t}dt$$

and this is convergent, since we can choose $c \in [a, a+\varepsilon)$. Therefore, $f \in E_{a+\varepsilon}$ for every $\varepsilon \in \mathbb{R}_+$.

Step 3. Finally we have to prove that for every $p_o \in H_a$ it holds that

$$\mathcal{L}[f](p_0) = F(p_0). \tag{1.4.38}$$

By use of the absolute and uniform convergence of the integral (1.4.36) and Fubini's theorem we have for $a < c < Re(p_0)$

$$\int_0^{\infty} e^{-p_o t}f(t)dt = (2\pi i)^{-1}\int_o^{\infty} e^{-p_o t}\int_{(c)} e^{pt}F(p)dpdt$$

$$= (2\pi i)^{-1}\int_{(c)} F(p)\int_o^{\infty} e^{(p-p_o)t}dtdp \tag{1.4.39}$$

$$= (2\pi i)^{-1}\int_{(c)} F(p)(p_o - p)^{-1}dp.$$

Choosing the numbers β, γ, c' such that p_0 is lying inside the rectangle $\mathfrak{R}$ of Figure 1 by means of Cauchy's integral formula we obtain

$$F(p_0) = (2\pi i)^{-1}\oint_{\mathfrak{R}_+} \frac{F(p)}{p-p_0}dp.$$

As β, γ and c' tend to $+\infty$ the integrals on the vertical segments and also the integral on the vertical segment through c' tend to zero and, therefore,

$$F(p_0) = (2\pi i)^{-1}\int_{c+i\infty}^{c-i\infty} \frac{F(p)}{p-p_0}\,dp = (2\pi i)^{-1}\int_{(c)} \frac{F(p)}{p_0-p}dp.$$

Together with formula (1.4.39) we obtain (1.4.38). ⬜

Remark 1.4.25 *The condition on the absolute convergence of $\int_{(c)} F(p)\,dp$ is, for example, fulfilled, if*

$$|F(p)| \leq C|p|^{-2}, \qquad p \in \bar{H}_a.$$

1.4.5   Inversion Methods

A general method for the calculation of the original f from a given image F is the application of the theory of residues on the complex inversion formula 1.4.4, (1.4.35). We first prove the following.

Lemma 1.4.1 (Jordan's Lemma) *Let*

$$\mathfrak{C}_n = \{p: \quad p \in \mathcal{C}, \quad |p - p_0| = R_n, \quad R_1 < R_2 < \ldots, \lim_{n \to \infty} R_n = \infty,$$
$$Re(t(p - p_0)) \le 0, \quad 0 \ne t \in \mathbb{R}\}$$

be half-circles (see Figure 3). Let F be continuous on $\mathfrak{C}_n$, $n \in \mathbb{N}$ and $F(p)$ tend to zero uniformly on $\mathfrak{C}_n$ as n tends to infinity. Then we have

$$\lim_{n \to \infty} \int_{\mathfrak{C}_n} e^{tp} F(p)\, dp = 0, \qquad t \ne 0. \tag{1.4.40}$$

Proof. From our assumptions we know that for every $\varepsilon \in \mathbb{R}_+$ there exists a number $n_0 = n_0(\varepsilon)$ such that

$$|F(p)| < \varepsilon, \qquad z \in \mathfrak{C}_n, \qquad n > n_0.$$

Step 1. $t < 0$.

Here we have

$$p = p_0 + R_n e^{i\varphi}, \quad -\pi/2 \le \varphi \le \pi/2.$$

Therefore, we get

$$\left| \int_{\mathfrak{C}_n} F(p) e^{pt}\, dp \right| < \varepsilon R_n e^{\sigma_0 t} \int_{-\pi/2}^{\pi/2} e^{t R_n \cos\varphi}\, d\varphi = 2\varepsilon R_n e^{\sigma_0 t} \int_{o}^{\pi/2} e^{t R_n \cos\varphi}\, d\varphi.$$

Set $\psi = \frac{\pi}{2} - \varphi$. Then $\sin\psi \ge \frac{2}{\pi}\psi$, $0 \le \psi \le \pi/2$. It follows because of $t < 0$

$$\int_{o}^{\pi/2} e^{t R_n \cos\varphi}\, d\varphi = \int_{o}^{\pi/2} e^{t R_n \sin\psi}\, d\psi \le \int_{o}^{\pi/2} e^{\frac{2t}{\pi} R_n \psi}\, d\psi$$

and therefore

$$\left| \int_{\mathfrak{C}_n} F(p) e^{pt}\, dp \right| \le 2\varepsilon R_n e^{\sigma_0 t} \int_{o}^{\pi/2} e^{\frac{2t}{\pi} R_n \psi}\, d\psi = \frac{\varepsilon\pi}{t} e^{\sigma_0 t}(e^{R_n t} - 1) < \frac{\varepsilon\pi}{(-t)}$$

as $n > n_0$.

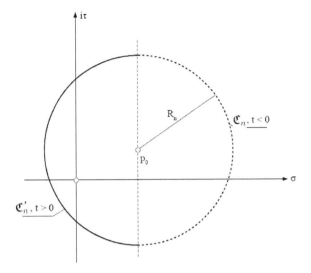

Figure 3

Step 2. $t > 0$. The proof runs analogously. Here we are setting $\psi = \varphi - \pi/2$ and we have

$$\int_{\pi/2}^{\pi} e^{t\,R_n \cos\varphi}\,d\varphi = \int_0^{\pi/2} e^{(-t)R_n \sin\psi}\,d\psi.$$

We proceed as in step 1 with $(-t)$ instead of t. $\qquad\qquad$ □

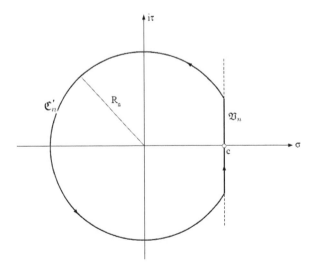

Figure 4

By means of Jordan's lemma we are going to prove the following.

Theorem 1.4.14 *Let $F(p) = \mathcal{L}[f](p)$ be a meromorphic function, analytical on the half-plane H_{σ_0}. Let $\mathfrak{C}_n = \{p : |p| = R_n\}$, $R_1 < R_2 < \cdots < R_n \to +\infty$ as $n \to +\infty$ a sequence of circles and let $F(p)$ tend to zero uniformly with respect to $\arg(p)$ as $p \to \infty$, $p \in \mathfrak{C}_n$. Furthermore, let $\int_{(c)} F(p)\,dp$ be absolute convergent for every $c > \sigma_0$. Then it holds that*

$$f(t) = \sum_{p \in \mathbb{C}} Res(F(p)e^{pt}) \tag{1.4.41}$$

where the summands have to be arranged according to ascending absolute values of the singularities.

Proof. Let $\Gamma_n = \mathfrak{C}'_n \cup \mathfrak{V}_n$ (see Figure 4). Then by means of the theorem of residues we have

$$\frac{1}{2\pi i} \oint_{\Gamma_n} e^{pt} F(p)\,dp = \sum_{|p|<R_n} Res[e^{pt}F(p)] = \frac{1}{2\pi i}\left(\int_{\mathfrak{V}_n} + \int_{\mathfrak{C}'_n}\right)e^{pt}F(p)\,dp.$$

The second integral vanishes (see Jordan's lemma) as $n \to \infty$. By means of the complex inversion formula (1.4.35) we obtain

$$f(t) = \frac{1}{2\pi i}\int_{(c)} e^{pt}F(p)\,dp = \lim_{n\to\infty}\frac{1}{2\pi i}\int_{\mathfrak{V}_n} e^{pt}F(p)\,dp = \lim_{n\to\infty}\sum_{|p|<R_n} Res[e^{pt}F(p)]$$

and this is formula (1.4.41). $\quad\square$

Example 1.4.27 *Let $F(p) = P(p)/Q(p)$ be a rational function, where $\deg(P) < \deg(Q)$. Then the assumptions of Theorem 1.4.14 are fulfilled. Let p_k be a pole of $F(p)$ with the multiplicity n_k. By the well-known formulas of the theory of residues we obtain*

$$f(t) = \sum_k \frac{1}{(n_k-1)!} \lim_{p\to p_k} \frac{d^{n_k-1}}{dp^{n_k-1}}[F(p)e^{pt}(p-p_k)^{n_k}]. \tag{1.4.42}$$

In particular, if $n_k = 1$ for every k we have

$$\mathcal{L}^{-1}\left[\frac{P(p)}{Q(p)}\right](t) = \sum_k \frac{P(p_k)}{Q'(p_k)}e^{p_k t}. \tag{1.4.43}$$

In practice one does not use the general formulas (1.4.42) and (1.4.43), but one performs the partial-fraction expansion of $F(p)$ and with help of formula 1.4.3, (1.4.20) one can evaluate the original $f(t)$. For example, if

$$F(p) = \frac{2p-1}{p^2-p} = \frac{1}{p} + \frac{1}{p-1},$$

then from (1.4.43) or directly from 1.4.3, (1.4.20) we obtain

$$f(t) = 1 + e^t, \qquad t \geq 0.$$

Another method is the inversion by means of the development in series. If $F(p)$ is analytical at ∞ and $F(\infty) = 0$ then the Laurent expansion with the center ∞ can be inverted term by term, using 1.4.2, formula (1.4.13).

Theorem 1.4.15 *Let $F(p)$ be analytical at ∞ and $F(\infty) = 0$,*

$$F(p) = \sum_{k=0}^{\infty} c_k p^{-k-1}, \qquad |p| > R. \tag{1.4.44}$$

Then $F(p) = \mathcal{L}[f](p)$ and

$$f(t) = 1_+(t) \sum_{k=0}^{\infty} \frac{c_k}{k!} t^k \tag{1.4.45}$$

and the series is convergent for every $t \in \mathbb{C}$.

Proof. Let $p \to p^{-1}$ and $\phi(p) = F(p^{-1})$. Then

$$\phi(p) = \sum_{k=0}^{\infty} c_k p^{k+1}$$

is analytical on $|p| \leq \rho < R^{-1}$. By means of the Cauchy inequality for the Taylor coefficients we have

$$|c_k| \leq M\rho^{k+1}.$$

Setting $s_n(t) = \sum_{k=1}^{n} c_k \frac{t^k}{k!}$ we get

$$|s_n(t)| \leq \sum_{k=1}^{n} |c_k| \frac{|t|^k}{k!} \leq M\rho \sum_{k=0}^{\infty} \frac{(\rho|t|)^k}{k!} = M\rho e^{s|t|}.$$

From the Weierstrass convergence theorem we conclude that the series

$$\lim_{n \to \infty} s_n(t) = \lim_{n \to \infty} \sum_{k=0}^{n} c_k \frac{t^k}{k!} = f(t)$$

is uniformly convergent on $\mathbb{C}$ and it defines an entire function f of exponential order, i.e.,

$$|f(t)| \leq Ce^{\rho|t|}.$$

Multiplication of the series of $f(t)$ with e^{-pt} and integration on $\mathbb{R}_+$ leads to

$$\mathcal{L}[f](p) = \sum_{k=0}^{\infty} c_k p^{-k-1} = F(p).$$

$\square$

Remark 1.4.26 *One can prove that conversely if f is an entire function of exponential type then there exists $F(p) = \mathcal{L}[f](p)$, $F(p)$ is analytical at ∞, $F(\infty) = 0$ and the formulas (1.4.44), (1.4.45) are valid (see [Doe. 3], Th. 30).*

Example 1.4.28 *Let*

$$F(p) = \frac{1}{\sqrt{p^2 + 1}} = \sum_{n=0}^{\infty} \binom{-\frac{1}{2}}{n} p^{-2n-1}, \qquad |p| > 1.$$

Since

$$\binom{-\frac{1}{2}}{n} = (-1)^n / \sqrt{\pi} n! \Gamma(n + 1/2)$$

wc obtain from Theorem 1.4.15 *and formula* (1.4.13)

$$f(t) = \mathcal{L}^{-1}[F](t) = \sum_{n=0}^{\infty} \frac{\Gamma(n+1/2)(-1)^n}{\sqrt{\pi} n!} \frac{t^{2n}}{(2n)!}.$$

By means of Legendre's formula for the Gamma function we have

$$\sqrt{\pi}(2n)! = 2^{2n} n! \Gamma(n + 1/2)$$

and this leads to

$$f(t) = \sum_{n=0}^{\infty} \frac{(-1)^n}{(n!)^2} \left(\frac{t}{2}\right)^{2n} = J_0(t), \tag{1.4.46}$$

where J_0 is the Bessel function of the first kind and zero order.

1.4.6   Asymptotic Behavior

Sometimes in applications of the *LT* we obtain the image $F(p)$ of an original $f(t)$ and we are not interested in the explicit form of the function $f(t)$ but only in their behavior as $t \to 0$ resp. $t \to +\infty$. Very simple results in this direction are the following two propositions.

Proposition 1.4.17 (Final-Value Theorem) *Let $F = \mathcal{L}[f]$ be analytical on $\bar{H}_0$ except possibly a simple pole at the origin and let there exists f' on $\mathbb{R}_+$, $f' \in E_0$, then we have*

$$\lim_{t \to +\infty} f(t) = \lim_{p \to 0} pF(p). \tag{1.4.47}$$

Proof.   From the differentiation rule 1.4.3, Proposition 1.4.16 we obtain

$$\lim_{p \to 0} \int_0^\infty f'(t)e^{-pt}dt = \lim_{p \to 0} pF(p) - f(+0) = \lim_{T \to +\infty} f(t) \mid_{+0}^\infty = \lim_{t \to \infty} f(t) - f(+0).$$

This is the desired result.   □

Analogous we have the following.

Proposition 1.4.18 (Initial-Value Theorem) *Let $F = \mathcal{L}[f]$ be analytical on $\bar{H}_0$ except possibly a simple pole at the origin and let there exist f' on $\mathbb{R}^+$, $f' \in E_0$, then it holds that*

$$\lim_{t \to +0} f(t) = \lim_{p \to \infty} pF(p). \tag{1.4.48}$$

Proof. Obviously, it holds for $p \in H_0$

$$p\,F(p) = \int_0^\infty f(t)p\,e^{-pt}dt = -f(t)e^{-pt}\,|_{+0}^\infty + \mathcal{L}[f'](p) = f(+0) + \mathcal{L}[f'](p).$$

From $f' \in E_0$ and Theorem 1.4.10 we know that $\mathcal{L}[f'](p)$ tends to zero as p tends to infinity and this yields formula (1.4.48). $\square$

Example 1.4.29 *Let $f(t) = \sin t$. From formula (1.4.9) we have $F(p) = (p^2 + 1)^{-1}$ and therefore*

$$\lim_{p\to\infty} pF(p) = 0.$$

But $\lim_{t\to\infty} \sin t$ does not exist and therefore formula (1.4.48) is not fulfilled because the poles of $F(p)$ are at $p = \pm i$.

Now we are going to prepare a more general theorem for the behavior of a original $f(t)$ as t tends to $+\infty$. It appears that the singularities with the greatest real part of the image $F(p)$ are of importance for this behavior of $f(t)$.

Lemma 1.4.2 *Let $\int_{(c)} |F(p)|\,dp < \infty$ for a vertical line (c) and $f(t) = \mathcal{L}^{-1}[F](t)$; see formula (1.4.35). Then it holds that*

$$f(t) = O(e^{ct}), \qquad t \in \mathbb{R}.$$

Proof. Obviously, it holds that

$$|f(t)| \le (2\pi)^{-1}e^{ct} \int_{(c)} |F(p)|dp$$

and this is the assertion. $\square$

Lemma 1.4.3 *Let $f \in E_a$ and $F = \mathcal{L}[f]$. Then it holds that a.e. the inversion formula*

$$f(t) = (2\pi i)^{-1}\frac{d}{dt} \int_{(c)} e^{pt}F(p)p^{-1}dp, \qquad c > a. \tag{1.4.49}$$

Proof. Setting $\varphi(t) = \int_o^t f(\tau)d\tau$, then from the integration rule 1.4.3, Proposition 1.4.15 we know that $\varphi \in E_c$, where $c = a$ if $a \in \mathbb{R}_+$ and $c \in \mathbb{R}_+$, arbitrarily, if $a \le 0$ and it holds that

$$\varphi(t) = (2\pi i)^{-1} \int_{(c)} e^{pt}p^{-1}F(p)dp. \tag{1.4.50}$$

Furthermore, there exist the derivative φ' a.e. and $\varphi' = f$. From (1.4.50) we find

$$\varphi'(t) = f(t) = (2\pi i)^{-1} \frac{d}{dt} \int_{(c)} e^{pt} p^{-1} F(p) dp, \qquad \text{a.e.,}$$

i.e., formula (1.4.49).   □

Lemma 1.4.4 *Let* $F_\nu = \mathcal{L}[f_\nu]$, $f_\nu \in E_a$, $\nu = 0,1,2$ *and* $\phi(p) = F_0(p) - F_1(p) - F_2(p)$ *analytical and bounded on a half-plane* H_c, $c > a$, *and moreover*

$$\int_{(c)} |\phi(p)| dp < \infty \tag{1.4.51}$$

and

$$f_2(t) = O(e^{ct}), \qquad t \to +\infty. \tag{1.4.52}$$

Then it holds that

$$f_0(t) = f_1(1) + O(e^{ct}), \qquad t \to +\infty. \tag{1.4.53}$$

Proof. From Lemma 1.4.3 we deduce

$$\varphi(t) = f_0(t) - f_1(t) - f_2(t) = (2\pi i)^{-1} \frac{d}{dt} \int_{(c)} e^{pt} p^{-1} \phi(p) dp.$$

Because of condition (1.4.51) we can differentiate under the integral and Lemma 1.4.2 leads to

$$f(t) = O(e^{ct}), \qquad t \to +\infty.$$

Taking into account that f_2 has the same behavior, we have (1.4.53).   □

Lemma 1.4.4 is the base of the proof of the following.

Theorem 1.4.16 *Let*

$$F(p) = O(|p|^{-2}), \qquad p \to \infty, \ p \in \bar{H}_a, \tag{1.4.54}$$

analytical on $\bar{H}_a$ *with the exception of a finite number of poles in the points*

$$p_\nu = \sigma_0 + i\tau_\nu, \ \nu = 1,2,\ldots,n, \ \sigma_0 > a.$$

If

$$F(p) \sim A_\nu (p - p_\nu)^{-k-1}, \qquad p \to p_\nu, \ k \in \mathbb{N}_0,$$

then it holds that

$$f(t) = \frac{t^k}{k!} e^{\sigma_0 t} \left(\sum_{\nu=1}^n A_\nu e^{i\tau_\nu t} + O(t^{-1}) \right), \qquad t \to +\infty. \tag{1.4.55}$$

Remark 1.4.27 *We would like to remark that the p_ν are the poles with the greatest real part of $F(p)$ (they all have the same real part $\mathrm{Re}(p_\nu) = \sigma_0$, $\nu = 1, 2, \ldots, n$) and these poles have the same order $k + 1$.*

Proof. We choose

$$F_1(p) = \sum_{\nu=1}^{n} \sum_{\kappa=1}^{k+1} \frac{a_{\kappa,\nu}}{(p - p_\nu)^\kappa}, \qquad a_{k+1,\nu} = A_\nu,$$

i.e., the sum of the singular parts of the Laurent expansion at the poles p_ν, $\nu = 1, 2, \ldots, n$ and $F_2 = 0$. Then the conditions of Lemma 1.4.4 are fulfilled and we have

$$f(t) = f_1(t) + O(e^{ct}), \qquad c \geq \sigma_0,$$

i.e.,

$$f(t) = \sum_{\kappa=1}^{k+1} \sum_{\nu=1}^{n} \left(a_{\kappa,\nu} \frac{t^{\kappa-1}}{(\kappa-1)!} e^{s_\nu t} \right) + O(e^{ct}) = \frac{t^k e^{\sigma_0 t}}{k!} \left(\sum_{\nu=1}^{n} A_\nu e^{i\tau_\nu t} + O(t^{-1}) \right),$$

since $O(e^{ct}) = O(e^{\sigma_0 t}) = e^{\sigma_0 t} O(1)$. $\qquad\qquad\qquad\qquad\qquad\qquad\qquad\qquad\qquad\square$

Example 1.4.30 *Let $F(p) = \frac{e^{-2\alpha/(p+1)}}{p^2+1}$, $\alpha > 0$. It has poles of the order 1 at $p = \pm i$, i.e., we have $\sigma_0 = k = 0$. From*

$$F(p) = \frac{1}{2i} e^{-2\alpha/(p+1)} \left(\frac{1}{p-i} - \frac{1}{p+i} \right) = \frac{1}{2i} \left(\frac{e^{\alpha(i-1)}}{p-i} - \frac{e^{-2\alpha(1+i)}}{p+i} \right) + O(1)$$

we have $A_1 = \frac{1}{2i} e^{-\alpha(1-i)}$, $A_2 = \overline{A}_1$. Theorem 1.4.16, formula (1.4.55) yields

$$f(t) = e^{-\alpha} \sin(t + \alpha) + O(t^{-1}) \qquad t \to +\infty.$$

Analogously one can prove a theorem for branching points instead of poles. For this and many other results in this direction we refer to [Doe.1], [Doe.3], [Me], and [Be.3].

1.4.7 Remarks on the Bilateral Laplace Transform

As stated in the introduction to section 1.4.1 we give the following definition.

Definition 1.4.5 *The bilateral (or two-sided) Laplace transform (BLT) of a function $f : \mathbb{R} \to \mathbb{C}$ is the function F defined by*

$$F^{(II)}(p) = \mathcal{L}^{(II)}[f](p) = \int_{-\infty}^{\infty} f(t) e^{-pt} dt, \qquad\qquad (1.4.56)$$

provided that the integral exists.

As suitable space of originals we choose:

Definition 1.4.6 E_a^b, $-\infty < a < b < \infty$, is the linear space of functions $f : \mathbb{R} \to \mathbb{C}$, $f \in L_1^{loc}(\mathbb{R})$, belonging to $L_1(\mathbb{R}; e^{-at})$ as well as to $L_1(\mathbb{R}; e^{-bt})$.

Similar to the one-sided case (section 1.4.1) one can prove:

Theorem 1.4.17 *Let $f \in E_a^b$. Then the bilateral Laplace integral (1.4.56) is absolutely and uniformly convergent on $\bar{H}_a^b = \{p : p \in \mathbb{C}, a \le Re(p) \le b\}$. The BLT is analytical on $H_a^b = \{p : p \in \mathbb{C}, a < Re(p) < b\}$ and it holds that*

$$D^k F^{(II)}(p) = \mathcal{L}^{(II)}[(-t)^k f(t)](p), \qquad k \in \mathbb{N}. \tag{1.4.57}$$

Furthermore, it is a linear transformation, i.e.,

$$\mathcal{L}^{(II)}[\alpha f + \beta g] = \alpha F + \beta G, \qquad \alpha, \beta \in \mathbb{C}, \ f, g \in E_a^b.$$

Since $e^{-\sigma t} f(t) \in L_1$, $a \le \sigma \le b$, provided that $f \in E_a^b$ we have analogously to 1.4.1, Corollary 1.4.5:

Corollary 1.4.8 *Let $f \in E_a^b$. Then for $p = \sigma + i\tau$ it holds that*

$$F^{(II)}(p) = \mathcal{F}[e^{-\sigma t} f(t)](\tau), \qquad p \in \bar{H}_a^b. \tag{1.4.58}$$

The complex inversion formula and Theorem 1.4.12, Theorem 1.4.13 are valid for the BLT, too. One has only to substitute $\mathbb{R}_+ \to \mathbb{R}$, $F \to F^{(II)}$, $E_a \to E_a^b$, $c \in \mathbb{R} : a < c < b$, $H_a \to H_a^b$.

The operational properties of section 1.4.3 hold analogously. One has only to change the assumptions and the conditions of the validity of the formulas in an easily understandable manner. In the case of the convolution theorem one must of course choose the Fourier convolution 1.3.3, (1.3.27). The differentiation rule becomes simpler:

Proposition 1.4.19 (Differentiation Rule) *Let $f \in C^n$, $n \in \mathbb{N}$ and let f and its derivatives up to the order n belong to E_a^b. Then there exist $\mathcal{L}^{(II)}[D^n f]$ and*

$$\mathcal{L}^{(II)}[D^n f](p) = p^n F^{(II)}(p), \qquad p \in H_a^b. \tag{1.4.59}$$

Because of the relation (1.4.58) the determination of the original f of a given $F = \mathcal{L}^{(II)}[f]$ can be made by the use of Fourier transform tables (see [O.1] and [EMOT], vol. I).

1.4.8   Applications

Similar to section 1.3.5 we would like to give some examples for the application of the Laplace transformation. We will consider integral equations and differential equations.

Example 1.4.31 *Let us consider the Volterra integral equation (IGL),*

$$\int_0^t f(x)k(t-x)dx = g(t), \quad g(0) = 0, \tag{1.4.60}$$

where g and k are known functions and f is the function we would like to determine.

Applying the LT on equation (1.4.60) leads quite formally to

$$F(p)K(p) = G(p)$$

or

$$F(p) = G(p) \cdot \frac{1}{K(p)}.$$

The application of the convolution theorem leads to

$$f(t) = g(t)\mathcal{L}^{-1}[1/K(p)](t).$$

Very often $\mathcal{L}^{-1}[1/K]$ does not exist. If there exist $\mathcal{L}^{-1}[1/pK(p)](t)$ and $L(p) = 1/pK(p)$ we have

$$F(p) = pG(p)L(p).$$

Setting $l := \mathcal{L}^{-1}[L]$ by means of the convolution theorem and with the help of the differentiation rule we obtain

$$f(x) = \frac{d}{dx}\int_0^x g(t)l(x-t)dt. \tag{1.4.61}$$

Let, for example,

$$k(x) = x^{-\alpha}, \quad 0 < \alpha < 1.$$

Then from formula (1.4.12) we have

$$K(p) = \Gamma(1-\alpha)p^{\alpha-1}$$

and, therefore,

$$L(p) = 1/pK(p) = \frac{p^{-\alpha}}{\Gamma(1-\alpha)}$$

and again by formula (1.4.12)

$$l(x) = \frac{x^{\alpha-1}}{\Gamma(1-\alpha)\Gamma(\alpha)} = \frac{\sin(\pi\alpha)}{\pi}x^{\alpha-1}.$$

Hence we have in this case as the solution (1.4.61) of (1.4.60)

$$f(x) = \pi^{-1} \sin(\pi\alpha) \frac{d}{dx} \int_0^x \frac{g(t)}{(x-t)^{1-\alpha}} dt. \qquad (1.4.62)$$

By simple substitutions one can transform this to integrals with other limit points. For example, let us consider the IGL

$$\int_t^b \frac{f(x)}{(x-t)^\alpha} dx = g(t), \quad 0 < t < b, \quad 0 < \alpha < 1, \quad g(b) = 0.$$

Substituting $t \to b - t$ and then $x := b - u$ we have from (1.4.62)

$$f(b-u) = \pi^{-1} \sin(\pi\alpha) \frac{d}{du} \int_0^u \frac{g(b-t)}{(u-t)^{1-\alpha}} dt.$$

Substituting conversely $u := b - x$, and then $b - t = z$ and replacing at the end $z \to t$ we have

$$f(x) = -\pi^{-1} \sin(\pi\alpha) \frac{d}{dx} \int_x^b \frac{g(t)}{(t-x)^{1-\alpha}} dt.$$

This result can be generalized in substituting $x \to \varphi(x)$, where $\varphi(x)$ is monotonic increasing and there exist φ' and $\varphi'(x) \neq 0$. Then the IGL is

$$\int_t^b \frac{f(x)}{(\varphi(x) - \varphi(t))^\alpha} dx = g(t), \quad a < t < b, \quad 0 < \alpha < 1, \quad g(b) = 0.$$

Substituting

$$\tau := \varphi(t), \quad \xi := \varphi(x), \quad \varphi(b) = \beta, \quad f(x)/\varphi'(x) =: h(x), \quad g(t) = g_o(\tau)$$

we have

$$\int_\tau^\beta \frac{h(\xi)}{(\xi - \tau)^\alpha} d\xi = g_o(\tau).$$

Using the last result we have the solution

$$h(x) = -\pi^{-1} \sin(\pi\alpha) \frac{d}{dx} \int_x^b \frac{g(t)\varphi'(t)}{(\varphi(t) - \varphi(x))^{1-\alpha}} dt.$$

Let, for example,

$$\varphi = \cosh, \quad x = \tau, \quad b = \infty, \quad t = \xi, \quad \alpha = 1/2.$$

Then, obviously,

$$h(\tau) = -\pi^{-1} \frac{d}{d\tau} \int_\tau^\infty \frac{g(\xi) \sinh \xi}{\left(\cosh \xi - \cosh \tau \right)^{1/2}} d\xi \qquad (1.4.63)$$

and therefore

$$g(\xi) = \int\limits_{\xi}^{\infty} \frac{h(\tau)}{\left(\cosh \tau - \cosh \xi\right)^{1/2}} d\tau. \tag{1.4.64}$$

For further examples of the solution of linear integral equations we refer to [Sn. 2], [Me], [Doe. 3], [Be. 2], and [De. 6].

Example 1.4.32 *Let us consider a general initial value problem (IVP) for a linear ordinary differential equation of order n, which has the form*

$$L[x](t) = D^n x(t) + a_1 D^{n-1} x(t) + \cdots + a_n x(t) = f(t) \tag{1.4.65}$$

with the initial value conditions at $t = 0+$

$$x(0+) = x_0, \quad x'(0+) = x_1, \ldots, x^{(n-1)}(0+) = x_{n-1}, \tag{1.4.66}$$

where $a_j \in \mathbb{C}$, $j = 1, 2, \ldots, n$, $x_k \in \mathbb{C}$, $k = 0, 1, \ldots, n-1$.

By means of the differentiation rule 1.4.3, Proposition 1.4.16 we obtain the equation in the domain of images

$$L(p)X(p) = F(p) + P(p), \tag{1.4.67}$$

or

$$X(p) = \frac{F(p)}{L(p)} + \frac{P(p)}{L(p)}. \tag{1.4.68}$$

Here $L(p)$ is the characteristic polynomial of $L[x]$, i.e.,

$$L(p) = p^n + a_1 p^{n-1} + \cdots + a_n$$

and $P(p)$ is a polynomial of degree $n-1$, which contains the initial values $x_0, x_1, \ldots, x_{n-1}$ in the coefficients. The original is

$$x(t) = \mathcal{L}^{-1}\left[\frac{F(p)}{L(p)}\right](t) + \mathcal{L}^{-1}\left[\frac{P(p)}{L(p)}\right](t) = x_s(t) + x_h(t). \tag{1.4.69}$$

Obviously, x_s is a special solution of the inhomogeneous equation with initial values zero and x_h is the general solution of the homogeneous equation with arbitrary initial values. The determination of x_h is possible, since $P(p)/L(p)$ is a rational function and the degree of $L(p)$ is greater than the degree of $P(p)$. So one obtains according to formula 1.4.3, (1.4.20) a linear combination of terms of the form $t^k e^{\lambda t}$. The inversion of $F(p)/L(p)$ is possible, if we assume that $f \in E_a$. So such functions as $f(t) = e^{t^2}$ are not enclosed. To close this gap we consider the IVP

$$L[\tilde{x}](p) = 1_+(t), \qquad t > 0, \tag{1.4.70}$$

with

$$\tilde{x}(0+) = \tilde{x}'(0+) = \cdots = \tilde{x}^{(n-1)}(0+) = 0.$$

The LT leads to

$$L(p)\tilde{X}(p) = p^{-1}. \tag{1.4.71}$$

For the solution of the inhomogeneous equation with right-hand side f and vanishing initial values at zero we have

$$L(p)X(p) = F(p)$$

and together with (1.4.71) we obtain

$$X(p) = (p\tilde{X}(p))F(p).$$

By inversion we get

$$x(t) = \int_0^t f(\tau)\tilde{x}'(t - \tau)d\tau = (f * \tilde{x}')(t). \tag{1.4.72}$$

This formula is the well-known Duhamel formula. It allows the calculation of the solution of equation (1.4.65) with vanishing initial values also if $\mathcal{L}[f]$ does not exist. Let us consider the equation

$$x''(t) + x(t) = e^{t^2}, \qquad x(+0) = x'(+0) = 0, \ t > 0. \tag{1.4.73}$$

Following Duhamel's method we consider first

$$\tilde{x}''(t) + \tilde{x} = 1, \qquad \tilde{x}(+0) = \tilde{x}(0+) = 0, \ t > 0.$$

The differentiation rule 1.4.3, Proposition 1.4.16 yields

$$(p^2 + 1)\tilde{X}(p) = p^{-1},$$

or

$$\tilde{X}(p) = \frac{1}{p(p^2 + 1)} = \frac{1}{p} - \frac{p}{p^2 + 1}.$$

The application of $\mathcal{L}^{-1}$ using 1.4.2, (1.4.8), (1.4.9) yields

$$\tilde{x}(t) = 1 - \cos t.$$

For the solution of (1.4.73) following Duhamel's formula (1.4.72) we obtain

$$x(t) = \int_0^t e^{\tau^2} \sin(t - \tau)d\tau.$$

Sometimes these methods also can be used for linear ordinary differential equations of the form (1.4.60), where the coefficients a_j are polynomials of a degree less than or equal to $m \in \mathbb{N}$. Using the multiplication and the differentiation rules in 1.4.3, Proposition 1.4.12, and Proposition 1.4.16, we obtain in the image domain an ordinary linear differential equation of order m with polynomial coefficients of a degree less than or equal to $n \in \mathbb{N}_0$. If $m < n$

then this differential equation can perhaps be solved more easily than the original. As an example we consider the Laguerre differential equation:

$$tx''(t) + (1-t)x'(t) + nx(t) = 0, \qquad n \in \mathbb{N}_0, \ x(0) = 1.$$

The application of the LT yields

$$-[\mathcal{L}[x''](p)]' + \mathcal{L}[x'](p) - [\mathcal{L}[x'](p)]' + nx(p) = 0,$$

or

$$-[p^2 X(p) - px_0 - \chi_1]' + pX(p) - x_0 + (pX(p) - x_0)' + nX(p) = 0,$$

i.e.,

$$p(1-p)X'(p) + (n+1-p)X(p) = 0.$$

The separation of the variables yields

$$\frac{dX}{X} = \frac{p-n-1}{p(1-p)}dp = \left(\frac{n}{p-1} - \frac{n+1}{p}\right)dp$$

and it follows that

$$X(p) = Cp^{-1}(1 - p^{-1})^n = C\sum_{k=0}^{n}\binom{n}{k}\frac{(-1)^k}{p^{k+1}}. \tag{1.4.74}$$

The application of $\mathcal{L}^{-1}$, using formula (1.4.13) and $x(0) = 1$ (i.e., $C = 1$) yields

$$x(t) = \sum_{k=0}^{n}\binom{n}{k}\frac{(-1)^k}{k!}t^k = L_n(t). \tag{1.4.75}$$

These are the well-known Laguerre polynomials. Another representation can be derived in the following manner. By means of the damping rule, Proposition 1.4.11, and formula (1.4.74) with $C = 1$ we get

$$\mathcal{L}[e^{-t}x(t)](p) = \frac{p^n}{(p+1)^{n+1}}. \tag{1.4.76}$$

From formula (1.4.20) we know

$$\mathcal{L}\left[e^{-t}\frac{t^n}{n!}\right](p) = \frac{1}{(p+1)^{n+1}}.$$

By means of the differentiation rule, Proposition 1.4.16 we obtain from the last equations (the initial values $x_0, x_1, \ldots, x_{n-1}$ are equal to zero)

$$\mathcal{L}\left[D^n e^{-t}\frac{t^n}{n!}\right](p) = \frac{p^n}{(p+1)^{n+1}},$$

and together with equation (1.4.76) and Theorem 1.4.9 we have the Rodrigues formula for the Laguerre polynomials

$$L_n(t) = \frac{e^t}{n!}D^n(e^{-t}t^n), \qquad n \in \mathbb{N}_0. \tag{1.4.77}$$

Remark 1.4.28 *One obtains at most $n - m$ solutions of equation (1.4.60) with polynomial coefficients of a degree less than or equal to m.*

Remark 1.4.29 *The transfer to systems of ordinary linear differential equations with constant coefficients can be done in an easily understandable manner.*

Remark 1.4.30 *For further examples we refer to [Da], [De.6], [Doe.3], [Fö], [Me], and other books on Laplace transforms.*

Example 1.4.33 *Now we are going to derive the solution of a linear partial differential equation with initial and boundary conditions. The LT transfers a linear ordinary differential equation of the form (1.4.60) to an algebraical equation (1.4.67). A partial differential equation for $\mathbb{R}^2 \to \mathbb{C}$ functions leads after application of the LT with respect to one variable to an ordinary differential equation. As an example we investigate the heat conduction equation in a semiinfinite linear medium.*

$$u_{xx}(x,t) - u_t(x,t) = 0, \qquad 0 < x, t < \infty \tag{1.4.78}$$

$$u(x,0) = 0, \qquad 0 < x < \infty \tag{1.4.79}$$

$$u(0,t) = u_0(t), \qquad 0 < t < \infty. \tag{1.4.80}$$

As usual, we apply the LT (with respect to the variable t) quite formally. Let $U(x,p) = \mathcal{L}[u(x,\cdot)](p)$ by means of the differentiation rule Proposition 1.4.16, because of $U(x,0) = 0$ we get

$$\frac{d^2 U(x,p)}{dx^2} + pU(x,p) = 0 \tag{1.4.81}$$

and

$$U(0,p) = U_0(p). \tag{1.4.82}$$

The solution of the ordinary linear differential equation (1.4.81) for the function U under the initial condition (1.4.82) is possible by means of the classical methods. One has

$$U(x,p) = Ae^{-\sqrt{p}x} + Be^{\sqrt{p}x},$$

where $\sqrt{p}$ is that branch of the square-root function, which is positive when p is positive. Since every LT of our spaces of originals tends to zero as p tends to $+\infty$ we obtain $B = 0$. The initial condition (1.4.82) yields $= U_0(p)$ such that the solution of (1.4.78)–(1.4.79) in the domain of images yields

$$U(x,p) = U_0(p)e^{-x\sqrt{p}}. \tag{1.4.83}$$

By means of the convolution theorem and formula 1.4.3, (1.4.24) we obtain

$$u(x,t) = \frac{x}{2\sqrt{\pi}} \int_0^t u_o(\tau) \frac{e^{-x^2/4(t-\tau)}}{(t-\tau)^{3/2}} d\tau = (u_0 * \psi(x,\cdot))(t), \tag{1.4.84}$$

with

$$\psi(x,t) = \frac{x}{2\sqrt{\pi}}t^{-3/2}e^{-x^2/4t}. \tag{1.4.85}$$

One can prove that (1.4.84) is the solution of the problem (1.4.78) through (1.4.81) provided that

$u_0 \in \mathcal{C}(\mathbb{R}_+).$

For further examples of solutions of partial differential equations by means of the *LT* we refer to [Da], [De. 6], [Doe. 3], [Me], [Ob], and [Sn. 2].

1.5 The Mellin Transform

1.5.1 Definition and Basic Properties

The Mellin transform (*MT*) is closely connected with the *FT* as well as with the two-sided *LT*. It is defined as follows:

Definition 1.5.7 *The MT of a function $f : \mathbb{R}_+ \to \mathbb{C}$ is the function f^* defined by*

$$f^*(s) = \mathcal{M}[f](s) = \int_0^\infty x^{s-1}f(x)dx, \tag{1.5.1}$$

where $s = \sigma + \iota\tau \in \mathbb{C}$, provided that the integral exists.

As space of originals we choose:

Definition 1.5.8 *The space P_a^b, $-\infty < a < b < \infty$, is the linear space of $\mathbb{R}_+ \to \mathbb{C}$ functions such that $x^{s-1}f(x) \in L_1(\mathbb{R}_+)$ for every $s \in \bar{H}_a^b$.*

Remark 1.5.31 *Functions $f \in L_1^{loc}(\mathbb{R}_+)$ with the estimate*

$$|f(x)| \le C \begin{cases} x^{-a}, & x \in (0,1] \\ x^{-b}, & x \in (1,\infty) \end{cases}$$

belong to $P_{a+\varepsilon}^{b-\varepsilon}$ for every $\varepsilon \in (0, (b-a)/2)$.

Theorem 1.5.18 *Let $f \in P_a^b$. Then the Mellin integral (1.5.1) converges absolutely and uniformly on $\bar{H}_a^b$. The MT f^* is an analytic function on H_a^b. If $k \in \mathbb{N}$ then we have*

$$D^k f^*(p) = \mathcal{M}[\log x)^k f(x)](p). \tag{1.5.2}$$

Furthermore, it is a linear transformation, i.e.,

$$\mathcal{M}[\alpha f + \beta g] = \alpha f^* + \beta g^*, \qquad \alpha, \beta \in \mathbb{C}, \quad f, g \in P_a^b.$$

Proof. The integral (1.5.1) converges (absolutely) if $f \in P_a^b$. This follows directly from the definition of the space. From

$$|x^{s-1}| < \begin{cases} x^{a-1}, & x \in (0,1] \\ x^{b-1}, & x \in (1, \infty) \end{cases}$$

and

$$\left| \int\limits_0^\infty x^{s-1} f(x) dx \right| \leq \int\limits_0^1 x^{a-1} |f(x)| dx + \int\limits_1^\infty x^{b-1} |f(x)| dx$$

it follows that the integral (1.5.1) is uniformly convergent on $\bar{H}_a^b$. Since the integral is an analytic function (with respect to s) and from

$$\left| \frac{d}{ds} x^{s-1} f(x) \right| = |(\log x) x^{s-1} f(x)| \leq \begin{cases} c x^{a-\varepsilon-1} |f(x)|, & x \in (0,1] \\ c x^{b+\delta-1} |f(x)|, & x \in (1, \infty), \end{cases}$$

where $\varepsilon, \delta \in \mathbb{R}_+$, arbitrarily, we deduce that the integral (1.5.1) after differentiation with respect to s under the integral sign is also uniformly convergent on $\bar{H}_{a+\varepsilon}^{b-\delta}$ and therefore we have (1.5.2) with $k = 1$. The general case follows by induction. The linearity is obviously true.

A connection between the Fourier and the Mellin transforms can be derived as follows. Substituting $x = e^{-t}$, $t \in (-\infty, \infty)$ in the integral (1.5.1) we have

$$f^*(\sigma + i\tau) = \int\limits_{-\infty}^\infty e^{-st} f(e^{-t}) dt = \mathcal{F}[e^{-\sigma t} f(e^{-t})](\tau), \qquad (1.5.3)$$

i.e., the following: □

Theorem 1.5.19 *Let $f \in P_a^b$. Then $e^{-\sigma t} f(e^{-t}) \in L_1(\mathbb{R})$, $a \leq \sigma \leq b$, and it holds that*

$$f^*(s) = \mathcal{F}[e^{-\sigma t} f(e^{-t})](\tau). \qquad (1.5.4)$$

By virtue of Theorem 1.3.1, and Theorem 1.3.2 we have two corollaries:

Corollary 1.5.9 *If $f \in P_a^b$, then $f^*(s)$ is bounded for each fixed σ, $a \leq \sigma \leq b$.*

Corollary 1.5.10 *If $f \in P_a^b$, then for each fixed σ, $a \leq \sigma \leq b$ it holds that*

$$\lim_{\tau \to \pm\infty} f^*(\sigma + i\tau) = 0.$$

Analogously from (1.5.4) one can obtain a connection between the MT and the bilateral LT.

By means of Definition 1.4.6 we have:

Theorem 1.5.20 *If $f \in P_a^b$, then $f(e^{-t}) \in E_a^b$ and it holds that*

$$f^*(s) = \mathcal{L}^{(II)}[f(e^{-t})](s). \tag{1.5.5}$$

Finally, we would like to derive a connection between the *MT* and the (one-sided) *LT*.

Theorem 1.5.21 *If $f \subset P_0^1$, then it holds that*

$$\mathcal{M}[\mathcal{L}[f]](s) = \Gamma(s)\mathcal{M}[f](1-s), \qquad s \in H_0^1. \tag{1.5.6}$$

Proof. Using formula (1.4.11), and the definition of the *MT* we obtain (using Fubini's theorem)

$$\Gamma(s)\mathcal{M}[f](1-s) = \int_0^\infty \tau^{s-1} e^{-\tau} d\tau \int_0^\infty t^{-s} f(t) dt$$

$$= \int_0^\infty f(t) \int_0^\infty x^{s-1} e^{-xt} dx dt = \int_0^\infty x^{s-1} \int_0^\infty e^{-xt} f(t) dt dx$$

$$= \mathcal{M}[\mathcal{L}[f]](s),$$

i.e., (1.5.6). □

By means of Theorem 1.5.19 and section 1.3.4, Corollary 1.3.4 we obtain

Theorem 1.5.22 *Let $f \in P_a^b$, $g \in P_c^d$ and $\alpha = \max(a,c) < \beta = \min(b,d)$. If*

$$f^*(s) = g^*(s), \qquad s \in H_\alpha^\beta,$$

then $f(x) = g(x)$ a.e. on $\mathbb{R}_+$.

Next we give some examples of Mellin transforms of elementary functions.

Example 1.5.34 *From formula (1.4.12) we have in another formulation*

$$\mathcal{M}[e^{-\alpha x}](s) = \alpha^{-s}\Gamma(s), \qquad Re(\alpha), \sigma \in \mathbb{R}_+. \tag{1.5.7}$$

Putting $\alpha = i\beta$, $\beta > 0$ we have for $0 < \sigma < 1$ from (1.5.7), in the sense of ordinary convergence

$$\mathcal{M}[e^{-i\beta x}](s) = \int_0^\infty x^{s-1}(\cos\beta x - i\sin\beta x)dx = e^{-\pi i s/2}\beta^{-s}\Gamma(s)$$

$$= \beta^{-s}\Gamma(s)\big(\cos(\pi s/2) - i\sin(\pi s/2)\big).$$

Comparing the real respectively imaginary part we get by means of analytical continuation with respect to s

$$\mathcal{M}[\cos\beta x](s) = \beta^{-s}\Gamma(s)\cos\pi s/2, \qquad \beta\in\mathbb{R}_+,\ 0<\sigma<1, \tag{1.5.8}$$

and

$$\mathcal{M}[\sin\beta x](s) = \beta^{-s}\Gamma(s)\sin\pi s/2, \qquad \beta\in\mathbb{R}_+,\ -1<\sigma<1. \tag{1.5.9}$$

Putting $\alpha=e^{-i\varphi}$, $\varphi\in(-\pi/2,\pi/2)$ *from* (1.5.7) *we obtain*

$$\mathcal{M}[e^{-ex^{-i\varphi}x}](s) = \Gamma(s)e^{i\varphi s} = \Gamma(s)[\cos\varphi s + i\sin\varphi s].$$

Comparing the real and the imaginary part we have

$$\mathcal{M}[e^{-x\cos\varphi}\cos(x\sin\varphi)](s) = \Gamma(s)\cos\varphi s, \qquad \sigma>0$$

$$\mathcal{M}[e^{-x\sin\varphi}\cos(x\sin\varphi)](s) = \Gamma(s)\sin\varphi s, \qquad \sigma>-1,\ |\varphi|<\pi/2. \tag{1.5.10}$$

Example 1.5.35 *Substituting* $u=s$, $v=\rho-s$, *and* $x=(t+1)^{-1}$ *in the definition of the Beta function, see equation* (1.4.30), *we obtain*

$$B(s,\rho-s) = \int_0^\infty (1+t)^{-\rho}t^{s-1}dt, \qquad 0<\sigma<Re(\rho),$$

or equivalently

$$\mathcal{M}[(1+x)^{-\rho}](s) = B(s,\rho-s), \qquad 0<\sigma<Re(\rho). \tag{1.5.11}$$

For further examples of Mellin transforms and inverse Mellin transforms we refer to [EMOT], vol. I, [O.3], and [M].

1.5.2 Operational Properties

Analogously to the investigation in sections 1.3.3 and 1.4.3 we obtain rules of operational calculus for the *MT*. By straightforward calculation we obtain the following elementary rules.

Proposition 1.5.20 *Let* $f\in P_a^b$ *and* s_0, $\alpha\in\mathbb{R}$, $\alpha\neq 0$, $\beta\in\mathbb{R}_+$. *Then it holds that*

$$\mathcal{M}[x^{s_0}f(x)](s) = f^*(s+s_0), \qquad s\in\bar{H}_{a-s_0}^{b-s_0}, \tag{1.5.12}$$

$$\mathcal{M}[f(x^\alpha)](s) = |\alpha|^{-1}f^*(s/\alpha), \qquad s/\alpha\in\bar{H}_a^b, \tag{1.5.13}$$

and

$$\mathcal{M}[f(\beta x)](s) = \beta^{-s}f^*(s), \qquad s\in\bar{H}_a^b. \tag{1.5.14}$$

Example 1.5.36 *From formulas* (1.5.8), (1.5.9), *and* Proposition 1.5.20, *formula* (1.5.12) *we have for* $\alpha \in \mathbb{R}$, $\beta \in \mathbb{R}_+$

$$\mathfrak{M}[x^\alpha \cos \beta x](s) = \beta^{-s-\alpha} \Gamma(s+\alpha) \cos \frac{\pi(s+\alpha)}{2}, \qquad -\alpha < \sigma < 1 - \alpha \qquad (1.5.15)$$

$$\mathfrak{M}[x^\alpha \sin \beta x](s) = \beta^{-s-\alpha} \Gamma(s+\alpha) \sin \frac{\pi(s+\alpha)}{2}, \qquad -1 - \alpha < \sigma < 1 - \alpha. \qquad (1.5.16)$$

Example 1.5.37 *From formula* (1.5.10) *with* $\rho = 1$ *and rule* (1.5.12) *we obtain*

$$\mathcal{M}\left[\frac{x^{1/2}}{1+x}\right](s) = B\left(s + \frac{1}{2}, \frac{1}{2} - s\right) = \frac{\pi}{\sin \pi(s + \frac{1}{2})},$$

i.e.,

$$\mathcal{M}\left[\frac{x^{1/2}}{1+x}\right](s) = \frac{\pi}{\cos \pi s}, \qquad -\frac{1}{2} < \sigma < \frac{1}{2}. \qquad (1.5.17)$$

Here the reflection law of the Gamma function

$$\Gamma(z)\Gamma(1-z) = \frac{\pi}{\sin \pi z} \qquad (1.5.18)$$

was used. By means of rule (1.5.13) *from* (1.5.17) *we obtain, putting* $\alpha \to 1/\alpha$,

$$\mathcal{M}\left[\frac{x^{1/2\alpha}}{1+x^{1/\alpha}}\right](s) = \frac{\pi\alpha}{\cos \alpha s}, \qquad \alpha \in \mathbb{R}_+, \quad -1/2\alpha < \sigma < 1/2\alpha. \qquad (1.5.19)$$

Now we are going to derive differentiation rules in the domain of images as well as in the space of originals. In the domain of images we have only to formulate 1.5.1, Theorem 1.5.18 in a new manner. Since the absolute value of $\log x$ for $x \in \mathbb{R}_+$ together with its powers is less than x^ε resp. $x^{-\varepsilon}$, $\varepsilon \in \mathbb{R}_+$, arbitrary small, as x is sufficient large resp. small we have:

Proposition 1.5.21 *Let* $f \in P_a^b$, $k \in \mathbb{N}$. *Then* $(\log x)^k f(x) \in P_{a+\varepsilon}^{b-\varepsilon}$, $\varepsilon \in \mathbb{R}_+$ *arbitrary small and it holds that*

$$\mathcal{M}[(\log x)^k f(x)](s) = D^k f^*(s), \qquad s \in H_a^b. \qquad (1.5.20)$$

By means of integration by parts one easily can prove:

Proposition 1.5.22 *Let* $f \in P_a^b$ *and there exists* $Df \in L_1^{loc}(\mathbb{R}_+)$. *Let there exist numbers* $a', b' \in \mathbb{R}$, *with* $a + 1 \le a' \le b' \le b + 1$ *such that*

$$\lim_{x \to +0} x^{a'-1} f(x) = \lim_{x \to +\infty} x^{b'-1} f(x) = 0.$$

Then there exists

$$\mathcal{M}[Df](s) = -(s-1)f^*(s-1), \qquad s \in H_{a'}^{b'}. \qquad (1.5.21)$$

By induction and using

$$(s-1)(s-2)\ldots(s-n) = \frac{\Gamma(s)}{\Gamma(s-n)}, \qquad n \in \mathbb{N}$$

we obtain

Corollary 1.5.11 *If $f^{(k)}$, $k = 0, 1, \ldots, n-1$, satisfy the conditions of* Proposition 1.4.11 *then*

$$\mathcal{M}[D^n f](s) = (-1)^n \frac{\Gamma(s)}{\Gamma(s-n)} f^*(s-n). \tag{1.5.22}$$

Now we are going to derive some further differentiation rules (without an explicit formulation of the conditions for their validity). From (1.5.22) and (1.5.12) we obtain

$$\mathcal{M}[x^n D^n f(x)](s) = (-1)^n \frac{\Gamma(s+n)}{\Gamma(s)} f^*(s). \tag{1.5.23}$$

By means of (1.5.21) and (1.5.12) and further by induction it follows that

$$\mathcal{M}[(-xD)^n f(x)](s) = s^n f^*(s) \tag{1.5.24}$$

and

$$\mathcal{M}[(-Dx)^n f(x)](s) = (s-1)^n f^*(s). \tag{1.5.25}$$

Replacing $f(x)$ with $\int_0^x f(t)\,dt$ from formula (1.5.21) we obtain

$$\mathcal{M}\left[\int_0^x f(t)dt\right](s) = -s^{-1} f^*(s+1). \tag{1.5.26}$$

Applying formula (1.5.21) to $\int_x^\infty f(t)\,dt$ we obtain by straightforward calculation

$$\mathcal{M}\left[\int_x^\infty f(t)dt\right](s) = s^{-1} f^*(s+1). \tag{1.5.27}$$

Analogously by straightforward calculation we obtain

$$\mathcal{M}\left[\int_0^\infty f(x,u)g(u)du\right](s) = \int_0^\infty x^{s-1} \int_0^\infty f(xu)g(u)du\,dx$$

$$= \int_0^\infty g(u) \int_0^\infty x^{s-1} f(xu)dx\,du = \int_0^\infty u^{-s}g(u)du \int_0^\infty t^{s-1}f(t)dt,$$

and therefore we have

$$\mathcal{M}\left[\int_0^\infty f(xu)g(u)du\right](s) = f^*(s)g^*(1-s). \tag{1.5.28}$$

Example 1.5.38 *From formulas (1.5.7) and (1.5.21) by means of (1.5.28) we obtain*

$$\mathcal{M}\left[\int_0^\infty e^{-(x\cos\varphi+1)u} \cos(xu\sin\varphi)du\right](s) = \Gamma(1-s)\,\Gamma(s)\cos\varphi s$$

$$= \mathcal{M}\left[\mathcal{L}[\cos(x\sin\varphi)u](1+x\cos\varphi)\right](s) = \mathcal{M}\left[\frac{1+x\cos\varphi}{1+2x\cos\varphi+x^2}\right](s),$$

where the LT (1.4.10) with $\lambda = x \sin\varphi$ was used. By means of the reflection law of the Gamma function (1.5.18) we obtain

$$\mathcal{M}\left[\pi^{-1}\frac{1 + x\cos\varphi}{1 + 2x\cos\varphi + x^2}\right](s) = \frac{\cos\varphi s}{\sin\pi s}, \qquad |\varphi| < \pi/2. \tag{1.5.29}$$

Replacing $\varphi \to \nu\varphi$ and then using (1.5.13) with $\alpha = \nu$ we have

$$\mathcal{M}\left[\pi^{-1}\nu\frac{1 + x^\nu\cos\nu\varphi}{1 + 2x^\nu\cos\nu\varphi + x^{2\nu}}\right](s) = \frac{\cos\varphi s}{\sin\pi s/\nu}, \tag{1.5.30}$$

$$\nu > 0, \ -\pi/2\nu < \varphi < \pi/2\nu, \ 0 < \sigma < \nu\pi.$$

Analogously we have

$$\mathcal{M}\left[\pi^{-1}\nu\frac{x^\nu\sin\nu\varphi}{1 + 2x^\nu\cos\nu\varphi + x^{2\nu}}\right](s) = \frac{\sin\varphi s}{\sin\pi s/\nu}, \tag{1.5.31}$$

$$\nu > 0, \ -\pi/2\nu < \varphi < \pi/2\nu, \ 0 < \sigma < \nu\pi.$$

Finally we define the Mellin convolution:

Definition 1.5.9 *Let $f, g : \mathbb{R}_+ \to \mathbb{C}$. The function $f \vee g$ defined by means of*

$$(f \vee g)(x) = \int_0^\infty f(t)g(x/t)t^{-1}dt \tag{1.5.32}$$

is called the Mellin convolution of f and g, provided that the integral exists.

Now we have:

Theorem 1.5.23 (Convolution Theorem) *Let $f, g \in P_a^b$. Then $f \vee g \in P_a^b$ and the Mellin convolution is commutative and associative. Furthermore, it holds that*

$$\mathcal{M}[f \vee g] = f^*g^*. \tag{1.5.33}$$

Proof.

$$|(f \vee g)^*(s)| \le \int_0^\infty x^{\sigma-1}|(f \vee g)(x)|dx = \int_0^\infty x^{\sigma-1}|\int_0^\infty f(t)(x/t)t^{-1}dt|dx$$

$$\le \int_0^\infty\int_0^\infty x^{\sigma-1}|g(x/t)|t^{-1}dt = \int_0^\infty t^{\sigma-1}|f(t)|dt\int_0^\infty u^{\sigma-1}|g(u)|dt.$$

Since the integrals on the right-hand side exist if $\sigma \in [a, b]$, we conclude the existence of the left-hand side, i.e., $f \vee g \in P_a^b$.

For the proof of formula (1.5.33) we have

$$\mathcal{M}[f \vee g](s) = \int_0^\infty\int_0^\infty f(t)g(x/t)(x/t)^{s-1}t^{s-2}dtdx$$

$$= \int_0^\infty t^{s-1}f(t)dt\int_0^\infty g(u)u^{s-1}\,du = f^*(s)g^*(s),$$

and this is the result (1.5.33). The commutativity and the associativity can be obtained by straightforward calculation or by taking the MT on both sides of the equations and making use of the commutativity resp. associativity of the ordinary product in the domain of images and of the uniqueness theorem; see Theorem 1.5.22. ◻

Example 1.5.39 *The convolution theorem is very useful for the computation of integrals. For details we refer to [M]. As an example we consider the integral representation of the MacDonald function K_0.*

$$K_0(2\sqrt{x}) = \frac{1}{2}\int_{-\infty}^{\infty} e^{-2\sqrt{x}ch\xi}d\xi. \tag{1.5.34}$$

Substituting $e^\xi = u$ and afterward $u\sqrt{x} = t$ we obtain

$$2K_0(2\sqrt{x}) = \int_0^\infty e^{-(u+u^{-1})\sqrt{x}}u^{-1}du = \int_0^\infty e^{-t-x/t}t^{-1}dt. \tag{1.5.35}$$

From the convolution theorem with $f(x) = g(x) = e^{-x}$ and 1.5.1, (1.5.7) with $\alpha = 1$ we obtain

$$\mathcal{M}[\int_0^\infty e^{-t-x/t}t^{-1}dt](s) = [\Gamma(s)]^2.$$

Together with formula (1.5.35) we have

$$2\mathcal{M}[K_0(2\sqrt{x})](s) = [\Gamma(s)]^2. \tag{1.5.36}$$

1.5.3 The Complex Inversion Formula

From the connection (1.5.4) between the MT and the FT we obtain by straightforward calculation

$$e^{\sigma t}f(e^t) = \mathcal{F}^{-1}[f^*(\sigma - i\cdot)](t).$$

Substituting $e^t = x$ by virtue of formula (1.3.31), and Remark 1.3.11 we have after the substitution $\tau \to -\tau$ in formula (1.3.31)

$$f(x) = (2\pi)^{-1}\int_{-\infty}^{\infty} x^{-(\sigma+i\tau)}f^*(\sigma + i\tau)d\tau.$$

With $s = \sigma + i\tau$ and using Theorem 1.3.5 and Remark 1.3.11 we obtain:

Theorem 1.5.24 *Let $f \in P_a^b$, $f^* = \mathcal{M}[f]$, $a \le c \le b$ and $f^*(c+i\tau) \in L_1(\mathbb{R})$ with respect to τ. Then at all points of continuity of f the complex inversion formula holds:*

$$f|x) = (2\pi i)^{-1}\int_{(c)} x^{-s}f^*(s)ds = \mathcal{M}^{-1}[f^*](s). \tag{1.5.37}$$

In this theorem we put conditions on the original f as well as on the image f^*. There exist sufficient conditions for f such that $f(c+i\cdot)$ belongs to $L_1(\mathbb{R})$. So from 1.3.4, Corollary 1.3.3 we have:

Corollary 1.5.12 *Let $f \in C(\mathbb{R}_+) \cap P_a^b$. Furthermore let there exist Df and $D(x^\sigma f(x)) \in L_1(\mathbb{R}_+)$, $a \le \sigma \le b$. Then the inversion formula (1.5.37) is valid for every $x \in \mathbb{R}_+$.*

Analogous to the inversion theorem for the *LT* respectively the bilateral *LT* (see Theorem 1.4.13 and section 1.4.7) one can formulate an inversion theorem for the *MT* using Theorem 1.5.20.

Theorem 1.5.25 *Let $f^*(s)$, $s = \sigma + i\tau$ be analytic in the infinite strip H_a^b and let $f^*(\sigma + i\tau)$ tend to zero as $\tau \to \pm\infty$ uniformly with respect to σ, $\sigma \in [a + \varepsilon, b - \varepsilon]$, $\varepsilon \in \mathbb{R}_+$, arbitrary small. Furthermore, let*

$$\int\limits_{-\infty}^{\infty} |f^*(\sigma + i\tau)| d\tau < \infty.$$

Then the function f defined on $\mathbb{R}_+$ by means of formula (1.5.37) with $c \in (a,b)$ belongs to $P_{a+\varepsilon}^{b-\varepsilon}$ and $f^* = \mathcal{M}[f]$.

For a direct proof we refer to [Sn.2], section 4.3.

1.5.4 Applications

Example 1.5.40 *We first apply the MT to the summation of (convergent) series. As usual quite formally we have by means of 1.5.2, (1.5.14) and 1.5.3, (1.5.37)*

$$f(nx) = \mathcal{M}^{-1}[n^{-s} f^*(s)](x) = (2\pi i)^{-1} \int\limits_{(c)} x^{-s} n^{-s} f^*(s) ds,$$

where $c \in (a,b)$ if $f^*(s)$ is analytical in H_a^b. Denoting Riemann's Zeta function by

$$\zeta(s) = \sum_{n=1}^{\infty} n^{-s} \qquad \sigma > 1 \tag{1.5.38}$$

we have

$$\sum_{n=1}^{\infty} f(nx) = (2\pi i)^{-1} \int\limits_{(c)} x^{-s} \zeta(s) f^*(s) ds = \mathcal{M}^{-1}[\zeta(s) f^*(s)](x). \tag{1.5.39}$$

Let us consider the sum

$$S(\beta) = \sum_{n=1}^{\infty} n^{-2} \cos \beta n, \qquad 0 \le \beta < 2\pi.$$

From formula (1.5.8) and with the help of formula (1.5.12) we have for $2 < \sigma < 3$

$$\mathcal{M}[x^{-2}\cos\beta x](s) = -\beta^{2-s}\cos(\pi s/2)\Gamma(s-2)$$

and therefore from (1.5.39) with $x = 1$ and $2 < c < 3$

$$S(\beta) = -(2\pi i)^{-1}\int\limits_{(c)}\beta^{2-s}\Gamma(s-2)\cos(\pi s/2)\zeta(s)ds.$$

By means of the theorem of residues by left-shifting of the path (c) of integration, $S(\beta)$ appears to be the sum of the residues of $-\beta^{2-s}\Gamma(s-2)\cos(\pi s/2)\zeta(s)$ at their three (!) simple poles at $s = 2$, $s = 1$ and $s = 0$. An easy calculation leads to

$$S(\beta) = \frac{\pi^2}{6} - \frac{\pi\beta}{2} + \frac{\beta^2}{4}.$$

Example 1.5.41 *The FT and the LT are very well suited for operational calculus with respect to the operator D and they can be applied to the solution of linear differential equations with respect to this operator. Such equations are transformed into algebraic equations. Because of rule (1.5.22) such ordinary differential equations are transformed by the MT into linear difference equations and their solution is not easier than the solution of the original differential equations. But from the rules (1.5.23) through (1.5.25) we see that linear ordinary differential equations with respect to the operator $x^n D^n$ respectively xD and Dx and its powers are transformed in the domain of images into algebraical equations. We use Euler's differential equation (we confine ourselves to the order 2)*

$$x^2 D^2 y(x) + pDy(x) + qy(x) = f(x),$$

where $p, q \in \mathbb{C}$ can be written in the form

$$P[xD]y(x) = ((xD)^2 + (p-1)xD + q)y(x) = f(x), \tag{1.5.40}$$

i.e., it is a linear differential equation with constant coefficients with respect to the operator xD. Application of the MT using rule (1.5.24) yields

$$P(-s)y^*(s) = f^*(s). \tag{1.5.41}$$

If $P(-s) = s^2 + (1-p)s + q$ has no zeros in a strip H_a^b we have

$$y(x) = \mathcal{M}^{-1}\Big[\frac{f^*(s)}{P^*(-s)}\Big](x). \tag{1.5.42}$$

Let, for example,

$$x^2 D^2 y(x) + 4xDy(x) + 2y(x) = e^{-x}. \tag{1.5.43}$$

The MT leads to, see formula (1.5.7),

$$(s^2 - 3s + 2)y^*(s) = \Gamma(s).$$

From that we obtain

$$y^*(s) = \frac{\Gamma(s)}{(s-1)(s-2)} = \Gamma(s-2).$$

By means of formulas (1.5.7), and (1.5.12) we have a particular solution of equation (1.5.43), namely

$$y(x) = x^{-2}e^{-x}. \tag{1.5.44}$$

Example 1.5.42 *Now we would like to find the potential $u(r, \varphi)$ in an infinite wedge, i.e., we will solve the potential equation in polar coordinates (r, φ)*

$$r^2 u_{rr}(r, \varphi) + r u_r(r, \varphi) + u_{\varphi\varphi}(r, \varphi) = 0 \tag{1.5.45}$$

in the infinite wedge $0 < r < \infty$, $-\alpha < \varphi < \alpha$, $\alpha \in (0, \pi/2)$, with the boundary conditions

$$\begin{cases} u(r, \alpha) & = u_+(r), \quad 0 \le r < \infty \\ u(r, -\alpha) & = u_-(r), \quad 0 \le r < \infty \end{cases} \tag{1.5.46}$$

$$u(r, \varphi) \to 0 \quad as \quad r \to \infty, \quad \varphi \in (-\alpha, \alpha). \tag{1.5.47}$$

Applying, as usual quite formally, the MT with respect to the variable r we have from 1.5.2, (1.5.23) the ordinary differential equation

$$\frac{\partial^2 u^*(s, \varphi)}{\partial \varphi^2} + s^2 u^*(s, \varphi) = 0 \tag{1.5.48}$$

with the boundary conditions

$$\begin{cases} u^*(s, \alpha) & = u_+^*(s) \\ u^*(s, -\alpha) & = u_-^*(s). \end{cases} \tag{1.5.49}$$

The solution of (1.5.48) is

$$u^*(s, \varphi) = A(s) \cos \varphi s + B(s) \sin \varphi s.$$

With the help of the boundary conditions (1.5.49) we get

$$A(s) = \frac{u_+^*(s) + u_-^*(s)}{2 \cos \alpha s}, \qquad B(s) = \frac{u_+^*(s) - u_-^*(s)}{2 \sin \alpha s},$$

i.e.,

$$u^*(s, \varphi) = u_+^*(s) \frac{\sin(\alpha + \varphi)s}{\sin 2\alpha s} + u_-^*(s) \frac{\sin(\alpha - \varphi)s}{\sin 2\alpha s}. \tag{1.5.50}$$

Putting

$$h^*(s, \varphi) = \frac{\sin \varphi s}{\sin 2\alpha s}$$

we have

$$u^*(s, \varphi) = u_+^*(s) h^*(s, \alpha + \varphi) + u_-^*(s) h^*(s, \alpha - \varphi)$$

with $\nu = \pi/2\alpha$, $\nu \in (-1, 1)$. By means of formula (1.5.31), namely,

$$h(r, \varphi) = \frac{\nu}{\pi} \frac{r^\nu \sin \nu\varphi}{1 + 2r^\nu \cos \nu\varphi + r^{2\nu}}, \tag{1.5.51}$$

and with the help of the convolution theorem of the MT we obtain the formal solution of our problem after a simple calculation:

$$u(r, \varphi) = u_+ \vee h(\cdot, \alpha + \varphi) + u_- \vee h(\cdot, \alpha - \varphi)$$

$$= \frac{\nu r^\nu \cos \nu \rho}{\pi} \left[\int_0^\infty \frac{\rho^{n-1} u_+(\varphi)}{\rho^{2\nu} - 2(r\rho)^\nu \sin \nu \varphi + r^{2\nu}} \, d\rho \right.$$

$$\left. + \int_0^\infty \frac{\rho^{n-1} u_-(\rho)}{\rho^{2\nu} + 2(r\rho)^\nu \sin \nu \varphi + r^{2\nu}} \, d\rho \right], \quad -1 < \nu < 1. \tag{1.5.52}$$

For further applications of the MT we refer to [Sn.2], [Tra.2] and [De.6].

Example 1.5.43 *Now we consider Fourier-type integral transforms. Let*

$$F(y) = \int_0^\infty f(x) k(xy) dx \tag{1.5.53}$$

exist and let there exist an inversion formula of the type

$$f(x) = \int_0^\infty F(y) h(xy) dy \tag{1.5.54}$$

with some kernel h. Then we have from (1.5.53) (provided that all calculations in the following can be justified)

$$\mathcal{M}[F](s) = \int_0^\infty y^{s-1} F(y) dy = \int_0^\infty f(x) \left(\int_0^\infty y^{s-1} k(xy) dy \right) dx.$$

Substituting $y \to u$ by $xy = u$ we have

$$\mathcal{M}[F](s) = \int_0^\infty x^{-s} f(x) dx \int_0^\infty u^{s-1} k(u) du,$$

that is,

$$\mathcal{M}[F](s) = \mathcal{M}[f](1-s) \cdot \mathcal{M}[k](s). \tag{1.5.55}$$

Analogously from (1.5.54) we obtain

$$\mathcal{M}[f](s) = \mathcal{M}[F](1-s) \cdot \mathcal{M}[h](s). \tag{1.5.56}$$

From (1.5.55) and (1.5.56) we have:

Proposition 1.5.23 *Let k and h be the kernels of a Fourier-type transform (1.5.53) and its inverse (1.5.54), respectively, then it holds that*

$$\mathcal{M}[k](1-s) \cdot \mathcal{M}[h](s) = 1. \tag{1.5.57}$$

The kernel h of the inversion formula (1.5.54) can therefore be calculated by means of

$$h(x) = \mathcal{M}^{-1}[\mathcal{M}[h]](x) = \mathcal{M}^{-1}\left[\frac{1}{\mathcal{M}[k](1-s)}\right](x). \tag{1.5.58}$$

These calculations can (under appropriate conditions) be done in inverse direction, that is, if two kernels h, k fulfill the equation (1.5.57) then the transform (1.5.53) has the inversion formula (1.5.54).

1.6 The Stieltjes Transform

1.6.1 Definition and Basic Properties

Definition 1.6.10 *The Stieltjes transform of a function $f : \mathbb{R}_+ \to \mathbb{C}$ is defined by means of*

$$\mathcal{S}[f](z) = \int_0^\infty \frac{f(t)}{t+z} dt, \tag{1.6.1}$$

provided that the integral exists.

For the existence we have the following three theorems.

Theorem 1.6.26 *If the integral in (1.6.1) converges for a point $z = z_0 \in \mathbb{C} \setminus (-\infty, 0]$ then it converges for every such point $z \in \mathbb{C} \setminus (-\infty, 0]$.*

Proof. Set

$$f_0(t) = \int_0^t \frac{f(u)}{u+z_0} du, \qquad t \in [0, \infty). \tag{1.6.2}$$

Then for any $z \in \mathbb{C} \setminus (-\infty, 0]$ and for any $R \in \mathbb{R}_+$ we have

$$\int_0^R \frac{f(t)}{t+z} dt = \int_0^R \frac{t+z_0}{t+z} f_0'(t) dt = f_0(R)\frac{R+z_0}{R+z} + (z_0 - z)\int_0^R \frac{f_0(t)}{(t+z)^2} dt.$$

Since $f_0(R)$ tends to $\mathcal{S}[f](z_0)$ as R tends to $+\infty$ the last integral converges absolutely as $R \to +\infty$. Therefore, (1.6.1) converges and we have: □

Corollary 1.6.13 *Under the conditions of* Theorem 1.6.26 *it holds that*

$$\mathcal{S}[f](z) = \mathcal{S}[f](z_0) + (z_0 - z) \int_0^\infty \frac{f_0(t)}{(t+z)^2} dt, \tag{1.6.3}$$

the integral being absolutely convergent.

Theorem 1.6.27 *If the integral in (1.6.1) converges in some point $z_0 \in \mathbb{C} \setminus (-\infty, 0]$ it converges uniformly in any compact subset K of $\mathbb{C}$ not containing points of the negative real axis $(-\infty, 0]$.*

Proof. Let

$$M = \max_{z \in K} |z|.$$

Then we have for any $R > M$ (with the notations of the latter proof) and (1.6.3)

$$\int_R^\infty \frac{f(t)}{t+z}dt = \mathcal{S}[f](z) - \int_0^R \frac{f(t)}{t+z}dt = \mathcal{S}[f](z_0) - f_0(R)\frac{R+z_0}{R+z} + (z_0 - z)\int_R^\infty \frac{f_0(t)}{(t+z)^2}dt,$$

and therefore

$$\left| \int_R^\infty \frac{f(t)}{t+z}dt \right| \leq |\mathcal{S}[f](z_0) - f_0(R)| + |f_0(R)|\frac{|z-z_0|}{|z+R|} + |z_0 - z|\int_R^\infty \frac{|f_0(t)|}{(|t+z|)^2}dt$$

$$\leq |\mathcal{S}[f](z_0) - f_0(R)| + |f_0(R)|\frac{M+|z_0|}{R-M} + (M+|z_0|)\int_R^\infty \frac{|f_0(t)|}{(t-M)^2}dt.$$

The right-hand side is independent of z and it tends to zero as R tends to $+\infty$. This completes the proof.   ☐

From Theorem 1.6.27 we have:

Theorem 1.6.28 *If the integral in (1.6.1) converges then the Stieltjes transform $\mathcal{S}[f](z)$ represents an analytic function in the complex plane cut along $(-\infty, 0]$ and*

$$D^k \mathcal{S}[f](z) = (-1)^k k! \int_0^\infty \frac{f(t)}{(t+z)^{k+1}}dt, \quad k \in \mathbb{N}_0.$$

Sufficient conditions for the existence of the Stieltjes transform (1.6.1) are given in the following:

Theorem 1.6.29 *Let $f \in L_1^{loc}(\mathbb{R}_+)$ and for some positive δ*

$$f(t) = 0(t^{-\delta}), \qquad t \to +\infty.$$

Then the Stieltjes transform $\mathcal{S}[f](z)$ exists on $\mathbb{C} \setminus (-\infty, 0]$.

Remark 1.6.32 *Stieltjes considered more generally the transform*

$$\alpha \to \int_0^\infty \frac{d\alpha(t)}{t+z}dt, \tag{1.6.4}$$

where α must be of bounded variation in $[0, R]$ for every positive R. The theorems 1.6.26 through 1.6.28 are valid also in this case and in Theorem 1.6.29 one has to replace the condition on the behavior at $+\infty$ by

$$\alpha(t) = 0(t^{1-\delta}), \qquad t \to +\infty.$$

For details and proofs we refer to [Wi.1] and [Wi.2].

Remark 1.6.33 *Still more general sometimes one considers the transforms*

$$\mathcal{S}_\varrho[f](z) = \int\limits_0^\infty \frac{f(t)}{(t+z)^\varrho} dt, \qquad (1.6.5)$$

resp.

$$\alpha \to \int\limits_0^\infty \frac{d\alpha(t)}{(t+z)^\varrho}, \qquad (1.6.6)$$

where $\varrho \subset \mathbb{R}_+$. The theorems 1.6.26 through 1.6.28 are again valid in those cases. In Theorem 1.6.29 one has to replace the condition on the behavior at $+\infty$ by

$$f(t) = 0(t^{\varrho-1-\delta}), \qquad t \to +\infty,$$

resp.

$$\alpha(t) = 0(t^{\varrho-\delta}), \qquad t \to +\infty.$$

For details see again [Wi.1] and [Wi.2].

Now we are going to derive the connection between the Stieltjes and the Laplace transforms.

Proposition 1.6.24 *Under the conditions of Theorem 1.6.29, but with $\delta > 1$, it holds that*

$$\mathcal{S}[f](z) = \mathcal{L}[\mathcal{L}[f](x)](z), \qquad Re(z) > 0. \qquad (1.6.7)$$

Proof. From

$$\frac{1}{t+z} = \int\limits_0^\infty e^{-(t+z)x} dx$$

it follows that

$$\mathcal{S}[f](z) = \int\limits_0^\infty f(t) \int\limits_0^\infty e^{-(t+z)x} dx dt = \int\limits_0^\infty e^{-zx} \int\limits_0^\infty f(t) e^{-xt} dt dx,$$

and the interchanging of the integration is permissible under our conditions. ☐

By means of the connection (1.6.7) and of the uniqueness theorem for the Laplace transform (see Theorem 1.4.9) we obtain

Theorem 1.6.30 *If the Stieltjes transform of f and g are absolutely convergent and* $S[f] = S[g]$, *then* $f = g$.

For examples of Stieltjes transforms we refer to the tables EMOT, vol. 2, Chapter XIV or the tables of Laplace transforms [PBM], vol. IV, [EMOT], vol. 1, or [OB] together with Proposition 1, and to the textbooks [De.6] and [Sn.2].

1.6.2   Operational Properties

The operational properties of the Stieltjes transform are valid if all the expressions appearing in the formulas exist, but we add sufficient conditions for the validity of the formulas. By straightforward calculation we have

Proposition 1.6.25 *Let f fulfill the conditions of 1.6.1, Theorem 1.6.29 and let* $f(t) = 0$ *if* $t < 0$. *Then for* $a, b \in \mathbb{R}_+$ *we have the translation rule*

$$S[f(t-a)](z) = S[f](z+a), \qquad z \in \mathbb{C} \setminus (-\infty, -a] \tag{1.6.8}$$

and the similarity rule

$$S[f(bt)](z) = S[f](bz), \qquad z \in \mathbb{C} \setminus (-\infty, 0]. \tag{1.6.9}$$

From the identity

$$\frac{t}{t+z} = 1 - \frac{z}{t+z}$$

we obtain the multiplication rule.

Proposition 1.6.26 *Let* $f \in L_1(\mathbb{R}_+)$. *Then it holds that*

$$S[tf(t)](z) = \int_0^\infty f(t)dt - zS[f](z). \tag{1.6.10}$$

Integrating by parts we deduce the differentiation rule (using Theorem 1.6.28).

Proposition 1.6.27 *Let there exist* f' *a.e.   and let* f, f' *fulfill the conditions of 1.6.1, Theorem 1.6.29. Then we have*

$$S[f'](z) = -\frac{d}{dz}S[f](z) - z^{-1}f(0). \tag{1.6.11}$$

More generally, we have:

Corollary 1.6.14

$$S[D^n f](z) = (-1)^n D^n S[f](z) - \left[\frac{(n-1)!}{z^n}f(0) + \cdots + \frac{f^{(n-2)}(0)}{z^2} + \frac{f^{(n-1)}(0)}{z}\right]. \quad (1.6.12)$$

For the iteration of two Stieltjes transforms we have quite formally

$$S[S[f]](z) = \int_0^\infty (x+z)^{-1}\left(\int_0^\infty \frac{f(t)}{t+x}dt\right)dx = \int_0^\infty f(t)\left(\int_0^\infty \frac{dx}{(x+z)(x+t)}\right)dt$$

$$= \int_0^\infty \frac{\log(t/z)f(t)}{t-z}dt.$$

Therefore we have under the assumptions that $S[f]$ and $S[S[f]]$ exist the connection

$$S[S[f]](z) = \int_0^\infty \frac{\log(t/z)}{t-z}f(t)dt. \quad (1.6.13)$$

The operational properties can also be formulated for the Stieltjes transform of index $\varrho \in \mathbb{R}_+$; see formula (1.6.5). Without proof we give only the results:

$$S_\varrho[f(t-a)](z) = S_\varrho[f](z+a), \qquad a \in \mathbb{R}_+, \quad (1.6.14)$$

$$S_\varrho[f(bt)](z) = b^{\varrho-1}S_\varrho[f](bz), \qquad b \in \mathbb{R}_+, \quad (1.6.15)$$

$$S_\varrho[tf(t)](z) = S_{\varrho-1}[f](z) - zS_\varrho[f](z), \quad (1.6.16)$$

$$S_\varrho[f'](z) = \varrho S_{\varrho+1}[f](z) - z^{-1}f(0), \quad (1.6.17)$$

$$S_\mu[S_\varrho f](z) = B(1, \mu+\varrho-1)z^{-\mu}\int_0^\infty t^{1-\varrho}\,{}_2F_1\left(\mu; 1; \mu+\varrho; 1-\frac{t}{z}\right)dt, \quad (1.6.18)$$

where B is the Beta function, see equation (1.4.31), and ${}_2F_1$ is Gauss' hypergeometric function. In this case one also can easily prove an integration rule. Setting

$$g(t) = \frac{1}{\Gamma(\mu)}\int_0^t f(u)(t-u)^{\mu-1}du, \qquad \mu > 0,$$

one can derive

$$S_\varrho[g](z) = \frac{\Gamma(\varrho-\mu)}{\Gamma(\varrho)}S_{\varrho-\mu}[f](z), \qquad \varrho > \mu. \quad (1.6.19)$$

In this case $\mu = 1$ this leads to

$$S_\varrho\left[\int_0^t f(u)du\right](z) = (\varrho-1)^{-1}S_{\varrho-1}[f](z), \qquad \varrho > 1. \quad (1.6.20)$$

For details we refer to [Za], Chapter VIII.

Following [SV] we prepare the proof of a convolution theorem for the Stieltjes transform.

Definition 1.6.11 *As convolution* $h = f \otimes g$ *of two functions* f *and* $g : \mathbb{R}_+ \to \mathbb{C}$ *we define*

$$h(t) = (f \otimes g)(t) = f(t) \int\limits_0^\infty \frac{g(u)}{u-t} du + g(t) \int\limits_0^\infty \frac{f(u)}{u-t} du, \qquad (1.6.21)$$

provided that the integrals exist.

Remark 1.6.34 *Setting*

$$f^\sim(t) = \int\limits_0^\infty \frac{f(u)}{u-t} du, \qquad g^\sim(t) = \int\limits_0^\infty \frac{g(u)}{u-t} du \qquad (1.6.22)$$

one has $f^\sim = \mathcal{H}[1_+(t)f(t)]$, $g^\sim = \mathcal{H}[1_+(t)g(t)]$, *where* $1_+(t)$ *is the Heaviside function and* $\mathcal{H}$ *is the Hilbert transform, investigated in section 1.7, which follows. From this section we know that* $f^\sim$, $g^\sim$ *exist if* $f, g \in L_1(\mathbb{R}_+)$, *provided that the integrals are understood in the sense of Cauchy's principal value* (PV).

So we can prove:

Theorem 1.6.31 (Convolution Theorem) *Let* $f, g \in L_1(\mathbb{R}_+)$ *and let the Stieltjes transforms of* $fg^\sim$ *and* $f^\sim g$ *be absolutely convergent. Then there exists the Stieltjes transform of the convolution* $f \otimes g$ *and it holds that*

$$\mathcal{S}[f \otimes g] = \mathcal{S}[f] \cdot \mathcal{S}[g]. \qquad (1.6.23)$$

Proof.   Under our assumptions we have

$$\mathcal{S}[f \otimes g](z) = \int\limits_0^\infty \frac{f(t)}{t+z} \Big(\int\limits_0^\infty \frac{g(u)}{u-t}, \ du \Big) dt + \int\limits_0^\infty \frac{g(t)}{t+z} \Big(\int\limits_0^\infty \frac{f(u)}{u-t} du \Big) dt.$$

Interchanging the order of integration we obtain

$$\mathcal{S}[f \otimes g](z) = \int\limits_0^\infty f(t) \Big(\int\limits_0^\infty \frac{g(u)}{(t+z)(u-t)} du \Big) dt + \int\limits_0^\infty f(t) \Big(\int\limits_0^\infty \frac{g(u)}{(u+z)(t-u)} du \Big) dt$$

$$= \int\limits_0^\infty f(t) \Big(\int\limits_0^\infty \frac{g(u)}{u-t} \Big[\frac{1}{t+z} - \frac{1}{u+z} \Big] du \Big) dt = \Big(\int\limits_0^\infty \frac{f(t)}{t+z} dt \Big) \Big(\int\limits_0^\infty \frac{g(u)}{u+z} du \Big)$$

$$= \mathcal{S}[f](z) \cdot \mathcal{S}[g](z).$$

$\square$

Remark 1.6.35 *In [SV] H.M. Silvastrava and Vu Kim Tuan have proved that for* $f \in L_p, g \in L_q$, $p, q > 1$ *and* $p^{-1} + q^{-1} =: r^{-1} < 1$ *it holds that* $h \in L_r$ *and* (1.6.23) *is valid.*

1.6.3 Asymptotics

We first would like to investigate the behavior of the originals at $+\infty$.

Theorem 1.6.32 *Let $f \geq 0$ (or $f \leq 0$) on $\mathbb{R}_+$ and let the Stieltjes transform 1.6.1, (1.6.1) exist. Then it holds that*

$$f(t) = o(1), \qquad t \to +\infty. \qquad (1.6.24)$$

Proof. Let $f \geq 0$ and set

$$f_0(t) = \int_0^t f(u)du, \qquad t \in \mathbb{R}_+,$$

and

$$\varphi(t) = \int_1^t \frac{f(u)}{u+1}du, \qquad t \in [1, \infty).$$

Then we have integrating by parts

$$f_0(t) - f_0(1) = \int_1^t f(u)du = \int_1^t (u+1)d\varphi(u) = (t+1)\varphi(t) - \int_1^t \varphi(u)du.$$

Therefore,

$$\frac{f_0(t)}{t} = \frac{f_0(1)}{t} + \left(1 + \frac{1}{t}\right)\varphi(t) - \frac{1}{t}\int_1^t \varphi(u)du$$

and it follows by the mean value theorem for the integral

$$\frac{f_0(t)}{t} \sim \varphi(\infty) - \varphi(\infty) = 0,$$

so

$$f_0(t) = o(t), \qquad t \to +\infty. \qquad (1.6.25)$$

By differentiation, which is permissible since f_0 is nondecreasing, we obtain (1.6.24). In the case of $f \leq 0$ we consider $g = -f$. □

Now we are going to derive the asymptotics of the images at $+\infty$.

Theorem 1.6.33 *If the Stieltjes transform (1.6.1) exists, then*

$$D^n \mathcal{S}[f](x) = o(x^{-n}), \qquad x \to +\infty, \ n \in \mathbb{N}_0. \qquad (1.6.26)$$

Proof.

1. $n \geq 1$: Setting $F(x) = \mathcal{S}[f](x)$, $x \in \mathbb{R}_+$ we have from 1.6.1, Theorem 1.6.28 integrating by parts

$$D^n F(x) = (-1)^n n! \int_0^\infty \frac{f(t)}{(t+x)^{n+1}}dx = (-1)^n (n+1)! \int_0^\infty \frac{f_0(t)}{(t+x)^{n+2}}dt, \qquad (1.6.27)$$

where the notations of Theorem 1.6.32 and the result (1.6.26) have been used. Again using Theorem 1.6.32 for arbitrary $\varepsilon \in \mathbb{R}_+$ we determine a positive number R so that

$$|f_0(t)| < \varepsilon t, \qquad t \in [R, \infty).$$

From (1.6.27) we obtain for $x > R$

$$|x^n D^n F(x)| \leq x^n (n+1)! \int_0^R \frac{|f_0(t)|}{(t+x)^{n+2}} dt + x^n \varepsilon (n+1)! \int_R^\infty \frac{t}{(t+x)^{n+2}} dt$$

$$\leq \frac{(n+1)!}{x^2} \int_0^R |f_0(t)| dt + \varepsilon \frac{(n+1)!}{n},$$

where in the latter integral the nominator was written as $(t+x) - x$. As x tends to $+\infty$ we obtain (1.6.26) in the case of $n \geq 1$.

2. $n = 0$: Using 1.6.1, Corollary 1.6.13 with $z_0 = 1$ we have

$$F(x) = F(1) + (1-x) \int_0^\infty \frac{f_1(1)}{(t+x)^2} dt,$$

where

$$f_1(t) = \int_0^t \frac{f(u)}{u+1} du.$$

For the proof of (1.6.26) in the case of $n = 0$ it is sufficient to show:

$$\lim_{x \to +\infty} x \int_0^\infty \frac{f_1(t)}{(t+x)^2} dt = F(1) = f_1(\infty).$$

From

$$\Delta := x \int_0^\infty \frac{f_1(t)}{(t+x)^2} dt - f_1(\infty) = x \int_0^\infty \frac{f_1(t) - f_1(\infty)}{(t+x)^2} dt$$

and since for arbitrary $\varepsilon \in \mathbb{R}_+$ there exists a number $R \in \mathbb{R}_+$ with

$$|f_1(t) - f_1(\infty)| < \varepsilon/2, \qquad t \in [R, \infty)$$

we have

$$|\Delta| < x \int_0^R \frac{|f_1(t) - f_1(\infty)|}{(t+x)^2} dt + \varepsilon/2 < \varepsilon$$

if x is sufficiently large.   ☐

Finally, we consider the behavior of $\mathcal{S}[f](z)$ at $z = 0$.

Theorem 1.6.34 *Under the conditions of* Theorem 1.6.32, *the Stieltjes transform* $F(z) = \mathcal{S}[f](z)$ *has a singularity at* $z = 0$.

Proof. We consider again only the case that f is nonnegative. Let us assume that $z = 0$ is no singularity of $F(z)$. Then the radius of convergence of the series

$$F(z) = \sum_{n=0}^{\infty} f^{(n)}(1) \frac{(z-1)^n}{n!}$$

is greater than one, i.e., the series

$$F(-\varepsilon) = \sum_{n=0}^{\infty} (-1)^n f^{(n)}(1) \frac{(\varepsilon+1)^n}{n!}$$

is convergent for some $\varepsilon \in \mathbb{R}_+$ which is taken such that $f(\varepsilon) > 0$. Then by means of Theorem 1.6.29 and Theorem 1.6.26 we get

$$F(-\varepsilon) = \sum_{n=0}^{\infty} \int_0^{\infty} \frac{(\varepsilon+1)^n}{(t+1)^{n+1}} f(t)dt = \sum_{n=0}^{\infty}(n+1) \int_0^{\infty} \frac{(\varepsilon+1)^n}{(t+1)^{n+2}} f(t)dt.$$

Therefore, the series

$$\sum_{n=0}^{\infty}(n+1) \int_{\varepsilon}^{\infty} \frac{(\varepsilon+1)^n}{(t+1)^{n+2}} f(t)dt \qquad (1.6.28)$$

is also convergent. Since the expression under the integral is nonnegative and since the series

$$\sum_{n=0}^{\infty}(n+1) \frac{(\varepsilon+1)^n}{(t+1)^{n+2}} = \frac{1}{(t-\varepsilon)^2}$$

converges for $t > \varepsilon$ one may interchange integration and summation in (1.6.28) and we have the convergent integral

$$\int_{\varepsilon}^{\infty} \frac{f(t)}{(t-\varepsilon)^2} dt.$$

But this integral is divergent because of

$$\lim_{t \to \varepsilon_+} f(t)/(t-\varepsilon) = +\infty.$$

Therefore the assumption that $z = 0$ is no singularity of $F(s)$ is wrong. □

1.6.4 Inversion and Application

Now we are going to derive a complex inversion formula for the Stieltjes transform. Following Widder [Wi.1], Chapter VIII, §7 we need two preliminary results.

Lemma 1.6.5 *Let $f \in L_1(0, R)$ and let $f(0+)$ exist. Then*

$$\lim_{y \to +0} \frac{y}{\pi} \int_0^R \frac{f(t)}{t^2 + y^2} \, dt = \frac{f(0+)}{2} \tag{1.6.29}$$

for every $R \in \mathbb{R}_+$.

Proof. We can assume that $f(0+) = 0$ since

$$\lim_{R \to +\infty} y \int_0^R \frac{dt}{t^2 + y^2} = \lim_{R \to +\infty} \operatorname{arctg} \frac{R}{y} = \frac{\pi}{2}.$$

For arbitrary $\varepsilon \in \mathbb{R}_+$ we determine a positive $\delta < R$ so that $|f(t)| < \varepsilon$ for $t \in [0, \delta]$. Then

$$\left| y \int_0^R \frac{f(t)}{t^2 + y^2} \, dt \right| \le \varepsilon y \int_0^\delta \frac{dt}{t^2 + y^2} + y \int_\delta^R \frac{|f(t)|}{t^2} \, dt$$

and it follows that

$$\overline{\lim_{y \to 0+}} \left| y \int_0^R \frac{f(t)}{t^2 + y^2} \, dt \right| \le \frac{\pi}{2} \varepsilon$$

and this is (1.6.29) in the case of $f(0+) = 0$.   □

Lemma 1.6.6 *Let $f \in L_1(0, R)$ and $x \in (0, R)$. If $f(x+)$ and $f(x-)$ exist, then*

$$\lim_{y \to 0+} \frac{y}{\pi} \int_0^R \frac{f(t)}{(t-x)^2 + y^2} \, dt = \frac{1}{2} \Big(f(x+) + f(x-) \Big). \tag{1.6.30}$$

Proof. The result follows from Lemma 1.6.5 by writing the integral in (1.6.30) as the sum of two integrals corresponding to the intervals $(0, x)$ and (x, R).   □

Now we are able to prove:

Theorem 1.6.35 *Let $f \in L_1^{loc}(\mathbb{R}_+)$ such that the Stieltjes transform $F(z) = \mathcal{S}[f](z)$ converges, then*

$$\lim_{y \to 0+} \frac{F(-x - iy) - F(-x + iy)}{2\pi i} = \frac{1}{2} \Big(f(x+) + f(x-) \Big) \tag{1.6.31}$$

for any $x \in \mathbb{R}_+$ at which $f(x+)$ and $f(x-)$ exist.

Proof. By straightforward calculation we have for $R > x$

$$\frac{F(-x - iy) - F(-x + iy)}{2\pi i} = \frac{y}{\pi} \int_0^\infty \frac{f(t)}{(t-x)^2 + y^2} \, dt = \frac{y}{\pi} \left(\int_0^R + \int_R^\infty \right) \frac{f(t)}{(t-x)^2 + y^2} \, dt = I_1 + I_2.$$

From Lemma 1.6.6 we know that

$$\lim_{y \to 0+} I_1 = \frac{f(x+) + f(x-)}{2}$$

and we have only to show that

$$\lim_{y \to 0+} I_2 = 0.$$

Set

$$f_0(t) = \int_0^t f(u)du, \qquad 0 \le t < \infty.$$

Then integration by parts leads to

$$I_2 = \frac{-y f_0(R)}{\pi[(R-x)^2 + y^2]} + \frac{2y}{\pi} \int_R^\infty \frac{f_0(t)(t-x)}{[(t-x)^2 + y^2]^2} dt.$$

The first term on the right-hand side tends to zero as $y \to 0+$. Denoting the second term by I_3 and using

$$|f_0(t)| < Mt, \qquad 0 \le f < \infty,$$

see formula (1.6.25), we have

$$|I_3| < \frac{2yM}{\pi} \int_R^\infty \frac{t}{(t-x)^3} dt.$$

The integral converges and, therefore, I_3 tends to zero as $y \to 0+$ and, hence,

$$\lim_{y \to 0+} I_2 = 0,$$

and the proof is completed. ▢

Remark 1.6.36 *The formula (1.6.31) gives the inversion only for the original f at points of the positive real axis.*

Remark 1.6.37 *Formula (1.6.31) may be written symbolically as*

$$f(x) = (2\pi i)^{-1} \left(F(xe^{-\pi i}) - F(xe^{\pi i}) \right) \tag{1.6.32}$$

in an easily understandable manner.

As an application we consider the Stieltjes integral equation

$$f(x) = \lambda \int_0^\infty \frac{u(t)}{t+x} dt, \qquad \lambda, x \in \mathbb{R}_+, \tag{1.6.33}$$

where f is a given function and u is the solution of the integral equation we are looking for. Quite formally applying formula (1.6.32) we have

$$u(x) = (2\pi i \lambda)^{-1}\big(f(xe^{-\pi i}) - f(xe^{\pi i})\big). \tag{1.6.34}$$

One can prove, for example, that a necessary and sufficient condition for (1.6.33) to have a solution (1.6.34) of $L^2(\mathbb{R}_+)$ is that $f(z)$ should be analytic on $\mathbb{C}\setminus(-\infty,0]$ and that

$$\int_0^\infty |f(re^{i\varphi})|^2 dr$$

should be bounded for $\varphi \in (-\pi,\pi)$ (see [T.2], 11.8). For further applications refer to [De.6], section 7.10; [Wi.1], Chapter VIII, sections 25–27; and [Za], 8.10.

1.7 The Hilbert Transform

1.7.1 Definition and Basic Properties

Definition 1.7.12 *The Hilbert transform of a function $f : \mathbb{R} \to \mathbb{C}$ is the function $f^\sim$ defined by*

$$f^\sim(x) = \mathcal{H}[f](x) = \frac{1}{\pi}\int_{-\infty}^{\infty}\frac{f(t)}{t-x}dt, \tag{1.7.1}$$

provided that the integral exists.

As space of originals we choose the space $L_1(\mathbb{R}) = L_1$. One can prove:

Theorem 1.7.36 *Let $f \in L_1$. Then the Hilbert transform $f^\sim$ exists a.e., provided that the integral in (1) is understood in the sense of Cauchy's principal value (PV), i.e.,*

$$f^\sim(x) = \frac{1}{\pi}(PV)\int_{-\infty}^{\infty}\frac{f(t)}{t-x}dt = \frac{1}{\pi}\lim_{\delta\to+0}\left(\int_{-\infty}^{x-\delta}+\int_{x+\delta}^{\infty}\right)\frac{f(t)}{t-x}dt. \tag{1.7.2}$$

The proof is too lengthy to be presented here. We refer to the functional analytic proof in [BuN], section 8, and a proof by means of the continuation into the complex domain $x \to z = x + iy$, $y > 0$ and by use of theorems on the existence of the limit of analytic functions defined in the upper half-plane as y tends to $+0$; see [T.2], section 5.10.

Remark 1.7.38 *From (1.7.2) we derive the following form:*

$$f^\sim(x) = \frac{1}{\pi}\lim_{\delta\to+0}\int_\delta^\infty\frac{f(x+t)-f(x-t)}{t}dt = \frac{1}{\pi}(PV)\int_{-\infty}^{\infty}\frac{f(x+t)}{t}dt. \tag{1.7.3}$$

Remark 1.7.39 *Similar as in the case of the FT, see 1.3.1, Example 1.3.2, the Hilbert transform $f^\sim$ of a L_1-function f is (in general) not a L_1-function. Let f be the L_1-function defined by*

$$f(t) = (1+t^2)^{-1}, \qquad t \in \mathbb{R}.$$

Then by fractional decomposition we have

$$f^\sim(x) = \frac{1}{\pi}(1+x^2)^{-1}(PV)\int_{-\infty}^{\infty}\left[\frac{-x}{t^2+1} - \frac{t}{t^2+1} + \frac{1}{t-x}\right]dt.$$

Considering the integral as (PV) not only with respect to x but also with respect to $\pm\infty$ the last two integrals vanish and we obtain

$$f^\sim(x) = -\frac{x}{1+x^2},$$

such that $f^\sim \notin L_1$.

Remark 1.7.40 *If $f \in L_p(\mathbb{R}) = L_p$, $p > 1$, then we have a stronger result. The Hilbert transform also exists a.e. and $f^\sim \in L_p$. In the case of $p = 2$ it holds moreover that*

$$\|f^\sim\|_2 = \|f\|_2.$$

For the proofs we refer to [BuN], section 8, or [T.2], section 5.10.

Now we derive the connection between the Hilbert transform on the one hand and the FT and the Stieltjes transform on the other hand.

Proposition 1.7.28 *Let $f, f^\sim \in L_1$. Then it holds that*

$$\mathcal{F}[\mathcal{H}[f]](x) = isgn(x)\mathcal{F}[f](x). \tag{1.7.4}$$

Proof. We have

$$(f^\sim)^\wedge(x) = frac1\pi \int_{-\infty}^{\infty} e^{-ixt}\left(\int_{-\infty}^{\infty}\frac{f(u)}{u-t}du\right)dt = \frac{1}{\pi}\int_{-\infty}^{\infty} f(u)\left(\int_{-\infty}^{\infty}\frac{e^{-ixt}}{u-t}dt\right)du$$
$$= \frac{1}{\pi}\int_{-\infty}^{\infty} f(u)e^{-ixu}\left(\int_{-\infty}^{\infty}\frac{e^{ixv}}{v}dv\right)du. \tag{1.7.5}$$

Now we have

$$\pi^{-1}\int_{-\infty}^{\infty} v^{-1}e^{ixv}dv = 2\mathcal{F}^{-1}[v^{-1}](x) = \frac{2i}{\pi}sgn(x)\int_{0}^{\infty}\frac{\sin y}{y}dy = isgn(x),$$

because of (1.4.23). Replacing x by $-x$ we get

$$\mathcal{F}[(\pi t)^{-1}](x) = -isgn(x). \tag{1.7.6}$$

Together with (1.7.5) we obtain the result (1.7.4).

By quite formal calculation we obtain the connection between the Hilbert transform and the Stieltjes transform; see formula (1.6.1). We have for $x > 0$

$$\mathcal{H}[f](x) = \frac{1}{\pi} \int_{-\infty}^{\infty} \frac{f(t)}{t-x} dt = \frac{1}{\pi} \left[\int_0^{\infty} \frac{f(t)}{t-x} dt - \int_0^{\infty} \frac{f(-t)}{t+x} dt \right]$$

$$= \frac{1}{2\pi} \left[\int_0^{\infty} \frac{f(t)}{t+xe^{\pi i}} dt + \int_0^{\infty} \frac{f(t)}{t+xe^{-\pi i}} dt \right] - \frac{1}{\pi} \int_0^{\infty} \frac{f(-t)}{t+x} dt.$$

Similarly we obtain a result in the case of $x < 0$. So we have   ▯

Proposition 1.7.29 *Let $f \in L_1$. Then it holds for $x > 0$ that*

$$\mathcal{H}[f](x) = \frac{1}{2\pi} \left[\mathcal{S}[f](xe^{\pi i}) + \mathcal{S}[f](xe^{-\pi i}) \right] - \frac{1}{\pi} \mathcal{S}[f(-t)](x), \qquad (1.7.7)$$

and for $x < 0$,

$$\mathcal{H}[f](x) = \frac{1}{2\pi} \mathcal{S}[f](-x) - \frac{1}{2\pi} \left[\mathcal{S}[f(-t)](|x|e^{\pi i}) + \mathcal{S}[f(-t)](|x|)e^{-\pi i}) \right], \qquad (1.7.8)$$

where the integral has to be taken as (PV).

For examples of Hilbert transforms we refer to the tables [EMOT], vol. 2, Chapter XV or to the tables of Stieltjes transforms [EMOT], vol. 2, Chapter XIV, together with Proposition 1.7.29.

Now we derive an inversion formula, but only quite formally. The Hilbert transform (1.7.1) can be written as the Fourier convolution 1.3.3, (1.3.11) of the functions f and g, where

$$g(t) = \frac{1}{\pi t}.$$

Using (1.7.6) and the convolution theorem (1.3.29) of the FT we have

$$(f^\sim)^\wedge(x) = isgn(x)f^\wedge(x),$$

or equivalently,

$$f^\wedge(x) = -isgn(x)(f^\sim)^\wedge(x).$$

Again using the convolution theorem of the FT (see 1.3.4, Remark 1.3.12) we have

$$f(x) = -\frac{1}{\pi} \int_{-\infty}^{\infty} \frac{f^\sim(x)}{t-x} dt. \qquad (1.7.9)$$

The result is:

Theorem 1.7.37 *Let f, $f^\sim \in L_1$. Then it holds a.e., the inversion formula (1.7.9) or equivalently*

$$f = -(f^\sim)^\sim. \tag{1.7.10}$$

Remark 1.7.41 *The inversion formula (1.7.9), (1.7.10) is also valid a.e. if $f \in L_p$, $p > 1$. For the proofs of this result and of* Theorem 1.7.37 *we refer to [BuN], Proposition 8.2.10 or [T.2], Th. 101.*

1.7.2 Operational Properties

By straightforward calculation we obtain the following elementary results of operational calculus of the Hilbert transform.

Proposition 1.7.30 *Let $a \in \mathbb{R}$, $0 \neq b \in \mathbb{R}$ and $f \in L_1$. Then it holds that*

$$\mathcal{H}[f(t+a)](x) = f^\sim(x+a), \tag{1.7.11}$$

$$\mathcal{H}[f(bt)](x) = sgn(b)f^\sim(bx), \tag{1.7.12}$$

and

$$\mathcal{H}[tf(t)](x) = xf^\sim(x) + \frac{1}{\pi} \int_{-\infty}^{\infty} f(t)dt. \tag{1.7.13}$$

Integrating by parts we obtain a differentiation rule.

Proposition 1.7.31 *Let there exist f' a.e. and let $f, f' \in L_1$. Then it holds that*

$$\mathcal{H}[f'](x) = \frac{d}{dx}f^\sim(x). \tag{1.7.14}$$

A convolution theorem for the Hilbert transform was published by Tricomi; see [Tri], section 4.2. Following [GlV] we have

Theorem 1.7.38 *Let f, g, $fg \in L_1$ such that their Hilbert transforms $f^\sim, g^\sim$, $(fg)^\sim$ belong also to L_1. Setting*

$$(f \otimes g)(x) = \pi^{-1} \int_{-\infty}^{\infty} [f(x)g(t) + g(x)f(t) - f(t)g(t)]\frac{dt}{t-x} \tag{1.7.15}$$

(convolution of the Hilbert transform), then $(f \otimes g)^\sim \in L_1$ and it holds that

$$(f \otimes g)^\sim = f^\sim g^\sim. \tag{1.7.16}$$

Proof. Under our conditions obviously $(f \otimes g)^\sim \in L_1$. Applying the Hilbert transform on equation (1.7.15) and using the inversion formula 1.7.1, (1.7.10) in the last summand of (1.7.15) we have

$$\tilde{h} = (f \otimes g)^\sim = (fg^\sim)^\sim + (f^\sim g)^\sim + fg.$$

Using the connection (1.7.4) between the FT and the Hilbert transform, the definition (1.3.27) of the convolution of the FT and 1.3.4, Remark 1.3.13, namely,

$$f^\wedge * g^\wedge = 2\pi(fg)^\wedge,$$

which holds under our conditions, we have

$$
\begin{aligned}
2\pi\mathcal{F}[h^\sim](x) &= 2\pi\mathcal{F}[(fg^\sim)^\sim + (f^\sim g)^\sim + fg](x) \\
&= -2\pi i sgn(x)\mathcal{F}[fg^\sim + f^\sim g](x) + 2\pi\mathcal{F}[fg](x) \\
&= -i sgn(x)[(f^\wedge * (g^\sim)^\wedge)(x) + ((f^\sim)^\wedge * g^\wedge)(x)] + (f^\wedge * g^\wedge)(x) \\
&= -sgn(x)[(f^\wedge * sgn(x)g^\wedge)(x) + (sgn(x)f^\wedge * g^\wedge)(x)] + (f^\wedge * g^\wedge)(x) \\
&= [(isgn(x))f^\wedge * (isgn(x))g^\wedge](x) \\
&= [(f^\sim)^\wedge * (g^\sim)^\wedge](x).
\end{aligned}
$$

Consequently,

$$h^\sim = f^\sim g^\sim,$$

and this is formula (1.7.16). ⬜

Remark 1.7.42 *One can also prove:*

Let $f \in L_p$, $g \in L_q$, $1 < p, q < \infty$, $p^{-1} + q^{-1} < 1$. Then $f \otimes g \in L_r$, $r^{-1} = p^{-1} + q^{-1}$ and formula (1.7.16) is valid.

A connection between the Hilbert transform and the Fourier convolution is given in:

Theorem 1.7.39 *If $f, g \in L_1$ and $f^\sim, g^\sim \in L_1$ then*

$$\mathcal{H}[f * g] = f^\sim * g = f * g^\sim \tag{1.7.17}$$

and

$$f * g = -(f^\sim * g^\sim). \tag{1.7.18}$$

Proof. Following [Za] section 14.5.4 we have

$$
\begin{aligned}
\mathcal{H}[f * g](x) &= \frac{1}{\pi}\int_{-\infty}^{\infty}(t-x)^{-1}\left(\int_{-\infty}^{\infty}f(y)g(t-y)dy\right)dt = \frac{1}{\pi}\int_{-\infty}^{\infty}f(y)\left(\int_{-\infty}^{\infty}\frac{g(z)}{z-(x-y)}dz\right)dy \\
&= \int_{-\infty}^{\infty}f(y)g^\sim(x-y)\,dy = (f * g^\sim)(x).
\end{aligned}
$$

The second half of (1.7.17) is proved analogously. Formula (1.7.18) follows from (1.7.17) by replacing g with $g^\sim$ and using formula (1.7.10). ⬜

1.7.3 Applications

Example 1.7.44 *We would like to explain a method for the calculation of Cauchy integrals. Let $f, f^\sim \in L_1$ and set $g = f^\sim$ in 1.7.2, (1.7.15). By means of the inversion formula 1.7.1, (1.7.10) this leads to*

$$f \otimes f^\sim = -f^2 + (f^\sim)^2 - (ff^\sim)^\sim. \tag{1.7.19}$$

But from the convolution theorem (1.7.16) we obtain in the special case $g = f^\sim$

$$(f \otimes f^\sim)^\sim = -ff^\sim.$$

Therefore,

$$(f \otimes f^\sim)^\sim = (ff^\sim)^\sim.$$

Together with (1.7.19) this can be written in the form

$$\left(f^\sim(x)\right)^2 - \left(f(x)\right)^2 = \frac{2}{\pi} \int\limits_{-\infty}^{\infty} \frac{f(t)f^\sim(t)}{t-x} dt. \tag{1.7.20}$$

Formula (1.7.20) can be applied to evaluate Cauchy integrals in the following manner: If the Hilbert transform $f^\sim$ of f is known, then the Hilbert transform of $ff^\sim$ is $[(f^\sim)^2 - f^2]/2$. For example, let

$$f(t) = \frac{\pi}{2} \exp(-|t|)I_0(t),$$

where I_0 is the modified Bessel function of the first kind and order zero. Then from [EMOT], vol. II, section XV, 15.3, (48) we have

$$f^\sim(x) = -\sinh(x)K_0(|x|),$$

where K_0 is the modified Bessel function of the second kind or the MacDonald function of order zero. Therefore,

$$\int\limits_{-\infty}^{\infty} \frac{\exp(-|t|)\sinh(t)K_0(|t|)I_0(t)}{x-t} dt = \sinh^2(x)K_0^2(|x|) - \left(\frac{\pi}{2}\right)^2 e^{-2|x|}I_0^2(x). \tag{1.7.21}$$

Example 1.7.45 *Let us look for the solution f of the nonlinear singular integral equation*

$$\frac{2}{\pi}f(x) \int\limits_{-\infty}^{\infty} \frac{f(t)}{t-x} dt - \frac{1}{\pi} \int\limits_{-\infty}^{\infty} \frac{f^2(t)}{t-x} dt = g(x) \tag{1.7.22}$$

on the real line, where g is a given function. As usual quite formally it can be rewritten in the equivalent form

$$f \otimes f = g.$$

Applying the Hilbert transform to this equation by means of the convolution theorem (1.7.16) we get

$$(f^\sim)^2 = g^\sim.$$

Therefore,

$$f^{\sim} = \pm\sqrt{g^{\sim}},$$

where $\sqrt{g^{\sim}}$ denotes that branch of the square root for which $Re(\sqrt{g^{\sim}})$ is nonnegative. Let $g \in L_p$, $p > 2$. Then $g^{\sim} \in L_p$ and $f^{\sim} \in L_{p/2}$. Furthermore, let Ω be any measurable subset of $\mathbb{R}$. Then one can prove that the function f_Ω with the Hilbert transform

$$f_{\widehat{\Omega}}(x) = \begin{cases} \sqrt{g^{\sim}(x)} & \text{if } x \in \Omega \\ -\sqrt{g^{\sim}(x)} & \text{elsewhere} \end{cases}$$

consists of all solutions of the integral equation (1.7.22); see also [GlV].

1.8   Bessel Transforms

There exist many integral transforms with Bessel functions as J_ν, Y_ν, $H_\nu^{(1)}$, $H_\nu^{(2)}$, K_ν, and others in the kernel.

We restrict ourselves to three different types of examples. In section 1.8.1 we deal with the so-called Hankel transform, an integral transform with the Bessel functions of the first kind and order ν, J_ν, in the kernel, ν being fixed. It is closely connected to the n–dimensional Fourier transform of circular symmetric functions. We point out that the Hankel transform is a generalization of the Fourier transform.

In section 1.8.2 we investigate the Meijer- or K-transform, a transform with the modified Bessel functions or MacDonald functions K_ν in the kernel, ν being again fixed. While the Hankel transforms depend on a real variable, in case of the K-transform the "image variable" is a complex number. It was first considered by C.S. Meijer in 1940. It is proved to be a generalization of the Laplace transform.

Another type of transform is investigated in section 1.8.3. Again there are MacDonald functions in the kernel, but they are of the type K_{it} and t is the variable of the transforms. Such transforms are called index transforms; see Yakubovich [Ya]. The Kontorovich–Lebedev transform was first investigated by M.J. Kontorovich and N.N. Lebedev in 1938–1939 and by Lebedev in 1946.

All the transforms considered in section 1.8 can be used for the solution of boundary value problems in cylindrical coordinates; see, for example, [Sn.2], [Za], and [Ze.2].

1.8.1 The Hankel Transform

Definition 1.8.13 *The Hankel transform (HT) of order ν of a function $f : \mathbb{R}_+ \to \mathbb{C}$ is defined by means of*

$$\mathcal{H}_\nu[f](y) = f_\nu^\wedge(y) = \int_0^\infty \sqrt{xy}\, J_\nu(xy)\, f(x)dx, \quad y \in \mathbb{R}_+, \quad \nu > -1/2, \tag{1.8.1}$$

provided that the integral exists.

Here J_ν is the Bessel function of the first kind and order ν, defined by means of

$$J_\nu(z) = \sum_{k=0}^\infty \frac{(-1)^k (z/2)^{2k+\nu}}{k!\,\Gamma(k+\nu+1)}, \quad \arg(z) < \pi, \tag{1.8.2}$$

see [E.1], vol. 2, 7.2.1, (2).

From (1.8.2) we conclude that

$$J_\nu(x) \sim (x/2)^\nu, \quad x \to 0+, \tag{1.8.3}$$

and from [W.2], 7.21, (1) we have

$$J_\nu(x) \sim (2/\pi x)^{1/2} \cos(x - y\nu/2 - \pi/4), \quad x \to +\infty. \tag{1.8.4}$$

From (1.8.3) and (1.8.4) we see that $\sqrt{x}J_\nu(x)$ is bounded on $\mathbb{R}_+$ if $\nu > -1/2$,

$$|\sqrt{x}J_\nu(x)| \le C_\nu, \quad x \in \mathbb{R}_+, \quad \nu > -1/2. \tag{1.8.5}$$

Therefore, we choose as space of originals of the *HT* the space $L_1(\mathbb{R}_+)$ of measurable functions on $\mathbb{R}_+$ with the norm

$$\|f\|_1 = \int_0^\infty |f(x)|dx.$$

Then we have:

Theorem 1.8.40 *Let $f \in L_1(\mathbb{R}_+)$ and $\nu > -1/2$. Then $\mathcal{H}_\nu[f]$ exists. It is a linear transform and*

$$|\mathcal{H}_\nu[f]| \le C_\nu\|f\|_1. \tag{1.8.6}$$

Remark 1.8.43 *From [W.2], 3.4, (3), or directly from (1.8.2) we have*

$$J_{1/2}(z) = (2/\pi z)^{1/2} \sin z$$

and hence, by means of formula (1.3.6) we have

$$\mathcal{H}_{1/2}[f](y) = (2/\pi)^{1/2} \int_0^\infty f(x) \sin(xy)dx = (2/\pi)^{1/2} \mathcal{F}_s[f](y). \tag{1.8.7}$$

Therefore, the HT is a generalization of the Fourier sine transform.

Remark 1.8.44 *There are also other definitions of the HT, for example,*

$$F_\nu(y) = \int_0^\infty x J_\nu f(xy) dx; \tag{1.8.8}$$

see [Sn.2], (1.1.4).

Remark 1.8.45 *The HT is closely connected with the n-dimensional Fourier transform of radially symmetric functions. Let* $\mathbf{y} = (y_1, y_2, \cdots, y_n)$ *and* $(y_1^2 + y_2^2 + \cdots + y_n^2)^{1/2} =: \rho$ *and let* $f(\mathbf{y}) = F(\rho)$. *Then the n-dimensional Fourier transform*

$$\mathcal{F}_n[f](\mathbf{x}) = \int_{\mathbb{R}^n} e^{-i(\mathbf{x},\mathbf{y})} f(\mathbf{y}) dy_1 \cdots dy_n,$$

$\mathbf{x} = (x_1, x_2, \ldots, x_n)$, $(\mathbf{x,y}) = (x_1 y_1 + x_2 y_2 + \ldots x_n y_n)$, *depends also only on* $r = (x_1^2 + x_2^2 + \ldots + x_n^2)^{1/2}$, $\mathcal{F}_n[f](\mathbf{x}) =: F^\wedge(r)$ *and*

$$F^\wedge(r) = (2\pi)^{(1-n)/2} \mathcal{H}_{\frac{n}{2}-1}[\rho^{(n-1)/2} F(\rho)].$$

For details see, for example, [BGPV], 1.1.3.

Remark 1.8.46 *For tables of Hankel transforms we refer to [EMOT], vol. 2, Chapter VIII.*

For a (quite formal) derivation of an inversion formula for the *HT* (1.8.1) we write the Mellin transform of the kernel of (1.8.1),

$$\mathcal{M}[\sqrt{x}\, J_\nu(x)](s) = \frac{2^{s-1/2} \Gamma\left(\frac{3}{2} + \frac{\nu}{2} + \frac{1}{4}\right)}{\Gamma\left(\frac{\nu}{2} - \frac{s}{2} + \frac{3}{4}\right)};$$

see [EMOT], vol. 1, section 6.8, formula (1) and section 1.5.2, Proposition 1.5.20, formula (1.5.12). Obviously,

$$\mathcal{M}[\sqrt{x}J_\nu(x)](1-s) = \frac{1}{\mathcal{M}[\sqrt{x}J_\nu(x)](s)},$$

and hence the *HT* is a Fourier type transform in the sense of 1.5.4, Example 1.5.43. So we have the inversion formula

$$f(x) = \int_0^\infty \sqrt{xy} J_\nu(xy) f_\nu^\wedge(y) \, dy =: \mathcal{H}_\nu^{-1}[f_\nu^\wedge](x). \tag{1.8.9}$$

More exactly we have:

Theorem 1.8.41 (Inversion Theorem) *Let* $f \in L_1(\mathbb{R}_+)$ *and of bounded variation in a neighborhood of a point* x *of continuity of* f. *Then for* $\nu \geq -1/2$ *the inversion formula* (1.8.9) *holds.*

For a rigorous proof we refer to [Sn.2], section 5–3 or [W.2], 14.12.

Remark 1.8.47 *It holds that* $\mathcal{H}_\nu = \mathcal{H}_\nu^{-1}$.

Now we prove a Parseval equation for the *HT*.

Theorem 1.8.42 *Let* $f, g^\wedge \in L_1(\mathbb{R}_+)$, $\nu \geq -1/2$ *and* $f^\wedge = \mathcal{H}_\nu[f]$, $g = \mathcal{H}_\nu^{-1}[g^\wedge]$. *Then*

$$\int_0^\infty f(x)g(x)dx = \int_0^\infty f^\wedge(y)g^\wedge(y)dy \qquad (1.8.10)$$

Proof. By means of Fubini's theorem we have

$$\int_0^\infty f(x)g(x)dx = \int_0^\infty f(x)\Big(\int_0^\infty g^\wedge(y)\sqrt{xy}J_\nu(xy)dy\Big)dx = \int_0^\infty g^\wedge(y)\Big(\int_0^\infty f(x)\sqrt{xy}J_\nu(xy)dx\Big)dy$$

$$= \int_0^\infty f^\wedge(y)\,g^\wedge(y)dy.$$

☐

Now we are going to derive some operational rules of the *HT*. In the following we assume that the transforms under consideration exist.

Proposition 1.8.32 *Let* $f_\nu^\wedge = \mathcal{H}_\nu[f]$. *Then it holds that*

$$\mathcal{H}_\nu[f(ax)](y) = a^{-1}f_\nu^\wedge(y/a), \qquad a \in \mathbb{R}_+, \qquad (1.8.11)$$

$$\mathcal{H}_\nu[x^n f(x)](y) = y^{1/2-\nu}\Big(\frac{1}{y}\frac{d}{dy}\Big)^n[y^{\nu+n-1/2}f_{\nu+n}^\wedge(y)], \qquad n \in \mathbb{N}_o, \qquad (1.8.12)$$

$$\mathcal{H}_\nu[2\nu x^{-1}f(x)](y) = y[f_{\nu-1}^\wedge(y) + f_{\nu+1}^\wedge(y)], \qquad (1.8.13)$$

and

$$\mathcal{H}_\nu[2\nu f'](y) = (\nu - 1/2)yf_{\nu+1}^\wedge(y) - (\nu + 1/2)yf_{\nu-1}^\wedge(y). \qquad (1.8.14)$$

Proof. The rule (1.8.11) is proved by substitution.

For the proof of (1.8.12) we use

$$\Big(z^{-1}\frac{d}{dz}\Big)^n\Big(z^{\nu+n}J_{\nu+n}(z)\Big) = z^\nu J_\nu(z), \qquad n \in \mathbb{N}_o; \qquad (1.8.15)$$

see [E.1], vol. 2, 7.2.8, (52). Starting with $n = 1$ and setting $z = xy$ we obtain

$$\frac{d}{dy}\Big[y^{\nu+1/2}f_{\nu+n}^\wedge(y)\Big] = y^{\nu+1/2}\int_0^\infty xf(x)\sqrt{xy}J_\nu(xy)dx,$$

and this is (1.8.12) in the case of $n = 1$. The general case $n \geq 1$ is proved inductively making use of (1.8.15).

For the proof of formula (1.8.13) we refer to

$$2\nu J_\nu(z) = z\Big(J_{\nu-1}(z) + J_{\nu+1}(z)\Big); \tag{1.8.16}$$

see [E.2], vol. 2, 7.2.8, (56) and perform the *HT* on both side of (1.8.16).

Formula (1.8.14) is proved integrating by parts on the left-hand side and using

$$2J_\nu' = J_{\nu-1} - J_{\nu+1}, \tag{1.8.17}$$

and

$$J_\nu'(z) = J_{\nu-1}(z) - \frac{x}{z}J_\nu(z); \tag{1.8.18}$$

see [E.1], vol. 2, 7.2.8, (57), and (54). We omit the straightforward calculation. ☐

Preparing the proof of a differentiation rule we need some preparation (following [Ze.2], section 5–4).

Lemma 1.8.7 *Let M_ν, N_ν be differential operators defined by means of*

$$(M_\nu f)(x) = x^{-\nu-1/2} D_x x^{\nu+1/2} f(x), \qquad D_x = \frac{d}{dx} \tag{1.8.19}$$

and

$$(N_\nu f)(x) = x^{\nu+1/2} D_x x^{-\nu-1/2} f(x), \tag{1.8.20}$$

where $\nu \geq -1/2$. Furthermore, let

$$f(x) = o(x^{-\nu-1/2}), \qquad x \to 0+$$

in the case of (1.8.19) *and*

$$f(x) = o(x^{-\nu-1/2}), \qquad x \to 0+$$

in the case of (1.8.20) *and*

$$f(x) = o(1), \qquad x \to +\infty.$$

Then it holds that

$$\mathcal{H}_\nu[M_\nu f](y) = y f_{\nu+1}^\wedge(y), \tag{1.8.21}$$

and

$$\mathcal{H}_{\nu+1}[N_\nu f](y) = -y f_\nu^\wedge. \tag{1.8.22}$$

Proof. For the proof of (1.8.21) we use [E.1], vol. 2, 7.2.8, (51)

$$D_z\Big[z^{-\nu} J_\nu(z)\Big] = -z^{-\nu} J_{\nu+1}(z). \tag{1.8.23}$$

It follows that

$$D_x\Big[x^{-\nu} J_\nu(xy)\Big] = -yx^{-\nu} J_{\nu+1}(xy).$$

Integrating $\mathcal{H}_\nu[M_\nu f](y)$ by parts we have

$$\mathcal{H}_\nu[M_\nu f](y) = (xy)^{1/2} J_\nu(xy) f(x)\Big|_0^\infty - \int_0^\infty (xy)^{1/2} x^\nu D_x[x^{-\nu} J_\nu(xy)] f(x) dx = -y f_{\nu+1}^\wedge(y),$$

because the first term vanishes under our assumptions.

Formula (1.8.22) can be proved in the same manner using

$$D_z[z^{\nu+1} J_{\nu+1}(z)] = z^{\nu+1} J_\nu(z); \qquad (1.8.24)$$

see [E.1], vol. 2, 8.2.8, (50). ☐

Combining the two differential operators of Lemma 1.8.7 we obtain a differentiation rule for a second-order differential operator. We consider the differential operator S_ν defined by means of

$$(S_\nu f)(x) = x^{-\nu-1/2} D_x[x^{2\nu+1} D_x x^{-\nu-1/2} f(x)]. \qquad (1.8.25)$$

Obviously, from (1.8.19), (1.8.20) we have

$$S_\nu = M_\nu N_\nu, \qquad (1.8.26)$$

and in another form,

$$(S_\nu f)(x) = f''(x) - (\nu^2 - 1/4) x^{-2} f(x). \qquad (1.8.27)$$

Combining (1.8.21) and (1.8.2) we obtain:

Proposition 1.8.33 (Differentiation Rule) *Let* $f \in L_1(\mathbb{R}_+) \cap C^2(\mathbb{R}_+)$ *and*

$$f(x) = o(x^{-\nu-3/2}), \quad (N_\nu f)(x) \in o(x^{-\nu-1/2}), \quad x \to 0+$$
$$f(x), (N_\nu f)(x) = o(1), \quad x \to +\infty.$$

Then

$$\mathcal{H}_\nu[S_\nu f](y) = -y^2 f_\nu^\wedge(y). \qquad (1.8.28)$$

Proof. From (1.8.26) and (1.8.21), (1.8.22) we obtain

$$\mathcal{H}_\nu[S_\nu](y) = y \mathcal{H}_{\nu+1}[N_\nu f](y) = -y^2 f_\nu^\wedge(y).$$

☐

Now we are going to derive a convolution for the *HT*. Following [Za], 21.6 we choose a slightly modified form of the *HT* (1.8.1).

Let

$$\tilde{J}_\nu(x) = 2^\nu \Gamma(\nu+1) x^{-\nu} J_\nu(x) \qquad (1.8.29)$$

and

$$d\rho_\nu = \frac{x^{2\nu+1}}{2^\nu \Gamma(\nu+1)} dx. \qquad (1.8.30)$$

We define the modified Hankel transform $\tilde{\mathcal{H}}_\nu$ by

$$\tilde{\mathcal{H}}_\nu[f](t) = \tilde{f}_\nu(t) = \int_0^\infty f(x)\tilde{J}_\nu(tx)d\rho_\nu(x). \qquad (1.8.31)$$

We obtain the connection with the *HT* $\mathcal{H}_\nu$:

$$\mathcal{H}_\nu[x^{\nu+1/2}f(x)](t) = t^{\nu+1/2}\tilde{\mathcal{H}}_\nu[f](t). \qquad (1.8.32)$$

From the inversion formula (1.8.9) of the Hankel transform we obtain an inversion formula for the modified version (1.8.31),

$$f(x) = \int_0^\infty \tilde{f}_\nu(t)\tilde{J}_\nu(xt)d\rho_\nu(t), \qquad (1.8.33)$$

which is valid, provided that $f_\nu^\wedge \in L_1(\mathbb{R}_+)$. It is derived from the integral in (1.8.33) by straightforward calculation using (1.8.32), (1.8.29), and (1.8.30) and taking note that $\mathcal{H}_a^{-1} = \mathcal{H}_a$.

Definition 1.8.14 *The convolution $f * g$ of two functions f, g associated with the Hankel transform $\tilde{\mathcal{H}}_\nu$ is defined by means of*

$$(f * g)(x) = \int_0^\infty \int_0^\infty f(y)g(z)D_\nu(x,y,z)d\rho_\nu(y)d\rho_\nu(z), \qquad (1.8.34)$$

provided that the integral exists.

The kernel D_ν is defined as

$$D_\nu(x,y,z) = \frac{2^{3\nu-1}\Gamma^2(\nu+1)A^{2\nu-1}(x,y,z)}{\sqrt{\pi}\Gamma(\nu+1/2)(xyz)^{2\nu}}, \qquad (1.8.35)$$

where $A(x,y,z)$ is the area of the triangle whose sides are x, y, z if there is a triangle with these sides and zero otherwise. The expression for the nonzero part of A is given by

$$4A(x,y,z) = [2(x^2y^2 + y^2z^2 + z^2x^2) - x^4 - y^4 - z^4]^{1/2}. \qquad (1.8.36)$$

From [Hi] we have the product formula

$$\int_0^\infty \tilde{J}_\nu(tx)D_\nu(x,y,z)d\rho_\nu(x) = \tilde{J}_\nu(ty)\tilde{J}_\nu(tz). \qquad (1.8.37)$$

For the kernel D_ν we have because of (1.8.37) and $\tilde{J}_\nu(0) = 1$, see (1.8.29) and (1.8.2):

Lemma 1.8.8 *The kernel D_ν of the product formula (1.8.37) is nonnegative, symmetrical with respect to its variables and it holds that*

$$\int_0^\infty D_\nu(x,y,z)d\rho_\nu(x) = 1. \qquad (1.8.38)$$

Remark 1.8.48 *Because of the symmetry of $D_\nu(x, y, z)$ with respect to its variables the integration in (1.8.38) can be done with respect to the variable y or z.*

Let $L_{1,\rho_\nu}(\mathbb{R}_+)$ be the space of measurable functions on $\mathbb{R}_+$ with respect to the measure $d\rho_\nu$ and let

$$\|f\|_{1,\rho_\nu} = \int_0^\infty |f(x)| d\rho_\nu(x)$$

be the norm in $L_{1,\rho_\nu}(\mathbb{R}_+)$. Then we have:

Theorem 1.8.43 (Convolution Theorem) *Let $f, g \in L_{1,\rho_\nu}(\mathbb{R}_+)$. Then $(f*g) \in L_{1,\rho_\nu}(\mathbb{R}_+)$ and*

$$\|f * g\|_{1,\rho_\nu} \le \|f\|_{1,\rho_\nu} \cdot |g|_{1,\rho_\nu}. \tag{1.8.39}$$

Furthermore it holds that

$$\tilde{\mathcal{H}}_\nu[f * g] = \tilde{f}_\nu \cdot \tilde{g}_\nu. \tag{1.8.40}$$

Proof. For the proof of (1.8.39) we have by means of (1.8.38)

$$\|f * g\|_{1,\rho_\nu} = \int_0^\infty |(f * g)(x)| \, d\rho_\nu(x)$$

$$= \int_0^\infty \left(\int_0^\infty \left(\int_0^\infty |f(y)\, g(z)|\, D_\nu(x, y, z)\, d\rho_\nu(y) \right) d\rho_\nu(z) \right) d\rho_\nu(x)$$

$$= \int_0^\infty |f(y)| d\rho_\nu(y) \int_0^\infty |g(z)|\, d\rho_\nu(z) = \|f\|_{1,\rho_\nu} \cdot \|g\|_{1,\rho_\nu}.$$

For the proof of (1.8.40) we obtain using (1.8.37):

$$\tilde{\mathcal{H}}_\nu[f * g](t) = \int_0^\infty (f * g)(x)\, \tilde{J}_\nu(tx)\, d\rho_\nu(x)$$

$$= \int_0^\infty \tilde{J}_\nu(tx) \left(\int_0^\infty \int_0^\infty f(y)\, g(z)\, D_\nu(x, y, z)\, d\rho_\nu(y)\, d\rho_\nu(z) \right) d\rho_\nu(x)$$

$$= \int_0^\infty \int_0^\infty f(y) g(z) \left(\int_0^\infty \tilde{J}_\nu(tx)\, D_\nu(x, y, z)\, d\rho_\nu(x) \right) d\rho_\nu(y)\, d\rho_\nu(z)$$

$$= \int_0^\infty f(y)\tilde{J}_\nu(ty)\, d\rho_\nu(y) \int_0^\infty g(z)\, \tilde{J}_\nu(tz)\, d\rho_\nu(z) = \tilde{f}_\nu(t) \cdot \tilde{g}_\nu(t).$$

□

Remark 1.8.49 *To complete the above we followed [Za].*

Remark 1.8.50 *In [Hi] a more general version is proved. Let $1 \leq r, s \leq \infty$ and $p^{-1} = r^{-1} + s^{-1} - 1$. If $F \in L_{r,\rho_\nu}(\mathbb{R}_+)$, $g \in L_{s,\rho_\nu}(\mathbb{R}_+)$ then $h = f * g$ exists on $\mathbb{R}_+$ and*

$$\|h\|_{p,\rho_\nu} \leq \|f\|_{r,\rho_\nu} \cdot \|g\|_{s,\rho_\nu}.$$

Moreover, if $p = \infty$ then $h \in C(\mathbb{R}_+)$. Furthermore, (1.8.40) holds.

Finally we consider an application of the *HT*. We look for a solution of an axially symmetric Dirichlet problem for a half-space. We look for a solution of Laplace's equation in cylindrical coordinates (r, φ, z) independent of the polar angle φ:

$$\frac{\partial^2 v(r, \varphi)}{\partial r^2} + \frac{1}{r}\frac{\partial v}{\partial r} + \frac{\partial^2 v}{\partial z^2} = 0, \quad r, z \in \mathbb{R}_+ \tag{1.8.41}$$

with the boundary condition

$$v(r, 0) = f(r) \tag{1.8.42}$$

and with the asymptotic behavior

$$v(r, z) \to 0 \quad \text{as} \quad \sqrt{r^2 + z^2} \to +\infty. \tag{1.8.43}$$

Substituting

$$u(r, z) = \sqrt{r}\, v(r, z), \qquad g(r) = \sqrt{r}\, f(r)$$

we have from (1.8.41)

$$\frac{\partial^2 u}{\partial r^2} + \frac{1}{4r^2}u + \frac{\partial^2 u}{\partial z^2} = 0, \tag{1.8.44}$$

with

$$u(r, 0) = g(r), \tag{1.8.45}$$

and

$$u(r, z) = o(\sqrt{r}), \quad \sqrt{r^2 + z^2} \to +\infty. \tag{1.8.46}$$

Applying the *HT* of order $\nu = 0$ to (1.8.44) we obtain

$$-g^2 u_o^\wedge(\rho, z) + \frac{\partial^2 u_o^\wedge(\rho, z)}{\partial z^2} = 0.$$

The solution of this ordinary differential equation with the growth (1.8.46) is

$$u_o^\wedge(\rho, z) = A(\rho)e^{\rho z}.$$

Because of (1.8.45) we have $A(\rho) = g_o^\wedge(\rho)$ and therefore

$$u_o^\wedge(\rho, z) = g_o^\wedge(\rho)e^{-\rho z}.$$

Making use of the inversion formula (1.8.9) of *HT* we obtain the solution of the problem (1.8.41) through (1.8.43):

$$v(r, z) = r^{-1/2} \int_0^\infty g_o^\wedge(\rho)\sqrt{r}\rho e^{-\rho z} J_o(r\rho)\,d\rho. \tag{1.8.47}$$

1.8.2   The Meijer (K-) Transform

Definition 1.8.15 *The Meijer- or K-transform (KT) of a function $f : \mathbb{R}_+ \to \mathbb{C}$ is defined as*

$$\mathcal{K}_\nu[f](s) = f_\nu(s) = \int_0^\infty f(t)\sqrt{st}\mathcal{K}_\nu(st)dt, \quad Re(\nu) \geq 0, \qquad (1.8.48)$$

provided that the integral exists.

Here $\mathcal{K}_\nu$ is the modified Bessel function of the third kind and order ν or MacDonald function, defined by means of

$$K_\nu(z) = \int_0^\infty e^{-z\cosh u}\cosh(\nu u)du, \qquad Re(z) > 0; \qquad (1.8.49)$$

see [E.1], vol. 2, 7.12, (21).

Remark 1.8.51 *The condition $Re(\nu) \geq 0$ is no less of generality since from (1.8.49) it follows that $K_\nu = K_{-\nu}$.*

Remark 1.8.52 *From [E.1], 7.2.6, (43) we have*

$$K_{1/2}(z) = (\pi/2z)^{1/2}e^{-z}.$$

Therefore, it follows that

$$\mathcal{K}_{1/2}[f](s) = (\pi/2)^{1/2}\mathcal{L}[f](s), \qquad (1.8.50)$$

and therefore the KT is a generalization of the Laplace transform (LT). It will be pointed out that the KT has operational properties similar to the LT.

Remark 1.8.53 *For tables of K-transforms we refer to [EMOT], vol. 2, Chapter X.*

The kernel K_ν has the following asymptotic behavior: From [E.1], vol. 2, 7.2.2, (12), and (13) we deduce that

$$K_\nu(z) = 2^{\nu-1}\Gamma(\nu)[1 + 0(1)], \qquad z \to 0, \quad |\arg(z)| < \pi/2, \quad Re(\nu) > 0, \qquad (1.8.51)$$

and from 7.2.4, (38) we have

$$K_o(z) = -\log(z/2)[1 + 0(1)], \qquad z \to 0, \quad |\arg(z)| < \pi/2. \qquad (1.8.52)$$

From the same source, section 7.4.1, (1) we know

$$K_\nu(z) = (\pi/2z)^{1/2}e^{-z}[1 + 0(1)], \qquad z \to 0, \quad |\arg(z)| < \pi/2, \quad Re(\nu) \geq 0. \qquad (1.8.53)$$

From these estimates we obtain

Theorem 1.8.44 *Let $f \in L_1^{loc}(\mathbb{R}_+)$ and $f(t) = 0(t^\alpha)$ as $t \to 0+$ where $\alpha > \nu - 3/2$ if $\nu \in \mathbb{R}_+$ and $\alpha > -1$ if $\nu = 0$. Furthermore let $f(t) = 0(e^{at})$ as $t \to +\infty$. Then its KT of order ν exists a.e. for $Re(s) > a$.*

Now we are going to prove an inversion theorem for the KT.

Theorem 1.8.45 (Inversion Theorem) *Let $-1/2 < Re(\nu) < 1/2$ and $F(s)$ analytic on the half-plane H_a, $a \leq 0$ and $s^{\nu-1/2}F(s) \to 0$, $|s| \to +\infty$, uniformly with respect to $\arg(s)$. For any number c, $c > a$ set*

$$f(t) = \frac{1}{\pi i} \int_{(c)} F(z)\sqrt{zt}I_\nu(zt)dz. \tag{1.8.54}$$

Then f does not depend on the choice of c and

$$F(s) = \mathcal{K}_\nu[f](s), \qquad s \in H_c. \tag{1.8.55}$$

Here I_ν is the modified Bessel function of the first kind and order ν; see [E.1], vol. 2, 7.2.2.

Proof. From Cauchy's integral theorem we have

$$F(s) = \frac{1}{2\pi i} \int_I \frac{(z/s)^{\nu-1/2}}{z-s}F(z)dz,$$

where I is the contour of Figure 5. Because of

$$\frac{2z}{z^2-s^2} = \frac{1}{z-s} + \frac{1}{z+s}$$

we have

$$F(s) = \frac{s^{-\nu+1/2}}{\pi i} \int_I \frac{z^{\nu+1/2}}{z^2-s^2}F(z)dz,$$

for the integral with the denominator $z+s$ vanishes because $z = -s$ is outside of the contour I. Writing

$$F(s) = \frac{s^{-\nu+1/2}}{\pi i}\left[\int_\mathcal{H} \cdots + \int_{c+iR}^{c-iR} \cdots\right]$$

and tending $R \to +\infty$ we see that $\int_\mathcal{H}$ vanishes as $R \to +\infty$ under our assumptions and so we obtain

$$F(s) = \frac{s^{-\nu+1/2}}{\pi i} \int_{(c)} \frac{z^{\nu+1/2}}{s^2-z^2}F(z)dz.$$

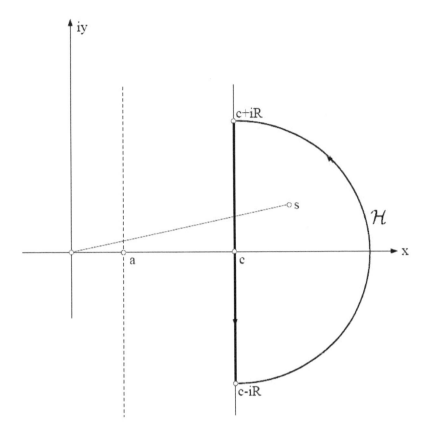

Figure 5

Now from [PBM], vol. 2, 2.16.28, 1. we have

$$\mathcal{K}_\nu[t^{1/2}I_\nu(zt)](s) = z^\nu s^{-\nu+1/2}/(s^2 - z^2)$$

and, therefore,

$$F(s) = \frac{1}{\pi i}\int_{(c)} \sqrt{z}F(z)\Big(\int_0^\infty \sqrt{ts}K_\nu(st)t^{1/2}I_\nu(zt)dt\Big)dz$$

$$= \frac{1}{\pi i}\int_{(c)} \sqrt{zs}F(z)\Big(\int_0^\infty tK_\nu(st)I_\nu(zt)dt\Big)dz.$$

Interchanging the order of integration we obtain

$$F(s) = \int_0^\infty \sqrt{st}K_\nu(st)\Big(\frac{1}{\pi i}\int_{(c)} \sqrt{zt}F(z)I_\nu(zt)dz\Big)dt$$

and this is (1.8.54), (1.8.55). □

Remark 1.8.54 *Formula (1.8.54) is the inversion formula for the KT (1.8.48):*

$$f(t) = \frac{1}{\pi i} \int\limits_{(c)} \sqrt{st} I_\nu(st) f_\nu(s) ds =: \mathcal{K}_\nu^{-1}[f_\nu](t). \tag{1.8.56}$$

Now we are going to derive some operational rule for the KT. In the following we assume that the transforms under consideration exist.

Proposition 1.8.34 *Let $f_\nu = \mathcal{K}_\nu[f]$. Then it holds that*

$$\mathcal{K}_\nu[f(at)](s) = a^{-1} f_\nu(s/a), \qquad a \in \mathbb{R}_+, \tag{1.8.57}$$

$$\mathcal{K}_\nu[t f(t)](s) = -s^{-\nu - 1/2} \frac{d}{ds} [s^{\nu + 1/2} f_{\nu+1}(s)], \tag{1.8.58}$$

and

$$\mathcal{K}_\nu[2\nu t^{-1} f(t)](s) = s[f_{\nu+1}(s) - f_{\nu-1}(s)]. \tag{1.8.59}$$

Proof. Formula (1.8.57) is proved by straightforward calculation substituting $at \to t$.

The basis for the proof of (1.8.58) is

$$\frac{d}{dz}\left(z^\nu K_\nu(z)\right) = -z^\nu K_{\nu-1}(z);$$

see [E.1], vol. 2, 7.11, (25). Setting $z \to st$, $\nu \to \nu + 1$, multiplying both sides by $\sqrt{st} f(t)$ and integrating over $\mathbb{R}_+$ one easily has (1.8.58).

Formula (1.8.59) is proved by means of

$$K_{\nu+1}(z) - K_{\nu-1}(z) = 2\nu z^{-1} K_\nu(z);$$

see [E.1], vol. 2, 7.11, (25). With $z = st$, multiplying by $\sqrt{st} f(t)$ and integrating on $\mathbb{R}_+$ we have immediately (1.8.59).   ⬚

The derivation of a differentiation rule is similar to the process for the Hankel transform. Again as in 1.8.1, Lemma 1.8.7 we consider diferential operators M_ν, N_ν (see formulas 1.8.19 and 1.8.20), and we obtain the formulas

$$\mathcal{K}_\nu[M_\nu f](s) = s f_{\nu+1}(s) \tag{1.8.60}$$

$$\mathcal{K}_{\nu+1}[N_\nu f](s) = s f_\nu(s). \tag{1.8.61}$$

This is proved analogously to the proof of Lemma 1.8.7, using the formulas

$$D_z[z^{-\nu} K_\nu(zx)] = -z^{-\nu} K_{\nu+1}(z) \tag{1.8.62}$$

and

$$D_z[z^{\nu+1} K_{\nu+1}(z)] = -z^{\nu+1} K_\nu(z); \tag{1.8.63}$$

see [E.1], vol. 2, (22), (21). The only difference between (1.8.23), (1.8.24) of 1.8.1, with $J_\nu \to K_\nu$ and (1.8.62), (1.8.63) is that the second formula has also the sign "$-$" on the right-hand side. Therefore, in (1.8.61) we have the sign "$+$" instead of "$-$" in 1.8.1, (1.8.22).

So we can use the same differential operator as in the case of the Hankel transform and we obtain:

Proposition 1.8.35 (Differentiation Rule) *Let S_ν be the differential operator 1.8.1, (1.8.25). Furthermore, let*

$$f(t), N_\nu f(t) = o(t^{\nu-1/2}), \qquad t \to 0+$$

$$f(t), N_\nu f(t) = O(e^{at}), \qquad t \to +\infty, a < 1.$$

Then it holds that

$$\mathcal{K}_\nu[S_\nu f](s) = s^2 f_\nu(s). \tag{1.8.64}$$

Now we define a convolution for the KT and similar to the case of the Hankel transform we will use a slightly different form of the KT (1.8.48), following Krätzel [K].

Instead of $\mathcal{K}_\nu$ according to (1.8.48) we consider the transform $\tilde{\mathcal{K}}_\nu$ with

$$\tilde{\mathcal{K}}_\nu[f](s) = \varphi_\nu(s) = 2\int_0^\infty (st)^{\nu/2} K_\nu(2\sqrt{st}) f(t) dt, \qquad Re(\nu) \geq 0. \tag{1.8.65}$$

The connection with the transform $\mathcal{K}_\nu$ can be calculated as follows:

$$\tilde{\mathcal{K}}_\nu[f](s) = 2^{2-\nu} s^{\nu-1/2} \mathcal{K}_\nu[t^{\nu-1/2} f(t^2)](s). \tag{1.8.66}$$

The transform $\tilde{\mathcal{K}}_\nu$ can be factorized by means of Laplace transforms. From [W.2], 6.22, (15) we have

$$K_\nu(z) = z^{-\nu-1} z^\nu \int_0^\infty \tau^{-\nu-1} \exp(-\tau - z^2/4\tau) d\tau, \qquad Re(z^2) > 0. \tag{1.8.67}$$

Putting K_ν according to (1.8.67) into (1.8.65) we have the factorization formula

$$\mathcal{K}_\nu[f](s) = \mathcal{L}\left[\tau^{-\nu-1}\mathcal{L}[t^\nu f(t)](1/\tau)\right](s). \tag{1.8.68}$$

Analogously substituting $\nu \to -\nu$ in (1.8.67) and using $K_{-\nu} = K_\nu$ we have

$$\tilde{\mathcal{K}}_\nu[f](s) = s^\nu \mathcal{L}\left[\tau^{\nu-1}\mathcal{L}[f](1/\tau)\right](s). \tag{1.8.69}$$

Now we are going to derive a convolution for KT $\tilde{\mathcal{K}}_\nu$.

Definition 1.8.16 *As convolution $f*g$ of two functions f, g with respect to the K-transform $\tilde{\mathcal{K}}_\nu$ we define $f * g$ by*

$(f * g)(t) =$

$$\frac{1}{\Gamma(n-\nu)} D^n \int_0^t (t-\sigma)^{n-1-\nu} \left(\int_0^\infty \xi^\nu \left(\int_0^1 (1-\eta)^\nu f(\xi\eta) g[(\eta-\xi)(1-\eta)] d\eta\right) d\xi\right) d\sigma, \tag{1.8.70}$$

where $D = d/dt$ and

$$Re(\nu) < n \le Re(\nu) + 1, \quad n \in \mathbb{N}. \tag{1.8.71}$$

Using the factorization formula (1.8.68) and then the differentiation rule and the convolution theorem of the Laplace transform we obtain quite formally

$$\tilde{\mathcal{K}}_\nu[f * g] = s^\nu \int_0^\infty e^{-s\tau} \tau^{\nu-1} \left(\int_0^\infty e^{-t/\tau} (f * g)(t) dt \right) d\tau$$

$$= \frac{s^\nu}{\Gamma(n-\nu)} \int_0^\infty e^{-\xi\tau} \tau^{\nu-1} \left(\int_0^\infty e^{-t/\tau} D^n \left(\int_0^t (t-\sigma)^{n-1-\nu} \right. \right.$$

$$\left. \left. \cdot \left(\int_0^\tau \xi^\nu \left(\int_0^1 (1-\eta)^\nu f(\xi\eta) g[(\sigma-\xi)(1-\eta)] d\eta \right) d\xi \right) d\sigma \right) dt \right) d\tau.$$

By means of the differentiation rule of the Laplace transform (1.4.34) applied on the inner Laplace transform we obtain

$$\tilde{\mathcal{K}}_\nu[f * g](s) = \frac{s^\nu}{\Gamma(n-\nu)} \int_0^\infty e^{-s\tau} \tau^{\nu-n-1} \left(\int_0^\infty e^{-t/\tau} \left(\int_0^t (t-\sigma)^{n-1-\nu} \right. \right.$$

$$\left. \left. \cdot \left(\int_0^\sigma \xi^\nu \left(\int_0^1 (1-\eta)^\nu f(\eta) g[(\sigma-\xi)(1-\eta)] d\eta \right) d\xi \right) d\sigma \right) dt \right) d\tau$$

Now applying the convolution theorem of the Laplace transform, Theorem 1.4.11, again on the inner Laplace integral by means of (1.4.12) we obtain

$$\tilde{\mathcal{K}}_\nu[f * g](s) = s^\nu \int_0^\infty e^{-s\tau} \tau^{-1} \left(\int_0^\infty e^{-t/\tau} \left(\int_0^t \xi^\nu \cdot \left(\int_0^1 (1-\eta)^\nu f(\xi\eta) g[(t-\xi)(1-\eta)] d\eta \right) d\xi \right) dt \right) d\tau.$$

Substituting $\xi \to t\xi, \quad \eta \to t^{-1}\eta$ we have

$$\tilde{\mathcal{K}}_\nu[f * g](s) = s^\nu \int_0^\infty e^{-s\tau} \tau^{-1} \left(\int_0^1 \xi^\nu \left(\int_0^\infty e^{-t/\tau} \cdot \left(\int_0^t (t-\eta)^\nu f(\xi\eta) g[(1-\xi)(t-\eta)] d\eta \right) dt \right) d\xi \right) d\tau.$$

From the similarity rule of the Laplace transform (see formula 1.4.18) we obtain

$$\mathcal{L}[f(\xi t)](1/\tau) = \xi^{-1} \mathcal{L}[f](1/\xi\tau),$$

and analogously

$$\mathcal{L}\left[t^\nu g((1-\xi)t) \right](1/\tau) = (1-\xi)^{-\nu-1} \mathcal{L}[t^\nu g(t)] \left(1/(1-\xi)\tau \right).$$

Again applying the convolution theorem of the Laplace transform we have

$$\tilde{\mathcal{K}}_\nu[f * g](s) = s^\nu \int_0^\infty e^{-s\tau} \tau^{-1} \left(\int_0^1 \xi^{\nu-1} [f](1/\xi\tau)(1-\xi)^{-\nu-1} \cdot \mathcal{L}[u^\nu g(u)] \left(1/(1-\xi)\tau \right) d\xi \right) d\tau.$$

Substituting $u = \tau\xi$ we have

$$\tilde{\mathcal{K}}_\nu[f * g](s) = s^\nu \mathcal{L}\left[t^{\nu-1}\mathcal{L}[f](1/t)\right](s) \quad \mathcal{L}\left[t^{-\nu-1}\mathcal{L}[u^\nu g(u)](1/t)\right](s)$$

and by means of the factorization formulas (1.8.69), (1.8.68) we have

$$\tilde{\mathcal{K}}_\nu[f * g] = \tilde{\mathcal{K}}_\nu[f] \cdot \tilde{\mathcal{K}}_\nu[g]. \tag{1.8.72}$$

Because of the connection between the Laplace and the $\tilde{\mathcal{K}}_\nu$ transform the calculations above can be justified if $\tilde{\mathcal{K}}_\nu[f]$ and $\tilde{\mathcal{K}}_\nu[g]$ are absolutely convergent in the point s. Because of the definition of the convolution (1.8.70) f, g must be n-time differentiable. Therefore, we proved:

Theorem 1.8.46 (Convolution Theorem) *Let f, g be n-time differentiable on $\mathbb{R}_+$, $n \in \mathbb{N}_o$ and according to (1.8.71), and let $\tilde{\mathcal{K}}_\nu[f]$ and $\tilde{\mathcal{K}}_\nu[g]$ be absolutely convergent in the point s. Then $\mathcal{K}_\nu[f * g](s)$ is absolutely convergent and (1.8.72) holds.*

Remark 1.8.55 *For details we refer to Krätzel [K].*

Remark 1.8.56 *As usual, one can prove that the convolution is commutative.*

Now we are going to consider an application of the KT. As usual, the KT can be used for the solution of boundary value problems with respect to the differential operator S_ν; see (1.8.25). We refer to [Za], 23.8, [Ze.2], 6.8, 6.9, and [K]. Here, we give an application for special functions (see [K]) and we will derive an addition theorem for the Bessel functions of the first kind with respect to the order. By means of the factorization formula (1.8.69) for the transform $\tilde{\mathcal{K}}_\nu$ we have with $\nu = 0$

$$\tilde{\mathcal{K}}_o[t^{\rho/2}J_s(2\sqrt{t})](s) = \mathcal{L}\left[\tau^{-1}\mathcal{L}[t^{\rho/2}J_\rho(2\sqrt{t})](1/\tau)\right](s).$$

From [PBM], vol. 4, 3.12.3, 8, we know

$$\mathcal{L}[t^{\rho/2}J_s(2\sqrt{t})](\tau^{-1}) = \tau^{\rho+1}e^{-\tau}.$$

Therefore,

$$\mathcal{K}_o\left[t^{(\mu+\nu+1)/2}J_{\mu+\nu+1}(2\sqrt{t})\right](s) = \frac{\Gamma(\mu+\nu+2)}{(s+1)^{\mu+\nu+2}}.$$

Otherwise

$$\begin{aligned}
\mathcal{K}_o\left[t^{\mu/2}J_\mu(2\sqrt{t}) * t^{\nu/2}J_\nu(2\sqrt{t})\right](s) &= \frac{\Gamma(\mu+1)\Gamma(\nu+1)}{(s+1)^{\mu+\nu+2}} \\
&= \frac{\Gamma(\mu+1)\Gamma(\nu+1)}{\Gamma(\mu+\nu+1)}\mathcal{K}_o\left[t^{(\mu+\nu+1)/2}J_{\mu+\nu+1}(2\sqrt{t})\right](s) \\
&= B(\mu+1,\nu+1)\mathcal{K}_o\left[t^{(\mu+\nu+1)/2}J_{\mu+\nu+1}(2\sqrt{t})\right](s),
\end{aligned}$$

where B is the Beta function; see equation (1.4.21). Applying the inverse $\mathcal{K}_o$ transform on both sides of this equation we have:

Proposition 1.8.36 *Let $\mu, \nu > -1$. Then the Bessel functions J_ν fulfill the addition theorem*

$$t^{\mu/2}J_\mu(2\sqrt{t}) * t^{\nu/2}J_\nu(2\sqrt{t}) = B(\mu+1,\nu+1)t^{(\mu+\nu+1)/2}J_{\mu+\nu+1}(2\sqrt{t}). \qquad (1.8.73)$$

1.8.3 The Kontorovich–Lebedev Transform

Definition 1.8.17 *The Kontorovich–Lebedev Transform (KLT) of a function $f : \mathbb{R}_+ \to \mathbb{C}$ is defined by means of*

$$\mathcal{KL}[f](t) = \int_0^\infty f(x)K_{it}(x)\frac{dx}{x}, \quad t \in \mathbb{R}_+, \qquad (1.8.74)$$

provided that the integral exists.

Here K_{it} is the modified Bessel function of the second kind or MacDonald function, defined by, see equation (1.8.48)

$$K_{it}(x) = \int_0^\infty e^{-x\cosh u}\cos(t)u\,du, \qquad x \in \mathbb{R}_+. \qquad (1.8.75)$$

From (1.8.75) we have $|K_{it}(x)| \leq K_o(x)$ and hence,

$$|\mathcal{KL}[f](t)| \leq \int_0^\infty |f(x)|K_o(x)\frac{dx}{x}.$$

If the integral above converges then the integral in (1.8.74) converges absolutely and uniformly and it defines a continuous function.

From

$$K_o(x) \sim -\log x/2, \qquad x \to 0+, \qquad (1.8.76)$$

see formula (1.8.52), and

$$K_o(x) \sim \left(\frac{\pi}{2x}\right)^{1/2}e^{-x}, \qquad x \to +\infty, \qquad (1.8.77)$$

(see formula 1.8.53) we deduce sufficient conditions for the existence of the *KLT*.

Theorem 1.8.47 *Let $f(x)/x \in L_1^{loc}(\mathbb{R}_+)$ and $f(x) = 0(e^{\alpha x}), x \to +\infty, 0 \leq \alpha \leq 1$. Then the integral (1.8.74) converges absolutely and uniformly and therefore $\mathcal{KL}[f] \in \mathcal{C}(\mathbb{R})$.*

In the following we consider the space $L_{-1,1}(\mathbb{R}_+) = L_{-1,1}$ of measurable functions on $\mathbb{R}_+$ such that

$$\|f\|_{-1,1} = \int_0^\infty |f(x)|\frac{dx}{x^2} \qquad (1.8.78)$$

is finite. Then we have:

Theorem 1.8.48 *Let $f \in L_{-1,1}$. Then $\mathcal{KL}[f]$ exists and*

$$|\mathcal{KL}[f](t)| \le C_o\|f\|_{-1,1}.$$

Moreover, $\mathcal{KL}[f] \in \mathcal{C}^\infty(\mathbb{R}_+)$.

Proof. From (1.8.76) and (1.8.77) we have $x\,K_o(x) \le C_o$, $x \in \mathbb{R}_+$, with some constant C_o. Therefore

$$|\mathcal{KL}[f](t)| \le \int_0^\infty |f(x)|x K_o(x)\frac{dx}{x^2} \le C_o\|f\|_{-1,1}.$$

Differentiating (1.8.74) under the sign of integral we obtain $\mathcal{KL}[f] \in \mathcal{C}^\infty(\mathbb{R}_+)$. $\qquad\square$

Remark 1.8.57 *For the definition of the KLT in various spaces of measurable functions on $\mathbb{R}_+$ see [YaL] and [Ya].*

Remark 1.8.58 *Sometimes instead of* Definition 1.8.17 *the KLT is defined by means of the integral*

$$\int_0^\infty f(x)K_{it}(x)dx;$$

see, for example, [Ya] and [Za], sometimes as

$$\int_0^\infty f(x)K_{it}(x)\frac{dx}{\sqrt{x}}.$$

Remark 1.8.59 *Another version is the transform with respect to the index*

$$\mathcal{KL}_{ind}[f](x) = \int_0^\infty f(t)K_{it}(x)dt;$$

see [Za], section 24.

Now we are going to derive an inversion formula for the *KLT*, but we do it in an operational manner, following [Sn.2], section 6.2, or [Za], section 24.4.

Using the Fourier cosine transform (see formula 1.3.5) the integral representation (1.8.75) can be written as

$$K_{it}(x) = \mathcal{F}_c\Big[e^{-x\cosh u}\Big](t).$$

By means of the inversion formula of the Fourier-cosine transform (see 1.3.4, Remark 1.3.14) we obtain

$$\mathcal{F}_c\Big[\mathcal{KL}[f]\Big](u) = \frac{\pi}{2}\int_0^\infty e^{-x\cosh u}f(x)\frac{dx}{x}$$

and this can be written as

$$\mathcal{L}[x^{-1}f(x)](\cosh u) = \frac{2}{\pi}\mathcal{F}_c\Big[\mathcal{KL}[f]\Big](u).$$

Putting $p = \cosh u$ we have

$$\mathcal{L}[x^{-1}f(x)](p) = \frac{2}{\pi}\int_0^\infty \mathcal{KL}[f](t)\cos(t\cosh^{-1}p)dt, \qquad (1.8.79)$$

where $\cosh^{-1}$ is the inverse of the function cosh. Making use of the division rule for the Laplace transform, Proposition 1.4.13, namely,

$$\mathcal{L}[t^{-1}f(t)](p) = \int_p^\infty \mathcal{L}[f](u)du,$$

by differentiation with respect to p we have

$$\mathcal{L}[f](p) = -\frac{d}{dp}\mathcal{L}[t^{-1}f(t)](p). \qquad (1.8.80)$$

Applying this result to (1.8.79) we get

$$\mathcal{L}[f](p) = \frac{2}{\pi}\int_0^\infty t\mathcal{KL}[f](t)\frac{\sin(t\cosh^{-1}p)}{\sqrt{p^2-1}}dt. \qquad (1.8.81)$$

From [PBM], vol. IV, section 3.16.1, 1. we know that

$$\mathcal{L}^{-1}\Big[\frac{\sin(t\cosh^{-1}p)}{\sqrt{p^2-1}}\Big](x) = \pi^{-1}\sinh(\pi t)K_{it}(x),$$

where $i\cos^{-1}x = \cosh^{-1}x$ has to be used and $\cos^{-1}$ is the inverse function of cos.

Therefore, from (1.8.81) it follows (performing the inverse Laplace transform on both sides of the equation) that

$$f(x) = \frac{2}{\pi^2}\int_0^\infty t\sinh(\pi t)K_{it}(x)\mathcal{KL}[f](t)dt =: \mathcal{KL}^{-1}\Big[\mathcal{KL}[f]\Big](x). \qquad (1.8.82)$$

With a more rigorous proof (see [Sn.2], 6-2) we have:

Theorem 1.8.49 (Inversion Theorem) *Let $x^{-1}f(x) \in \mathcal{C}(\mathbb{R}_+)$ and $xf(x)$, $x\frac{d}{dx}\big(x^{-1}f(x)\big) \in L_1(\mathbb{R}_+)$. Then with $\mathcal{KL}[f] =: F$ we have*

$$f(x) = \frac{2}{\pi^2}\int_0^\infty t\sinh(\pi t)K_{it}(x)F(t)dt =: \mathcal{KL}^{-1}[F](x).$$

Applying the inversion formula leads to a Parseval relation for the KLT. Assuming that the integrals in the following exist we have

$$\frac{2}{\pi^2}\int_0^\infty t\sinh(\pi t)\mathcal{KL}[f](t)\mathcal{KL}[g](t)dt = \frac{2}{\pi^2}\int_0^\infty t\sinh(\pi t)\mathcal{KL}[f](t)\left(\int_0^\infty g(x)K_{it}(x)\frac{dx}{x}\right)dt$$

$$= \frac{2}{\pi^2}\int_0^\infty g(x)\left(\int_0^\infty t\sinh(\pi t)K_{it}(x)\mathcal{KL}[f](t)dt\right)\frac{dx}{x}$$

$$= \int_0^\infty f(x)g(x)\frac{dx}{x}.$$

Among applications we look for a differentiation rule of the KLT. Let

$$A_x := x^2 D^2 + xD - x^2, \qquad D = \frac{d}{dx}. \tag{1.8.83}$$

From [E.1], vol. 2, 7.2.2, (11) we have

$$A_x K_{it}(x) = -t^2 K_{it}(x). \tag{1.8.84}$$

Applying the KLT on $A_x f$ and integrating by parts we require that the terms outside of the integral signs vanish at $0+$ and at $+\infty$. After straightforward calculations this leads to:

Proposition 1.8.37 (Differentiation Rule) *Let f be such that the KLT of f and $A_x f$ exist and furthermore*

$$\lim_{x\to 0^+(\infty)} K_{it}(x)f(x) = \lim_{x\to 0^+(+\infty)} xK_{it}(x)f'(x) = \lim_{x\to 0^+(+\infty)} x\left(DK_{it}(x)\right)f(x) = 0.$$

Then it holds that

$$\mathcal{KL}[A_x f](t) = -t^2\mathcal{KL}[f](t). \tag{1.8.85}$$

Now we derive a convolution theorem for the KLT. First we determine a linearization formula for the product of two MacDonald functions. From [W.2], 13.71, (1) we have

$$K_\nu(x)K_\nu(y) = \frac{1}{2}\int_0^\infty exp\left[-\frac{1}{2}(u+\frac{x^2+y^2}{u})\right]K_\nu\left(\frac{xy}{u}\right)\frac{du}{u}. \tag{1.8.86}$$

Substituting $\nu = it$, $xy/u = z$ after a straightforward calculation we have the kernel form of the product formula.

Lemma 1.8.9 *For $x,y\in\mathbb{R}_+$, $t\in\mathbb{R}$ it holds that*

$$K_{it}(x)K_{it}(y) = \int_0^\infty K(x,y,z)K_{it}(z)\frac{dz}{z} = \mathcal{KL}[K(x,y,\cdot)](t). \tag{1.8.87}$$

with the kernel

$$K(x, y, z) = \frac{1}{2}\exp\left[-\frac{1}{2}\left(\frac{xz}{y} + \frac{xy}{z} + \frac{yz}{x}\right)\right].\qquad(1.8.88)$$

By means of the product formula one defines a generalized translation operator $(GT0)\,T_x$.

Definition 1.8.18 *As GTO for the KLT one defines*

$$(T_x f)(y) = \int_0^\infty K(x, y, z)f(z)\frac{dz}{z}, \qquad x, y \in \mathbb{R}_+ \qquad(1.8.89)$$

provided that it exists.

Proposition 1.8.38 *Let $f \in L_{-1,1}$. Then the GTO $T_x, x \in \mathbb{R}_+$ exists, it is a positive linear operator of $L_{-1,-1}$ into itself and it holds that*

(i) $\qquad \|T_x f\|_{-1,1} \leq x^{-1}\|f\|_{-1,-1},$

(ii) $\qquad (T_x f)(y) = (T_y f)(x),$

(iii) $\qquad (T_x K_{it})(y) = K_{it}(x) \cdot K_{it}(y),$

and

(iv) $\qquad \mathcal{KL}[T_x f](t) = K_{it}(x) \cdot \mathcal{KL}[f](t).$

Proof. T_x is obviously a linear operator and it is positive, since the kernel K from (1.8.88) is positive. From (1.8.89) we have because of $\exp\left[-y\frac{x^2+z^2}{2xz}\right] \leq e^{-y} \leq 1$

$$\|T_x\|_{-1,1} = \int_0^\infty \left|\int_0^\infty K(x, y, z)f(z)\frac{dz}{z}\right|\frac{dy}{y^2} \leq \frac{1}{2}\int_0^\infty |f(z)|\left(\int_0^\infty e^{-\frac{xz}{2y}}\frac{dy}{y^2}\right)\frac{dz}{z}.$$

Putting $u := \frac{xz}{2y}$, $du = -\frac{xz}{2y^2}dy$, i.e., $\frac{dy}{y^2} = -\frac{2}{xz}du$ we have because of $\int_0^\infty e^{-u}du = 1$ the estimate

$$\|T_x\|_{-1,1} \leq x^{-1}\int_0^\infty |f(z)|\frac{dz}{z^2} = x^{-1}\|f\|_{-1,1}$$

and this is (i).

Formula (ii) follows from the symmetry of $K(x, y, z)$ in its variables and (iii) is nothing other than the product formula (1.8.87).

Formula (iv) follows by straightforward calculation:

$$\mathcal{KL}[T_x f](t) = \int_0^\infty K_{it}(y)\left(\int_0^\infty K(x, y, z)f(z)\frac{dz}{z}\right)\frac{dy}{y} = \int_0^\infty f(z)\left(\int_0^\infty K_{it}(y)K(x, y, z)\frac{dy}{y}\right)\frac{dz}{z}$$

$$= \int_0^\infty f(z)K_{it}(x)K_{it}(z)\frac{dz}{z} = K_{it}(x) \cdot \mathcal{KL}[f](t),$$

where again the symmetry of $K(x, y, z)$ and the product formula (1.8.87) were used. ☐

Definition 1.8.19 *The convolution $f * g$ of the KLT is defined by means of*

$$(f * g)(x) = \int\limits_0^\infty f(x)(T_x g)(y)\frac{dy}{y},$$ (1.8.90)

provided that it exists.

Then one can prove:

Theorem 1.8.50 *Let $f, g \in L_{-1,1}$. Then $f * g$ exists and it holds that*

(i) $\qquad \|f * g\|_{-1,1} \leq \|f\|_{-1,1} \cdot \|g\|_{-1,1}$

and

(ii) $\qquad \mathcal{KL}[f * g] = \mathcal{KL}[f] \cdot \mathcal{KL}[g].$

Proof. By means of Proposition 1.8.38, (i) we obtain

$$\|f * g\|_{-1,1} \leq \int\limits_0^\infty \Big(\int\limits_0^\infty |f(y)||T_x g(y)|\frac{dy}{y}\Big)\frac{dx}{x^2} = \int\limits_0^\infty |f(y)|\Big(\int\limits_0^\infty |T_y g(x)|\frac{dx}{x^2}\Big)\frac{dy}{y}$$

$$\leq \|g\|_{-1,1}\int\limits_0^\infty |f(y)|\frac{dy}{y^2} = \|f\|_{-1,1} \cdot \|g\|_{-1,1},$$

and this is (i).

The formula (ii) can be derived by straightforward calculation using Proposition 1.8.38, (ii) and (iv). ☐

Corollary 1.8.15 *The convolution of the KLT is commutative,*

$$f * g = g * f.$$

This can easily be proved taking the *KLT* of both sides of this equation and interchanging the factors in the product of the images.

Remark 1.8.60 *For an extensive investigation of the convolution of the KLT in various spaces of measurable functions on $\mathbb{R}_+$ we refer to [YaL] and [Ya].*

1.8.4   Application

We consider a boundary value problem for the Laplace equation in cylindrical coordinates.

In cylindrical coordinates $\{(r, \theta, z) : \quad 0 < r < \infty, \quad 0 \le \theta < 2\pi, \quad 0 \le z < \infty\}$ the Laplace equation can be written

$$\frac{\partial^2 u}{\partial r^2} + \frac{1}{r}\frac{\partial u}{\partial r} \quad \frac{1}{r^2}\frac{\partial^2 u}{\partial \theta^2} + \frac{\partial^2 u}{\partial z^2} - 0 \qquad (1.8.91)$$

We look for a solution of (1.8.91) in the wedge

$$S_o^\alpha = \{(r, \varphi) : \quad 0 < r < \infty, \quad 0 \le \varphi \le \alpha, \quad \alpha \le \frac{\pi}{2}\}$$

with the boundary conditions

$$u(r, 0) = 0, \qquad u(r, \alpha) = f(r), \qquad (1.8.92)$$

which additionally has the form $u(r, \theta, z) = e^{i\sigma z} u(r, \theta)$, $\sigma > 0$. Putting $\rho = \sigma r$ from (1.8.91) it follows that $v(\rho, \theta) = u(\rho/\sigma, \theta)$

$$A_\rho v(\rho, \theta) + \frac{\partial^2 v(\rho, \theta)}{\partial \theta^2} = 0,$$

where A_ρ is taken from (1.8.83). Performing the KLT on the equation by means of Proposition 1.8.37 we obtain setting $\mathcal{KL}[v(\cdot, \theta)](t) = V(t, \theta)$

$$\frac{\partial^2 V}{\partial \theta^2} - t^2 V = 0.$$

The solution of this (ordinary) differential equation is

$$V(t, \theta) = A(t)\cosh(t\theta) + B(t)\sinh(t\theta).$$

Applying the KLT to the boundary values (1.8.92) we have

$$V(t, 0) = 0, \quad V(t, \alpha) = F(t) = \mathcal{KL}[\{](\sqcup),$$

and, therefore,

$$V(t, \theta) = \frac{F(t)}{\sinh(\alpha t)} \sinh(\theta t).$$

By means of the inversion formula (1.8.82) we obtain

$$v(\rho, \theta) = \frac{2}{\pi^2} \int\limits_0^\infty \frac{t\sinh(\pi t)}{\sinh(\alpha t)} K_{it}(\rho) F(t) \sinh(\theta t) dt.$$

Therefore

$$u(r, \theta) = \frac{2}{\pi^2} \int\limits_0^\infty \frac{t\sinh(\pi t)}{\sinh(\alpha t)} K_{it}(\sigma r) F(t) \sinh(\theta t) dt. \qquad (1.8.93)$$

This is the solution of the boundary value problem (1.8.91), (1.8.92) provided that f has a sufficient "good" behavior.

1.9 The Mehler–Fock Transform

The Mehler–Fock transform (MFT) was first considered by Mehler, see [Meh], and later on by Fock, see [Fo]. It was extensively investigated by Lebedev, see [Le] and the original papers cited there. It has applications for the solution of integral equations and of boundary value problems, especially of axial symmetrical problems and of problems in torodial coordinates in the theory of elasticity, see [Sn.2], [Le], and [U].

Definition 1.9.20 *The Mehler–Fock transform of a function $f : (1; \infty) \to \mathbb{C}$ is defined as*

$$\mathcal{MF}[f](t) = F_o(t) = \int_1^\infty f(x) P_{it-1/2}(x) dx, \tag{1.9.1}$$

provided that the integral exists.

Here P_ν are the Legendre functions of order ν, connected with Gauss' hypergeometric function $_2F_1$ by

$$P_\nu(x) =_2 F_1(-\nu, \nu+1; 1; \frac{1-x}{2}); \tag{1.9.2}$$

see [Le], (7.3.4). Sometimes the Legendre functions P_ν with index $\nu = it - 1/2$ are called cone functions; see [Er.1], vol. 1, section 3.14.

Putting $x = \cosh \xi$, $\xi \in \mathbb{R}_+$ from [Le], (7.4.1), we have the integral representation

$$P_{it-1/2}(\cosh \xi) = \frac{\sqrt{2}}{\pi} \int_0^\xi \frac{\cos(t\tau)}{(\cosh \xi - \cosh \tau)^{1/2}} d\tau, \tag{1.9.3}$$

and from [Le], (7.4.7) we obtain

$$P_{it-1/2}(\cosh \xi) = \frac{\sqrt{2}}{\pi} \coth(\pi t) \int_\xi^\infty \frac{\sin(t\tau)}{(\cosh \tau - \cosh \xi)^{1/2}} d\tau. \tag{1.9.4}$$

From (1.9.2) we have

$$P_{it-1/2}(1) = 1. \tag{1.9.5}$$

By means of connection formulas for Gauss' hypergeometric functions (see [E.1], vol. 1, section 3.2, formulas 9 and 23) we obtain for $x > 1$

$$\begin{aligned} P_{it-1/2}(x) = &\frac{\Gamma(it)}{\sqrt{\pi}\Gamma(it+1/2)} (2x)_2^{it-1/2} F_1\left(\frac{3}{4} - \frac{it}{2}, 1 - \frac{it}{2}; 1 - it; x^{-2}\right) \\ &+ \frac{\Gamma(-it)}{\sqrt{\pi}\Gamma(1/2-it)} (2x)_2^{-it-1/2} F_1\left(\frac{it}{2} + \frac{3}{4}, \frac{it}{2} + \frac{1}{4}; 1 + it; x^{-2}\right). \end{aligned} \tag{1.9.6}$$

Formula (1.9.6) leads to

$$P_{it-1/2}(x) = 0(x^{-1/2}), \qquad x \to +\infty. \tag{1.9.7}$$

From here we deduce sufficient conditions for the existence of the *MFT*.

Theorem 1.9.51 *Let $f \in L_1^{loc}(1,\infty)$ and $f(x) = 0(x^\alpha), x \to +\infty$ for some $\alpha < -1/2$, then the MFT F_o of f exists.*

From (1.9.5) and (1.9.7) we conclude that F_o exists too if $f(x)/\sqrt{x} \in L_1(1,\infty)$. This is certainly fulfilled if $f \in L_1(1,\infty)$, the space of measurable functions on $(1,\infty)$ with the norm

$$\|f\|_1 = \int_1^\infty |f(x)|dx < \infty.$$

In the following we deal with functions belonging to this space. Then we have:

Theorem 1.9.52 *Let $f \in L_1(1,\infty)$. Then $\mathcal{MF}[f]$ exists and with some constant C*

$$|\mathcal{MF}[f]| \le C\|f\|_1.$$

Proof. From (1.9.5) though (1.9.7) we know that $|P_{it-1/2}(x)|$ is bounded by some constant C on $[1,\infty)$ and therefore

$$|\mathcal{MF}[f](t)| \le C \int_1^\infty |f(x)|dx = C\|f\|_1.$$

$\Box$

Remark 1.9.61 *For the investigation of the MFT in several spaces of measurable functions on $(1,\infty)$ we refer to [Ya], section 3.*

Remark 1.9.62 *Sometimes instead of the cone functions (of order zero) $P_{it-1/2}$ the cone functions of order $n, n \in \mathbb{N}_o$ are used as the kernel in (1.9.1). They are defined by*

$$P_\nu^{-n}(x) = \frac{(x^2-1)^{n/2}}{2^n\, n!}\, {}_2F_1\Big(n-\nu, n+\nu+1; n+1; \frac{1-x}{2}\Big),$$

with $\nu = it - 1/2$, $P_\nu^o = P_\nu$.

The *MFT* is closely connected with the *KLT*; see section 1.8.3, (1.8.74). From [PBM], vol. III, section 2.17.7, formula 1, we know that

$$K_{\nu+1/2}(ap) = \Big(\frac{\pi p}{2a}\Big)^{1/2} \int_a^\infty e^{-py}\, P_\nu(y/a)dy, \quad a,p \in \mathbb{R}_+.$$

With $a = 1$, $p = x$, $\nu = it - 1/2$ we obtain

$$K_{it}(x) = (\pi x/2)^{1/2} \int_1^\infty e^{-xy} P_{it-1/2}((y)dy. \tag{1.9.8}$$

From 1.8.3, (1.8.74) we deduce

$$\mathcal{KL}[f](t) = \int_0^\infty f(x)K_{it}(x)\frac{dx}{x} = (\pi/2)^{1/2}\int_0^\infty x^{-1/2}f(x)\left(\int_1^\infty e^{-xy}P_{it-1/2}(y)dy\right)dx$$

$$= (\pi/2)^{1/2}\int_1^\infty P_{it-1/2}(y)\left(\int_0^\infty e^{-xy}x^{-1/2}f(x)dx\right)dy$$

$$= (\pi/2)^{1/2}\int_1^\infty P_{it-1/2}(y)\mathcal{L}[x^{-1/2}f(x)](y)dy.$$

Therefore, we have the relation

$$\mathcal{KL}[f](t) = \sqrt{\frac{\pi}{2}}\mathcal{MF}\left[\mathcal{L}\left[\frac{f(x)}{\sqrt{x}}\right]\right](t), \qquad (1.9.9)$$

where $\mathcal{L}$ is the Laplace transform.

In the same manner one can derive the formula

$$\mathcal{MF}[f](t) = \frac{\sqrt{2}\cosh(\pi t)}{\pi\sqrt{\pi}}\mathcal{KL}[\sqrt{u}\mathcal{L}[(x-1)_+f(x)](u)](t). \qquad (1.9.10)$$

Here the integral representation

$$P_{it-1/2}(x) = \frac{\sqrt{2}\cosh(\pi t)}{\pi\sqrt{\pi}}\int_0^\infty K_{it}(u)e^{-xu}\frac{du}{\sqrt{u}}, \qquad x > 1,$$

(see [PBM], vol. 2, section 2.16.6, formula 3) is used. As usual we use the notation

$$h(x)_+ = \begin{cases} h(x), & h(x) > 0, \\ 0, & h(x) \le 0. \end{cases}$$

Now we are going to derive an inversion formula for the *MFT*. We do so in a quite formal, operational manner, following [Sn.2], section 7.5 or [Za], section 25.5.

Substituting in (1.9.1) $x := \cosh\xi$, $f(x) = g(\xi)$ we have

$$F_o(t) = \int_0^\infty P_{it-1/2}(\cosh\xi)g(\xi)\sinh\xi d\xi.$$

By means of formula (1.9.3) we obtain

$$F_o(t) = \int_0^\infty g(\xi)\left(\frac{\sqrt{2}}{\pi}\int_0^\xi \frac{\cos(t\tau)}{(\cosh\xi - \cosh\tau)^{1/2}}d\tau\right)\sinh\xi d\xi$$

$$= \frac{\sqrt{2}}{\pi}\int_0^\infty \cos(t\tau)\left(\int_\tau^\infty \frac{g(\xi)\sinh\xi}{(\cosh\xi - \cosh\tau)^{1/2}}d\xi\right)d\tau$$

$$= \frac{\sqrt{2}}{\pi}\mathcal{F}_c\left[\int_\tau^\infty \frac{g(\xi)\sinh\xi}{(\cosh\xi - \cosh\tau)^{1/2}}d\xi\right](t),$$

where $\mathcal{F}_c$ is the Fourier cosine transform; see formula (1.3.5).

By means of the well-known formulas for the Fourier cosine and the Fourier sine transforms (see 1.3.3, Remark 1.3.7) namely

$$\mathcal{F}_s[h'](t) = -t\mathcal{F}_c[h](t)$$

we have

$$tF_o(t) = -\frac{\sqrt{2}}{\pi}\mathcal{F}_s\left[\frac{d}{d\tau}\int_\tau^\infty \frac{g(\xi)\sinh\xi}{(\cosh\xi - \cosh\tau)^{1/2}}d\xi\right](t).$$

Using the inversion formula for the Fourier sine transform (see 1.3.4, Remark 1.3.14) we obtain

$$\mathcal{F}_s[tF_o(t)](\tau) = -\frac{1}{\sqrt{2}}\frac{d}{d\tau}\int_\tau^\infty \frac{g(\xi)\sinh\xi}{(\cosh\xi - \cosh\tau)^{1/2}}d\xi.$$

This is an integral equation for the function g. By means of 1.4.8, formulas (1.4.64), (1.4.63) with $h = \frac{\sqrt{2}}{\pi}tF_o(t)$ we have the solution

$$g(\xi) = \frac{\sqrt{2}}{\pi}\int_\xi^\infty \mathcal{F}_s[tF_o(t)](\tau)\frac{d\tau}{(\cosh\tau - \cosh\xi)^{1/2}}$$

$$= \frac{\sqrt{2}}{\pi}\int_\xi^\infty \frac{1}{(\cosh\tau - \cosh\xi)^{1/2}}\left(\int_0^\infty tF_o(t)\sin(t\tau)dt\right)d\tau$$

$$= \frac{\sqrt{2}}{\pi}\int_0^\infty tF_o(t)\left(\int_\xi^\infty \frac{\sin(t\tau)}{(\cosh\tau - \cosh\xi)^{1/2}}d\tau\right)dt.$$

Substituting the inner integral by means of formula (1.9.4) we have

$$g(\xi) = \int_0^\infty t\tanh(\pi t)\,P_{it-1/2}(\cosh\xi)F_o(t)dt.$$

Resubstituting $x := \cosh\xi$, $g(\xi) = f(x)$ we have

$$f(x) = \int_0^\infty t\tanh(\pi t)P_{it-1/2}(x)F_o(t)dt =: \mathcal{MF}^{-1}[F_o](x). \qquad (1.9.11)$$

Conditions for the validity of (1.9.16) can be given as follows. Because of

$$P_{it-1/2}(\cosh\alpha) \approx \frac{\sqrt{2}}{(\pi t\sinh\alpha)^{1/2}}\sin(\alpha t + \pi/4)$$

as $t \to +\infty$, $\delta \le \alpha \le a < \infty$; see [Le], formula (7.11.8), and Problem 14 of Chapter 7. Therefore,

$$P_{it-1/2}(x) = 0(t^{-1/2}), \quad t \to +\infty$$

and consequently the integral (1.9.11) exists if $t^{1/2}F_o(t) \in L_1(\mathbb{R}^+)$. Taking into consideration the asymptotic behavior of $P_{it-1/2}(x)$ as x tends to $+\infty$, see formula (1.9.7), we have the following result:

Theorem 1.9.53 (Inversion Theorem) *Let $f \in L_1^{loc}(1, \infty)$ such that $f(x)/\sqrt{x} \in L_1(1, \infty)$. Furthermore, let $\sqrt{t}\mathcal{MF}[f](t) \in L_1(\mathbb{R}_+)$. Then (a.e.) we have the connection (1.9.1), (1.9.11) between the MFT of an original f and f itself.*

Remark 1.9.63 *For further conditions for the validity of the inversion formula (1.9.10) we refer to [Le], section 8.9, [Sn.2], section 7.5, and [Ya], section 3.1.*

Remark 1.9.64 *The pair of formulas (1.9.1), (1.9.11) is often called the Mehler–Fock theorem.*

By means of the Mehler–Fock theorem one can easily derive a Parseval-type relation for the MFT. If F_o, G_o, are the Mehler–Fock transforms of f and g, respectively, then we obtain from (1.9.1) and (1.9.10)

$$\int_0^\infty t\tanh(\pi t)F_o(t)\left(\int_1^\infty g(x)P_{it-1/2}(x)dx\right)dt = \int_1^\infty g(x)\left(\int_0^\infty t\tanh(\pi t)F_o(t)P_{it-1/2}(x)\right)dx$$

$$= \int_1^\infty f(x)\,g(x)dx.$$

Therefore, we have:

Theorem 1.9.54 *Let $F_o = \mathcal{MF}[f]$ and $G_o = \mathcal{MF}[g]$ and furthermore f and g fulfill the conditions of Theorem 1.9.53. Then*

$$\int_0^\infty t\tanh(\pi t)F_o(t)G_o(t)dt = \int_1^\infty f(x)g(x)dx. \tag{1.9.12}$$

In view of applications we look for a differentiation rule of the MFT.

Let

$$B = D(x^2 - 1)D + 1/4, \qquad D = d/dx. \tag{1.9.13}$$

From [E.1], vol. 1, 3.2, (1) we have

$$B(P_{it-1/2}(x)) = -t^2 P_{it-1/2}(x). \tag{1.9.14}$$

Applying the MFT on Bf and integrating by parts we have to look for conditions on f such that the terms outside the integral signs vanish at $1+$ and $+\infty$. After straightforward calculation we obtain:

Proposition 1.9.39 (Differentiation Rule) *Let f be such that the MFT of f and of Bf exist and let furthermore*

$$\lim_{x \to 1+, +\infty} (x^2 - 1) f'(x) P_{it-1/2}(x) = \lim_{x \to 1+, +\infty} (x^2 - 1) \Big(D P_{it-1/2}(x) \Big) f(x) = 0.$$

Then it holds that

$$\mathcal{MF}[Bf](t) = -t^2 \mathcal{MF}[f](t). \tag{1.9.15}$$

The investigation of the convolution structure of the MFT starts as usual with a linearization formula for the product of two cone functions, see [Vi], Chapter VI, 4., formula (2), which after some substitutions can be written as

$$P_{it-1/2}(x) \, P_{it-1/2}(y) = \pi^{-1} \int_0^{\pi} P_{it-1/2}(z(x, y, \theta)) \, d\theta, \tag{1.9.16}$$

with

$$z(x, y, \theta) = xy + [(x^2 - 1)(y^2 - 1)]^{1/2} \cos \theta. \tag{1.9.17}$$

Substituting θ by z we obtain the kernel form of (1.9.16). We have

$$dz = -[(x^2 - 1)(y^2 - 1)]^{1/2} \sin \theta d\theta$$

or

$$\begin{aligned}
d\theta &= -[(x^2 - 1)(y^2 - 1)]^{-1/2}(1 - \cos^2 \theta)^{-1/2} d\theta \\
&= -[(x^2 - 1)(y^2 - 1)]^{-1/2} \Big[1 - \frac{(z - xy)^2}{(x^2 - 1)(y^2 - 1)} \Big]^{-1/2} dz \\
&= -(2xyz + 1 - x^2 - y^2 - z^2)^{-1/2} dz.
\end{aligned}$$

Here

$$z \in \Big(xy - [(x^2 - 1)(y^2 - 1)]^{1/2}, xy + [(x^2 - 1)(y^2 - 1)]^{1/2} \Big) =: I_{x,y}. \tag{1.9.18}$$

Since the left-hand point of this interval is greater than one, we have:

Lemma 1.9.10 *For $x, y > 1$, $t \in \mathbb{R}$ it holds that*

$$P_{it-1/2}(x) \, P_{it-1/2}(y) = \int_1^{\infty} K(x, y, z) P_{it-1/2}(z) dz = \mathcal{MF}[K(x, y, \cdot)](t) \tag{1.9.19}$$

with the kernel

$$K(x, y, z) = \begin{cases} \pi^{-1}(2xyz + 1 - x^2 - y^2 - z^2)^{-1/2}, & z \in I_{x,y}, \\ 0, & \text{otherwise.} \end{cases} \tag{1.9.20}$$

Lemma 1.9.11 *The kernel $K(x, y, z)$ is positive and symmetrical with respect to x, y, z. Furthermore, it holds that*

$$\int_1^\infty K(x, y, z)dz = 1. \tag{1.9.21}$$

Proof. The positivity and the symmetry follow directly from (1.9.20). Putting $t = -\frac{i}{2}$ in (1.9.19), because of $P_o = 1$, see (1.9.2), we obtain (1.9.21). ☐

By means of the product formulas one defines a generalized translation operator (GTO) T_x.

Definition 1.9.21 *As GTO for the MFT one defines the operator T_x by means of*

$$(T_x f)(y) = \int_1^\infty K(x, y, z)f(z)dz, \tag{1.9.22}$$

provided that it exists.

Proposition 1.9.40 *Let $f \in L_1(1, \infty)$. Then the GTO T_x, $x > 1$ exists, it is a positive linear operator from $L_1(1, \infty)$ into itself and it holds that*

(i) $\|T_x f\|_1 \leq \|f\|_1,$

(ii) $(T_x f)(y) = (T_y f)(x),$

(iii) $(T_x P_{it-1/2})(y) = P_{it-1/2}(x) \cdot P_{it-1/2}(y)$

and

(iv) $\mathcal{MF}[T_x f](t) = P_{it-1/2}(x) \cdot \mathcal{MF}[f](t).$

Proof. Obviously, T_x is a linear operator and it is positive since K is positive. For the proof of estimate (i) we have by means of (1.9.21)

$$\|T_x f(\cdot)\|_1 = \int_1^\infty \left| \int_1^\infty K(x, y, z)f(z)dz \right| dy \leq \int_1^\infty |f(z)| \left(\int_1^\infty K(x, y, z)dy \right)dz = \|f\|_1.$$

The results (ii)–(iv) follow in the same manner as derived in the case of the KLT in section 1.8.3. ☐

Definition 1.9.22 *The convolution $f * g$ of the MFT is defined by means of*

$$(f * g)(x) = \int_1^\infty f(y)(T_x g)(y)dy, \tag{1.9.23}$$

provided that it exists.

Then one can prove:

Theorem 1.9.55 (Convolution Theorem) *Let $f, g \in L_1(1, \infty)$. Then $f * g \in L_1(1, \infty)$ and it holds that*

(i) $\|f * g\|_1 \leq \|f\|_1 \cdot \|g\|_1$

and

(ii) $\mathcal{MF}[f * g] - \mathcal{MF}[f] \cdot \mathcal{MF}[g]$.

Proof.   By means of (1.9.23) and Proposition 1.9.40, (ii) and (i) we obtain

$$\|f * g\|_1 = \int\limits_1^\infty |(f * g)(x)| dx = \int\limits_1^\infty \left| \int\limits_1^\infty f(y) \, T_x g(y) dy \right| dx$$

$$\leq \int\limits_1^\infty |f(y)| \left(\int\limits_1^\infty |T_y g(x)| dx \right) dy = \int\limits_1^\infty |f(y)| \|T_y g\|_1 dy$$

$$\leq \|g\|_1 \int\limits_1^\infty |f(y)| dy = \|f\|_1 \cdot \|g\|_1.$$

Property (ii) can easily be derived by straightforward calculation using Proposition 1.9.40, (ii) and (iv).   □

As usual one has:

Corollary 1.9.16 *The convolution is commutative, i.e.,*

$$f * g = g * f.$$

For application of the MFT we refer to [Sn.2], sections 7.8 through 7.12. There boundary value problems for partial differential equations and dual integral equations are solved. We restrict ourselves to a simple application of the Parseval relation; see Theorem 1.9.54 formula (1.9.12). Taking

$$f(x) = e^{-ax}, \quad g(x) = e^{-bx}, \quad a, b \in \mathbb{R}_+$$

from equation (1.9.8) we obtain

$$F_o(t) = \sqrt{\frac{2}{\pi a}} \, K_{it}(a), \quad G_o(t) = \sqrt{\frac{2}{\pi b}} \, K_{it}(b).$$

Now

$$\int\limits_1^\infty f(x) \, g(x) \, dx = \int\limits_1^\infty e^{-(a+b)x} dx = \frac{e^{-(a+b)}}{a+b}.$$

So from (1.9.12) we obtain

$$\int\limits_0^\infty t \, \tanh(\pi t) \, K_{it}(a) K_{it}(b) dt = \frac{\pi \sqrt{ab}}{2(a+b)} e^{-(a+b)}.$$

1.10   Finite Integral Transforms

1.10.1   Introduction

In the preceding section we investigated integral transforms where the images (transforms) were functions defined on some interval of the real axis or on some domain of the complex plane. They are sometimes called continuous integral transforms.

Now we deal with integral transforms, where the images are functions defined on an (infinite) subset of the set $\mathbb{Z}$ of integers. They are sometimes called finite integral transforms (see Sneddon [Sn.2], Chapter 8, Churchill [Ch.2]) sometimes discrete integral transforms (see Zayed [Za], Definition 4.2). So the concept of finite or discrete integral transforms is not uniform.

The kernels of the transforms investigated in the following subsections are polynomials of a complete orthogonal system in some Hilbert space of square-integrable functions (with some weight) on some interval of the real line. The transforms are the (standardized) Fourier coefficients with respect to this orthogonal system and quite formally one has an inversion formula, namely the Fourier series with respect to the orthogonal system in consideration. We will not develop the L_2-theory. The originals of our transforms are L_1-functions on some interval with some weight.

Such integral transforms were investigated first by Scott; see [Sc], (Jacobi transform, 1953); Churchill, see [Ch.1], Churchill and Dolph [ChD], (Legendre transform, 1954); Conte, see [Co], (Gegenbauer transform, 1955); McCully, see [MC], (Laguerre transform of order zero, 1960); Debnath, see [De.1], [De.5], [De.7], (Laguerre transform of arbitrary order, 1960, 1961, 1969); [De.3], [De.4], (Hermite transform, 1964, 1968); and others.

We consider the case of a finite interval, as usual standardized as $(-1, 1)$, and this leads to the Jacobi transform and its special cases, the Chebyshev transform, Legendre transform and Gegenbauer transform, the case of a semi-infinite interval, standardized as $(0, \infty)$ with the Laguerre polynomials and the case of the interval $(-\infty, +\infty)$, which leads to the Hermite polynomials.

The reader interested in some other discrete integral transforms is referred to Churchill [Ch.2], Firth [Fi], Jerry [Je], Sneddon [Sn.2], Zayed [Za], and others.

Many examples of transforms of concrete functions are given by Debnath, [De.6].

The subsections are organized as follows:

- Foundation (Definition, Spaces of originals, Inversion formula,...)

- Operational Rules (Differentiation, Integration, Convolution theorem)

- Applications (Solution of boundary value problems).

The properties of the kernels – the classical orthogonal polynomials — are assumed to be known to the reader and they can be found in textbooks of special functions and orthogonal polynomials. We refer mostly to Erdeyi, [E.1], vol. 2.

Designation: The transform $\mathfrak{T}[f]$ of a function f is written in simplified form as $f^\wedge$, also if the transform depends additionally on parameters

$$\alpha, \beta, (\alpha; \beta), \cdots : \qquad \mathfrak{T}[f] = (f^\alpha)^\wedge = f^\wedge.$$

So $f^\wedge$ means another transform in each subsection. The designation $(f^\alpha)^\wedge$, – is only used if in some formula transforms with different parameters appear.

1.10.2   The Chebyshev Transform

Definition 1.10.23 *The Chebyshev transform (CT) of a function $f : [-1; 1] \to \mathbb{C}$ is defined by means of*

$$\mathfrak{T}[f](n) = f^\wedge(n) = \frac{1}{\pi} \int_{-1}^{1} f(x) T_n(x)(1-x^2)^{-1/2} dx, \qquad n \in \mathbb{N}_o, \qquad (1.10.1)$$

provided that the integral exists. Here T_n are the Chebyshev polynomials (of the first kind), defined by

$$T_n(x) = \cos(n \arccos x), \qquad n \in \mathbb{N}_o. \qquad (1.10.2)$$

Remark 1.10.65 *For the properties of the Chebyshev polynomials we refer to [R] or tables of special functions, for example [E.1], vol. II, 10.11.*

As space of originals we choose the space $L_1^0(-1; 1)$, written in abbreviated form as L_1^0, of measurable functions on $(-1; 1)$ such that

$$\|f\|_{1,0} = \frac{1}{\pi} \int_{-1}^{1} |f(x)|(1-x^2)^{-1/2} dx \qquad (1.10.3)$$

is finite. It is well known that it is a Banach space with the norm (1.10.3).

Remark 1.10.66 *The considerations in the following also can be extended for originals in the space L_p^0 of measurable functions on $(-1; 1)$, $1 \leq p < \infty$ with the norm*

$$\|f\|_{p,0} = \left[\frac{1}{\pi} \int_{-1}^{1} |f(x)|^p (1-x^2)^{-1/2} dx \right]^{1/p} \qquad (1.10.4)$$

or in the space $\mathcal{C}[-1; 1]$ of continuous functions on $[-1; 1]$ with the norm $\|f\| = \sup |f(x)|$, $x \in [-1, 1]$, see [BuS].

Theorem 1.10.56 *Let $f \in L_1^0$ and $k, n \in \mathbb{N}_o$. Then the Chebyshev transform $\mathfrak{T}$ is a linear transform and moreover it holds that*

(i) $\qquad |\mathfrak{T}[f](n)| \leq \|f\|_{1,0}, \quad n \in \mathbb{N}_o,$

(ii) $\qquad \lim_{n \to \infty} \mathfrak{T}[f](n) = 0,$

(iii) $\qquad \mathfrak{T}[f](n) = 0, \quad n \in \mathbb{N}_o$ *if and only if* $f(x) = 0$ *(a.e)*,

(iv) $\qquad \mathfrak{T}[T_k](n) = \left\{ \begin{array}{ll} 1, & n = k = 0 \\ \dfrac{1}{2}, & n = k \neq 0 \\ 0, & n \neq k \end{array} \right\}, \quad n, k \in \mathbb{N}_o.$

Proof. The linearity follows from (1.10.1), as does the estimate (i), taking into account that $|T_n(x)| \leq 1$ (see 1.10.2).

Putting $x = \cos\theta, \quad \theta \in [0; \pi]$ the CT takes the form

$$\mathfrak{T}[f](n) = \frac{1}{\pi} \int_0^\pi (f \circ \cos)(\theta) \cos n\theta d\theta \qquad (1.10.5)$$

and this are up to a constant factor the Fourier coefficients of the function $\varphi = f \circ \cos$, which belongs to $L_1(0; \pi)$ if $f \in L_1^0$.

The properties (ii) and (iii) then follow from well-known results of Fourier series theory.

Formula (iv) is nothing other than the orthogonality relation of the Chebyshev polynomials (see [E.1], vol. II, 10–11, (7)). $\qquad \square$

The result (iii) can be formulated as a uniqueness theorem for the CT.

Theorem 1.10.57 (Uniqueness Theorem) *Let $f, g \in L_1^0$ and $\mathfrak{T}[f](n) = \mathfrak{T}[g](n)$ for every $n \in \mathbb{N}_o$. Then $f = g$ (a.e.).*

An inversion formula for the CT can be easily derived by means of Lebesgue's dominated convergence theorem.

Theorem 1.10.58 *If $f \in L_1^0$ can be expanded into a series of the form*

$$f(x) = a_o + 2 \sum_{k=1}^\infty a_k T_k(x) \qquad (a.e.), \qquad (1.10.6)$$

the series being dominatedly convergent, i.e., for each $m \in \mathbb{N}_o$ it holds that

$$\left| \sum_{k=0}^m a_k T_k(x) \right| \leq g(x) \qquad (a.e.),$$

where $g \in L_1^0$, then $a_k = \mathfrak{T}[f](k)$.

So, under the conditions of Theorem 1.10.58 we have:

Corollary 1.10.17 (Inversion Formula)

$$f(x) = f^\wedge(0) + 2 \sum_{k=1}^{\infty} f^\wedge(k) T_k(x) =: \mathfrak{T}^{-1}[f^\wedge](x). \tag{1.10.7}$$

Now we are going to formulate some rules of operational calculus.

Proposition 1.10.41 *Let* $f \in L_1^0$ *and* $k, m, n \in \mathbb{N}_o$. *Then it holds that*

$$\mathfrak{T}[T_m f](n) = \frac{1}{2} \left[f^\wedge(m+n) + f^\wedge(|m-n|) \right], \tag{1.10.8}$$

$$\mathfrak{T}[x f(x)](n) = \frac{1}{2} \left[f^\wedge(n+1) + f^\wedge(n-1) \right], \tag{1.10.9}$$

and, more generally,

$$\mathfrak{T}[x^k f(k)](n) = 2^{-k} \sum_{l=0}^{k} \binom{k}{l} f^\wedge(n + 2l - k). \tag{1.10.10}$$

Proof. Formula (1.10.8) follows directly from

$$T_m T_n = \frac{1}{2} \left[T_{m+n} + T_{|m-n|} \right], \tag{1.10.11}$$

which itself follows from (1.10.2) by means of the addition theorem of the cosine function (see also [E.1], vol. II, 10.11, (34)). The result (1.10.9) follows from the three-term recurrence for Chebyshev polynomials

$$T_{n+1}(x) = 2x T_n(x) - T_{n-1}(x), \tag{1.10.12}$$

see [E.1], vol. II, 10.11, (16). Finally, formula (1.10.10) can easily be proved by mathematical induction. □

The Chebyshev polynomials are eigenfunctions of the differential operator T defined by means of

$$(Ty)(x) = -(1-x^2)^{1/2} D(1-x^2)^{1/2} D, \qquad D = \frac{d}{dx} \tag{1.10.13}$$

with respect to the eigenvalues n^2, i.e.,

$$T T_n = n^2 T_n, \qquad n \in \mathbb{N}_o. \tag{1.10.14}$$

Therefore, we obtain:

Proposition 1.10.42 (Differentiation Rule) *Let* $f \in L_1^o$ *be two times differentiable (a.e.) on* $(-1; 1)$ *and*

$$\lim_{x \to \pm 1} (1-x^2)^{1/2} f(x) = \lim_{x \to \pm 1} (1-x^2)^{1/2} f'(x) = 0.$$

Then it holds that

$$\mathfrak{T}[Tf](n) = n^2 f^\wedge(n). \tag{1.10.15}$$

Proof. We have by means of integration by parts

$$\mathfrak{T}[Tf](n) = -\frac{1}{\pi}\int\limits_{-1}^{1} D[(1-x^2)^{1/2}f'(x)]T_n(x)dx = \frac{1}{\pi}\int\limits_{-1}^{1} f'(x)(1-x^2)^{1/2}T_n'(x)dx$$

$$= -\frac{1}{\pi}\int\limits_{-1}^{1} f(x)[(1-x^2)^{1/2}T_n'(x)]'dx = \frac{1}{\pi}\int\limits_{-1}^{1} f(x)(TT_n)(x)(1-x^2)^{-1/2}\,dx$$

$$= n^2 f^\wedge(n).$$

$\square$

Corollary 1.10.18
$$\mathfrak{T}[T^k f](n) = n^{2k}\mathfrak{T}[f](n). \tag{1.10.16}$$

Now let
$$Tf = g.$$

Then
$$(1-x^2)^{1/2}f'(x) = -\int\limits_{-1}^{x}(1-v^2)^{-1/2}g(v)dv$$

or
$$f(x) = -\int\limits_{-1}^{x}(1-u^2)^{-1/2}\int\limits_{-1}^{u}(1-v^2)^{-1/2}g(v)dvdu =: (T^{-1}g)(x). \tag{1.10.17}$$

Because of $\mathfrak{T}[Tf](n) = n^2 f^\wedge(n) = g^\wedge(n)$ we have to assume that $g^\wedge(0) = 0$.

Performing the Chebyshev transform on (1.10.17) we have

$$f^\wedge(n) = \mathfrak{T}[T^{-1}g](n) = n^{-2}g^\wedge(n),$$

or:

Proposition 1.10.43 (Integration Rule) *Let $g \in L_1^0$ and $g^\wedge(0) = 0$. Then it holds that*

$$\mathfrak{T}[T^{-1}g](n) = n^{-2}g^\wedge(n), \qquad n \in \mathbb{N}, \tag{1.10.18}$$

where T^{-1} is defined by means of (1.10.17).

Now we are going to derive a convolution theorem for the Chebyshev transform. First of all we note a linearization formula for the product of two Chebyshev polynomials which easily can be proved by (1.10.2) and by means of the addition theorem of the cosine function.

Lemma 1.10.12 *It holds that*

$$T_n(x)T_n(y) = \frac{1}{2}\left[T_n\left(xy + \sqrt{(1-x^2)(1-y^2)}\right) + T_n\left(xy - \sqrt{(1-x^2)(1-y^2)}\right)\right]. \tag{1.10.19}$$

Definition 1.10.24 *As the generalized translation operator (GTO) we denote the operator* τ_x^o, $x \in [-1; 1]$, *defined by*

$$(\tau_x^o f)(y) = \frac{1}{2}\left[f\left(xy + \sqrt{(1-x^2)(1-y^2)}\right) + f\left(xy - \sqrt{(1-x^2)(1-y^2)}\right)\right], \qquad (1.10.20)$$

$y \in [-1; 1]$.

Then we have:

Proposition 1.10.44 *Let* $f \in L_1^0$, $x \in [-1; 1]$. *Then we have*

(i) τ_x^o *is a bounded linear operator of* L_1^0 *into itself and* $\|\tau_x^o f\|_{1,0} \le \|f\|_{1,0}$,

(ii) $(\tau_x^o f)(y) = (\tau_y^o f)(x)$,

(iii) $\mathfrak{T}[\tau_x^o f](n) = T_n(x) f^{\wedge}(n)$,

(iv) $(\tau_x^o T_n)(y) = T_n(x) T_n(y)$.

Proof. Substituting $x = \cos\theta$, $0 \le \theta \le \pi$, we obtain

$$\|\tau_x^o f\|_{1,0} = \frac{1}{2\pi} \int\limits_{-1}^{1} \left| f\left(y\cos\theta + (1-y^2)^{1/2}\sin\theta\right)\right.$$
$$\left. + f\left(y\cos\theta - (1-y^2)^{1/2}\sin\theta\right)\right| (1-y^2)^{-1/2}\, dy.$$

Putting $y = \cos\varphi$, $0 \le \varphi \le \pi$, after a short calculation leads to

$$\|\tau_x^o f\|_{1,0} = \frac{1}{2\pi} \int\limits_{0}^{\pi} \left| f\left(\cos(\varphi-\theta)\right) + f\left(\cos(\varphi+\theta)\right)\right| d\varphi$$

$$= \frac{1}{4\pi} \int\limits_{-\pi}^{\pi} \left| f\left(\cos(\varphi-\theta)\right) + f\left(\cos(\varphi+\theta)\right)\right| d\varphi$$

$$\le \frac{1}{2\pi} \int\limits_{-\pi}^{\pi} \left| f\left(\cos(\varphi-\theta)\right)\right| d\varphi = \frac{1}{2\pi} \int\limits_{-\pi}^{\pi} |f(\cos\varphi)| d\varphi$$

$$= \frac{1}{\pi} \int\limits_{0}^{\pi} |f(\cos\varphi)|\, d\varphi = \frac{1}{\pi} \int\limits_{-1}^{1} |f(y)|(1-y^2)^{-1/2} dy = \|f\|_{1,0},$$

and this is (i). The symmetry (ii) follows from the definition (1.10.20). With the same substitutions as in part (i) we obtain easily (iii), and (iv) is another formulation of the product formula (1.10.19).   □

Definition 1.10.25 *As the convolution* $f * g$ *of* f *and* g *we denote*

$$(f * g)(x) = \frac{1}{\pi} \int\limits_{-1}^{1} f(y)(\tau_x^o g)(y)(1-y^2)^{-1/2} dy, \qquad (1.10.21)$$

provided that it exists.

Theorem 1.10.59 (Convolution Theorem) *Let $f, g \in L_1^0$. Then $f * g$ exists (a.e.), it belongs to L_1^0 and*

(i) $\qquad \|f * g\|_{1,0} \leq \|f\|_{1,0} \cdot \|g\|_{1,0}$,

(ii) $\qquad \mathfrak{T}[f * g] = f^\wedge g^\wedge$,

(iii) $\qquad f * g = g * f$.

Proof. First we are going to prove (i). Obviously, by Proposition 1.10.44, (ii) it holds that

$$\|f * g\|_{1,0} = \frac{1}{\pi^2} \int_{-1}^{1} \left| \int_{-1}^{1} f(y)(\tau_x^o g)(y)(1 - y^2)^{-1/2}(1 - x^2)^{-1/2} dy \right| dx$$

$$\leq \frac{1}{\pi} \int_{-1}^{1} |f(y)|(1 - y^2)^{-1/2} \int_{-1}^{1} |(\tau_y^o g)(x)|(1 - x^2)^{-1/2} dx dy$$

$$= \|f\|_{1,0} \cdot \|\tau_y^o g\|_{1,0} \leq \|f\|_{1,0} \cdot \|g\|_{1,0},$$

where for the latter Proposition 1.10.44 was used, and this is (i). Furthermore, by means of Proposition 1.10.44, (iii) we have

$$\mathfrak{T}[f * g](n) = \frac{1}{\pi^2} \int_{-1}^{1} f(y)(1 - y^2)^{-1/2} \int_{-1}^{1} (\tau_y^o g)(x) T_n(x)(1 - x^2)^{-1/2} dx dy$$

$$= \left(\frac{1}{\pi} \int_{-1}^{1} f(y) T_n(y)(1 - y^2)^{-1/2} dy \right) g^\wedge(n) = f^\wedge(n) \cdot g^\wedge(n).$$

The commutativity (iii) follows directly from (ii) by means of the uniqueness theorem, Theorem 1.10.57. $\qquad\qquad\qquad\qquad\qquad\qquad\qquad\qquad\qquad\qquad\qquad\qquad\qquad\quad$ ☐

Remark 1.10.67 *The results can be generalized to originals of the spaces L_p^0; see Remark 1.10.66 and [BuS], Theorem 1.10.56. Here we have to consider functions $f \in L_p^0$ and $g \in L_1^0$. The convolution belongs to L_p^0 and (ii) takes the form*

$$\|f * g\|_{p,0} \leq \|f\|_{p,0} \cdot \|g\|_{1,0}.$$

Remark 1.10.68 *All the considerations of this section are valid also in the space $C[-1; 1]$ (with minor changes).*

Now we are going to consider an application. We look for a solution of the initial value problem

$$(1 - x^2) u_{xx}(x, t) - x u_x(x, t) = (1 - t^2) u_{tt}(x, t) - t u_t(x, t),$$

$$u(x, 1) = u_o(x), \qquad x, t \in [-1; 1].$$

The partial differential equation can be written as

$$T_x u(x,t) = T_t u(x,t),$$

where T_x resp. T_t is the operation T defined in (1.10.19) considered as a partial differential operator with respect to x resp. t. The application of the Chebyshev transform with respect to x and the use of the differentiation rule (1.10.15) lead to

$$n^2 \mathfrak{T}[u(\cdot,t)](n) = T_t \mathfrak{T}[u(\cdot,t)](n).$$

This is the eigenvalue equation (1.10.14) and, therefore, we have

$$\mathfrak{T}[u(\cdot,t)](n) = a_n T_n(t), \qquad n \in \mathbb{N}_o,$$

where $a_n \in \mathbb{R}$ are constants. By means of the boundary conditions we obtain

$$\mathfrak{T}[u(\cdot,1)](n) = \mathfrak{T}[u_o](n).$$

From Proposition 1.10.44, (iv) and Theorem 1.10.57 we conclude

$$a_n = \mathfrak{T}[u_o](n)$$

and Proposition 1.10.44, (iii) leads to

$$\mathfrak{T}[u(\cdot,t)](n) = \mathfrak{T}[u_o](n) T_n(t) = \mathfrak{T}[\tau_t^o u_o](n)$$

and by inversion we have

$$u(x,t) = (\tau_t^o u_o)(x) = \frac{1}{2}\left[u_o\left(xt + \sqrt{(1-x^2)(1-t^2)} \right) + u_o\left(xt - \sqrt{(1-x^2)(1-t^2)} \right) \right],$$

which is the solution of the initial value problem, if u_o is sufficiently smooth.

1.10.3 The Legendre Transform

Definition 1.10.26 *The Legendre transform (LeT) of a function $f : [-1;1] \to \mathbb{C}$ is defined by means of*

$$\mathfrak{P}[f](n) = f^\wedge(n) = \frac{1}{2}\int_{-1}^{1} f(x) P_n(x)dx, \qquad n \in \mathbb{N}_o, \tag{1.10.22}$$

provided that the integral exists. Here P_n are the Legendre polynomials, defined by

$$P_n(x) = \frac{(-1)^n}{2^n n!}\frac{d^n}{dx^n}(1-x^2)^n, \quad n \in \mathbb{N}_o. \tag{1.10.23}$$

Remark 1.10.69 *For the properties of the Legendre polynomials we refer to [E.1], vol. II, 10.10. We note especially that $|P_n(x)| \le 1$, $-1 \le x \le 1$ and $P_n(1) = 1$, $n \in \mathbb{N}_o$.*

As space of originals we choose the space $L_1(-1;1) = L_1$ of measurable functions on $(-1;1)$ such that

$$\|f\|_1 = \frac{1}{2} \int\limits_{-1}^{1} |f(x)| dx \tag{1.10.24}$$

is finite. It is a Banach space with the norm (1.10.24).

Remark 1.10.70 *The considerations in the following also can be extended to originals in the space L_p, $1 \leq p < \infty$, of measurable functions on $(-1;1)$ with the norm*

$$\|f\|_p = \left[\frac{1}{2} \int\limits_{-1}^{1} |f(x)|^p dx \right]^{1/p} \tag{1.10.25}$$

or in the space $C[-1;1]$ of continuous functions on $[-1;1]$ with the sup-norm; see [StW].

Theorem 1.10.60 *Let $f \in L_1$ and $k,n \in \mathbb{N}_o$. Then the Legendre transform $\mathfrak{P}$ is a linear transform and moreover it holds that*

(i) $\qquad |\mathfrak{P}[f](n)| \leq \|f\|_1$,

(ii) $\qquad \mathfrak{P}[P_k](n) = \frac{1}{2n+1}\, \delta_{nk}$,

(iii) $\qquad \mathfrak{P}[f](n) = 0$, $n \in \mathbb{N}_o$ *if and only if $f(x) = 0$ (a.e.).*

Proof. From (1.10.22) we have the linearity of the transform and also the estimate (i), since $|P_n(x)| \leq 1$, $x \in [-1;1]$; see [E.1], vol. II, 10.18, (1).

Property (ii) is nothing other than the orthogonality property of the Legendre polynomials; see [E.1], vol. II, 10.10, (4).

For the proof of property (iii) we consider the mapping

$$d\mu_f(x) = \frac{1}{2} f(x) dx.$$

It assigns a bounded measure μ_f on $[-1;1]$ to every $f \in L_1$. Let $g \in C[-1;1]$. It is well known that every continuous linear functional on $C[-1;1]$ can be determined in unique manner by a bounded measure, i.e.,

$$F_f(g) = \int\limits_{-1}^{1} g(x) d\mu_f(x).$$

From our assumption we have $\mathfrak{P}[f](n) = F_f(P_n) = 0$ for every $n \in \mathbb{N}_o$. Since every polynomial is a linear combination of the set of Legendre polynomials, we have $F_f(p) = 0$ for every polynomial p. By means of the Weierstrass approximation theorem every continuous function g on $[-1;1]$ can be approximated by means of polynomials. Therefore, we conclude

$F_f(g) = 0$ for every $g \in \mathcal{C}[-1; 1]$ and so we have $d\mu_f(x) = 0$, i.e., $f(x) = 0$ a.e., This concludes the proof of one direction of the assertion (iii). The proof of the other one is trivial.   □

The result (iii) can be formulated as a uniqueness theorem for the LeT.

Theorem 1.10.61 (Uniqueness Theorem) *Let $f, g \in L_1$ and $\mathfrak{P}[f](n) = \mathfrak{P}[g](n)$ for every $n \in \mathbb{N}_o$. Then $f = g$ (a.e.).*

For the derivation of an inversion formula we assume that f can be expanded into a series

$$f(x) = \sum_{n=o}^{\infty} c_n P_n(x),$$

the series being uniformly convergent on $[-1; 1]$. Then, as usual in the theory of Fourier series, by means of Theorem 1.10.60, (ii) we conclude

$$c_n = (2n + 1)f^\wedge(n).$$

So we have quite formally an inversion formula for the LT:

$$f(x) = \sum_{n=o}^{\infty} (2n + 1)f^\wedge(n)P_n(x) =: \mathfrak{P}^{-1}[f^\wedge](x). \tag{1.10.26}$$

This formula is not valid for originals $f \in L_1$ or $f \in \mathcal{C}[-1; 1]$ but for $f \in L_p(-1; 1)$ with $p \in (4/3; 4)$; see [StW], section 2. Furthermore, we refer to conditions for the validity of (1.10.26) given in [NU], Paragraph 8, Theorem 1 for series expansions with respect to orthogonal polynomials. The proof is too lengthy to be given in this text. In particular, we have:

Theorem 1.10.62 *Let $f \in \mathcal{C}^1[-1; 1]$. Then the inversion formula (1.10.26) holds, the series being uniformly convergent on $[-1; 1]$.*

Now we are going to formulate some rules of operational calculus.

Proposition 1.10.45 *Let $f \in L_1$ and $n \in \mathbb{N}_o$. Then it holds that*

$$\mathfrak{P}[xf(x)](n) = \frac{1}{2n + 1}\Big[(n + 1)f^\wedge(n + 1) + nf^\wedge(n - 1)\Big], \tag{1.10.27}$$

Proof. Formula (1.10.27) follows immediately from (1.10.22) by means of the three-term recurrence for Legendre polynomials

$$(n + 1)P_{n+1}(x) = (2n + 1)xP_n(x) - nP_{n-1}(x); \tag{1.10.28}$$

see [E.1], vol. II, 10.10, (9).   □

The Legendre polynomials are eigenfunctions of the differential operator P defined by means of

$$(Py)(x) = -D(1-x^2)D, \qquad D = \frac{d}{dx} \qquad (1.10.29)$$

with respect to the eigenvalues $n(n+1)$, i.e.,

$$PP_n = n(n+1)F_n, \qquad n \in \mathbb{N}_o. \qquad (1.10.30)$$

Therefore, we have:

Proposition 1.10.46 (Differentiation Rule) *Let $f \in L_1$ be two times differentiable (a.e.) on $(-1;1)$ and*

$$\lim_{x \to \pm 1} (1-x^2)\, f(x) = \lim_{x \to \pm 1} (1-x^2)\, f'(x) = 0.$$

Then it holds that

$$\mathfrak{P}[Pf](n) = n(n+1)f^{\wedge}(n). \qquad (1.10.31)$$

Proof. The proof is straightforward by means of (1.10.29) and integration by parts, similar to the proof of 1.10.2, Proposition 1.10.42. ⬚

Corollary 1.10.19

$$\mathfrak{P}[P^k f](n) = [n(n+1)]^k f^{\wedge}(n), \qquad k \in \mathbb{N}_o. \qquad (1.10.32)$$

Now let

$$Pf = g.$$

Then

$$(1-x^2)f'(x) = -\int_{-1}^{x} g(v)dv$$

and

$$f(x) = -\int_{-1}^{x} (1-u^2)^{-1} \int_{-1}^{u} g(v)dvdu =: (P^{-1}g)(x). \qquad (1.10.33)$$

Because of $\mathfrak{P}[Pf](n) = n(n+1)f^{\wedge}(n) = g^{\wedge}(n)$ we have to assume that $g^{\wedge}(0) = 0$.

Performing the Legendre transform on (1.10.33) we have

$$f^{\wedge}(n) = \mathfrak{P}[P^{-1}g](n) = \frac{1}{n(n+1)}g^{\wedge}(n),$$

i.e.,

Proposition 1.10.47 (Integration Rule) *Let* $g \in L_1$ *and* $g^\wedge(0) = 0$. *Then it holds that*

$$\mathfrak{P}[P^{-1}g](n) = [n(n+1)]^{-1}g^\wedge(n), \qquad n \in \mathbb{N}, \tag{1.10.34}$$

where P^{-1} *is defined by means of* (1.10.33).

Now we are going to derive a convolution theorem for the Legendre transform. First we note a linearization formula for the product of two Legendre polynomials, which is a special case of a formula for spherical harmonics, see, for example, [Vi], Chapter III, Paragraph 4, formula (3).

Lemma 1.10.13 *For the Legendre poynomials it holds that*

$$P_n(x)P_n(y) = \frac{1}{\pi}\int_0^\pi P_n\Big(xy + \sqrt{(1-x^2)(1-y^2)}\cos\varphi\Big)d\varphi. \tag{1.10.35}$$

Substituting $t = \cos\varphi$ we have:

Corollary 1.10.20

$$P_n(x)P_n(y) = \frac{1}{\pi}\int_{-1}^1 P_n\Big(xy + t\sqrt{(1-x^2)(1-y^2)}\Big)(1-t^2)^{-1/2}dt. \tag{1.10.36}$$

Substituting $z = xy + \sqrt{(1-x^2)(1-y^2)}t$ in formula (1.10.36) we obtain by straightforward calculation:

Corollary 1.10.21 *Let* $z_j = xy + (-1)^j\sqrt{(1-x^2)(1-y^2)}$, $j = 1,2$, *then the product formula* (1.10.36) *can be written in the so-called kernel form*

$$P_n(x)P_n(y) = \frac{1}{2}\int_{-1}^1 K(x,y,z)P_n(z)dz = \mathfrak{P}[K(x,y,\cdot)](n), \tag{1.10.37}$$

where

$$K(x,y,z) = \begin{cases} \frac{2}{\pi}[1 - x^2 - y^2 - z^2 + 2xyz]^{-1/2}, & z_1 < z < z_2, \\ 0, & \text{otherwise.} \end{cases} \tag{1.10.38}$$

Lemma 1.10.14 $K(x,y,z)$ *is positive, symmetrical in* x,y,z *and it holds that*

$$\frac{1}{2}\int_{-1}^1 K(x,y,z)dz = 1. \tag{1.10.39}$$

By means of the product formula we define a generalized translation operator (GTO) for the Legendre transform.

Definition 1.10.27 *Let $f \in L_1$ and $x \in [-1; 1]$. Then a GTO τ_x is defined by means of*

$$(\tau_x f)(y) = \frac{1}{2} \int_{-1}^{1} K(x, y, z) f(z) dz. \tag{1.10.40}$$

We have:

Proposition 1.10.48 *The GTO τ_x is a positive bounded linear operator of L_1 into itself satisfying*

(i) $\qquad \|\tau_x f\|_1 \leq \|f\|_1,$

(ii) $\qquad (\tau_x f)(y) = (\tau_y f)(x),$

(iii) $\qquad (\tau_x P_n)(y) = P_n(x) \cdot P_n(y),$

(iv) $\qquad \mathfrak{P}[\tau_x f](n) = P_n(x) f^\wedge(n),$

(v) $\qquad \lim_{x \to 1^-} \|\tau_x f - f\|_1 = 0.$

Proof. Since K is positive (see Lemma 1.10.14) τ_x is a positive (linear) operator. Now let $f \in L_1$. Then we have

$$\|\tau_x f\|_1 = \frac{1}{2} \int_{-1}^{1} |(\tau_x f)(y)| \, dy = \frac{1}{2} \int_{-1}^{1} \left| \frac{1}{2} \int_{-1}^{1} f(z) K(x, y, z) dz \right| dy.$$

By means of the Hölder inequality, Fubini's theorem and formula (1.10.39) we conclude

$$\|\tau_x f\|_1 \leq \frac{1}{2} \int_{-1}^{1} \frac{1}{2} \int_{-1}^{1} |f(z)| K(x, y, z) dz dy \leq \frac{1}{2} \int_{-1}^{1} |f(z)| \frac{1}{2} \int_{-1}^{1} K(x, y, z) dy dz$$

$$= \frac{1}{2} \int_{-1}^{1} |f(z)| dz = \|f\|_1.$$

The symmetry relation follows from the symmetry of the kernel K; see Lemma 1.10.14. Formula (iii) is nothing other than the product formula (1.10.37).

For the proof of formula (iv) we use $|P_n(x)| \leq 1$, $x \in [-1; 1]$, formula (1.10.39) and Fubini's theorem:

$$(\tau_x f)^\wedge(n) = \frac{1}{4} \int_{-1}^{1} \int_{-1}^{1} f(z) K(x, y, z) dz P_n(y) dy = \frac{1}{4} \int_{-1}^{1} f(z) \int_{-1}^{1} P_n(y) K(x, y, z) dy dz$$

$$= \frac{P_n(x)}{2} \int_{-1}^{1} f(z) P(z) dz = P_n(x) \, f^\wedge(n),$$

where $K(x, y, z) = K(x, z, y)$ and the product formula (1.10.37) were used.

Preparing the proof of assertion (v) we consider the case $f = P_n$. From (iii) we have

$$|\tau_x P_n - P_n| = |P_n(x) - 1| |P_n|.$$

Since $|P_n(x)| \leq 1$ if $x \in [-1; 1]$ and $P_n(1) = 1$, $n \in \mathbb{N}_o$, see Remark 1.10.69, we obtain

$$\lim_{x \to 1^-} \|\tau_x P_n - P_n\|_1 = 0.$$

Since every polynomial is a linear combination of Legendre polynomials the assertion is valid for polynomials. Since the set of polynomials is dense in L_1 there exists for every $\epsilon > 0$ and $f \in L_1$ a polynomial p such that

$$\|f - p\|_1 < \epsilon.$$

Therefore, from Proposition 1.10.48, (i) we obtain

$$\|\tau_x f - f\|_1 \leq \|\tau_x p - p\|_1 + \|\tau_x f - \tau_x p\|_1 + \|f - p\|_1 < \epsilon + \|\tau_x (f - p)\|_1 + \epsilon < 3\epsilon.$$

$$\Box$$

Now we define the convolution of the Legendre transform.

Definition 1.10.28 *As the convolution of the Legendre transform we denote*

$$(f * g)(x) = \frac{1}{2} \int\limits_{-1}^{1} f(y)(\tau_x g)(y) dy, \qquad (1.10.41)$$

provided that it exists.

Theorem 1.10.63 (Convolution Theorem) *Let $f, g \in L_1$. Then there exists $f * g$ (a.e.), $f * g \in L_1$ and it holds that*

(i) $\|f * g\|_1 \leq \|f\|_1 \cdot \|g\|_1,$

(ii) $\mathfrak{P}[f * g] = f^\wedge g^\wedge,$

(iii) $f * g = g * f.$

Proof. By means of Proposition 1.10.48, (ii) and (i) we have

$$\|f * g\|_1 = \frac{1}{2} \int\limits_{-1}^{1} \frac{1}{2} \left| \int\limits_{-1}^{1} f(y)(\tau_x g)(y) dy \right| dx \leq \frac{1}{2} \int\limits_{-1}^{1} \frac{1}{2} \int\limits_{-1}^{1} |f(y)| |(\tau_y g)(x)| dx dy$$

$$= \frac{1}{2} \int\limits_{-1}^{1} |f(y)| \|\tau_y g\|_1 dy = \|f\|_1 \cdot \|g\|_1$$

and, therefore, $f * g \in L_1$ and (i) is valid.

The result (ii) follows by Fubini's theorem, using Proposition 1.10.48 and $|P_n| \leq 1$:

$$(f * g)^{\wedge}(n) = \frac{1}{2} \int_{-1}^{1} (f * g)(x) P_n(x) dx = \frac{1}{4} \int_{-1}^{1} \int_{-1}^{1} f(y)(\tau_y\, g)(x) P_n(x) dx dy$$

$$= \frac{1}{4} \int_{-1}^{1} f(y) \int_{-1}^{1} \tau_y g(x) P_n(x) dx dy = \frac{1}{2} \int_{-1}^{1} f(y)(\tau_y\, g)^{\wedge}(n) dy$$

$$= \frac{1}{2} g^{\wedge}(n) \int_{-1}^{1} f(y) P_n(y) dy = f^{\wedge}(n) \cdot g^{\wedge}(n)$$

and this is (ii). The commutativity follows directly from (ii) and Theorem 1.10.61, applying the Legendre transform to one side of equation (iii). □

Remark 1.10.71 *The results can be generalized to originals of the spaces L_p, $1 \leq p < \infty$; see Remark 1.10.70 and [StW], Lemma 3. For the convolution we have to consider functions $f \in L_p$ and $g \in L_1$. The convolution belongs to L_p and in place of (i) we have*

$$\|f * g\|_p \leq \|f\|_p \cdot \|g\|_1.$$

Remark 1.10.72 *All the convolutions of this section are valid also for functions of the space $\mathcal{C}[-1; 1]$.*

Finally, we prove a Riemann–Lebesgue type result for Legendre transforms.

Theorem 1.10.64 *Let $f \in L_1$. Then*

$$\lim_{n \to \infty} f^{\wedge}(n) = 0. \tag{1.10.42}$$

Proof. Let x_n be the largest root of P_n. From Proposition 1.10.48, (iv) we obtain

$$(\tau_{x_n}\, f)^{\wedge}(n) = P_n(x_n)\, f^{\wedge}(n) = 0$$

and, therefore, from Theorem 1.10.60, (i) it follows that

$$|f^{\wedge}(n)| = |(f - \tau_{x_n} f)^{\wedge}(n)| \leq \|f - \tau_{x_n} f\|_1. \tag{1.10.43}$$

From Bruns' inequality, see [Sz], formula (6.21.5), we know that

$$\lim_{n \to \infty} x_n = 1$$

and using Proposition 1.10.48, (v), from (1.10.43) with $n \to \infty$ it follows (1.10.42). □

Now we consider an application. We are looking for the solution of the Dirichlet problem for the Laplace equation in the unit ball of the three-dimensional Euclidian space. Let

$\mathcal{K}_1 = \{x, y, z : x^2 + y^2 + z^2 < 1\}$, $\bar{\mathcal{K}}_1 = \{x, y, z : x^2 + y^2 + z^2 \leq 1\}$ and $\partial\mathcal{K}_1 = \bar{\mathcal{K}}_1 \setminus \mathcal{K}_1$ the unit sphere. We look for a function $u \in \mathcal{C}^2(\mathcal{K}_1)$ such that

$$\Delta_3\, u(x, y, z) = u_{xx} + u_{yy} + u_{zz} = 0, \qquad x, y, z \in \mathcal{K}_1$$

with the boundary condition

$$\lim_{r \to 1^-} u(x, y, z) = u_o, \qquad r = (x^2 + y^2 + z^2)^{1/2}.$$

Introducing spherical coordinates

$$x = r \cos\vartheta \cos\varphi,$$
$$y = r \cos\vartheta \sin\varphi,$$
$$z = r \sin\vartheta, \quad \text{where} \quad 0 \leq r \leq 1,\ 0 \leq \vartheta \leq \pi,\ 0 \leq \varphi < 2\pi,$$

and assuming that the solution is independent of φ, with the notation

$$u(x, y, z) = U(r, \vartheta), \quad u_o = U_o(\vartheta)$$

after straightforward calculation we have

$$\frac{\partial^2 U}{\partial r^2} + 2r^{-1}\frac{\partial U}{\partial r} + \frac{1}{r^2 \sin^2\vartheta}\frac{\partial}{\partial\vartheta}(\sin\vartheta U_\vartheta) = 0.$$

Putting $t = \cos\vartheta$, $\quad -1 \leq t \leq 1$ and

$$U(r, \vartheta) = V(r, t), \quad U_o(\vartheta) = V_o(t)$$

after a short calculation we obtain

$$\frac{\partial}{\partial r}\left(r^2\frac{\partial V}{\partial r}\right) + \frac{\partial}{\partial t}\left[(1 - t^2)\frac{\partial V}{\partial t}\right] = 0$$

and the boundary condition is

$$\lim_{r \to 1^-} V(r, t) = V_o(t).$$

By means of the Legendre transform with respect to t and using formulas (1.10.29) through (1.10.31) we have quite formally

$$r^2\frac{\partial^2 V^\wedge(r, n)}{\partial r^2} + 2r\frac{\partial V^\wedge(r, n)}{\partial r} - n(n + 1)V^\wedge(r, n) = 0$$

and

$$\lim_{r \to 1^-} V^\wedge(r, n) = V_o^\wedge(n).$$

The differential equation is a Euler type and can be solved by means of $V^\wedge = r^\alpha$, and so we obtain

$$\alpha_1 = n \quad \text{or} \quad \alpha_2 = -(n + 1).$$

Because of the continuity of the solution at $r = 0$ we obtain

$$V^\wedge(r, n) = c(n)\, r^n$$

and the boundary condition yields $c(n) = V_o^\wedge(n)$.

Applying $\mathfrak{P}^{-1}$ according to (1.10.26) we have

$$V(r, t) = \sum_{n=0}^{\infty} (2n + 1)\, V_o^\wedge(n)\, P_n(t) r^n$$

or

$$U(r, \vartheta) = \sum_{n=0}^{\infty} (2n + 1) V_o^\wedge P_n(\cos \vartheta) r^n.$$

One can prove that this formal solution of the Dirichlet problem is the solution of our problem, if u_o is sufficiently smooth.

1.10.4 The Gegenbauer Transform

Definition 1.10.29 *The Gegenbauer transform (GT) of a function $f : [-1; 1] \to \mathbb{C}$ is defined by means of*

$$\mathfrak{P}^\lambda[f](n) = f^\wedge(n) = \int_{-1}^{1} f(x)\, P_n^\lambda(x) d\mu^\lambda(x), \qquad \lambda \in \mathbb{R}^+,\ n \in \mathbb{N}_o, \qquad (1.10.44)$$

provided that the integral exists. Here P_n^λ are the Gegenbauer polynomials, defined by

$$P_n^\lambda(x) = \frac{(-1)^n}{2^n (\lambda + \frac{1}{2})_n} (1 - x^2)^{\frac{1}{2} - \lambda} \frac{d^n}{dx^n} (1 - x^2)^{n + \lambda - 1/2}, \qquad n \in \mathbb{N}_o \qquad (1.10.45)$$

and

$$d\mu^\lambda(x) = \frac{\Gamma(\lambda + 1)}{\sqrt{\pi}\Gamma(\lambda + \frac{1}{2})} (1 - x^2)^{\lambda - 1/2} dx. \qquad (1.10.46)$$

Remark 1.10.73 *Sometimes the Gegenbauer polynomials are defined in another standardization and notation:*

$$C_n^\lambda = \frac{(2\lambda)_n}{n!} P_n^\lambda, \qquad (1.10.47)$$

see, for example, [E.1], vol. 2, 10.9.

Moreover, sometimes these C_n^λ are denoted by P_n^λ and are called ultraspherical polynomials; see, for example, [Sz], 4.7. Therefore, one has to look carefully at the definitions.

Remark 1.10.74 *For the properties of the Gegenbauer polynomials we refer to [E.1], vol. 2, 10.1. In particular, we note that, in our standardization (1.10.45) we have*

$$|P_n^\lambda(x)| \leq 1, \qquad -1 \leq x \leq 1 \quad and \quad P_n^\lambda(1) = 1, \quad n \in \mathbb{N}_o.$$

Remark 1.10.75 *In the case of $\lambda = \frac{1}{2}$ we obtain the Legendre polynomials and (1.10.44) is the Legendre transform with $d\mu^{1/2}(x) = \frac{1}{2}dx$, see formulas (1.10.22), (1.10.23).*

Since the proofs of many properties of the GT follow the same line as the proofs for the Legendre transform we will omit these proofs and make remarks only if there are differences.

As space of originals we choose the space $L_1^\lambda(-1;1) = L_1^\lambda$ of measurable functions on $(-1;1)$ such that

$$\|f\|_{1,\lambda} = \int_{-1}^{1} |f(x)|d\mu^\lambda(x) \tag{1.10.48}$$

is finite. It is a Banach space with the norm (1.10.48). We remark that L_1^0 is the space of section 1.10.2 and $L_1^{1/2} = L_1$; see section 1.10.3.

Lemma 1.10.15 *We have*

$$\|1\|_{1,\lambda} = \int_{-1}^{1} d_\mu^\lambda(x) = 1. \tag{1.10.49}$$

Proof. Using (1.10.46) and substituting $x = 1 - 2t$ we obtain

$$\|1\|_{1,\lambda} = \frac{\Gamma(\lambda+1)2^{2\lambda}}{\sqrt{\pi}\Gamma(\lambda+\frac{1}{2})} \int_{0}^{1} [t(1-t)]^{\lambda-1/2}dt = \frac{\Gamma(\lambda+1)\,2^{2\lambda}}{\sqrt{\pi}\Gamma(\lambda+\frac{1}{2})}B\left(\lambda+\frac{1}{2},\lambda+\frac{1}{2}\right),$$

where the Beta function is defined by equation (1.4.30). Using formula (1.4.31) and the duplication formula of the Gamma function, see [E.1], vol. I, 1.2, (15),

$$\Gamma(2z) = 2^{2z-1}\pi^{-1/2}\,\Gamma(z)\,\Gamma\left(z+\frac{1}{2}\right)$$

we obtain (1.10.49).   ▯

Remark 1.10.76 *The considerations in the following also can be extended to originals of the space L_p^λ, $1 \le p < \infty$ of measurable functions on $(-1;1)$ with the norm*

$$\|f\|_{p,\lambda} = \left[\int_{-1}^{1} |f(x)|^p d\mu^\lambda(x)\right]^{1/p} \tag{1.10.50}$$

or of the space $\mathcal{C}[-1;1]$ of continuous functions on $[-1;1]$ with the sup-norm; see [VP].

Theorem 1.10.65 *Let $f \in L_1^\lambda$ and $k,n \in \mathbb{N}_o$. Then the Gegenbauer transform $\mathfrak{P}^\lambda$ is a linear transform and moreover it holds that*

(i) $\quad |\mathfrak{P}^\lambda[f](n)| \le \|f\|_{1,\lambda}$,

(ii) $\quad \mathfrak{P}^\lambda[P_k^\lambda](n) = \begin{cases} \frac{n!\lambda}{(2\lambda)_n(n+\lambda)} =: h_n^\lambda, & k = n \\ 0, & k \ne n, \end{cases}$

(iii) $\quad \mathfrak{P}^\lambda[f](n) = 0, \qquad n \in \mathbb{N}_o$ *if and only if* $f(x) = 0$ *(a.e.)*.

Proof. For the proofs of (i) and (iii) look at the proofs of Theorem 1.10.60, (i), (iii) in section 1.10.3 for the Legendre transform. The formula (ii) is the orthogonality relation of the Gegenbauer polynomials; see, for example, [E.1], vol. II, 10.9, (7). $\qquad \square$

The result (iii) can be formulated as a uniqueness theorem for the GT.

Theorem 1.10.66 (Uniqueness Theorem) *Let* $f, g \in L_1^\lambda$ *and* $\mathfrak{P}^\lambda[f](n) = \mathfrak{P}^\lambda[g](n)$ *for every* $n \in \mathbb{N}_o$. *Then* $f = g$ *(a.e.)*.

Analogous to the derivation of an inversion formula for the Legendre transform we obtain quite formally an inversion formula for the GT:

$$f(x) = \sum_{n=o}^\infty \frac{1}{h_n^\lambda} f^\wedge(n) P_n^\lambda(x) =: \left(\mathfrak{P}^\lambda\right)^{-1}[f^\wedge](x). \qquad (1.10.51)$$

This formula is not valid for originals $f \in L_1^\lambda$ or $f \in \mathcal{C}[-1; 1]$ but for $f \in L_p^\lambda(-1; 1)$ with $p \in (2 - (1 + \lambda)^{-1}; 2 + \lambda^{-1})$; see [Po]. Similar to Theorem 1.10.62 we have

Theorem 1.10.67 *Let* $f \in \mathcal{C}^1[-1; 1]$. *Then the inversion formula* (1.10.51) *holds, the series being uniformly convergent on* $[-1; 1]$. *Here* $f^\wedge$ *is taken from* (1.10.44) *and* h_n^λ *is defined in* Theorem 1.10.62, *(ii)*.

Now we derive some rules of operational calculus.

Proposition 1.10.49 *Let* $f \in L_1^\lambda$ *and* $n \in \mathbb{N}_o$. *Then it holds that*

$$\mathfrak{P}^\lambda[xf(x)](n) = \frac{1}{2(n+\lambda)}\Big[(n+2\lambda)f^\wedge(n+1) + nf^\wedge(n-1)\Big]. \qquad (1.10.52)$$

Proof. Formula (1.10.52) follows directly from (1.10.44) by means of the tree-term-recurrence for Gegenbauer polynomials

$$(n+2\lambda)P_{n+1}^\lambda(x) = 2(n+\lambda)xP_n^\lambda(x) - nP_{n-1}^\lambda(x); \qquad (1.10.53)$$

see [E.1], vol. II, 10.9, (13). $\qquad \square$

The Gegenbauer polynomials are eigenfunctions of the differential operator P^λ defined by means of

$$(P^\lambda y)(x) = -(1 - x^2)^{\frac{1}{2} - \lambda} D(1 - x^2)^{\lambda + \frac{1}{2}} Dy(x) \qquad (1.10.54)$$

with respect to the eigenvalues $n(n+2\lambda)$, see [E.1], vol. II, 10.9, (14), i.e.,

$$P^\lambda P_n^\lambda = n(n+2\lambda)P_n^\lambda, \qquad n \in \mathbb{N}_o. \tag{1.10.55}$$

Therefore, we have:

Proposition 1.10.50 (Differentiation Rule) *Let* $f \in L_1^\lambda$ *be two times differentiable (a.e.) on* $(-1;1)$ *and*

$$\lim_{x\to\pm1}(1-x^2)^{\lambda+\frac{1}{2}}f(x) = \lim_{x\to\pm1}(1-x^2)^{\lambda+\frac{1}{2}}f'(x) = 0.$$

Then it holds that

$$\mathfrak{P}^\lambda[P^\lambda f](n) = n(n+2\lambda)f^\wedge(n). \tag{1.10.56}$$

Corollary 1.10.22

$$\mathfrak{P}^\lambda[(P^\lambda)^k f](n) = [n(n+2\lambda)]^k f^\wedge(n), \qquad k \in \mathbb{N}_o. \tag{1.10.57}$$

Now let

$$P^\lambda f = g.$$

Set

$$w_\lambda(x) = (1-x^2)^{\frac{1}{2}-\lambda}. \tag{1.10.58}$$

Then

$$(1-x^2)^{\lambda+\frac{1}{2}}Df = -\int_{-1}^{x}\frac{g(v)}{w_\lambda(v)}dv$$

and

$$f(x) = -\int_{-1}^{x}(1-u^2)^{-\lambda-\frac{1}{2}}\int_{-1}^{u}\frac{g(v)}{w_\lambda(v)}dvdu =: (P^\lambda)^{-1}g(x). \tag{1.10.59}$$

Because of $\mathfrak{P}^\lambda[P^\lambda f](n) = n(n+2\lambda)f^\wedge(n) = g^\wedge(n)$ then $g^\wedge(0) = 0$. By means of the *GT* from (1.10.59) we have

$$f^\wedge(n) = \mathfrak{P}^\lambda[(P^\lambda)^{-1}g](n) = \frac{1}{n(n+2\lambda)}g^\wedge(n),$$

i.e.,

Proposition 1.10.51 (Integration Rule) *Let* $g \in L_1^\lambda$ *and* $g^\wedge(0) = 0$. *Then it holds that*

$$\mathfrak{P}^\lambda[(P^\lambda)^{-1}g](n) = [n(n+2\lambda)]^{-1}g^\wedge(n), \quad n \in \mathbb{N}, \tag{1.10.60}$$

where $(P^\lambda)^{-1}$ *is defined by* (1.10.59).

Preparing the definition of the convolution for the GT we note a linearization formula for Gegenbauer polynomials, see [Vi], Chapter IX, Paragraph 4, (2).

Lemma 1.10.16 *For the Gegenbauer poynomials it holds that*

$$P_n^\lambda(x)P_n^\lambda(y) = \frac{\Gamma(\lambda+\frac{1}{2})}{\sqrt{\pi}\Gamma(\lambda)} \int_0^\pi P_n^\lambda\Big(xy + \sqrt{(1-x^2)(1-y^2)}\,\cos\varphi\Big)(\sin\varphi)^{2\lambda-1}\,d\varphi. \quad (1.10.61)$$

Substituting $t = \cos\varphi$ we have

Corollary 1.10.23

$$P_n^\lambda(x)P_n^\lambda(y) = \frac{\Gamma(\lambda+\frac{1}{2})}{\sqrt{\pi}\Gamma(\lambda)} \int_{-1}^1 P_n^\lambda\Big(xy + t\sqrt{(1-x^2)(1-y^2)}\Big)(1-t^2)^{\lambda-\frac{1}{2}}\,dt. \quad (1.10.62)$$

Substituting $z = xy + t\sqrt{(1-x^2)(1-y^2)}$ in formula (1.10.62) we obtain by straightforward calculation the kernel form of (1.10.61):

Corollary 1.10.24 *Let $z_j = xy + (-1)^j\sqrt{(1-x^2)(1-y^2)}$, $j = 1, 2$, then*

$$P_n^\lambda(x)P_n^\lambda(y) = \int_{-1}^1 K^\lambda(x,y,z)P_n^\lambda(z)d\mu^\lambda(z) = \mathfrak{P}[K^\lambda(x,y,\cdot)](n), \quad (1.10.63)$$

where

$$K^\lambda(x,y,z) = \begin{cases} \frac{\Gamma^2(\lambda+\frac{1}{2})}{\Gamma(\lambda)\Gamma(\lambda+1)} \frac{[(1-x^2)(1-y^2)(1-z^2)]^{1/2-\lambda}}{[1-x^2-y^2-z^2+2xyz]^{1-\lambda}}, & z_1 < z < z_2. \\ 0, & otherwise. \end{cases} \quad (1.10.64)$$

From (1.10.64) and (1.10.63) with $n = 1$ because of $P_0^\lambda = 1$ we obtain immediately:

Lemma 1.10.17 *$K^\lambda(x,y,z)$ is positive, symmetrical in x, y, z and it holds that*

$$\int_{-1}^1 K^\lambda(x,y,z)d\mu^\lambda(z) = 1. \quad (1.10.65)$$

In the same manner as in sections 1.10.2 and 1.10.3 we are able to define a generalized translation operator (GTO) for the Gegenbauer transform.

Definition 1.10.30 *Let $f \in L_1^\lambda$ and $x \in [-1;1]$. Then a GTO τ_x^λ for the GT is defined by means of*

$$(\tau_x^\lambda f)(y) = \int_{-1}^1 K^\lambda(x,y,z)f(z)d\mu^\lambda(z). \quad (1.10.66)$$

We have

Proposition 1.10.52 *The GTO τ_x^λ is a positive bounded linear operator of L_1^λ into itself satisfying*

(i) $\|\tau_x^\lambda f\|_{1,\lambda} \leq \|f\|_{1,\lambda}$,

(ii) $(\tau_x^\lambda f)(y) = (\tau_y^\lambda f)(x)$,

(iii) $(\tau_x^\lambda P_n^\lambda)(y) = P_n^\lambda(x) \cdot P_n^\lambda(y)$,

(iv) $\mathfrak{P}^\lambda[\tau_x^\lambda f](n) = P_n^\lambda(x) f^\wedge(n)$,

(v) $\lim\limits_{x \to 1^-} \|\tau_x^\lambda f - f\|_{1,\lambda} = 0$.

The proof follows the same line as the proof of Proposition 1.10.48 in section 1.10.3.

Definition 1.10.31 *As the convolution of the Gegenbauer transform (GT) we denote*

$$(f * g)(x) = \int\limits_{-1}^{1} f(y)(\tau_x^\lambda g)(y) d\mu^\lambda(y), \qquad (1.10.67)$$

provided that it exists.

Analogous to Theorem 1.10.63, section 1.10.3 one can prove:

Theorem 1.10.68 (Convolution Theorem) *Let $f, g \in L_1^\lambda$. Then there exists $f * g$ (a.e.), $f * g \in L_1^\lambda$, and it holds that*

(i) $\|f * g\|_{1,\lambda} \leq \|f\|_{1,\lambda} \cdot \|g\|_{1,\lambda}$,

(ii) $\mathfrak{P}^\lambda[f * g] = f^\wedge g^\wedge$,

(iii) $f * g = g * f$.

Remark 1.10.77 *The results can be generalized to originals of the spaces L_p^λ, $1 \leq p < \infty$; see Remark 1.10.76 and [VP]. For the convolution we have to consider functions $f \in L_p^\lambda$ and $g \in L_1^\lambda$. The convolution belongs to L_p^λ and in place of (i) we have*

$$\|f * g\|_{p,\lambda} \leq \|f\|_{p,\lambda} \cdot \|g\|_{1,\lambda}.$$

Remark 1.10.78 *All the considerations of this section also are valid for functions of the space $\mathcal{C}[-1; 1]$.*

Finally we have a Riemann–Lebesgue type result for Gegenbauer transforms.

Theorem 1.10.69 *Let $f \in L_1^\lambda$. Then*

$$\lim\limits_{n \to \infty} f^\wedge(n) = 0. \qquad (1.10.68)$$

The proof is analogous to the proof of Theorem 1.10.64 in section 1.10.3 in the case of the Legendre transform.

Proof. In the case of $0 \leq \lambda \leq \frac{1}{2}$ the proof follows the line of the proof of Theorem 1.10.64, section 1.10.3. Only the Bruns' inequality for the roots of the Gegenbauer polynomials is here; [Sz], formula (6.21.7). In the general case ($\lambda \in \mathbb{R}^+$) we refer to [VP].

For an application we refer to [De.6], section 13.6. ⬚

1.10.5   The Jacobi Transform

Definition 1.10.32 *The Jacobi transform* (JT) *of a function* $f : [-1; 1] \to \mathbb{C}$ *is defined by means of*

$$\mathfrak{P}^{(\alpha,\beta)}[f](n) = f^{\wedge}(n) = \int\limits_{-1}^{1} f(x) R_n^{(\alpha,\beta)}(x) d\mu^{(\alpha,\beta)}(x), \qquad (1.10.69)$$

where $\alpha \geq \beta \geq -1/2$, $n \in \mathbb{N}_o$, *provided that the integral exists. Here* $R_n^{(\alpha,\beta)}$ *are the Jacobi polynomials, standardized in such a manner that*

$$R_n^{(\alpha,\beta)}(x) = \frac{(-1)^n}{2^n (\alpha+1)_n} [(1-x)^{\alpha}(1+x)^{\beta}]^{-1} \frac{d^n}{dx^n} [(1-x)^{\alpha}(1+x)^{\beta}(1-x^2)^n], \quad (1.10.70)$$

and

$$d\mu^{(\alpha,\beta)}(x) = \frac{\Gamma(a+1)}{2^a \Gamma(\alpha+1)\Gamma(\beta+1)} (1-x)^{\alpha}(1+x)^{\beta} dx, \qquad (1.10.71)$$

where

$$a = \alpha + \beta + 1. \qquad (1.10.72)$$

Remark 1.10.79 *Sometimes the Jacobi polynomials are defined in another standardization and notation:*

$$P_n^{(\alpha,\beta)}(x) = \frac{(\alpha+1)_n}{n!} R_n^{(\alpha,\beta)}(x); \qquad (1.10.73)$$

see, for example, [E.1], vol. 2, 10.8. Here one also can find all important properties of these polynomials.

Remark 1.10.80 *In our standardization we have*

$$|R_n^{(\alpha,\beta)}(x)| \leq 1, \quad -1 \leq x \leq 1 \quad and \quad R_n^{(\alpha,\beta)}(1) = 1, n \in \mathbb{N}_o;$$

see [Sz], (7.32.2).

Remark 1.10.81 *If* $\alpha = \beta = \lambda - \frac{1}{2}$ *one gets the Gegenbauer polynomials* P_n^{λ}, *see 1.10.4, (1.10.45) in particular for* $\alpha = \beta = 0$, *one gets the Legendre polynomials, see formula (1.10.2) and for* $\alpha = \beta = -\frac{1}{2}$ *the Chebyshev polynomials, see formula (1.10.2).*

As the space of originals we choose the space $L_1^{(\alpha,\beta)}(-1;1) = L_1^{(\alpha,\beta)}$ of measurable functions on $(-1;1)$ such that

$$\|f\|_{1,(\alpha,\beta)} = \int_{-1}^{1} |f(x)| d\mu^{(\alpha,\beta)}(x) \tag{1.10.74}$$

is finite. It is a Banach space with the norm (1.10.74).

Lemma 1.10.18 *We have*

$$\|1\|_{1,(\alpha,\beta)} = \int_{-1}^{1} d_\mu^{(\alpha,\beta)}(x) = 1. \tag{1.10.75}$$

Proof. The proof follows the same line as the proof of Lemma 1 in section 1.10.4. Substituting $x = 1 - 2t$ and using the definition of Euler's Beta function and the duplication formula of the Gamma function the result (1.10.75) is derived. ▯

Remark 1.10.82 *The considerations in the following also can be extended to originals of the space $L_p^{(\alpha,\beta)}$, $1 \leq p < \infty$ of measurable functions on $(-1;1)$ with the norm*

$$\|f\|_{p,(\alpha,\beta)} = \left[\int_{-1}^{1} |f(x)|^p d\mu^{(\alpha,\beta)}(x) \right]^{1/p} \tag{1.10.76}$$

or of the space $\mathcal{C}[-1;1]$ of continuous functions on $[-1;1]$ with the sup-norm.

Theorem 1.10.70 *Let $f \in L_1^{(\alpha,\beta)}$ and $k, n \in \mathbb{N}_o$. Then the Jacobi transform $\mathfrak{P}^{(\alpha,\beta)}$ is a linear transform and moreover it holds that*

(i) $|\mathfrak{P}^{(\alpha,\beta)}[f](n)| \leq \|f\|_{1,(\alpha,\beta)},$

(ii) $\mathfrak{P}^{(\alpha,\beta)}[R_k^{(\alpha,\beta)}](n) = \begin{cases} \frac{n!\Gamma(a+1)\Gamma(\alpha+1)\Gamma(n+\beta+1)}{\Gamma(\beta+1)\Gamma(n+a)\Gamma(n+\alpha+1)(2n+a)} =: h_n^{(\alpha,\beta)}, & k = n \\ 0, & k \neq n, \end{cases}$

(iii) $\mathfrak{P}^{(\alpha,\beta)}[f](n) = 0$, $n \in \mathbb{N}_o$ *if and only if $f(x) = 0$ (a.e.).*

Proof. The property (i) follows directly from the definition (1.10.69) and $|R_n^{(\alpha,\beta)}| \leq 1$; see Remark 1.10.80. Property (ii) is the orthogonality relation of the Jacobi polynomials; see [E.1], vol. 2, 10.8,(4), taking note of the standardization (1.10.73). The proof of (iii) follows the same line as the proof of 1.10.3, Theorem 1.10.60, (iii). ▯

The result (iii) can again be formulated as a uniqueness theorem for the JT.

Theorem 1.10.71 (Uniqueness Theorem) *Let $f, g \in L_1^{(\alpha,\beta)}$ and $\mathfrak{P}^{(\alpha,\beta)}[f](n) = \mathfrak{P}^{(\alpha,\beta)}[g](n)$ for every $n \in \mathbb{N}_o$. Then $f = g$ (a.e.).*

For the derivation of an inversion formula we assume that f can be extended into a series

$$f(x) = \sum_{n=o}^{\infty} c_n R_n^{(\alpha,\beta)}(x),$$

the series being uniformly convergent on $[-1;1]$. Then as usual in Fourier series theory by means of Theorem 1.10.70, (ii) we have

$$c_n = \frac{1}{h_n^{(\alpha,\beta)}} f^{\wedge}(n).$$

So quite formally we have an inversion formula for the JT:

$$f(x) = \sum_{n=o}^{\infty} \frac{1}{h_n^{(\alpha,\beta)}} f^{\wedge}(n) R_n^{(\alpha,\beta)}(x) =: \left(\mathfrak{P}^{(\alpha,\beta)}\right)^{-1} [f^{\wedge}](x). \qquad (1.10.77)$$

Again, as in section 1.10.3, from [NU], § 8, Theorem 1 one has:

Theorem 1.10.72 *Let $f \in C^1[-1;1]$. Then the inversion formula (1.10.77) holds, the series being uniformly convergent on $[-1;1]$. Here $f^{\wedge}$ is taken from (1.10.69) and $h_n^{(\alpha,\beta)}$ is defined in* Theorem 1.10.70, *(ii).*

Now we are going to formulate some rules of operational calculus.

Proposition 1.10.53 *Let $f \in L_1^{(\alpha,\beta)}$ and $n \in \mathbb{N}_o$. Then it holds that*

$$\mathfrak{P}^{(\alpha,\beta)}[f(-x)](n) = (-1)^n \frac{P_n^{(\beta,\alpha)}(1)}{P_n^{(\alpha,\beta)}(1)} \mathfrak{P}^{(\beta,\alpha)}[f](n), \qquad (1.10.78)$$

$$\mathfrak{P}^{(\alpha,\beta)}[f](n+1) = \mathfrak{P}^{(\alpha,\beta)}[f](n) - \frac{2n+a+1}{a+1}\mathfrak{P}^{(\alpha+1,\beta)}[f](n), \qquad (1.10.79)$$

$$\mathfrak{P}^{(\alpha,\beta)}[f](n+1) = \frac{2n+a+1}{(a+1)(n+\alpha+1)}\mathfrak{P}^{(\alpha,\beta+1)}[f](n) - \frac{n+\beta+1}{n+\alpha+1}\mathfrak{P}^{(\alpha,\beta)}[f](n), \quad (1.10.80)$$

$$\mathfrak{P}^{(\alpha,\beta)}[f](n) = \frac{n+\alpha+1}{a+1}\mathfrak{P}^{(\alpha+1,\beta)}[f](n) + \frac{\beta+1}{a+1}\mathfrak{P}^{(\alpha,\beta+1)}[f](n), \qquad (1.10.81)$$

$$\begin{aligned}
\mathfrak{P}^{(\alpha,\beta)}[xf(x)](n) = {} & \frac{2(n+a)(n+\alpha+1)}{(2n+a)(2n+a+1)}\mathfrak{P}^{(\alpha,\beta)}[f](n+1) \\
& + \frac{\beta^2-\alpha^2}{(2n+a-1)(2n+a+1)}\mathfrak{P}^{(\alpha,\beta)}[f](n) \qquad (1.10.82) \\
& + \frac{2n(n+\beta)}{(2n+a)(2n+a-1)}\mathfrak{P}^{(\alpha,\beta)}[f](n),
\end{aligned}$$

where the value a is taken from (1.10.72).

Proof. The proof is straightforward using appropriate formulas for the Jacobi polynomials; see [E.1], vol. 2, 10.8, (13), (32), (23), (11), the proof of (1.10.78), (1.10.79), (1.10.80), and

(1.10.82), respectively. Formula (1.10.81) follows from (1.10.79) and (1.10.80) by subtraction. ▯

The Jacobi polynomials are eigenfunctions of the differential operator $P^{(\alpha,\beta)}$ defined by means of

$$\left(P^{(\alpha,\beta)}y\right)(x) = -(1-x)^{-\alpha}(1+x)^{-\beta}D\left[(1-x)^{\alpha+1}(1+x)^{\beta+1}D\right]y(x) \qquad (1.10.83)$$

with respect to the eigenvalues $n(n+a)$; see [E.1], vol. 2, 10.8, (14), i.e.,

$$P^{(\alpha,\beta)}P_n^{(\alpha,\beta)} = n(n+a)P_n^{(\alpha,\beta)}, \qquad n \in \mathbb{N}_o. \qquad (1.10.84)$$

Therefore, we have:

Proposition 1.10.54 (Differentiation Rule) *Let $f \in L_1^{(\alpha,\beta)}$ be two times differentiable (a.e.) on $(-1;1)$ and*

$$\lim_{x\to\pm1}(1-x)^{\alpha+1}(1+x)^{\beta+1}f(x) = \lim_{x\to\pm1}(1-x)^{\alpha+1}(1+x)^{\beta+1}f'(x) = 0.$$

Then it holds that

$$\mathfrak{P}^{(\alpha,\beta)}[P^{(\alpha,\beta)}f](n) = n(n+a)f^\wedge(n). \qquad (1.10.85)$$

Corollary 1.10.25

$$\mathfrak{P}^{(\alpha,\beta)}[(P^{(\alpha,\beta)})^k f](n) = [n(n+a)]^k f^\wedge(n), \qquad k \in \mathbb{N}_o. \qquad (1.10.86)$$

Now let

$$P^{(\alpha,\beta)}f = g,$$

and let

$$w(x) = (1-x)^\alpha(1+x)^\beta. \qquad (1.10.87)$$

Then we obtain

$$(1-x^2)w(x)Df(x) = -\int_{-1}^{x} w(v)\,g(v)dv$$

and

$$f(x) = -\int_{-1}^{x}[(1-u^2)w(u)]^{-1}\int_{-1}^{u}w(v)g(v)dvdu =: (P^{(\alpha,\beta)})^{-1}g(x). \qquad (1.10.88)$$

Because of

$$\mathfrak{P}^{(\alpha,\beta)}[P^{(\alpha,\beta)}f](n) = n(n+a)f^\wedge(n) = g^\wedge(n)$$

we require that $g^\wedge(0) = 0$ is valid. Applying the Jacobi transform to both sides of equation (1.10.88) we have:

Proposition 1.10.55 (Integration Rule) *Let* $g \in L_1^{(\alpha,\beta)}$ *and* $g^\wedge(0) = 0$. *Then it holds that*

$$\mathfrak{P}^{(\alpha,\beta)}[(P^{(\alpha,\beta)})^{-1}g](n) = [n(n+a)]^{-1}g^\wedge(n), \qquad n \in \mathbb{N}, \qquad (1.10.89)$$

where $(P^{(\alpha,\beta)})^{-1}$ *is defined by* (1.10.88).

Now we are going to define a convolution for the Jacobi transform. First, we note a product formula for Jacobi polynomials; see [Koo].

Lemma 1.10.19 *Let* $\alpha > \beta > -1/2$ *and* $x, y \in [-1;1]$. *Then for the Jacobi poynomials it holds that*

$$R_n^{(\alpha,\beta)}(x)R_n^{(\alpha,\beta)}(y) = \int_0^1 \int_0^\pi R_n^{(\alpha,\beta)}\left[\frac{1}{2}(1+x)(1+y) + \frac{1}{2}(1-x)(1-y)r^2\right.$$
$$\left. + (1-x^2)^{1/2}(1-y^2)^{1/2}r\cos\theta - 1\right]dm^{(\alpha,\beta)}(r,\theta), \qquad (1.10.90)$$

where

$$dm^{(\alpha,\beta)}(r,\theta) = \frac{2\Gamma(\alpha+1)}{\sqrt{\pi}\Gamma(\alpha-\beta)\Gamma(\beta+1/2)}(1-r^2)^{\alpha-\beta-1}r^{2\beta+1}(\sin\theta)^{2\beta}drd\theta. \qquad (1.10.91)$$

Substituting (r,θ) with (z,φ) by means of

$$^{1/2}r\cos\theta + [(1+x)(1+y)]^{1/2} = (2z)^{1/2}\cos\varphi$$
$$[(1-x)(1-y)]^{1/2}r\sin\theta = (2z)^{1/2}\sin\varphi$$

we obtain formula (1.10.90) in the so-called kernel form, first proved by Gasper; see [Ga.1], [Ga.2].

Lemma 1.10.20 *Let* $\alpha \geq \beta \geq -1/2$, $\alpha > -1/2$, $x \in (-1;1)$. *Then*

$$R_n^{(\alpha,\beta)}(x)R_n^{(\alpha,\beta)}(y) = \int_{-1}^1 K^{(\alpha,\beta)}(x,y,z)R_n^{(\alpha,\beta)}(z)\,d\mu^{(\alpha,\beta)}(z). \qquad (1.10.92)$$

The kernel $K^{(\alpha,\beta)}$ *is well defined. It is positive, symmetrical with respect to* x, y, z *and it holds that*

$$\int_{-1}^1 K^{(\alpha,\beta)}(x,y,z)d\mu^{(\alpha,\beta)}(z) = 1. \qquad (1.10.93)$$

Remark 1.10.83 *For details on the kernel* $K^{(\alpha,\beta)}$ *we refer to* [Ga.1], [Ga.2].

Remark 1.10.84 *Here and in the following we assume* $\alpha \geq \beta \geq 1/2$, $\alpha > -1/2$.

In the same manner as in sections 1.10.2, 1.10.3, and 1.10.4 we are able to define a generalized translation operator (GTO) for the Jacobi transform.

Definition 1.10.33 *Let* $f \in L_1^{(\alpha,\beta)}$. *Then a GTO* $\tau_x^{(\alpha,\beta)} =: \tau_x$ *is defined by means of*

$$(\tau_x f)(y) = \int\limits_{-1}^{1} K^{(\alpha,\beta)}(x,y,z)f(z)d\mu^{(\alpha,\beta)}(z). \tag{1.10.94}$$

Then we have

Proposition 1.10.56 *The GTO* τ_x *is a positive bounded linear operator of* $L_1^{(\alpha,\beta)}$ *into itself satisfying*

(i) $\|\tau_x f\|_{1,(\alpha,\beta)} \leq \|f\|_{1,(\alpha,\beta)}$,

(ii) $(\tau_x f)(y) = (\tau_y f)(x)$,

(iii) $\tau_x R_n^{(\alpha,\beta)}(y) = R_n^{(\alpha,\beta)}(x) \cdot R_n^{(\alpha,\beta)}(y)$,

(iv) $\mathfrak{F}^{(\alpha,\beta)}[\tau_x f](n) = R_n^{(\alpha,\beta)}(x)f^\wedge(n)$,

(v) $\lim\limits_{x \to 1^-} \|\tau_x f - f\|_{1,(\alpha,\beta)} = 0$.

Proof. The proof is analogous to the proof of Proposition 1.10.48 in section 1.10.3 in the case of the Legendre transform because the properties of the Legendre polynomials and of the kernel are the same in the case of the Jacobi polynomials. Therefore, we omit the proof.
☐

Definition 1.10.34 *As the convolution of the Jacobi transform we denote* $f * g$ *defined by*

$$(f * g)(x) = \int\limits_{-1}^{1} f(y)(\tau_x g)(y)d\mu^{(\alpha,\beta)}(y), \tag{1.10.95}$$

provided that it exists.

Analogous to Theorem 1.10.62, section 1.10.3 one can prove:

Theorem 1.10.73 (Convolution Theorem) *Let* $f, g \in L_1^{(\alpha,\beta)}$. *Then there exists* $f * g$ *(a.e.),* $f * g \in L_1^{(\alpha,\beta)}$ *and it holds that*

(i) $\|f * g\|_{1,(\alpha,\beta)} \leq \|f\|_{1,(\alpha,\beta)} \cdot \|g\|_{1,(\alpha,\beta)}$,

(ii) $\mathfrak{F}^{(\alpha,\beta)}[f * g] = f^\wedge g^\wedge$,

(iii) $f * g = g * f$.

Remark 1.10.85 *Because of the equivalence of* (1.10.90) *and* (1.10.92), (1.10.93) *for* $\alpha > \beta > -1/2$ *we have*

$$(\tau_x f)(y) = \int_0^1 \int_0^\pi f\Big[\frac{1}{2}(1+x)(1+y) + \frac{1}{2}(1-x)(1-y)r^2 \tag{1.10.96}$$
$$+ (1-x^2)^{1/2}(1-y^2)^{1/2}r\cos\theta - 1\Big]dm^{(\alpha,\beta)}(r,\theta)drd\theta,$$

where $dm^{(\alpha,\beta)}(r,\theta)$ *is taken from* (1.10.91). *Because the expression for the kernel* $K^{(\alpha,\beta)}$ *in formula* (1.10.92) *is very complicated and for the proof of* Proposition 1.10.56 *and* Theorem 1.10.73 *one needs only properties of the kernel proved in* [Ga.1], [Ga.2] *we have with* (1.10.96) *an explicit expression for* τ_x *and therefore also for the convolution* $f*g$ *in* (1.10.95).

Remark 1.10.86 *The results can be generalized; see* [Ga.2]. *Let* $f \in L_p^{(\alpha,\beta)}, g \in L_q^{(\alpha,\beta)}$ *and* $r^{-1} = p^{-1} + q^{-1} - 1$. *Then* $f * g \in L_r^{(\alpha,\beta)}$ *and*

$$\|f * g\|_{r,(\alpha,\beta)} \leq \|f\|_{p,(\alpha,\beta)} \cdot \|g\|_{q,(\alpha,\beta)}.$$

In particular, if $f \in L_p^{(\alpha,\beta)}$ *and* $g \in L_1^{(\alpha,\beta)}$, *then* $f * g \in L_p^{(\alpha,\beta)}$.

Remark 1.10.87 *All the considerations of this section are also valid in the space* $C[-1;1]$ *of continuous functions on* $[-1;1]$ *with the sup-norm.*

Finally we explain an application. Let us look for a solution of

$$(1-x^2)\frac{\partial^2 u(x,t)}{\partial x^2} - [\alpha + (\alpha+2)x]\frac{\partial u(x,t)}{\partial x} = \frac{\partial^2 u(x,t)}{\partial t^2}$$

where $\alpha > -1$, $x,t \in \mathbb{R}_+$ and with the initial conditions

$$u(x,0) = u_o(x), \quad \frac{\partial u(x,t)}{\partial t}\Big|_{t=0} = u_1(x).$$

Applying the Jacobi transform $\mathfrak{P}^{(\alpha,0)}$ with respect to x to the differential equation, by means of (1.10.83), (1.10.85) we get

$$\mathfrak{P}^{(\alpha,0)}\Big[\frac{\partial^2 u(x,t)}{\partial t^2}\Big](n) = -n(n+\alpha+1)\mathfrak{P}^{(\alpha,0)}[u(\cdot,t)](n)$$

or

$$\frac{\partial^2}{\partial t^2}\mathfrak{P}^{(\alpha,0)}[u(\cdot,t)](n) + n(n+\alpha+1)\mathfrak{P}^{(\alpha,0)}[u(\cdot,t)](n) = 0.$$

The solution of this ordinary differential equation (with respect to t) is

$$\mathfrak{P}^{(\alpha,0)}[u(\cdot,t)](n) = c_1 \cos\Big(\sqrt{n(n+\alpha+1)}t\Big) + c_2 \sin\Big(\sqrt{n(n+\alpha+1)}t\Big). \tag{1.10.97}$$

From the (transformed) initial value condition we obtain

$$c_1 = \mathfrak{P}^{(\alpha,0)}[u_o](n), \quad c_2 = \frac{\mathfrak{P}^{(\alpha,0)}[u_1](n)}{\sqrt{n(n+\alpha+1)}}. \tag{1.10.98}$$

Applying the inversion formula (1.10.77) to equation (1.10.97) we have

$$u(x,t) = \sum_{n=0}^{\infty} \frac{1}{h_n^{(\alpha,\beta)}} \left[c_1(n) \cos \sqrt{n(n+\alpha+1)}t + c_2(n) \sin \sqrt{n(n+\alpha+1)}t \right] R_n^{(\alpha,0)}(x),$$

where c_1, c_2 must be taken from (1.10.98).

This is the solution of the initial value problem above provided that u_o, u_1 are sufficiently smooth.

1.10.6 The Laguerre Transform

Definition 1.10.35 *The Laguerre transform (LaT) of a function $f : [0; \infty] \to \mathbb{C}$ is defined by means of*

$$\mathfrak{L}a^{\alpha}[f](n) = f^{\wedge}(n) = \int_0^{\infty} f(x) R_n^{\alpha}(x) dw^{\alpha}(x), \qquad (1.10.99)$$

where $\alpha > -1$, $n \in \mathbb{N}_o$, *provided that the integral exists. Here R_n^{α} are the Laguerre polynomials of order α and degree n, standardized in such a manner that*

$$R_n^{\alpha}(x) = \frac{L_n^{\alpha}(x)}{L_n^{\alpha}(0)}, \qquad (1.10.100)$$

where

$$L_n^{\alpha}(x) = \frac{e^x x^{-\alpha}}{n!} D^n(e^{-x} x^{n+\alpha}), \quad D = \frac{d}{dx} \qquad (1.10.101)$$

are the Laguerre polynomials in the usual designation and

$$L_n^{\alpha}(0) = \frac{(\alpha+1)_n}{n!}. \qquad (1.10.102)$$

Furthermore,

$$dw^{\alpha}(x) = \frac{e^{-x} x^{\alpha}}{2^{\alpha+1}\Gamma(\alpha+1)} dx. \qquad (1.10.103)$$

Remark 1.10.88 *From [E.1], vol. 2, 10.18, formula (14) we have with (1.10.100), (1.10.102)*

$$|R_n^{\alpha}(x)| \le e^{x/2}, \quad \alpha \ge 0 \qquad (1.10.104)$$

and

$$R_n^{\alpha}(0) = 1, \quad \alpha > -1.$$

Remark 1.10.89 *We follow Görlich and Markett throughout this section; see [GöM].*

As the space of originals we choose the space $L_{1,w^{\alpha}}(\mathbb{R}_+) = L_{1,w^{\alpha}}$ of measurable functions on $\mathbb{R}_+$ such that

$$\|f\|_{1,w^{\alpha}} = \int_0^{\infty} |f(x)|e^{x/2} dw^{\alpha}(x) \qquad (1.10.105)$$

is finite. It is a Banach space with the norm (1.10.105).

Lemma 1.10.21 *We have*

$$\|1\|_{1,w^\alpha} = 1.$$

Proof. Substituting $t = \frac{x}{2}$ in (1.10.105) and using the integral formula for the Gamma function, see formula (1.4.11), we obtain the result: ☐

Remark 1.10.90 *The considerations in the following also can be extended to originals of the space L_{p,w^α}, $1 \le p < \infty$ of measurable functions on $\mathbb{R}_+$ with the norm*

$$\|f\|_p = \frac{(p/2)^{\alpha+1}}{\Gamma(\alpha+1)} \left[\int_0^\infty |f(x)e^{-x/2}|^p x^\alpha dx \right]^{1/p};$$

see [GöM].

Theorem 1.10.74 *Let $F \in L_{1,w^\alpha}$ and $k, n \in \mathbb{N}_o$. Then the LaT is linear and moreover it holds that*

(i) $\qquad |\mathcal{L}a^\alpha[f](n)| \le \|f\|_{1,w^\alpha}, \quad \alpha \ge 0,$

(ii) $\qquad \mathcal{L}a^\alpha[R_k^\alpha](n) = \frac{n!}{2^{\alpha+1}(\alpha+1)_n} \delta_{kn} =: h_{n,\alpha}\delta_{kn}.$

Proof. The property (i) follows from (1.10.99), (1.10.105) and

$$|R_n^\alpha(x)| \le e^{x/2}, \quad \alpha \ge 0;$$

see Remark 1.10.90.

Property (ii) is the orthogonality relation of the Laguerre polynomials (see [E.1], vol. 2, 10.12, II) taking note of (1.10.100), (1.10.102), and (1.10.103). ☐

Now we formulate (without proof) an expansion theorem for series in Laguerre polynomials, particularly a general theorem for the classical orthogonal polynomials; see [NU], §8, Theorem 1.

Theorem 1.10.75 *Let $f \in C^1[0; \infty)$ and let furthermore the integrals*

$$\int_0^\infty [f(x)]^2 dw^\alpha(x) \quad and \quad \int_0^\infty [f'(x)]^2 x\, dw^\alpha(x)$$

be convergent. Then we have the inversion formula

$$f(x) = \sum_{n=0}^\infty c_n R_n^\alpha(x) =: (\mathcal{L}a^\alpha)^{-1}[f^\wedge](x)$$

with

$$c_n = f^\wedge(n)/h_{n,\alpha},$$

the series converging uniformly on every interval $[x_1, x_2] \subset \mathbb{R}_+$.

Directly we obtain:

Corollary 1.10.26 *It holds that* $\mathcal{L}a^\alpha[f](n) = 0$ *for every* $n \in \mathbb{N}_o$ *if and only if* $f(x) = 0$ *(a.e.).*

The result can be formulated as a uniqueness theorem for the *LaT*:

Theorem 1.10.76 (Uniqueness Theorem) *Let* f, g *fulfill the conditions of* Theorem 1.10.75 *and let*

$$\mathcal{L}a^\alpha[f](n) = \mathcal{L}a^\alpha[g](n) \quad \text{for every} \quad n \in \mathbb{N}_o.$$

Then $f(x) = g(x)$ *(a.e.), i.e.,* $f = g$.

Now we are going to formulate some rules of operational calculus. The Laguerre polynomials are eigenfunctions of the differential operator L^α defined by means of

$$(L^\alpha y)(x) = -e^x x^{-\alpha} D(e^{-x} x^{\alpha+1}) D, \qquad D = d/dx \tag{1.10.106}$$

with respect to the eigenvalues n, i.e.,

$$L^\alpha R_n^\alpha = n R_n^\alpha, \qquad n \in \mathbb{N}. \tag{1.10.107}$$

From (1.10.107) we obtain:

Proposition 1.10.57 (Differentiation Rule) *Let* $f, f' \in L_{1,w^\alpha}$ *and* f *be two times differentiable (a.e.) on* $\mathbb{R}_+$. *Then it holds that*

$$\mathcal{L}a^\alpha[L^\alpha f](n) = n f^\wedge(n), \qquad n \in \mathbb{N}. \tag{1.10.108}$$

Proof. By means of integration by parts we have from (1.10.99) and (1.10.103)

$$\mathcal{L}a^\alpha[L^\alpha f](n) = -\frac{1}{2^{\alpha+1}\Gamma(\alpha+1)} \int\limits_0^\infty [D(e^{-x} x^{\alpha+1}) D f(x)] R_n^\alpha(x) dx$$

$$= \frac{1}{2^{\alpha+1}\Gamma(\alpha+1)} \left\{ -e^{-x} x^{\alpha+1} f'(x) R_n^\alpha(x) \Big|_0^\infty + \int\limits_0^\infty f'(x) e^{-x} x^{\alpha+1} D R_n^\alpha(x) dx \right\}$$

$$= \frac{1}{2^{\alpha+1}\Gamma(\alpha+1)} \left\{ e^{-x} x^{\alpha+1} f(x) D R_n^\alpha(x) \Big|_0^\infty - \int\limits_0^\infty f(x) D\Big(e^{-x} x^{\alpha+1} D R_n^\alpha(x)\Big) dx \right\}$$

$$= \int\limits_0^\alpha f(x) L^\alpha R_n^\alpha(x) dw^\alpha(x) = n \mathcal{L}a^\alpha[f](n).$$

∎

Corollary 1.10.27

$$\mathfrak{La}^\alpha[(L^\alpha)^k f](x) = n^k f^\wedge(n), \qquad k, n \in \mathbb{N}. \tag{1.10.109}$$

Now let $L^\alpha f = g$. Then

$$e^{-x} x^{\alpha+1} f'(x) = -\int_0^x g(v) e^{-v} v^\alpha dv,$$

and

$$f(x) = -\int_0^x e^u u^{-\alpha-1} \int_0^u g(v) e^{-v} v^\alpha dv =: (L^\alpha)^{-1} g(x). \tag{1.10.110}$$

Now we apply the *LaT* (1.10.99) to both sides of (1.10.110). Because of (1.10.108) we have $g^\wedge(n) = n \, f^\wedge(n)$ and so we require that $g^\wedge(0) = 0$. This yields:

Proposition 1.10.58 (Integration Rule) *Let $g \in L_{1,w^\alpha}$ and $g^\wedge(0) = 0$. Then it holds that*

$$\mathfrak{La}^\alpha[(L^\alpha)^{-1} g](n) = n^{-1} g^\wedge(n), \quad n \in \mathbb{N}, \tag{1.10.111}$$

where $(L^\alpha)^{-1}$ is defined by means of (1.10.110).

Preparing the definition of a convolution for the *LaT* we note a product formula for Laguerre polynomials given by Watson; see [W.1].

Lemma 1.10.22 *Let $\alpha > -1/2$ and $x, y \in \mathbb{R}_+$, $n \in \mathbb{N}_o$. Then it holds that*

$$R_n^\alpha(x) R_n^\alpha(y) = \frac{2^\alpha \Gamma(\alpha+1)}{\sqrt{2\pi}} \int_0^\pi R_n^\alpha(x + y + 2\sqrt{xy} \, \cos \theta) e^{-\sqrt{xy} \cos \theta}$$

$$\cdot \frac{J_{\alpha-1/2}(\sqrt{xy} \, \sin \theta)}{(\sqrt{xy} \sin \theta)^{\alpha-1/2}} \sin^{2\alpha} \theta d\theta. \tag{1.10.112}$$

This formula can be extended to the case $\alpha = -1/2$. We follow a proof of Boersma, published by Markett; see [Ma.1], Lemma 3.

Lemma 1.10.23 *For every $x, y \in \mathbb{R}_+$, $n \in \mathbb{N}_o$ it holds that*

$$R_n^{-1/2}(x) R_n^{-1/2}(y) = \frac{1}{2} \Big\{ e^{-\sqrt{xy}} R_n^{-1/2}\left([\sqrt{x} + \sqrt{y}]^2\right) + e^{\sqrt{xy}} R_n^{-1/2}\left([\sqrt{x} - \sqrt{y}]^2\right)$$

$$- \sqrt{xy} \int_0^\pi R_n^{-1/2}(x + y + 2\sqrt{xy} \cos \theta) e^{-\sqrt{xy} \cos \theta} J_1(\sqrt{xy} \sin \theta) \, d\theta \Big\}. \tag{1.10.113}$$

Proof. We write the power series of $J_{\alpha-1/2}$ in the form

$$J_{\alpha-1/2}(\sqrt{xy}\sin\theta) = \frac{(\frac{1}{2}\sqrt{xy}\sin\theta)^{\alpha-1/2}}{\Gamma(\alpha+1/2)}$$

$$+ \sum_{k=1}^{\infty} \frac{(-1)^k}{k!\,\Gamma(k+\alpha+1/2)} \left(\frac{1}{2}\sqrt{xy}\sin\theta\right)^{2k+\alpha-1/2}, \quad \alpha > -1/2. \tag{1.10.114}$$

The series tends to $-J_1(\sqrt{xy}\sin\theta)$ as $\alpha \to -\frac{1}{2}$. Inserting this limit into (1.10.112) we obtain the integral term of (1.10.113).

Substituting the first term of the right-hand side of equation (1.10.114) into (1.10.112) we have the term

$$I := \frac{\Gamma(\alpha+1)}{\sqrt{\pi}\,\Gamma(\alpha+1/2)} \int_0^\pi R_n^\alpha(x+y+2\sqrt{xy}\cos\theta)e^{-\sqrt{xy}\cos\theta}\sin^{2\alpha}\theta d\theta.$$

The integral has nonintegrable singularities at $\theta = 0$ and at $\theta = \pi$ when $\alpha = -1/2$. To remove these singularities I is rewritten as

$$I = e^{-\sqrt{xy}} R_n^\alpha([\sqrt{x}+\sqrt{y}]^2) \frac{\Gamma(\alpha+1)}{\sqrt{\pi}\Gamma(\alpha+1/2)} \int_0^{\pi/2} \sin^{2\alpha}\theta\,d\theta$$

$$+ e^{\sqrt{xy}} R_n^\alpha([\sqrt{x}-\sqrt{y}]^2) \frac{\Gamma(\alpha+1)}{\sqrt{\pi}\Gamma(\alpha+1/2)} \int_{\pi/2}^\pi \sin^{2\alpha}\theta d\theta$$

$$+ \frac{\Gamma(\alpha+1)}{\sqrt{\pi}\Gamma(\alpha+1/2)} \int_0^{\pi/2} \left\{ e^{-\sqrt{xy}\cos\theta} R_n^\alpha(x+y+2\sqrt{xy}\cos\theta) \right.$$

$$\left. - e^{-\sqrt{xy}} R_n^\alpha([\sqrt{x}+\sqrt{y}]^2) \right\} \sin^{2\alpha}\theta d\theta$$

$$+ \frac{\Gamma(\alpha+1)}{\sqrt{\pi}\Gamma(\alpha+1/2)} \int_{\pi/2}^\pi \left\{ e^{-\sqrt{xy}\cos\theta} R_n^\alpha(x+y+2\sqrt{xy}\cos\theta) \right.$$

$$\left. - e^{\sqrt{xy}} R_n^\alpha([\sqrt{x}-\sqrt{y}]^2) \right\} \sin^{2\alpha}\theta d\theta.$$

The third and the fourth terms tend to zero as $\alpha \to -1/2$ since the integrals are convergent for $\alpha = -1/2$ and $1/\Gamma(\alpha+1/2)$ tends to zero. Furthermore, by means of [PBM], vol. I, section 2.5.3, formula 1., namely,

$$\int_0^\pi (\sin\theta)^{2\alpha} d\theta = B(\alpha+1/2, 1/2) = \frac{\Gamma(\alpha+1/2)\Gamma(1/2)}{\Gamma(\alpha+1)}, \quad \alpha > -1/2, \tag{1.10.115}$$

we have

$$\frac{\Gamma(\alpha+1)}{\sqrt{\pi}\Gamma(\alpha+1/2)} \int_0^{\pi/2} \sin^{2\alpha}\theta d\theta = \frac{\Gamma(\alpha+1)}{\sqrt{\pi}\,\Gamma(\alpha+1/2)} \int_{\pi/2}^\pi \sin^{2\alpha}\theta d\theta$$

$$= \frac{\Gamma(\alpha+1)}{\Gamma(1/2)\,\Gamma(\alpha+1/2)} \frac{\Gamma(\alpha+1/2)\Gamma(1/2)}{2\,\Gamma(\alpha+1)} \longrightarrow \frac{1}{2}$$

as $\alpha \to -\frac{1}{2}$. Thus, the sum of the first two terms in I tends to

$$\frac{1}{2}\left\{e^{-\sqrt{xy}}R_n^{-1/2}([\sqrt{x}+\sqrt{y}]^2) + e^{\sqrt{xy}}R_n^{-1/2}([\sqrt{x}-\sqrt{y}]^2)\right\}$$

as $\alpha \to -1/2$ and so we arrive at (1.10.113). $\qquad\qquad\qquad\qquad$ □

Substituting $z = z(\theta) = x + y + 2\sqrt{xy}\cos\theta$, $0 \le \theta \le \pi$ in (1.10.112) we have

$$\sqrt{xy}\sin\theta = \frac{1}{2}[2(xy+yz+zx) - x^2 - y^2 - z^2]^{1/2} =: \rho(x,y,z), \qquad (1.10.116)$$

and after a short calculation

$$R_n^\alpha(x)\,R_n^\alpha(y) = \frac{2^{\alpha-1}\Gamma(\alpha+1)}{\sqrt{2\pi}(xy)^\alpha} \int\limits_{(\sqrt{x}-\sqrt{y})^2}^{(\sqrt{x}+\sqrt{y})^2} R_n^\alpha(z)\exp\left(-\frac{z-x-y}{2}\right)J_{\alpha-1/2}(\rho)\rho^{\alpha-1/2}dz.$$

From formula (1.10.103) we have $dz = 2^{\alpha+1}\Gamma(\alpha+1)\,e^z\,z^{-\alpha}dw^\alpha(z)$ and setting

$$K_\alpha(x,y,z) = \begin{cases} \frac{2^{2\alpha}[\Gamma(\alpha+1)]^2}{\sqrt{2\pi}(xyz)^\alpha}\exp\left(\frac{x+y+z}{2}\right)J_{\alpha-1/2}(\rho)\rho^{\alpha-1/2}, \text{ for} \\ \qquad\qquad z \in ([\sqrt{x}-\sqrt{y}]^2, [\sqrt{x}+\sqrt{y}]^2) \\ 0, \qquad\qquad\qquad\qquad\qquad\qquad \text{elsewhere} \end{cases} \qquad (1.10.117)$$

we obtain the product formula (1.10.112) in the kernel form.

Lemma 1.10.24 *Let* $\alpha > -1/2$, $x, y \in \mathbb{R}_+, n \in \mathbb{N}_o$. *Then it holds that*

$$R_n^\alpha(x)R_n^\alpha(y) = \int\limits_0^\infty K_\alpha(x,y,z)R_n^\alpha(z)dw^\alpha(z) = \mathfrak{L}a^\alpha[K_\alpha(x,y,\cdot)](n). \qquad (1.10.118)$$

Remark 1.10.91 *Analogous to (1.10.118), the formula (1.10.113) can also be written in kernel form. From (1.10.118), (1.10.117), and (1.10.113) we obtain the following after a straightforward calculation*

$$R_n^{-1/2}(x)R_n^{-1/2}(y) = \frac{1}{2}\left\{e^{-\sqrt{xy}}R_n^{-1/2}([\sqrt{x}+\sqrt{y}]^2) + e^{\sqrt{xy}}R_n^{-1/2}([\sqrt{x}-\sqrt{y}]^2)\right\}$$

$$-\frac{1}{4}\int\limits_{(\sqrt{x}-\sqrt{y})^2}^{(\sqrt{x}+\sqrt{y})^2}\sqrt{xyz}R_n^{-1/2}(z)\exp\left(-\frac{z-x-y}{2}\right)J_1(\rho)\rho^{-1}dz.$$

$$(1.10.119)$$

Lemma 1.10.25 *The kernel* $K_\alpha(x,y,z)$ *is symmetrical in* x, y, z *and it holds that*

(i) $\qquad \int\limits_0^\infty K_\alpha(x,y,z)dw^\alpha(z) = 1$

and

(ii) $\qquad \|K_\alpha(x,y,\cdot)\|_{1,w^\alpha} \le e^{(x+y)/2}.$

Proof. Formula (i) follows directly from (1.10.117) with $n = 0$ and $R_0^\alpha = 1$; see (1.10.100).

For the proof of (ii) we quote that for $\alpha \geq 0$

$$|J_{\alpha-1/2}(t)| \leq \frac{1}{\Gamma(\alpha+1/2)}\left(\frac{t}{2}\right)^{\alpha-1/2}, \qquad t \in \mathbb{R}_+; \tag{1.10.120}$$

see [W.2], 3.31, (1).

Then substituting inversely z by θ we have

$$e^{-(x+y)/2}\int_0^\infty |K_\alpha(x,y,z)|e^{z/2}d\,w^\alpha(z) = \frac{2^{\alpha-1}\Gamma(\alpha+1)}{\sqrt{2\pi}\,(xy)^\alpha}\int_{(\sqrt{x}-\sqrt{y})^2}^{(\sqrt{x}+\sqrt{y})^2}\left|J_{\alpha-1/2}(\rho)\rho^{\alpha-1/2}\right|dz$$

$$\leq \frac{1}{2}\frac{\Gamma(\alpha+1)}{\Gamma(\frac{1}{2})\Gamma(\alpha+\frac{1}{2})}\frac{1}{(xy)^\alpha}\int_{(\sqrt{x}-\sqrt{y})^2}^{(\sqrt{x}+\sqrt{y})^2}\rho^{2\alpha-1}dz = \frac{\Gamma(\alpha+1)}{\Gamma(\frac{1}{2})\Gamma(\alpha+\frac{1}{2})}\int_0^\infty (\sin\theta)^{2\alpha}d\theta = 1,$$

using (1.10.115). ⬜

Remark 1.10.92 *Because of the alternating property of the Bessel function the kernel is not positive as in the case of the polynomials P_n, P_n^λ and $P_n^{(\alpha,\beta)}$.*

As usual in the preceding section now we are able to define a generalized translation operator (GTO) for the LaT.

Definition 1.10.36 *Let $f \in L_{1,w^\alpha}$. Then a GTO T_x^α for the LaT is defined by means of*

$$(T_x^\alpha f)(y) = \int_0^\infty K_\alpha(x,y,z)f(z)dw^\alpha(z). \tag{1.10.121}$$

Proposition 1.10.59 *Let $\alpha \geq 0$ and $f \in L_{1,w^\alpha}$. Then one has*

(i) $\|T_x^\alpha f\|_{1,w^\alpha} \leq e^{x/2}\|f\|_{1,w^\alpha}$,

(ii) $(T_x^\alpha f)(y) = (T_y^\alpha f)(x)$,

(iii) $T_x^\alpha R_n^\alpha(y) = R_n^\alpha(x)\cdot R_n^\alpha(y)$,

(iv) $\mathcal{L}a^\alpha[T_x^\alpha f](n) = R_n^\alpha(x)f^\wedge(n)$.

Proof. Let $x \in \mathbb{R}_+$. Then we have

$$\|T_x^\alpha f\|_{1,w^\alpha} \leq \int_0^\infty \left(\int_0^\infty |f(z)||K_\alpha(x,y,z)|dw_\alpha^\alpha(z)\right)e^{y/2}dw^\alpha(y)$$

$$\leq \int_0^\infty \left(\int_0^\infty |K_\alpha(x,y,z)|e^{y/2}dw^\alpha(y)\right)|f(z)|dw^\alpha(z).$$

Because of the symmetry of K and Lemma 1.10.25, (ii) applied to the inner integral we obtain

$$\|T_x^\alpha f\|_{1,w^\alpha} \le e^{x/2} \int\limits_0^\infty |f(z)|e^{z/2}dw^\alpha(z) = e^{x/2}\|f\|_{1,w^\alpha}.$$

The result (ii) follows from the symmetry of the kernel K_α and (iii) is the product formula (1.10.117). Finally, we have

$$\begin{aligned}
\mathcal{L}a^\alpha[T_x^\alpha f](n) &= \int\limits_0^\infty \left(\int\limits_0^\infty K_\alpha(x,y,z)f(z)dw^\alpha(z) \right) R_n^\alpha(y)dw^\alpha(y) \\
&= \int\limits_0^\infty \left(\int\limits_0^\infty K_\alpha(x,y,z)R_n^\alpha(y)dw^\alpha(y) \right) f(z)dw^\alpha(z) \\
&= R_n^\alpha(x)f^\wedge(n),
\end{aligned}$$

again using the symmetry of K and (iii). ◻

Definition 1.10.37 *As the convolution of the LaT we call $f*g$ defined by*

$$(f*g)(x) = \int\limits_0^\infty f(y)(T_x^\alpha g)(y)dw^\alpha(y), \qquad (1.10.122)$$

provided that it exists.

Theorem 1.10.77 (Convolution Theorem) *Let $\alpha \ge 0$ and $f,g \in L_{1,w^\alpha}$. Then there exists $f*g$ (a.e.), $f*g \in L_{1,w^\alpha}$ and it holds that*

(i) $\|f*g\|_{1,w^\alpha} \le \|f\|_{1,w^\alpha}\|g\|_{1,w^\alpha}$,

(ii) $\mathcal{L}a^\alpha[f*g] = f^\wedge g^\wedge$,

(iiii) $f*g = g*f$.

Proof. The estimate (i) follows from

$$\|f*g\|_{1,w^\alpha} = \|\int\limits_0^\infty f(y)(T_x^\alpha g)(y)dw^\alpha(y)\|_{1,w^\alpha} \le \int\limits_0^\infty |f(y)|\|T_y^\alpha g\|_{1,w^\alpha}dw^\alpha(y)$$

$$\le \|g\|_{1,w^\alpha} \int\limits_0^\infty |f(y)|e^{y/2}dw^\alpha(y) = \|f\|_{1,w^\alpha}\|g\|_{1,w^\alpha},$$

using the generalized Minkowski inequality

$$\int\limits_0^\infty |\int\limits_0^\infty f(x,y)dy|dx \le \int\limits_0^\infty \left(\int\limits_0^\infty |f(x,y)|dx \right) dy$$

and Proposition 1.10.59, (ii) and (i).

Now

$$(f * g)^{\wedge}(n) = \int_0^\infty \left(\int_0^\infty f(y)(T_x^\alpha g)(y)dw^\alpha(y) \right) R_n^\alpha(x)dw^\alpha(x)$$

$$= \int_0^\infty f(y) \int_0^\infty (T_y^\alpha g)(x)R_n^\alpha(x)dw^\alpha(x)dw^\alpha(y)$$

$$= g^{\wedge}(n) \int_0^\infty f(y)R_n^\alpha(y)dw^\alpha(y) = f^{\wedge}(n)g^{\wedge}(n),$$

where the properties of Proposition 1.10.59, (ii), (iii) of the *GTO* were used. ☐

Remark 1.10.93 *The results of Proposition 1.10.59 and Theorem 1.10.77 can also be generalized to spaces L_{p,w^α}; see Remark 1.10.90. For details we refer to [GöM]. It holds that if $f \in L_{p,w^\alpha}$, $g \in L_{q,w^\alpha}$, $r^{-1} = p^{-1} + q^{-1} - 1$, then $f * g \in L_{r,w^\alpha}$ and*

$$\|f * g\|_{r,w^\alpha} \le \|f\|_{p,w^\alpha}\|g\|_{q,w^\alpha}.$$

In particular,

$$\text{if} \quad p \in L_{p,w^\alpha} \quad \text{and} \quad g \in L_{1,w^\alpha}, \quad \text{then} \quad f * g \in L_{p,w^\alpha}.$$

Now we consider an application. Following Debnath [De.7] we investigate the problem of oscillations $u(x,t)$ of a very long and heavy chain with variable tension. The mathematical model of this problem is given by the differential equation

$$x\frac{\partial^2 u(x,t)}{\partial x^2} + (\alpha - x + 1)\frac{\partial u(x,t)}{\partial x} = \frac{\partial^2 u(x,t)}{\partial t^2}, \quad t,\, x \in \mathbb{R}_+, \tag{1.10.123}$$

with the initial value conditions

$$\begin{cases} u(x,0) = u_0(x), & x \ge 0 \\ \frac{\partial u(x,t)}{\partial t}\Big|_{t=0} = u_1(x), & x \ge 0. \end{cases} \tag{1.10.124}$$

The left-hand side of (1.10.123) can be written as $L^\alpha u(x,t)$, when the differentiation is taken with respect to the variable x. Therefore,

$$(L^\alpha u)(x,t) + u_{tt}(x,t) = 0. \tag{1.10.125}$$

Applying the *LaT* with respect to x, by means of the differentiation rule (1.10.108) we have

$$u_{tt}^{\wedge}(n,t) + n\,u^{\wedge}(n,t) = 0,$$

where $u^{\wedge}$ is the LaT $\mathcal{L}a^\alpha[u(\cdot,t)](n)$. The solution of this (ordinary) differential equation with respect to t is

$$u^{\wedge}(n,t) = A(n)\cos(\sqrt{n}t) + B(n)sin\sqrt{n}t. \tag{1.10.126}$$

Applying the *LaT* on the initial value conditions (1.10.124) we have

$$u^\wedge(n,0) = u_0^\wedge(n), \qquad n \in \mathbb{N}_o$$
$$u_t^\wedge(n,0) = u_1^\wedge(n), \qquad n \in \mathbb{N}.$$

Therefore, from (1.10.126) we conclude

$$u^\wedge(n,t) = \begin{cases} u_0^\wedge(0), & n = 0 \\ u_0^\wedge(n)\cos\sqrt{n}t + \frac{u_1^\wedge(n)}{\sqrt{n}}\sin\sqrt{n}t, & n \in \mathbb{N}. \end{cases}$$

Applying (quite formally) the inversion formula, Theorem 1.10.75, we get

$$u(x,t) = \frac{1}{2^{\alpha+1}}u_0^\wedge(0) + \sum_{n=1}^{\infty}\frac{1}{h_{n,\alpha}}\left[u_0^\wedge(n)\cos\sqrt{n}t + \frac{u_1^\wedge(n)}{\sqrt{n}}\sin\sqrt{n}t\right]R_n^\alpha(x), \qquad (1.10.127)$$

with $h_{n,\alpha}$ from Theorem 1.10.74, (ii). The series expansion (1.10.127) is the solution of the initial value problem (1.10.123), (1.10.124) provided that u_0, u_1 are sufficiently smooth.

1.10.7 The Hermite Transform

Definition 1.10.38 *The Hermite transform (HeT) of a function $f : \mathbb{R} \to \mathbb{C}$ is defined by means of*

$$\mathcal{H}e[f](n) = f^\wedge(n) = \frac{1}{\sqrt{2\pi}}\int_{-\infty}^{\infty}f(x)\tilde{H}_n(x)e^{-x^2}dx, \qquad n \in \mathbb{N}_o \qquad (1.10.128)$$

provided that the integral exists. Here $\tilde{H}_n$ are the Hermite polynomials of degree n standardized in such a manner that

$$\tilde{H}_n(x) = \begin{cases} \frac{H_{2k}(x)}{H_{2k}(0)}, & \text{if } n = 2k, \ k \in \mathbb{N}_o, \\ \frac{H_{2k+1}(x)}{(DH_{2k+1})(0)}, & \text{if } n = 2k+1, \end{cases} \qquad (1.10.129)$$

where

$$H_n(x) = (-1)^n e^{x^2}D^n(e^{-x^2}), \qquad D = d/dx \qquad (1.10.130)$$

are the Hermite polynomials in the usual designation and

$$\begin{cases} H_{2k}(0) = (-1)^k(2k)!/k! \\ (DH_{2k+1})(0) = 2(2k+1)H_{2k}(0) = (-1)^k\frac{2(2k+1)!}{k!}. \end{cases} \qquad (1.10.131)$$

For formulas for Hermite polynomials we refer to [E.1], vol. 2, 10.13.

Because of the connection between Laguerre and Hermite polynomials we have

$$\tilde{H}_n(x) = \begin{cases} R_k^{-1/2}(x^2), & \text{if } n = 2k, \\ xR_k^{1/2}(x^2), & \text{if } n = 2k+1, \end{cases} \qquad (1.10.132)$$

where R_n^α are the Laguerre polynomials; see section 1.10.6, (1.10.100), (1.10.101).

Remark 1.10.94 *From [E.1], vol. 2, 10.18, (19) we have*

$$e^{-x^2/2}|H_n(x)| < 2\sqrt{2^n n!}.$$

Therefore, in our standardization (1.10.129), (1.10.131) *we obtain*

$$e^{-x^2/2}|\tilde{H}_n(x)| \leq \left\{ \begin{array}{ll} 2^{k+1}\dfrac{k!}{\sqrt{(2k)!}}, & \text{if } n = 2k, \\[3mm] 2^{k+1/2}\dfrac{k!}{\sqrt{(2k+1)!}}, & \text{if } n = 2k+1 \end{array} \right\} =: C_n. \qquad (1.10.133)$$

As the space of originals we choose the space $L_{1,\exp}(\mathbb{R}) =: L_{1,\exp}$ of measurable functions on $\mathbb{R}$ such that

$$\|f\|_{1,\exp} = \frac{1}{\sqrt{2\pi}} \int_{-\infty}^{\infty} |f(x)|e^{-x^2/2}dx \qquad (1.10.134)$$

is finite. It is a Banach space with the norm (1.10.134). From

$$\int_{-\infty}^{\infty} e^{-x^2}dx = \sqrt{\pi}$$

we have

Lemma 1.10.26 *It holds that*

$$\|1\|_{1,\exp} = 1. \qquad (1.10.135)$$

Remark 1.10.95 *The investigations in the following also can be extended to the space $L_{p,\exp}$, $1 \leq p < \infty$ of measurable functions on $\mathbb{R}$ with the norm*

$$\|f\|_{p,\exp} = \frac{1}{\sqrt{2\pi}} \int_{-\infty}^{\infty} \left|f(x)e^{-x^2/2}\right|^p dx; \qquad (1.10.136)$$

see Markett [Ma.2]. This paper is the basis for many explanations in this section.

Theorem 1.10.78 *Let $f \in L_{1,\exp}$ and $k, n \in \mathbb{N}_o$. Then the HeT is a linear transform and moreover it holds that*

(i) $|\mathcal{H}e[f](n)| \leq C_n\|f\|_{1,\exp}$,

(ii) $\mathcal{H}e[\tilde{H}_k](n) = \tilde{h}_n \delta_{kn}$,

with C_n from (1.10.133) *and*

$$\tilde{h}_n = \left\{ \begin{array}{ll} \dfrac{2^{2k-1/2}(k!)^2}{(2k)!}, & \text{if } n = 2k, \ k \in \mathbb{N}_o, \\[3mm] \dfrac{2^{2k-3/2}(k!)^2}{(2k+1)!}, & \text{if } n = 2k+1. \end{array} \right. \qquad (1.10.137)$$

Proof. Part (i) follows directly from (1.10.128), (1.10.133), and (1.10.135). The formula (ii) is nothing other than the orthogonality relation of Hermite polynomials; see [E.1], vol. 2, 10.13, (4), taking note of (1.10.129) and (1.10.131). □

Now we formulate (without proof) an expansion theorem for Hermite polynomials; see [NU], §8, Theorem 1.

Theorem 1.10.79 *Let $f \in C^1(\mathbb{R})$ and furthermore let the integrals*

$$\int_{-\infty}^{\infty} [f(x)]^2 e^{-x^2}\,dx \quad and \quad \int_{-\infty}^{\infty} [f'(x)]^2 x e^{-x^2}\,dx$$

be convergent. Then we have the inversion formula

$$f(x) = \sum_{n=0}^{\infty} c_n \tilde{H}_n(x) =: (\mathcal{H}e)^{-1}[f^\wedge](x),$$

with

$$c_n = f^\wedge(n)/\ddot{h}_n,$$

the series being uniformly convergent on every interval $[x_1, x_2] \subset \mathbb{R}$.

Directly, we conclude:

Corollary 1.10.28 $\mathcal{H}e[f](n) = 0$ *for every* $n \in \mathbb{N}_o$ *if and only if* $f(x) = 0$ *(a.e.).*

This result can be formulated as the uniqueness theorem for the *HeT*.

Theorem 1.10.80 *Let f, g fulfill the conditions of Theorem 1.10.79 and let $\mathcal{H}e[f](n) = \mathcal{H}e[g](n)$ for every $n \in \mathbb{N}_o$. Then $f(x) = g(x)$ (a.e.), i.e., $f = g$.*

Now we are going to formulate some rules of operational calculus. The Hermite polynomials are eigenfunctions of the differential operator H defined by means of

$$(Hy)(x) = -e^{x^2} D(e^{-x^2})Dy, \qquad D = \frac{d}{dx} \tag{1.10.138}$$

with respect to the eigenvalues $2n$, i.e.,

$$H\tilde{H}_n = 2n\tilde{H}_n, \qquad n \in \mathbb{N}. \tag{1.10.139}$$

Therefore, we obtain:

Proposition 1.10.60 (Differentiation Rule) *Let $f, f' \in L_{1,\exp}$ and f be two times differentiable (a.e.) on $\mathbb{R}$. Then it holds that*

$$\mathcal{H}e[Hf](n) = 2nf^\wedge(n), \qquad n \in \mathbb{N}. \tag{1.10.140}$$

Proof. By means of integration by parts we have with (1.10.138), (1.10.139)

$$\sqrt{2\pi}\mathcal{H}e[Hf](n) = -\int_{-\infty}^{\infty} [D(e^{-x^2})Df(x)]\tilde{H}_n(x)dx$$

$$= -e^{-x^2}(Df(x))\tilde{H}_n(x)\Big|_{-\infty}^{\infty} + \int_{-\infty}^{\infty} f'(x)e^{-x^2}D\tilde{H}_n(x)dx$$

$$= f(x)e^{-x^2}D\tilde{H}_n(x)\Big|_{-\infty}^{\infty} - \int_{-\infty}^{\infty} f(x)D(e^{-x^2}D\tilde{H}_n(x))dx$$

$$= \int_{-\infty}^{\infty} f(x)(H\tilde{H}_n(x))e^{-x^2}dx = 2n\mathcal{H}e[f](n).$$

$$\square$$

Corollary 1.10.29

$$\mathcal{H}e[H^k f](n) = (2n)^k f^\wedge(n), \qquad k, n \in \mathbb{N}. \tag{1.10.141}$$

Now let $Hf = g$. Then

$$e^{-x^2}f'(x) = -\int_{-\infty}^{\infty} g(v)e^{-v^2}dv,$$

and

$$f(x) = -\int_{-\infty}^{x} e^{u^2}\int_{-\infty}^{u} g(v)e^{-v^2}dv =: (H^{-1}g)(x). \tag{1.10.142}$$

Now we apply the HeT on both sides of equation (1.10.142). Because of (1.10.140) we have $g^\wedge(n) = 2nf^\wedge(n)$ and, therefore, we require $g^\wedge(0) = 0$. Then the following holds:

Proposition 1.10.61 (Integration Rule) *Let $g \in L_{1,exp}$ and $g^\wedge(0) = 0$. Then it holds that*

$$\mathcal{H}e[H^{-1}g](n) = (2n)^{-1}g^\wedge(n), \qquad n \in \mathbb{N}, \tag{1.10.143}$$

where H^{-1} is defined by means of (1.10.142).

Preparing the definition of a convolution for the HeT we note a product formula for Hermite polynomials given by Markett; see [Ma.2].

Lemma 1.10.27 *Let $x, y \in \mathbb{R}$, $n \in \mathbb{N}_o$. Then*

$$\tilde{H}_n(x)\tilde{H}_n(y)$$
$$= \frac{1}{4}\left\{\left[\tilde{H}_n(-x-y) + \tilde{H}_n(x+y)\right]e^{-xy} + \left[\tilde{H}_n(y-x) + \tilde{H}_n(x-y)\right]e^{xy}\right\} \tag{1.10.144}$$
$$+ \mathcal{H}e[K(x,y,\cdot)](n).$$

Here

$$K(x,y,z) = \begin{cases} \sqrt{\frac{\pi}{8}}\left[J_o(\triangle) - \frac{xyz}{\triangle}J_1(\triangle)\right]sgn(xyz)e^{(x^2+y^2+z^2)/2}, & z \in S(x,y) \\ 0, & elsewhere, \end{cases} \quad (1.10.145)$$

with

$$S(x,y) = \left(-|x|-|y|, -||x|-|y||\right) \cup \left(||x|-|y||, |x|+|y|\right) \quad (1.10.146)$$

$$\triangle(x,y,z) = \frac{1}{2}\left[2(x^2y^2+y^2z^2+z^2x^2) - x^4 - y^4 - z^4\right]^{1/2} = \rho(x^2,y^2,z^2), \quad (1.10.147)$$

where ρ is defined in 1.10.6, (1.10.116).

Proof. Following [M.2] we use for convenience the functions

$$\mathfrak{L}_n^\alpha(x) = e^{-x^2/2}R_n^\alpha(x^2), \qquad x \geq 0 \quad (1.10.148)$$

and

$$\mathcal{H}_n(x) = e^{-x^2/2}\tilde{H}_n(x), \qquad x \in \mathbb{R}. \quad (1.10.149)$$

From (1.10.132) we obtain

$$\mathcal{H}_n(x) = \begin{cases} \mathfrak{L}_k^{-1/2}(|x|), & n = 2k, \\ & k \in \mathbb{N}_o, \\ x\mathfrak{L}_k^{1/2}(|x|), & n = 2k+1. \end{cases} \quad (1.10.150)$$

Case 1: $n = 2k$. From (1.10.150), (1.10.148), and 1.10.6, (1.10.119) we obtain (substituting $z \longrightarrow z^2$)

$$\mathcal{H}_{2k}(x)\mathcal{H}_{2k}(y) = \mathfrak{L}_k^{-1/2}(|x|)\mathfrak{L}_k^{-1/2}(|y|) = \frac{1}{2}\left[\mathfrak{L}_k^{-1/2}(|x|+|y|) + \mathfrak{L}_k^{-1/2}(|x-y|)\right]$$
$$- \frac{1}{2}\int_{||x|-|y||}^{|x|+|y|} \mathfrak{L}_k^{-1/2}(z)|xy|zJ_1(\triangle)\triangle^{-1}dz.$$

Substituting $z \longrightarrow -z$ we have

$$\mathcal{H}_{2k}(x)\mathcal{H}_{2k}(y) = \frac{1}{2}\left[\mathfrak{L}_k^{-1/2}(|x-y|) + \mathfrak{L}_k^{-1/2}(|x|+|y|)\right]$$
$$- \frac{1}{2}\int_{-(|x|+|y|)}^{-(|x|-|y|)} \mathfrak{L}_k^{-1/2}(|z|)|xyz| J_1(\triangle)\triangle^{-1}d z.$$

Taking the arithmetical mean of these two formulas we get (again using formula 1.10.150)

$$\mathcal{H}_{2k}(x)\mathcal{H}_{2k}(y) = \frac{1}{2}\left[\mathcal{H}_{2k}(x-y) + \mathcal{H}_{2k}(x+y)\right] - \frac{1}{8}\int_{S(x,y)} \mathcal{H}_{2k}(z)|xyz|J_1(\triangle)\triangle^{-1}dz. \quad (1.10.151)$$

Case 2: $n = 2k+1$. Now from (1.10.150), (1.10.118), and (1.10.117) we obtain (again substituting $z \longrightarrow z^2$)

$$\mathcal{H}_{2k+1}(x)\mathcal{H}_{2k+1}(y) = xy\mathfrak{L}_k^{1/2}(|x|)\mathfrak{L}_k^{1/2}(|y|) = \frac{1}{2}\int_{||x|-|y||}^{|x|+|y|} \mathfrak{L}_k^{1/2}(z)sgn(xyz)J_o(\triangle)dz.$$

Substituting $z \longrightarrow -z$ we have

$$\mathcal{H}_{2k+1}(x)\mathcal{H}_{2k+1}(y) = \frac{1}{2} \int\limits_{-|x|-|y|}^{-||x|-|y||} \mathfrak{L}_k^{1/2}(z)sgn(xyz)J_o(\triangle)dz.$$

Again we take the arithmetical mean of these two equations and we obtain

$$\mathcal{H}_{2k+1}(x)\mathcal{H}_{2k+1}(y) = \frac{1}{4} \int\limits_{S(x,y)} \mathcal{H}_{2k+1}(z)\, sgn(xyz)J_o(\triangle)dz. \qquad (1.10.152)$$

Since $\mathcal{H}_{2k}$ (respectively $\mathcal{H}_{2k+1}$) are even (respectively odd) functions we have

$$\int\limits_{-\infty}^{\infty} \mathcal{H}_{2k+1}(z)|z|d\,z = \int_{-\infty}^{\infty} \mathcal{H}_{2k}(z)sgn(z)dz = 0.$$

Therefore we have a unified form for (1.10.151) and (1.10.152):

$$\mathcal{H}_n(x)\mathcal{H}_n(y) = \frac{1}{4}\Big[\mathcal{H}_n(-x-y) + \mathcal{H}_n(x+y) + \mathcal{H}_n(x-y) + \mathcal{H}_n(y-x)\Big]$$
$$+ \frac{1}{4}\int\limits_{-\infty}^{\infty} \mathcal{H}_n(z)\Big[sgn(xyz)J_o(\triangle) - |xyz|J_1(\triangle)\triangle^{-1}\Big]dz. \qquad (1.10.153)$$

Because of $|xyz| = sgn(xyz)\, xyz$ we obtain, returning to the polynomials $\tilde{H}_n$ with the help of (1.10.149) the product formula (1.10.144) with the kernel (1.10.145). $\quad\square$

Lemma 1.10.28 *The kernel $K(x,y,z)$ is symmetrical in its variables. Furthermore, it holds that*

$$e^{-(y^2+z^2)/2}\|K(\cdot,y,z)\|_{1,\exp} \le M|yz|^{1/2}, \qquad yz \neq 0, \qquad (1.10.154)$$

where M is a constant, $M \ge 1$.

Proof. The symmetry of the kernel K follows directly from (1.10.145). For the proof of (1.10.154) we obtain from (1.10.145)

$$A := e^{-(y^2+z^2)/2}\|K(\cdot,y,z)\|_{1,\exp} = \frac{1}{4} \int\limits_{S(y,z)} \Big|-\frac{xyz}{\triangle}J_1(\triangle) + J_o(\triangle)\Big|dx$$

$$\le \frac{1}{4}\Big[\int\limits_{S(y,z)} |xyz|\frac{|J_1(\triangle)|}{\triangle}dx + \int\limits_{S(y,z)} |J_o(\triangle)|dx\Big].$$

$S(y,z)$ is (see formula 1.10.146) symmetrical with respect to the origin and the integrals are even functions (with respect to x). Therefore,

$$A \le \frac{|yz|}{z} \int\limits_{||y|-|x||}^{|y|+|z|} x\triangle^{-1}|J_1(\triangle)|dx + \frac{1}{2} \int\limits_{||y|-|z||}^{|y|+|z|} |J_o(\triangle)|dx. \qquad (1.10.155)$$

From Watson [W.2], 3.31 (1) and Szegö [Sz], Theorem 7.31.2 we have simultaneously for $x \in \mathbb{R}^+$

$$|J_o(x)| \leq 1 \quad \text{and} \quad |J_o(x)| \leq \sqrt{\frac{2}{\pi x}}$$

and

$$|J_1(x)| \leq x/2 \quad \text{and} \quad |J_1(x)| \leq M\sqrt{\frac{2}{\pi x}}$$

with some constant $M > 1$. Preparing the estimate of A we note Markett's lemma [Ma.2], Lemma 3.1:
⬜

Lemma 1.10.29 *Let* $b \leq 0$, $a + b > -1$ *and let* $\triangle$ *be defined as in* (1.10.147). *For all* $y, z \in \mathbb{R}^+$ *one has*

$$I(a,b) := \int\limits_{|y-z|}^{y+z} \triangle^{2a} x^{2b+1} dx = \gamma(yz)^{2a+1}(y+z)^{2b},$$

with some positive constant γ, $\gamma \geq \frac{\sqrt{\pi}\Gamma(a+1)}{\Gamma(a+3/2)}$.

Using the estimates for J_o and J_1 we see with the help of this lemma (after a short calculation) that the first (respectively second) expression on the right-hand side of equation (1.10.155) is less than $const \cdot \min(|yz|^2, |yz|^{1/2})$ (respectively $const \cdot \min(1, |yz|^{1/2})$). A careful discussion of the appeasing constants leads to (1.10.154).

Remark 1.10.96 *The kernel* K *(see equation* 1.10.145*) is not nonnegative for all* $x, y \in \mathbb{R}\backslash\{0\}$, $z \in S(x,y)$; *see [Ma.2], Corollary 2.4.*

Now we are able to define a *GTO* for the *HeT*.

Definition 1.10.39 *Let* $f \in L_{1,\exp}$. *Then a GTO* T_x *is defined by means of*

$$(T_x f)(y) = \frac{1}{4}\Big\{[f(x+y) + f(-x-y)]e^{-xy} + [f(x-y) + f(y-x)]e^{xy}\Big\}$$

$$+ \frac{1}{\sqrt{2\pi}} \int\limits_{-\infty}^{\infty} f(z)K(x,y,z)e^{-z^2} dz. \tag{1.10.156}$$

Proposition 1.10.62 *Let* $f, \sqrt{|x|}\, f \in L_{1,\exp}$. *Then it holds that*

(i) $\|T_x f\|_{1,\exp} \leq M e^{x^2/2} \left\{ \|f\|_{1,\exp} + \sqrt{|x|} \cdot \|\sqrt{y}\, f(y)\|_{1,\exp} \right\},$

(ii) $(T_x f)(y) = (T_y f)(x),$

(iii) $(T_x \tilde{H}_n)(y) = \tilde{H}_n(x)\tilde{H}_n(y),$

(iv) $\mathcal{H}e[T_x f](n) = \tilde{H}_n(x) f^\wedge(n).$

Proof. From (1.10.156) and Lemma 1.10.28, (1.10.154) we have

$$
\sqrt{2\pi}\|T_x f\|_{1,\exp} \leq \frac{1}{4} e^{x^2/2} \Big\{ \int_{-\infty}^{\infty} \Big[|f(x+y)| + |f(-x-y)| \Big] e^{-(x+y)^2/2} dy
$$

$$
+ \int_{-\infty}^{\infty} \Big[|f(x-y)| + |f(y-x)| \Big] e^{-(x-y)^2/2} dy \Big\}
$$

$$
+ \frac{1}{\sqrt{2\pi}} \int_{-\infty}^{\infty} |f(z)| \Big(\int_{-\infty}^{\infty} |K(x,y,z)| e^{-y^2/2} dy \Big) e^{-z^2} dz
$$

$$
\leq e^{x^2/2} \Big\{ \|f\|_{1,\exp} + M\sqrt{|x|} \cdot \|\sqrt{z}\, f(z)\|_{1,\exp} \Big\}
$$

and this is (i) (because of $M > 1$).

The result (ii) follows directly from the symmetry of $K(x,y,z)$ with respect to the variables x, y, z by means of (1.10.156).

Formula (iii) is nothing other than the product formula (1.10.144). Relation (iv) is proved straightforward, similar to the case of the *LaT*; see 1.10.6, Proof of Proposition 1.10.59. $\square$

Definition 1.10.40 *As convolution for the HeT we call* $f * g$ *defined by*

$$
(f * g)(x) = \frac{1}{\sqrt{2\pi}} \int_{-\infty}^{\infty} f(y)(T_x g)(y) e^{-y^2} dy, \tag{1.10.157}
$$

provided that it exists.

Theorem 1.10.81 (Convolution Theorem) *Let* $\sqrt{|x|}\, f,\ \sqrt{|x|}\, g \in L_{1,\exp}$. *Then* $f * g \in L_{1,\exp}$ *and it holds that*

(i) $\mathcal{H}e[f * g] = f^\wedge g^\wedge,$

(ii) $f * g = g * f.$

Proof. From (1.10.157) it follows by means of Proposition 1.10.62, (i), (ii)

$$\|f * g\|_{1,\exp} \leq \frac{1}{\sqrt{2\pi}} \int_{-\infty}^{\infty} \left(\frac{1}{\sqrt{2\pi}} \int_{-\infty}^{\infty} |f(y)| |T_x g(y)| e^{-y^2} dy \right) e^{-x^2/2} dx$$

$$= \frac{1}{\sqrt{2\pi}} \int_{-\infty}^{\infty} |f(y)| e^{-y^2} \left(\frac{1}{\sqrt{2\pi}} \int_{-\infty}^{\infty} |T_y g(x)| e^{-x^2/2} dx \right) dy$$

$$\leq \frac{M}{\sqrt{2\pi}} \int_{-\infty}^{\infty} |f(y)|^2 e^{-y^2/2} \left(\|g\|_{1,\exp} + \sqrt{|y|} \cdot \|\sqrt{z} g(z)\|_{1,\exp} \right) dy$$

$$= M \left\{ \|f\|_{1,\exp} \cdot \|g\|_{1,\exp} + \|\sqrt{y} f(y)\|_{1,\exp} \cdot \|\sqrt{z} g(z)\|_{1,\exp} \right\}$$

and therefore $f * g \in L_{1,\exp}$.

Formula (ii) follows by straightforward calculation analogous to the proof of 1.10.6, Theorem 1.10.77, (ii). The commutativity (ii) is proved in calculating the HeT of $f * g$ and of $g * f$, taking into account that $f^\wedge g^\wedge = g^\wedge f^\wedge$, and then performing the inverse transformation $(\mathcal{H}e)^{-1}$.

Now we consider an application. We are looking for the solution of the differential equation

$$\frac{\partial^2 u(x,t)}{\partial x^2} - 2x \frac{\partial u(x,t)}{\partial x} = x f(t), \qquad x \in \mathbb{R}, \ t \in \mathbb{R}^+, \tag{1.10.158}$$

under the condition

$$u(x,0) = 0. \tag{1.10.159}$$

Applying the HeT on both sides of equation (1.10.158) using the differential rule (see Proposition 1.10.60) we obtain

$$-2n \, u^\wedge(n,t) = f(t)(\tilde{H}_1)^\wedge(n) = \tilde{h}_1 \delta_{1n} f(t),$$

since $x = \tilde{H}_1(x)$. So we obtain

$$\tilde{u}^\wedge(n,t) = -\frac{\tilde{h}_1}{2n} \delta_{1n} f(t).$$

By means of the inversion theorem, Theorem 1.10.79, we have

$$u(x,t) = -\frac{1}{2} x f(t),$$

and this is the solution of (1.10.158) under the condition (1.10.159). □

Chapter 2

Operational Calculus

2.1 Introduction

In Chapter 1 we considered integral transforms, which can be used for the solution of linear differential equations with respect to a certain differential operator. For example, in section 1.4 we considered the Laplace transformation, which is fit for the solution of linear differential equations with respect to the operator of differentiation $D = \frac{d}{dx}$. The disadvantage of the use of integral transforms is that some integral has to be convergent. So one should look for a pure algebraic version of operational rules with respect to the operator D for the application to the solution of differential equations. This was first done by D. Heaviside [H.1] through [H.3] in a quite formal manner. In the 1950s the problem was solved by J. Mikusiński, see [Mi.7], who used elements of algebra to develop an operational calculus for the operator D in an elementary but perfect manner; see also [DP] and [Be.1] for similar representations of the same topic. Meanwhile, operational calculi for many other differential operators were developed; see, for example, [Di]. In this book we deal only with the classical one, i.e., operational calculus for the operator D.

The basis of the construction is the algebraic result that every commutative ring without divisors of zero can be extended to a field. Its elements are fractions of elements of the ring. So in a manner similar to how the field $\mathbb{Q}$ of rational numbers is constructed by means of the ring $\mathbb{Z}$ of integers, a field of operators can be constructed by means of a ring of functions, which is without divisors of zero with respect to a later defined multiplication.

We will now explain this in more detail. Let R be a commutative ring without a divisor of zero, i.e., that from the equation

$$fg = 0$$

where $f, g \in R$ and $g \neq 0$ it follows that $f = 0$. Two pairs (f, g) and (f_1, g_1) of elements of R with $g, g_1 \neq 0$ are called equivalent if $fg_1 = f_1g$, and if this condition is fulfilled we write $(f, g) \backsim (f_1, g_1)$. This relation is an equivalence relation and, therefore, it divides the set $R \times R$ into disjointed classes of pairs of equivalent elements. A pair (f, g) is called a representation of the class of all pairs equivalent to (f, g). This class is denoted by the symbol $\frac{f}{g}$. For this symbol we define the operators of addition and multiplication according to the rules of calculation with fractions in elementary arithmetics:

$$\frac{f}{g} + \frac{\tilde{f}}{\tilde{g}} = \frac{f\tilde{g} + \tilde{f}g}{g\tilde{g}},$$

$$\frac{f}{g} \cdot \frac{\tilde{f}}{\tilde{g}} = \frac{f\tilde{f}}{g\tilde{g}}.$$

These definitions are correct since $g\tilde{g} \neq 0$ and one can easily prove that result on the right-hand side does not depend on the representation of the classes $\frac{f}{g}$ and $\frac{f_1}{g_1}$. The set of symbols $\frac{f}{g}$ with the operations of addition and multiplication defined above is easily proved to be not only a commutative ring, but also a field, K. Its elements (the "symbols" $\frac{f}{g}$, $f, g \in R, g \neq 0$) are called operators. The field K is the quotient field of the ring R.

Sometimes operators are denoted only by one letter: $a = \frac{f}{g}$, $b = \frac{\tilde{f}}{\tilde{g}}$.

We denote

$$e = \frac{f}{f}, \quad 0 = \frac{0}{g}, \quad f, g \neq 0$$

and they are called the unit respectively zero element of K. Obviously, they are the unit and zero element of the field K:

$$0 + \frac{f}{g} = \frac{f}{g}, \quad 0 \cdot \frac{f}{g} = 0, \quad e \cdot \frac{f}{g} = \frac{f}{g}.$$

In the field K equations of the form $ax = b$, $a, b \in K$, $a \neq 0$ have a (unique) solution. Defining the inverse operator a^{-1} of $a = \frac{f}{g}$ by $a^{-1} = \frac{g}{f}$ (which exists since $a \neq 0$), we have $x = a^{-1}b$, which easily can be verified.

Operators of the form $\frac{f}{e}$ comprise a subfield of the field K since

$$\frac{f}{e} + \frac{g}{e} = \frac{f + g}{e} \qquad \frac{f}{e} \cdot \frac{g}{e} = \frac{fg}{e}.$$

One can easily prove that this subfield is isomorphic to the ring R, whose elements f can also be denoted by (f, e). In this sense the field K is an extension of the ring R to a field of quotients of elements of R and in K the usual calculations for fractions are valid. So we have given a short version of the construction of the field of quotients of the elements of a commutative ring without divisors of zero.

An example of this construction is the extension of the field $\mathbb{Z}$ of integers to the field $\mathbb{Q}$ of rational numbers, i.e., to the set of quotients $\frac{m}{n}$, $m, n \in \mathbb{Z}$, $n \neq 0$ with the usual calculation for fractions. The zero element is the number 0, the element e is identical with the number 1. This was the model for the general construction. A second example is the extension of the ring of polynomials on the real line to the field of rational functions, whose elements are quotients of polynomials.

In operational calculus one considers rings of operators generated by a differential operator. Here we choose the operator $D = \frac{d}{dx}$. We start with a ring of functions defined on the interval $[0, \infty)$ with the usual addition and multiplication of functions. We consider functions of the space L, which are measurable on every finite interval $[0, a]$, $a \in \mathbb{R}_+$. The zero element 0 is the function that is zero almost everywhere on $[0, \infty)$, the unit element is the function that has the value 1 on $[0, \infty)$ with the exception of a set of points of measure zero. The set L considered as a commutative ring contains the product λf as a usual product of a function f by a number λ or as the product of elements of the ring L. We will later define the product in such a manner that these two products coincide. (For another version see [Mi.7].) Then we have $1 \cdot f = f$, therefore $e = 1$. In the field of quotients of the ring L (which will be constructed later on) the quotient of operators $\frac{\lambda}{\mu}$, where λ and μ are numbers, can then be identified with the usual quotient of numbers. The operator of integration (denoted by $\frac{1}{p}$) of an element $f \in L$ will be defined by means of

$$\left(\frac{1}{p}f\right)(t) = \int_0^t f(u)du. \tag{2.1.1}$$

Later one can see that this somewhat strange notation makes sense. $\frac{1}{p}$ is an operator of L into L. Therefore, one can consider powers of this operator. Obviously,

$$\left(\left(\frac{1}{p}\right)^2 f\right)(t) = \int_0^t \int_0^u f(v)dvdu = \int_0^t (t-v)f(v)dv$$

and in general

$$\left(\frac{1}{p}\right)^n f = \frac{1}{p^n}f$$

and from elementary calculus we know that

$$\left(\frac{1}{p^n}f\right)(t) = \frac{1}{(n-1)!}\int_0^t (t-v)^{n-1}f(v)dv, \qquad n \in \mathbb{N},$$

and this can be written in the form

$$\left(\frac{1}{p^n}f\right)(t) = \frac{d}{dt}\int_0^t \frac{(t-v)^n}{n!}f(v)dv, \qquad n \in \mathbb{N}_0. \tag{2.1.2}$$

The product of $\frac{1}{p^n}$ with $f \in L$ is derivative of the convolution of the function $\frac{t^n}{n!}$ with the function f (see formula 1.4.25) and therefore sometimes one describes the relation between

$$\frac{1}{p^n} \quad \text{and} \quad \frac{t^n}{n!}$$

by

$$\frac{1}{p^n} = \frac{t^n}{n!}. \tag{2.1.3}$$

This notation will be understood later, see section 2.3.1.

Formula (2.1.3) can be considered as a particular case of formula (2.1.2), if $f(t) = 1$, $\quad 0 \le t < \infty$. Let us consider formula (2.1.3) in more detail. Obviously, according to (2.1.3), we have for $m, n \in \mathbb{N}_0$

$$\lambda \frac{1}{p^n} = \frac{\lambda t^n}{n!}, \qquad \frac{1}{p^m} + \frac{1}{p^n} = \frac{t^m}{m!} + \frac{t^n}{n!}. \tag{2.1.4}$$

On the one hand

$$\frac{1}{p^{m+n}} = \frac{t^{m+n}}{(m+n)!} \tag{2.1.5}$$

and on the other hand from (2.1.3) and (2.1.2) we obtain for an arbitrary function $f \in L$, $m, n \in \mathbb{N}$

$$\frac{1}{p^m}\left(\frac{1}{p^n}f\right)(t) = \frac{d}{dt}\int_0^t \frac{(t-v)^m}{m!}\frac{d}{dv}\int_0^v \frac{(v-u)^n}{n!}f(u)dudv$$

$$= \frac{d}{dt}\int_0^t \frac{(t-v)^m}{m!}\int_0^v \frac{(v-u)^{n-1}}{(n-1)!}f(u)dudv \tag{2.1.6}$$

$$= \frac{d}{dt}\int_0^t f(u)\int_u^t \frac{(t-v)^m(v-u)^{n-1}}{m!(n-1)!}dudv,$$

after interchanging the order of integration. Substituting $v - u = \xi$ we get

$$\frac{1}{p^m}\left(\frac{1}{p^n}f\right)(t) = \frac{d}{dt}\int_0^t f(u)\int_0^{t-u} \frac{(t-u-\xi)^m\xi^{n-1}}{m!(n-1)!}d\xi. \tag{2.1.7}$$

Taking into account that

$$\int_0^t \frac{(t-\xi)^m\xi^{n-1}}{m!(n-1)!}d\xi = \int_0^t \frac{(t-\xi)^{n-1}\xi^m}{m!(n-1)!}d\xi = \frac{t^{m+n}}{(m+n)!}$$

from (2.1.7) we obtain

$$\frac{1}{p^m}\left(\frac{1}{p^n}f\right)(t) = \frac{d}{dt}\int_0^t \frac{(t-u)^{m+n}}{(m+n)!}f(u)du.$$

Therefore, the product of operators $\frac{1}{p^m}$ and $\frac{1}{p^n}$ is adjoint the function

$$\frac{t^{m+n}}{(m+n)!}$$

and, therefore (see formula 2.1.5), we have

$$\frac{1}{p^m}\frac{1}{p^n} = \frac{1}{p^{m+n}}. \tag{2.1.8}$$

Formula (2.1.8) can be rewritten in the form

$$\frac{1}{p^m}\frac{1}{p^n} = \frac{d}{dt}\int_0^t \frac{(t-\xi)^m\xi^n}{m!n!}d\xi. \tag{2.1.9}$$

On the left is the product of operators, and on the right is some type of product of the adjointed function, in fact, it is the definition of the convolution of these functions. We denote this product by a bold star:

$$\frac{t^m}{m!} \star \frac{t^n}{n!} = \frac{d}{dt} \int_0^t \frac{(t-\xi)^m \xi^n}{m! n!} d\xi. \tag{2.1.10}$$

Later we will generalize this notation to the arbitrary functions $f, g \in L$. So we can include more general functions, than power functions $\frac{t^m}{m!}$, and therefore also more general functions of the operator $\frac{1}{p}$. Formula (2.1.10) by means of (2.1.9), (2.1.8), and (2.1.5) can be written as

$$t^m \star t^n = \frac{m! n!}{(m+n)!} t^{m+n}. \tag{2.1.11}$$

Finally in the next chapter we will start to construct the field of operators by the set L of functions. One also could choose some other ring, for example, the ring of continuous functions on $[0, \infty)$, and one arrives at the same field of operators (see Mikusiński [Mi.7]).

2.2   Titchmarsh's Theorem

First we recall the definition of *the convolution of two functions f and g, defined on $[0, \infty)$* (see section 1.4.3, formula 1.4.25) namely,

$$h(t) = (f * g)(t) = \int_0^t f(t-u)g(u)du. \tag{2.2.1}$$

If f and g are continuous in the interval $[0, \infty)$, then the function

$$\varphi(u) = f(t-u)g(u)$$

is continuous on the segment $0 \leq u \leq t$. Therefore, integral (2.2.1) exists. It is easy to prove that the function h is also continuous on $[0, \infty)$.

In addition to Example 1.4.25 and Example 1.4.26 in section 1.4.3 we consider three more examples.

Example 2.2.46 *Let $f(t) = e^{\alpha t}$, $g(t) = e^{\beta t}$. Then*

$$h(t) = \int_0^t e^{\alpha(t-u)} e^{\beta u} du = e^{\alpha t} \int_0^t e^{(\beta-\alpha)u} du,$$

and hence, we obtain

$$\int_0^t e^{\alpha(t-u)} e^{\beta u} du = \begin{cases} \frac{e^{\beta t} - e^{\alpha t}}{\beta - \alpha}, & \text{if } \alpha \neq \beta \\ t e^{\alpha t}, & \text{if } \alpha = \beta. \end{cases} \tag{2.2.2}$$

The operation of convolution (2.2.1) may be applied not only to continuous functions. It may be shown that the convolution of two locally integrable functions on $\mathbb{R}_+$ is also locally integrable on $\mathbb{R}_+$ and

$$\int_0^a |h(t)|dt \leq \int_0^a |f(t)|dt \int_0^a |g(t)|dt,$$

for any $a \in \mathbb{R}_+$.

Example 2.2.47 *Let $f(t) = g(t) = t^{-3/4}$. Obviously, at the point $t = 0$ these functions are discontinuous. From Example 1.4.26, section 1.4.3, it follows that their convolution has the form*

$$\int_0^t (t-u)^{-\frac{3}{4}} u^{-\frac{3}{4}} du = \frac{\Gamma^2\left(\frac{1}{4}\right)}{\Gamma\left(\frac{1}{2}\right)} \frac{1}{\sqrt{t}}$$

and it has also a discontinuity at the point $t = 0$.

Example 2.2.48 *Let us define a family of functions $\eta(t; \lambda)$ depending on the parameter λ by the condition*

$$\eta(t; \lambda) = \begin{cases} 0, & \text{if } 0 \leq t < \lambda, \\ 1, & \text{if } \lambda \leq t. \end{cases}$$

Obviously, the parameter λ varies in the bounds $0 \leq \lambda < \infty$. In the case of $\lambda = 0$ we have $\eta(t; 0) = \eta(t) = 1$ for all $t \geq 0$. Let us find the convolution of the functions $\eta(t; \lambda)$ and $\eta(t; \mu)$. We have

$$h(t) = \int_0^t \eta(t-u; \lambda)\eta(u; \mu)du.$$

Since $\eta(u; \mu) = 0$ for $u < \mu$ and $\eta(u; \mu) = 1$ for $u \geq \mu$, we have $h(t) = 0$ for $t < \mu$. For $t \geq \mu$ we have $h(t) = \int_\mu^t (t-u; \lambda)du$. On putting $t - u = \xi$, $du = -d\xi$, we find for $t > \mu$ the equality $h(t) = \int_0^{t-\mu} \eta(\xi; \lambda)d\xi$.

Reasoning similarly, we deduce $h(t) = 0$ for $t - \mu < \lambda$. For $t - \mu \geq \lambda$ we have $h(t) = \int_\lambda^{t-\mu} d\xi$. Thus, the desired convolution has the form

$$h(t) = \int_0^t \eta(t-u; \lambda)\eta(u; \mu)du = \begin{cases} 0, & \text{for } t < \lambda + \mu, \\ t - \lambda - \mu, & \text{for } \lambda + \mu \leq t, \end{cases}$$

or

$$\int_0^t \eta(t-u; \lambda)\eta(u; \mu)du = \int_0^t \eta(u; \lambda + \mu)du. \qquad (2.2.3)$$

Properties of the Convolution

1. *Commutativity:*

$$f * g = g * f.$$

To prove this we change the variable of integration in the first integral, assuming $t - u = \xi$, then $du = -d\xi$. We obtain the relation

$$\int_0^t f(t-u)g(u)du = -\int_t^0 f(\xi)g(t-\xi)d\xi = \int_0^t g(t-\xi)f(\xi)d\xi.$$

2. *Associativity:*

$$(f * g) * h = f * (g * h).$$

For the proof let us recall one more formula from the theory of multiple integrals. If $f(x, y)$ is an arbitrary function integrable in the triangular region T limited by the lines

$$y = a, \quad x = b \quad \text{and} \quad y = x,$$

then the following relation holds:

$$\int_a^b dx \int_a^x f(x,y)dy = \int_a^b dy \int_y^b f(x,y)dx, \tag{2.2.4}$$

which is often called the *Dirichlet formula* for double integrals. To prove formula (2.2.4) it is sufficient to note that both of the iterated integrals in (2.2.4) are equal to the double integral

$$\iint_T f(x,y)dxdy,$$

calculated on the triangular region T. The reader can easily check this fact, applying the formulae reducing the double integral to the iterated one.

Now we consider the integral

$$\int_0^{t-u} f(t-u-v)g(v)dv.$$

Making the change in the variable of integration by the formula $v = w - u$, $dv = dw$, we have

$$\int_0^{t-u} f(t-u-v)g(v)dv = \int_u^t f(t-w)g(w-u)dw;$$

hence,

$$\int_0^t \left[\int_0^{t-u} f(t-u-v)g(v)dv\right] h(u)du = \int_0^t h(u)du \int_u^t f(t-w)g(w-u)dv.$$

Applying Dirichlet's formula (2.2.4), where $a = 0$, $b = t$, $y = u$, $x = w$, we find

$$\int\limits_0^t \left[\int\limits_0^{t-u} f(t - u - v)g(v)dv \right] h(u)du = \int\limits_0^t f(t - w)dw \int\limits_0^w g(w - u)h(u)du,$$

as was to be proved.

3. *Distributivity:*

$$(f + g) * h = f * h + g * h.$$

4. *Multiplication of the convolution by a number:*

$$\lambda(f * g) = (\lambda f) * g = f * (\lambda g).$$

5. *If $f, g \in L$ and their convolution $f * g$ vanishes on $\mathbb{R}_+$, then at least one of these functions vanishes (a.e.) on $\mathbb{R}_+$, we can say: if the convolution of two functions is equal to zero, then at least one of these functions is equal to zero.* In the general case, without any additional conditions on the functions, property 5 was first proved by Titchmarsh; therefore, property 5 is often called Titchmarsh's theorem. We first prove property 5 for a special case, when the Laplace integrals for the functions f and g converge absolutely.

Let $(f * g)(t) = 0$ for all $t \in [0, \infty)$. By the assumption there exists a number $\gamma \in \mathbb{R}$ such that the Laplace integrals

$$F(z) = \int\limits_0^\infty f(t)e^{-zt}dt \quad \text{and} \quad G(z) = \int\limits_0^\infty g(t)e^{-zt}dt$$

converge absolutely in the region H_γ; see 1.4.1, Theorem 1.4.8. By virtue of the convolution theorem for the Laplace transform, section 1.4.3, Theorem 1.4.11, the product FG is the Laplace transform of the function $f * g$, which vanishes on $\mathbb{R}_+$, whence $F(z)G(z) = 0$ in the region H_γ. The functions F and G are analytic in the region H_γ; hence, if $G(z)$ does not vanish identically, then there exists a point $z_0 \in H_\gamma$ and $G(z_0) \neq 0$. Then in a sufficiently small neighborhood of z_0 the function $G(z)$ is not equal to zero, whence in this neighborhood $F(z)$ vanishes everywhere, and because $F(z)$ is an analytic function it follows that $F(z) = 0$ for all z, such that $z \in H_\gamma$. On applying Theorem 1.4.9 of section 1.4.1, we conclude that $f(t) = 0$ at every point t of the interval $(0, \infty)$, where $f(t)$ is continuous. In particular, if f is a continuous function in $[0, \infty)$, then $f(t) = 0$ for all $t \in [0, \infty)$.

Now we are going to formulate and prove Titchmarsh's theorem.

Theorem 2.2.82 *If $f, g \in C[0, \infty)$ and their convolution*

$$(f * g)(t) = 0, \qquad 0 \leq t < \infty, \tag{2.2.5}$$

then at least one of these functions vanishes everywhere in the interval $[0, \infty)$.

Remark 2.2.97 *The condition of continuity of the functions f and g on $[0,\infty)$ is not essential.*

Indeed, let

$$f_1(t) = \int\limits_0^t f(u)du, \quad g_1(t) = \int\limits_0^t g(u)du.$$

On integrating (2.2.5), we obtain

$$\int\limits_0^t \int\limits_0^\xi f(\xi - u)g(u)\, du\, d\xi = C.$$

But the left-hand side for $t = 0$ is equal to zero; hence,

$$\int\limits_0^t \int\limits_0^\xi f(\xi - u)g(u)\, du\, d\xi = 0.$$

From this, applying (2.2.4), we find

$$\int\limits_0^t g(u)du \int\limits_u^t f(\xi - u)d\xi = 0;$$

putting $\xi - u = \eta$, $d\xi = d\eta$, we have

$$\int\limits_u^t f(\xi - u)d\xi = \int\limits_0^{t-u} f(\eta)d\eta = f_1(t-u).$$

Hence, we have

$$\int\limits_0^t f_1(t-u)g(u)du = 0,$$

or

$$\int\limits_0^t g(t-u)f_1(u)du = 0.$$

Integrating the latter relation once more, we find

$$\int\limits_0^t f_1(t-u)g(u)du = 0, \qquad 0 \le t < \infty. \tag{2.2.6}$$

If the theorem holds for continuous functions, this immediately implies its validity for locally integrable functions on $\mathbb{R}_+$. In order to prove Titchmarsh's theorem we need a series of lemmas.

If a sequence of functions $f_n(x)$ converges uniformly for $a \le x \le b$, then it is known that

$$\lim_{n\to\infty} \int\limits_a^b f_n(x)dx = \int\limits_a^b \lim_{n\to\infty} f_n(x)dx.$$

However, the uniform convergence on $[a, b]$ is only a sufficient condition, supplying a possibility of permutation of the operations of integration and passage to the limit. If $f_n(x)$ are continuous functions and the limit $f(x) = \lim\limits_{n \to \infty} f_n(x)$ is discontinuous, then the convergence is necessary not uniformly and the question about the passage to the limit under the sign of the integral requires further investigation.

Lemma 2.2.30 *If*

$$\lim_{n \to \infty} f_n(x) = f(x), \quad a \le x \le b$$

and

1) *the function $f(x)$ has a finite number of points of discontinuity;*

2) *there exists a number Q such that for all $a \le x \le b$ and $n = 1, 2, 3, \ldots$*

$$|f_n(x)| \le Q;$$

3) *the sequence $(f_n(x))_{n\ in\mathbb{N}}$ converges to $f(x)$ uniformly on the segment $[a, b]$ except arbitrary small neighborhoods of points of discontinuity of the limit function $f(x)$, then the following relation holds:*

$$\lim_{n \to \infty} \int_a^b f_n(x)dx = \int_a^b f(x)dx. \tag{2.2.7}$$

Proof. It is sufficient to prove the lemma for the special case, when $[a, b]$ contains only one point of discontinuity of f. The general case may be reduced to this one by partition of the interval of integration into a finite number of intervals, each of which contains only one point of discontinuity of f. Thus, let the function $f(x)$ have a gap at $x = t$. Let $a < t < b$. The cases when $t = a$, or $t = b$ are considered similarly. Let $\varepsilon > 0$ be an arbitrary sufficiently small number and $\delta = \frac{\varepsilon}{8Q}$. Let us take as a neighborhood of the point t the interval $(t - \delta, t + \delta)$. We have

$$\int_a^b f_n(x)dx - \int_a^b f(x)dx = \int_a^{t-\delta} f_n(x)dx - \int_a^{t-\delta} f(x)dx + \int_{t+\delta}^b f_n(x)dx$$
$$- \int_{t+\delta}^b f(x)dx + \int_{t-\delta}^{t+\delta} f_n(x)dx - \int_{t-\delta}^{t+\delta} f(x)dx.$$

Under this assumption $|f(x)| \le Q$, therefore, we have the inequalities

$$\left| \int_{t-\delta}^{t+\delta} f(x)dx \right| \le 2\delta Q \quad \text{and} \quad \left| \int_{t-\delta}^{t+\delta} f_n(x)dx \right| \le 2\delta Q.$$

Under this assumption, the sequence $(f_n(x))_{n \in \mathbb{N}}$ converges to $f(x)$ uniformly in the intervals $a \leq x \leq t - \delta$ and $t + \delta \leq x \leq b$. The uniform convergence implies the relations

$$\lim_{n \to \infty} \int_a^{t-\delta} f_n(x)dx = \int_a^{t-\delta} f(x)dx;$$

$$\lim_{n \to \infty} \int_{t+\delta}^b f_n(x)dx = \int_{t+\delta}^b f(x)dx.$$

Therefore, there exists a number n_0 such that

$$\left| \int_a^{t-\delta} f_n(x)dx - \int_a^{t-\delta} f(x)dx \right| < \frac{\varepsilon}{4},$$

$$\left| \int_{t+\delta}^b f_n(x)dx - \int_{t+\delta}^b f(x)dx \right| < \frac{\varepsilon}{4}$$

for all $n \geq n_0$. Now for $n \geq n_0$ we have the inequality

$$\left| \int_a^b f_n(x)dx - \int_a^b f(x)dx \right| \leq \frac{\varepsilon}{4} + \frac{\varepsilon}{4} + 2\delta Q + 2\delta Q = \frac{\varepsilon}{2} + 4\delta Q.$$

On putting $\delta = \frac{\varepsilon}{8Q}$, we finally have

$$\left| \int_a^b f_n(x)dx - \int_a^b f(x)dx \right| \leq \varepsilon \quad \text{for all} \quad n \geq n_0.$$

$\square$

Lemma 2.2.31 *If $f \in C[0, T)$, then there exists the limit*

$$\lim_{n \to \infty} \sum_{k=0}^{\infty} \frac{(-1)^k}{k!} \int_0^T e^{-kn(t-u)} f(u)du = \int_0^t f(u)du \qquad (2.2.8)$$

for any $t \in [0, T)$.

Proof. Let $0 \leq t < T$ and

$$Q = \max_{0 \leq t \leq T} |f(t)|.$$

Let us consider the series

$$\sum_{k=0}^{\infty} \frac{(-1)^k}{k!} e^{-nk(t-u)} f(u). \qquad (2.2.9)$$

Let us fix the numbers $n > 0$ and t. The general term of the series satisfies the inequality

$$\left| \frac{(-1)^k}{k!} e^{-nk(t-u)} f(u) \right| \leq \frac{Q e^{nkT}}{k!}.$$

But the series

$$\sum_{k=0}^{\infty} \frac{e^{nkT}}{k!}$$

converges and its terms do not depend on the variable u; hence, series (2.2.9) converges uniformly in the region $0 \leq u \leq T$ and

$$\sum_{k=0}^{\infty} \frac{(-1)^k}{k!} \int_0^T e^{-nk(t-u)} f(u) du = \int_0^T \left(\sum_{k=0}^{\infty} \frac{(-1)^k}{k!} e^{-nk(t-u)} \right) f(u) du.$$

We have the relation

$$\sum_{k=0}^{\infty} \frac{(-1)^k}{k!} e^{-nk(t-u)} = e^{-e^{-n(t-u)}};$$

hence

$$\sum_{k=0}^{\infty} \frac{(-1)^k}{k!} \int_0^T e^{-nk(t-u)} f(u) du = \int_0^T e^{-e^{-n(t-u)}} f(u) du.$$

Taking account of the relation

$$\lim_{n \to \infty} e^{-e^{-n(t-u)}} = \begin{cases} 1 & \text{for } u < t, \\ e^{-1} & \text{for } u = t, \\ 0 & \text{for } u > t \end{cases}$$

for the sequence $f_n(u) = e^{-e^{-n(t-u)}} f(u)$ we have the relations

$$\lim_{n \to \infty} f_n(u) = f(u), \quad \text{if} \quad 0 \leq u < t,$$

and

$$\lim_{n \to \infty} f_n(u) = 0, \quad \text{for} \quad u > t.$$

Let $\varphi_n(u) = e^{-e^{-n(t-u)}}$; then, obviously, we have

$$\varphi_n'(u) = -n e^{-e^{-n(t-u)}} e^{-n(t-u)} < 0.$$

Therefore, the function $\varphi_n(u)$ decreases when the variable u increases. If $\delta > 0$ is sufficiently small, and $0 \leq u \leq t - \delta$, then $1 - \varphi_n(u) \leq 1 - \varphi_n(t - \delta)$. For $t + \delta \leq u \leq T$ we have $\varphi_n(u) \leq \varphi_n(t + \delta)$. Therefore, the inequalities follow:

$$|f(u) - f_n(u)| \leq Q(1 - \varphi_n(t - \delta)) \quad \text{for} \quad 0 \leq u \leq t - \delta,$$

$$|f_n(u)| \leq Q\varphi_n(t + \delta) \quad \text{for} \quad t + \delta \leq u \leq T.$$

These inequalities imply the uniform convergence of the sequence $(f_n(u))_{n \in \mathbb{N}}$ on the segments $[0, t - \delta]$ and $[t + \delta, T]$. Hence, the third condition of Lemma 2.2.30 is also fulfilled. Thus, we have the relation

$$\lim_{n \to \infty} \sum_{k=0}^{\infty} \frac{(-1)^k}{k!} \int_0^T e^{-nk(t-u)} f(u) \, du = \int_0^T \lim_{n \to \infty} e^{-e^{-n(t-u)}} f(u) \, du = \int_0^T f(u) \, du.$$

□

Lemma 2.2.32 *If $f \in C[0,T)$ and there exists a number Q such that the inequality*

$$\left| \int_0^T e^{nt} f(t)dt \right| \leq Q \tag{2.2.10}$$

holds for all $n \in \mathbb{N}_o$, then $f(t) = 0$ on the whole interval $[0,T]$.

Proof. Lemma 2.2.31 implies

$$\lim_{n \to \infty} \sum_{k=0}^{\infty} \frac{(-1)^k e^{-nkt}}{k!} \int_0^T e^{nku} f(u)du = \int_0^t f(u)du. \tag{2.2.11}$$

On the other hand, condition (2.2.10) implies

$$\left| \sum_{k=1}^{\infty} \frac{(-1)^k e^{-nkt}}{k!} \int_0^T e^{nku} f(u)du \right| \leq Q \sum_{k=1}^{\infty} \frac{e^{-nkt}}{k!} = Q(e^{e^{-nt}} - 1) \to 0$$

for $n \to \infty$. From (2.2.11) we have for $n \to \infty$

$$\int_0^T f(u)\,du = \int_0^t f(u)du, \qquad 0 \leq t < T,$$

therefore,

$$\int_0^t f(u)\,du = 0 \quad \text{for all} \quad t \in [0,T];$$

hence,

$$f(t) = 0 \qquad \text{for} \qquad t \in [0,T].$$

$\square$

Lemma 2.2.33 *If $f \in C[0,T]$ and*

$$\int_0^T t^n f(t)dt = 0; \qquad n = 0,1,2,\ldots,$$

then $f(t) = 0$ on the whole interval $[0,T]$.

Proof. Let $t = \alpha x$; we have

$$\alpha^{n+1} \int_0^{\frac{T}{\alpha}} x^n f(\alpha x)\,dx = 0.$$

Suppose that $\alpha > 0$ is small enough in order for $\frac{T}{\alpha} > 1$. Then the relation holds

$$\int_1^{\frac{T}{\alpha}} x^n f(\alpha x)dx = - \int_0^1 x^n f(\alpha x)dx,$$

therefore,

$$\left| \int_{1}^{\frac{T}{\alpha}} x^n f(\alpha x) dx \right| \leq \int_{0}^{1} |f(\alpha x)| dx = Q.$$

On putting $x = e^\xi$, $dx = e^\xi d\xi$, we find

$$\left| \int_{0}^{\ln \frac{T}{\alpha}} e^{n\xi} f(\alpha e^\xi) e^\xi d\xi \right| \leq Q, \qquad n = 0, 1, 2, \ldots.$$

Applying Lemma 2.2.32, we conclude that $f(\alpha e^\xi) = 0$ for all $0 \leq \xi \leq \ln \frac{T}{\alpha}$, or $f(t) = 0$ for $\alpha \leq t \leq T$. Since the function $f(t)$ is continuous on $[0, T]$ and $\alpha > 0$ is an arbitrary sufficiently small number, we have $f(t) = 0$ for $0 \leq t \leq T$.   ▯

Lemma 2.2.34 *If $h(t) = (f * g)(t)$, then the relation holds*

$$\int_{0}^{T} h(t)e^{-zt}dt = \int_{0}^{T} f(t)e^{-zt}dt \int_{0}^{T} g(t)e^{-zt}dt - e^{-Tz} \int_{0}^{T} e^{-zt}dt \int_{t}^{T} f(t+T-\xi)g(\xi)d\xi.$$

Proof. We have

$$H_T(z) = \int_{0}^{T} h(t)e^{-zt}dt = \int_{0}^{T} dt \int_{0}^{t} e^{-z(t-u)-zu} f(t-u)g(u)du.$$

Let us make the change of variables in the double intergral. On putting $t - u = x$, $u = y$, then $0 < x + y < T$, $y > 0$, $c > 0$. Hence, the new domain of integration is the triangle ABO (Figure 6) and

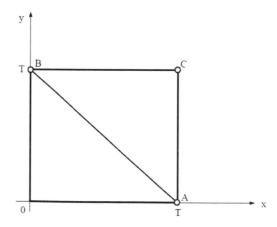

Figure 6

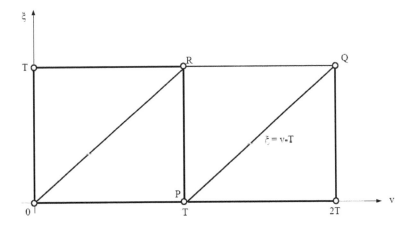

Figure 7

$$H_T(z) = \int_0^T e^{-zx} f(x)dx \int_0^{T-x} g(y)e^{-zy}dy = F_T(z)G_T(z) - \int_0^T e^{-zx} f(x)dx \int_{T-x}^T g(y)e^{-zy}dy.$$

Here

$$F_T(z) = \int_0^T f(t)e^{-zt}dt \quad \text{and} \quad G_T(z) = \int_0^T g(t)e^{-zt}dt,$$

and the domain of integration is the triangle ABC (see Figure 6). On changing in the last double integral the variables of integration by the formulae $x + y = v$, $y = \xi$, we find that $0 < v - \xi < T$, $v > T$, $\xi < T$, and the new domain of integration is the triangle PQR (Figure 7); therefore, the following formula holds:

$$H_T(z) = F_T(z)G_T(z) - \int_T^{2T} e^{-zv}dv \int_{v-T}^T f(v - \xi)g(\xi)d\xi.$$

Finally, on putting $v = t + T$, $dv = dt$, we have

$$H_T(z) = F_T(z)G_T(z) - \int_0^T e^{-z(t+T)}dt \int_t^T f(t + T - \xi)g(\xi)d\xi. \qquad (2.2.12)$$

□

Proof of Titchmarsh's theorem for the case: $f = g$

Let f be a continuous function in the region $0 \le t < \infty$ and

$$h(t) = (f * f)(t) = 0 \qquad \text{for} \quad 0 \le t < \infty. \qquad (2.2.13)$$

Applying Lemma 2.2.34, we conclude that

$$\left(\int_0^T f(t)e^{-zt}dt \right)^2 = e^{-Tz} \int_0^T e^{-zt}dt \int_t^T f(t+T-\xi)f(\xi)d\xi,$$

where $T > 0$ is an arbitrary fixed number. If we put $z = n$ and denote $\varphi(t) = \int_t^T f(t+T-\xi)f(\xi)d\xi$, then

$$e^{nT}\left(\int_0^T f(t)e^{-nt}dt \right)^2 = \int_0^T e^{-nt}\varphi(t)dt,$$

or

$$\left(\int_0^T f(t)e^{n(\frac{T}{2}-t)}dt \right)^2 = \int_0^T e^{-nt}\varphi(t)dt.$$

Hence,

$$\left| \int_0^T f(t)e^{n(\frac{T}{2}-t)}dt \right|^2 \leq \int_0^T e^{-nt}|\varphi(t)|\,dt \leq \int_0^T |\varphi(t)|dt;$$

therefore,

$$\left| \int_0^T f(t)e^{n(\frac{T}{2}-t)}dt \right| \leq \sqrt{\int_0^T |\varphi(t)|dt}.$$

From this inequality we find

$$\left| \int_0^{T/2} f(t)e^{n(\frac{T}{2}-t)}dt \right| - \left| \int_{T/2}^T f(t)e^{n(\frac{T}{2}-t)}dt \right| \leq \sqrt{\int_0^T |\varphi(t)|dt},$$

or taking into account that $e^{n(\frac{T}{2}-t)} \leq 1$, for $t \geq \frac{T}{2}$,

$$\left| \int_0^{T/2} f(t)e^{n(\frac{T}{2}-t)}dt \right| \leq \sqrt{\int_0^T |\varphi(t)|dt} + \int_{T/2}^T |f(t)|dt = Q.$$

Finally, on putting $\frac{T}{2} - t = \xi$, we find

$$\left| \int_0^{T/2} f\left(\frac{T}{2} - \xi\right)e^{n\xi}d\xi \right| \leq Q, \qquad n = 0,1,2,\ldots.$$

By virtue of Lemma 2.2.32 we deduce from here that

$$f\left(\frac{T}{2} - \xi\right) = 0 \qquad \text{for} \qquad 0 \leq \xi \leq \frac{T}{2},$$

or

$$f(t) = 0 \qquad \text{for} \qquad 0 \leq t \leq \frac{T}{2};$$

and since $T > 0$ is an arbitrary number, hence, $f(t) = 0$ for all $t \geq 0$.

Proof of Titchmarsh's theorem in the general case. Let f and g be continuous functions for $0 \leq t < \infty$ and

$$(f * g)(t) = 0 \qquad \text{for all} \qquad 0 \leq t < \infty. \tag{2.2.14}$$

Then we have

$$\int_0^t (t-u)f(t-u)g(u)du + \int_0^t f(t-u)ug(u)du = t\int_0^t f(t-u)g(u)du = 0.$$

On denoting $f_1(t) = tf(t)$ and $g_1(t) = tg(t)$, we rewrite the previous relation in the form

$$\int_0^t f_1(t-u)g(u)du + \int_0^t f(t-u)g_1(u)du = 0,$$

i.e.,

$$f_1 * g + f * g_1 = 0,$$

which implies

$$(f * g_1) * (f_1 * g + f * g_1) = 0,$$

or, using the properties of convolution, we obtain

$$f * g_1 * f_1 * g + f * g_1 * f * g_1 = (f * g) * (f_1 * g_1) + (f * g_1) * (f * g_1) = 0.$$

But $f * g = 0$, therefore, $(f * g_1) * (f * g_1) = 0$. By virtue of Titchmarsh's theorem proven for the case $f = g$, we have $f * g_1 = 0$, or

$$\int_0^t f(t-u)ug(u)du = 0 \tag{2.2.15}$$

for all $t \geq 0$. Thus, (2.2.14) implies (2.2.15).

Suppose that

$$\int_0^t f(t-u)u^n g(u)du = 0, \qquad t \geq 0. \tag{2.2.16}$$

In the same way that we obtained (2.2.15) from (2.2.14), we find from (2.2.16) that

$$\int_0^t f(t-u)u^{n+1}g(u)du = 0 \qquad \text{for all} \qquad t \geq 0. \tag{2.2.17}$$

Thus, (2.2.16) holds for all $n \geq 0$. By Lemma 2.2.33 we deduce from here that

$$f(t-u)g(u) = 0 \qquad \text{for all} \qquad 0 \leq u \leq t < \infty. \tag{2.2.18}$$

If there exists a point $u_0 \in [0, \infty)$ such that $g(u_0) \neq 0$, then (2.2.18) implies $f(t - u_0) = 0$, $u_0 \leq t < \infty$, i.e. $f(t) = 0$ for all $t \geq 0$.

2.3 Operators

2.3.1 Ring of Functions

Let us denote by M the set of all functions defined in the region $0 \le t < \infty$, differentiable in this region and whose derivative belongs to the set $L_1^{loc}(\mathbb{R}_+) = L$. Then every function F belonging to the set M may be represented in the form

$$F(t) = F(0) + \int_0^t f(u)du, \text{ where } f \in L.$$

Conversely, if $F(t)$ has the form

$$\int_0^t f(u)\,du + \lambda, \text{ where } f \in L \text{ and } \lambda \text{ is a number, then } F \in M.$$

Obviously, M is contained in the set L and is a linear set with respect to the ordinary operations of addition of functions and multiplication of a function by a number.

Lemma 2.3.35 *If $F \in M$ and $g \in L$, then the convolution H of these functions, defined by*

$$H(t) = \int_0^t F(t - u)g(u)\,du,$$

belongs to the set M.

Proof. The condition $F \in M$ implies that

$$F(t) = \int_0^t f(u)\,du + F(0), \text{ where } f \in L.$$

Let

$$h(t) = \int_0^t f(t - v)g(v)\,dv,$$

then

$$\int_0^t h(u)du = \int_0^t du \int_0^u f(u - v)g(v)dv.$$

On changing the order of integration in the double integral, we obtain

$$\int_0^t h(u)du = \int_0^t g(v)dv \int_v^t f(u - v)du,$$

or, on putting $u - v = \xi, \quad du = d\xi$, we have

$$\int\limits_0^t h(u)\, du = \int\limits_0^t g(v)dv \int\limits_0^{t-v} f(\xi)d\xi.$$

But $\int\limits_0^t f(\xi)d\xi = F(t) - F(0)$; therefore, the following equality holds

$$\int\limits_0^t h(u)du = \int\limits_0^t F(t-v)g(v)dv - F(0)\int\limits_0^t g(v)dv,$$

and hence,

$$H(t) = \int\limits_0^t F(t-v)g(v)dv = F(0)\int\limits_0^t g(v)dv + \int\limits_0^t h(u)du. \qquad (2.3.1)$$

Both functions on the right-hand side of the latter equation belong to M; hence, $H \in M$. $\Box$

Corollary 2.3.30 *If the functions $F \in M$ and $G \in M$, then there exists the derivative*

$$\frac{d}{dt}\int\limits_0^t F(t-u)G(u)du = H(t)$$

and H also belongs to M.

Proof. Indeed, replacing in (2.3.1) the function $g(t)$ by $G(t)$ we obtain

$$\int\limits_0^t F(t-v)G(v)dv = F(0)\int\limits_0^t G(v)dv + \int\limits_0^t H_1(u)du,$$

where

$$H_1(u) = \int\limits_0^u f(u-v)G(v)dv = \int\limits_0^u G(u-v)f(v)dv.$$

Lemma 2.3.35 implies that $H_1 \in M$, therefore, for all $t \geq 0$ there exists the derivative

$$H(t) = \frac{d}{dt}\int\limits_0^t F(t-v)G(v)dv = F(0)G(t) + H_1(t),$$

and obviously, $H \in M$. $\Box$

Let $F \in M$ and $g \in L$. According to Lemma 2.3.35 the convolution of these functions belongs to M; hence, this convolution is differentiable. Let us introduce the notation

$$h(t) = \frac{d}{dt}\int\limits_0^t F(t-u)g(u)du. \qquad (2.3.2)$$

The function h belongs to the set L. This follows from the definition of M. Obviously, (2.3.2) is a linear operator defined on the set L, whose range also belongs to L. This operator is uniquely defined by the choice of the function F. For instance, if $F(t) = t$, then we obtain

$$h(t) = \frac{d}{dt} \int_0^t (t-u)g(u)du = \int_0^t g(u)du. \tag{2.3.3}$$

Therefore, the function $F(t) = t$ is associated with the integration operator.

Let $G(t) \in M$. Consider two linear operators:

$$h(t) = \frac{d}{dt} \int_0^t F(t-u)g(u)du$$

and

$$q(t) = \frac{d}{dt} \int_0^t G(t-u)g(u)du. \tag{2.3.4}$$

Let us find the product of these operators. In order to do this, we have to compute $\frac{d}{dt} \int_0^t G(t-u)h(u)du$, where $h(u)$ is defined by (2.3.2). We have

$$\int_0^t G(t-u)h(u)du = \int_0^t G(t-u)du \frac{d}{du} \int_0^u F(u-\xi)g(\xi)d\xi$$

$$= \int_0^t G(t-u)\,du \left[F(0)g(u) + \int_0^u F'(u-\xi)g(\xi)d\xi \right]$$

$$= F(0) \int_0^t G(t-u)g(u)\,du + \int_0^t G(t-u)du \int_0^u F'(u-\xi)g(\xi)d\xi.$$

On changing the order of integration in the last integral we find

$$\int_0^t G(t-u)h(u)du = F(0) \int_0^t G(t-u)g(u)du + \int_0^t g(\xi)d\xi \int_\xi^t G(t-u)F'(u-\xi)du.$$

In the second integral we change the variables of integration by means of $u = t - \eta$, $du = -d\eta$:

$$\int_0^t G(t-u)h(u)du = F(0) \int_0^t G(t-u)g(u)du + \int_0^t g(\xi)d\xi \int_0^{t-\xi} F'(t-\xi-\eta)G(\eta)d\eta$$

$$= \int_0^t g(\xi)d\xi \left[F(0)G(t-\xi) + \int_0^{t-\xi} F'(t-\xi-\eta)G(\eta)d\eta \right].$$

Let us introduce the notation

$$K(t) = F(0)G(t) + \int_0^t F'(t-\eta)G(\eta)d\eta = \frac{d}{dt}\int_0^t F(t-\eta)G(\eta)d\eta,$$

then we obtain

$$\frac{d}{dt}\int_0^t G(t-u)h(u)du = \frac{d}{dt}\int_0^t K(t-\xi)g(\xi)d\xi.$$

This implies that the product of operators (2.3.4) is associated with the function

$$K(t) = \frac{d}{dt}\int_0^t F(t-u)G(u)du.$$

Let us introduce in the set M an operation of multiplication. We shall call the function

$$K(t) = F(t) \star G(t) = \frac{d}{dt}\int_0^t F(t-u)G(u)du \qquad (2.3.5)$$

by the product of the functions $F \in M$ and $G \in M$. It was proved (see Corollary 2.3.30) that K also belongs to M. It is easy to check on the basis of properties of the convolution (see 2.2) that the product (2.3.5) satisfies the following properties:

$$F \star G = G \star F \ (commutativity), \qquad (2.3.6)$$

$$F \star (G \star H) = (F \star G) \star H \ (associativity), \qquad (2.3.7)$$

$$F \star (G + H) = F \star G + F \star H \ (distributivity), \qquad (2.3.8)$$

$$\lambda F \star G = \lambda(F \star G). \qquad (2.3.9)$$

Therefore, the linear set M with the above-defined operation of multiplication (2.3.5) is a commutative ring; this ring is called Mikusiński's ring. It was noted that a linear operator (2.3.2) is associated with every function F belonging to M. The ring M is a ring of linear operators. The ring M also contains functions which are constant on $[0, \infty)$. We identify such a function with its constant value, say λ. So we can consider the product $\lambda \star F$, $F \in M$. In the ring M is also defined the product of a function $F \in M$ by a number λ: λF. In our case those two products coincide, i.e., the product (2.3.5) has also the property that

for any function $F \in M$ the relation (2.3.10)

$$\boxed{\lambda \star F = \lambda F},$$

holds, where λ is a number.

The proof follows from the equation

$$\lambda \star F(t) = \frac{d}{dt}\int_0^t \lambda F(u)du = \lambda F(t).$$

Property (2.3.10) in the special case $F(t) = \mu$ for every $t \in [0, \infty)$ yields

$$\lambda \star \mu = \lambda\mu,$$

i.e., the multiplication of numbers in the ring M satisfies the ordinary rules of the arithmetic.

Let us note once more that in the general case the product in this ring differs from the ordinary product of functions. In section 1.4.3, Example 1.4.26 we computed the convolution of the functions t^α and t^β with the result

$$t^\alpha * t^\beta = \frac{\Gamma(\alpha + 1)\Gamma(\beta + 1)}{\Gamma(\alpha + \beta + 2)} t^{\alpha+\beta+1}.$$

Differentiation leads to

$$t^\alpha \star t^\beta = \frac{\Gamma(\alpha + 1)\Gamma(\beta + 1)}{\Gamma(\alpha + \beta + 1)} t^{\alpha+\beta}. \tag{2.3.11}$$

We took here into account the formula

$$\Gamma(\alpha + \beta + 2) = (\alpha + \beta + 1)\Gamma(\alpha + \beta + 1).$$

Let us compute the product $L_n(t) \star L_m(t)$, where $L_n(t)$ and $L_m(t)$ are Laguerre polynomials of degree n and m. The *Laguerre polynomial* of degree n has the form, see formula (1.4.77),

$$L_n(t) = \sum_{k=0}^{n} (-1)^k \binom{n}{k} \frac{t^k}{k!}.$$

Taking into account (2.3.11), we have

$$L_n(t) \star L_m(t) = \left(\sum_{k=0}^{n} (-1)^k \binom{n}{k} \frac{t^n}{k!} \right) \star \left(\sum_{r=0}^{m} (-1)^r \binom{m}{r} \frac{t^r}{r!} \right)$$

$$= \sum_{k=0}^{n} \sum_{r=0}^{m} (-1)^{k+r} \binom{n}{k} \binom{m}{r} \frac{t^k}{k!} \star \frac{t^r}{r!}$$

$$= \sum_{k=0}^{n} \sum_{r=0}^{m} (-1)^{k+r} \binom{n}{k} \binom{m}{r} \frac{t^{k+r}}{(k+r)!}.$$

In the last line we make the change of indices of summation. On putting $k + r = \nu$, we obtain

$$L_n(t) \star L_m(t) = \sum_{\nu=0}^{m+n} (-1)^\nu \frac{t^\nu}{\nu!} \sum_{k=0}^{\nu} \binom{n}{k} \binom{m}{\nu - k}.$$

In order to compute the sum $\Sigma_{k=0}^{\nu} \binom{n}{k} \binom{m}{\nu-k}$ we compare the coefficients for the equal powers of x in the identity

$$\sum_{k=0}^{n} \binom{n}{k} x^k \sum_{r=0}^{m} \binom{m}{r} x^r = \sum_{\nu=0}^{m+n} \binom{m+n}{\nu} x^\nu.$$

We easily find

$$\sum_{k=0}^{\nu} \binom{n}{k} \binom{m}{\nu - k} = \binom{m+n}{\nu},$$

and, therefore, finally, we obtain

$$L_n(t) \star L_m(t) = L_{n+m}(t). \tag{2.3.12}$$

For a proof of (2.3.12) see also [KS.2].

2.3.2 The Field of Operators

In the previous section we introduced the ring M. Let us prove that M has no divisors of zero.

Let $F \in M$, $G \in M$ and let

$$F(t) \star G(t) = \frac{d}{dt} \int_0^t F(t-u)G(u)du = 0 \tag{2.3.13}$$

on $[0, \infty)$. Then the convolution $\int_0^t F(t-u)G(u)\,du$ is equal to a constant for all $t \geq 0$. Setting $t = 0$ we conclude that this constant has to be zero. Applying Titchmarsh's theorem (see 2.2.), we conclude that at least one of the functions F or G vanishes in the interval $[0, \infty)$. Thus, M has no divisors of zero. From algebra it is well known that every commutative ring without divisors of zero may be extended to a field of quotients (see, for example, [BL], Chapter II, 2, Theorem 7). So M may be extended to the quotient field. This field we denote by $\mathfrak{M}(M)$. Below we often shall write simply $\mathfrak{M}$ instead of $\mathfrak{M}(M)$. The elements of $\mathfrak{M}$ we shall call *operators*.

Recall that elements of the field are sets. Every such set consists of mutually equivalent pairs (F, G), $G \neq 0$. An element of the field is denoted by $\frac{F}{G}$. Two pairs (F, G) (F_1, G_1) are called *equivalent* if $F \star G_1 = F_1 \star G$, $\frac{F}{G} = \frac{F_1}{G_1}$ if and only if $F \star G_1 = F_1 \star G$. The sum and product of operators satisfy the usual rules of arithmetic, only the product is computed by formula (2.3.12). Hence,

$$\frac{F}{G} + \frac{F_1}{G_1} = \frac{F \star G_1 + F_1 \star G}{G \star G_1},$$

$$\frac{F}{G} \star \frac{F_1}{G_1} = \frac{F \star F_1}{G \star G_1}.$$

The set of all operators of the field, which may be represented by the form $\frac{F}{1}$, forms a subring of this field isomorphic to the given ring M. Therefore, we shall write F instead of $\frac{F}{1}$, i.e., $\frac{F}{1} = F$. If $F = \lambda$, where λ is a number, then $\frac{\lambda}{1} = \lambda$. In particular, $\frac{1}{1} = 1$ and $\frac{0}{1} = 0$. An expression $\frac{F}{G}$ may be considered as an operation of division in the field $\mathfrak{M}$. The latter is essentially different from the usual operation of division. Only in the case when F and G are constants, $F = \lambda$, $G = \mu$, will the relation $\frac{\lambda}{\mu}$ be equal to the usual fraction of constants, i.e., only for constants does the operation of division coincide with the usual division.

Consider all operators in $\mathfrak{M}$ which may be represented by the form $\frac{F}{I}$, $F(0) = 0$, where I is the function defined by $I(t) = t$, $t \in [0, \infty)$. Obviously, the collection of these elements is a linear set. The function F is differentiable and $F' = f$. Let us associate with the operator $\frac{F}{I}$ the function $F' = f$. If two operators $\frac{F}{I}$, $F(0) = 0$, $\frac{G}{I}$, $G(0) = 0$, are different, then the associated functions, $f = F'$, $g = G'$, are different too. Indeed, if f coincides with g, i.e., $\int_0^t f(u)\,du = \int_0^t g(u)\,du$ for all $t \geq 0$, then $F(t) = G(t)$ for all $t \geq 0$. Then we obtain $\frac{F}{I} = \frac{G}{I}$, and this contradicts the assumption $\frac{F}{I} \neq \frac{G}{I}$. Thus, the correspondence between the set of all operators of the form $\frac{F}{I}$, $F(0) = 0$, and the set of all functions $f = F'$ (F is an arbitrary function of the original ring) is one to one (bijective). This correspondence maps a sum of operators $\frac{F}{I} + \frac{G}{I}$ onto the sum of functions $F' + G'$. This follows from the equations $\frac{F}{I} + \frac{G}{I} = \frac{F+G}{I}$ and $F' + G' = (F + G)'$.

The product of an operator $\frac{F}{I}$ by a number λ is associated with the product of λ by the function $f = F'$. Indeed, $\lambda \frac{F}{I} = \frac{\lambda F}{I}$ and $(\lambda F)' = \lambda F'$. Thus, the linear set of all operators of the form $\frac{F}{I}$, $F(0) = 0$ is isomorphic to the set of all functions $f = F'$. Operators of the field $\mathfrak{M}$ reducible to the form $\frac{F}{I}$, $F(0) = 0$ are called functions and we shall write f instead of $\frac{F}{I}$. Thus,

$$\frac{F}{I} = F' = f, \qquad \text{if} \qquad F(0) = 0.$$

Not every operator is reducible to a function. For instance, the operator $\frac{1}{I}$, obviously, cannot be reduced to the form $\frac{F}{I}$, $F(0) = 0$. Hence, the operator $\frac{1}{I}$ is not reducible to a function. It was said that a sum of functions is always a function. A simple example shows that a product of functions will not always be a function, i.e., that the product of operators $\frac{F}{I}$ and $\frac{G}{I}$, $F(0) = 0$, $G(0) = 0$ is not always reducible to the form $\frac{H(t)}{t}$, $H(0) = 0$. Indeed, let $F(t) = G(t) = \sqrt{t}$; then the following formula holds:

$$\frac{F(t)}{t} \star \frac{G(t)}{t} = \frac{\sqrt{t} \star \sqrt{t}}{t \star t}.$$

By virtue of 2.3.1, (2.3.11) we have

$$\frac{\sqrt{t}}{t} \star \frac{\sqrt{t}}{t} = \frac{\sqrt{t} \star \sqrt{t}}{t \star t} = \frac{\Gamma^2\left(\frac{3}{2}\right)t}{\Gamma(2)t \star t} = \frac{\pi}{4t},$$

because $\Gamma\left(\frac{3}{2}\right) = \Gamma\left(1 + \frac{1}{2}\right) = \frac{1}{2}\Gamma\left(\frac{1}{2}\right) = \frac{\sqrt{\pi}}{2}$ and $\Gamma(2) = 1$. Consequently, in general a product of functions is an operator.

Theorem 2.3.83 *The product of functions $\frac{F}{I} = f$ and $\frac{G}{I} = g$ is a function if and only if the convolution of the functions f and g belongs to the original ring M and vanishes at the origin.*

Proof. Let the product $\frac{F}{I} \star \frac{G}{I}$ be a function. In this case the formula holds

$$\frac{F}{I} \star \frac{G}{I} = \frac{H}{I}, \qquad \text{where} \qquad H \in M \qquad \text{and} \qquad H(0) = 0.$$

By the definition we have

$$\frac{F(t)}{I(t)} \star \frac{G(t)}{I(t)} = \frac{\frac{d}{dt} \int\limits_0^t F(t-u)G(u)du}{t \star t} = \frac{F(0)G(t) + \int\limits_0^t F'(t-u)G(u)du}{t \star t}$$

$$= \frac{\int\limits_0^t f(t-u)G(u)\,du}{t \star t} = \frac{H(t)}{t},$$

which implies

$$\frac{\int\limits_0^t f(t-u)G(u)du}{t} = H(t),$$

or

$$\int\limits_0^t G(t-u)f(u)du = t \star H(t) = \int\limits_0^t H(u)du.$$

Differentiating and taking into account that $G(0) = 0$, we find

$$\int\limits_0^t g(t-u)f(u)du = H(t),$$

and therefore $\int\limits_0^t f(t-u)g(u)du \in M$.

Conversely, if the convolution $\int\limits_0^t f(t-u)g(u)du = H(t) \in M$ and $H(0) = 0$, then from

$$\frac{F(t)}{t} \star \frac{G(t)}{t} = \frac{\int\limits_0^t f(t-u)G(u)du}{t \star t} = \frac{\frac{d}{dt} \int\limits_0^t f(t-u)G(u)du}{t} = \frac{H(t)}{t}$$

we conclude that the product of the functions $\frac{F(t)}{t}$ and $\frac{G(t)}{t}$ is a function, too; obviously, $f \star g = \frac{d}{dt} \int\limits_0^t f(t-u)g(u)du$. $\qquad \square$

Corollary 2.3.31 *The product of a function $F \in M$ with an arbitrary function $g \in L$ is again a function.*

Indeed, the convolution $\int\limits_0^t F(t-u)g(u)du$ (2.3.1, Lemma 2.3.35) belongs to the ring M and at $t = 0$ is equal to zero.

In section 2.2. we introduced the function $\eta(t; \lambda) = 0$, if $t < \lambda$ and $\eta(t; \lambda) = 1$ if $t \geq \lambda$. It was proved that the convolution of $\eta(t; \lambda)$ and $\eta(t; \mu)$ is equal to

$$\int\limits_0^t \eta(t-u; \lambda)\eta(u; \mu)\,du = \int\limits_0^t \eta(u; \lambda+\mu)du.$$

This implies that the convolution of this functions belongs to M. Hence, the product $\eta(t; \lambda) \star \eta(t; \mu)$ is a function, and, obviously,

$$\eta(t; \lambda) \star \eta(t; \mu) = \eta(t; \lambda + \mu). \tag{2.3.14}$$

Thus, the field $\mathfrak{M}$ contains all locally integrable functions on $\mathbb{R}_+$. Complex numbers also lie in $\mathfrak{M}$, and the product of numbers in $\mathfrak{M}$ coincides with the ordinary product of complex numbers. Thus, an operator is a generalization of the notion of a function and a complex number; the elements of the field $\mathfrak{M}$ could be called *generalized functions*. However, taking into account established operational calculus terminology, we consider as the best name for an element of $\mathfrak{M}$ the term "operator." An operator is essentially different from a function. In contrast to functions, one cannot speak about the value of an operator at a point.

On defining the ring of operators M we departed from the relation

$$h(t) = \frac{d}{dt} \int_0^t F(t - u)g(u)\, du,$$

where the operator $F \in M$ acts on the function $g \in L$. The range of this operator lies in L. And both of the functions F and g belong to the field $\mathfrak{M}$. Therefore, one can compute the product of the operators $F = \frac{F}{1}$ and $g = \frac{G}{I}$. This product is equal to

$$F(t) \star g(t) = \frac{F(t) \star G(t)}{t} = \frac{\int_0^t F(t-u)g(u)du}{t} = \frac{d}{dt} \int_0^t F(t-u)g(u)du.$$

Thus, the operation of application of the operator F to the function g coincides with the product of the operators F and $\frac{G}{I} = g$, where $G(t) = \int_0^t g(u)du$.

The operator $\frac{F}{I}$ may be considered as the product of the operator $\frac{1}{I}$ by the function F: $\frac{F}{I} = \frac{1}{I} \star \frac{F}{1}$.

The operator $\frac{1}{I}$ plays a fundamental role in the operational calculus. A special notation is introduced for it:

$$\boxed{p = \frac{1}{I}}. \tag{2.3.15}$$

In this case formula (2.3.12) takes the form

$$p \star F = F'; \quad F(0) = 0, \tag{2.3.16}$$

hence, in the case $F(0) = 0$ the multiplication of the function F, belonging to M, by the operator $p = \frac{1}{I}$ denotes the differentiation of the function F. The operator p may be multiplied by any operator $\frac{F}{G}$, i.e., the product $p \star \frac{F}{G}$ has a meaning for any operator $\frac{F}{G} \in \mathfrak{M}$. In the general case the product $p \star \frac{F}{G}$ is an operator. The operator p is called the *differentiation operator*. If F is an arbitrary function belonging to M, then (2.3.16) implies

$$p \star (F - F(0)) = F',$$

$$p \star F = F' + pF(0). \tag{2.3.17}$$

If $F' \in M$, then (2.3.17) implies

$$p \star (p \star F) = p \star F' + p^2 F(0),$$

or

$$p^2 \star F = F'' + pF'(0) + p^2 F(0). \tag{2.3.18}$$

In the general case, when F has an nth order derivative $F^{(n)}$ belonging to the set L, successive application of (2.3.17) yields

$$p^n \star F = F^{(n)} + pF^{(n-1)}(0) + p^2 F^{(n-2)}(0) + \cdots + p^n F(0), \tag{2.3.19}$$

where p^n denotes the product $p \star p \cdots \star p$ of n operators. The inverse of the operator p is obviously equal to $\frac{1}{p} = I$. The function I belongs to the set M. It therefore follows from (2.3.15) that

$$\frac{1}{p} \star f(t) = \frac{d}{dt} \int_0^t (t-u)f(u)du = \int_0^t f(u)du, \tag{2.3.20}$$

hence $\frac{1}{p}$ is the operator of integration. On applying the operator $\frac{1}{p}$ to both sides of the relation

$$\frac{1}{p} = t, \tag{2.3.21}$$

we find from (2.3.20) that $\frac{1}{p} \star \frac{1}{p} = \left(\frac{1}{p}\right)^2 = \frac{t^2}{2!}$ and $\left(\frac{1}{p}\right)^n = \frac{1}{p}\left(\frac{1}{p}\right)^{n-1} = \frac{t^n}{n!}$.

Then taking into account (2.3.15), we have

$$\left(\frac{1}{p}\right)^n \star f(t) = \frac{d}{dt} \int_0^t \frac{(t-u)^n}{n!} f(u)du;$$

but

$$\left(\frac{1}{p}\right)^n = \frac{1}{p} \star \frac{1}{p} \star \cdots \star \frac{1}{p} = \frac{1}{p^n},$$

therefore,

$$\frac{1}{p^n} = \frac{t^n}{n!} \tag{2.3.22}$$

and

$$\frac{1}{p^n} \star f(t) = \frac{d}{dt} \int_0^t \frac{(t-u)^n}{n!} f(u)du. \tag{2.3.23}$$

Formula (2.3.22) is proved for positive integer n; however, it can be extended to arbitrary values $n = \nu$, where $\nu \geq 0$. Let us denote the function $\frac{t^\nu}{\Gamma(1+\nu)}$, where $\nu \geq 0$, by $\frac{1}{p^\nu}$. It follows from formula (2.3.23) that in such notation we have

$$\frac{1}{p^\nu} \star \frac{1}{p^\mu} = \frac{1}{p^{\mu+\nu}},$$

hence for all $\nu \geq 0$ the following formula holds:

$$\boxed{\frac{1}{p^\nu} = \frac{t^\nu}{\Gamma(1+\nu)}}. \tag{2.3.24}$$

Let us agree below, for the sake of simplicity of notations, to omit the asterisk in the product of operators, if it will not cause ambiguities. Hence, we often shall write $\frac{F}{G} \frac{F_1}{G_1}$ instead of $\frac{F}{G} \star \frac{F_1}{G_1}$. Thus, formula (2.3.23) may be represented in the form

$$\frac{1}{p^n} f(t) = \frac{d}{dt} \int_0^t \frac{(t-u)^n}{n!} f(u)\,du; \tag{2.3.25}$$

and formulae (2.3.17), (2.3.19) take the form

$$F' = pF - pF(0), \tag{2.3.26}$$

$$F^{(n)} = p^n F - p^n F(0) - p^{n-1} F'(0) - \cdots - p F^{(n-1)}(0). \tag{2.3.27}$$

Often we shall denote operators of the field $\mathfrak{M}$ by a single letter, for instance, $\frac{F}{G} = a$, $\frac{H}{R} = b$, etc.

2.3.3   Finite Parts of Divergent Integrals

It was proved that the field $\mathfrak{M}$ contains all locally integrable functions on $\mathbb{R}_+$. The question arises if is it possible to prove in the same way that some nonintegrable functions also belong to the field $\mathfrak{M}$. Let us consider nonintegrable functions with power or logarithmic singularities at $t = 0$. Let N_0 denote the set of all functions satisfying the following conditions:

1. In some neighborhood of $t = 0$, i.e., for $0 < t \leq \delta$ the function $f(t)$ may be represented in the form

$$f(t) = \sum_{i=0}^{n} \sum_{k=0}^{m} \beta_{ik} t^{\alpha_{ik}} \log^k t + h(t), \tag{2.3.28}$$

where α_{ik} and β_{ik} are arbitrary real or complex numbers, and the function h is absolutely integrable on the interval $(0, \delta)$ and bounded for $t \to +0$.

2. The function f is absolutely integrable on the interval $\delta \leq t < T$.

Obviously, N_0 is a linear set.

Let us find the indefinite integrals of the functions $t^\alpha \log^k t$, $k = 0, 1, 2, \ldots$, which appear in the right-hand side of (2.3.28). If $\alpha = -1$, then obviously

$$\int t^{-1} \log^k t\,dt = \frac{\log^{k+1} t}{k+1} + C.$$

If $\alpha \neq -1$, then integrating by parts we find that

$$\int t^\alpha \log^k t\, dt = \frac{t^{\alpha+1}}{\alpha+1} \log^k t - \frac{k}{\alpha+1} \int t^\alpha \log^{k-1} t\, dt.$$

Repeatedly applying this formula, we obtain the relation

$$\int t^\alpha \log^k t\, dt = \frac{t^{\alpha+1}}{\alpha+1} \log^k t - \frac{kt^{\alpha+1}}{(\alpha+1)^2} \log^{k-1} t + \cdots + (-1)^k \frac{k!}{(\alpha+1)^k} \int t^\alpha dt,$$

or

$$\int t^\alpha \log^k t\, dt = \frac{t^{\alpha+1}}{\alpha+1} \left(\log^k t - \frac{k}{\alpha+1} \log^{k-1} t \right.$$
$$\left. + \frac{k(k-1)}{(\alpha+1)^2} \log^{k-2} t - \cdots + (-1)^k \frac{k!}{(\alpha+1)^k} \right) + C.$$

Let us define the function $\Phi_k(t;\alpha)$, $k = 0,1,2,\ldots$, on putting

$$\Phi_k(t;\alpha) = \frac{t^{\alpha+1}}{\alpha+1} \left(\log^k t - \frac{k}{\alpha+1} \log^{k-1} t + \cdots + \frac{k!}{(\alpha+1)^k} \right) \quad \text{if} \quad \alpha \neq -1,$$

$$\Phi_k(t;\alpha) = \frac{\log^{k+1} t}{k+1} \quad \text{if} \quad \alpha = -1.$$

In this case we can write for all α

$$\int t^\alpha \log^k t\, dt = \Phi_k(t;\alpha) + C.$$

Let $f \in N_0$. Consider the integral

$$J(\epsilon) = \int_\epsilon^t f(u)du,$$

where $0 < \epsilon < \delta < t$. Taking into account (2.3.28) we have

$$J(\epsilon) = \int_\epsilon^\delta f(u)du + \int_\delta^t f(u)du = \sum_{i=0}^n \sum_{k=0}^m \beta_{ik} \int_\epsilon^\delta u^{\alpha_{ik}} \log^k u\, du + \int_\epsilon^\delta h(u)\, du + \int_0^t f(u)du$$

$$= \sum_{i,k} \beta_{ik} \Phi_k(\delta;\alpha_{ik}) - \sum_{i,k} \beta_{ik} \Phi_k(\epsilon;\alpha_{ik}) + \int_\epsilon^\delta h(u)\, du + \int_\delta^t f(u)du,$$

or

$$\int_\epsilon^t f(u)\, du + \sum_{i,k} \beta_{ik} \Phi_k(\epsilon;\alpha_{ik}) = \sum_{i,k} \beta_{ik} \Phi_k(\delta;\alpha_{ik}) + \int_\epsilon^\delta h(u)\, du + \int_\delta^t f(u)du.$$

We see from the last relation that as $\epsilon \to +0$ there exists the limit

$$\lim_{\epsilon \to +0} \left[\sum_{i,k} \beta_{ik} \Phi_k(\epsilon;\alpha_{ik}) + \int_\epsilon^t f(u)du \right]$$

$$= \sum_{i,k} \beta_{ik} \Phi_k(\delta;\alpha_{ik}) + \int_0^\delta h(u)\, du + \int_\delta^t f(u)du.$$

(2.3.29)

This limit is called the *finite part* of the (in general divergent) integral $\int\limits_0^t f(u)du$ and is denoted by

$$\overline{\left|\int\limits_0^t f(u)du.\right.}$$

Thus, if we introduce the notation

$$\Phi(\epsilon) = \sum_{ik} \beta_{ik}\Phi_k(\epsilon;\alpha_{ik}),$$

then we obtain for all $t > 0$

$$\lim_{\epsilon \to +0}\left[\Phi(\epsilon) + \int\limits_\epsilon^t f(u)du\right] = \overline{\left|\int\limits_0^t f(u)du.\right.} \qquad (2.3.30)$$

Let us give, for instance, the following finite parts of integrals:

1) $\overline{\left|\int\limits_0^t u^\alpha du\right.} = \frac{t^{\alpha+1}}{\alpha+1}, \quad \alpha \neq -1;$

2) $\overline{\left|\int\limits_0^t \frac{du}{u}\right.} = \log t;$

3) It follows from (2.3.29) and the definition of the finite part that

$$\overline{\left|\int\limits_0^t u^\alpha \log^k u\, du\right.} = \Phi_k(t;\alpha). \qquad (2.3.31)$$

Properties of Finite Parts of Integrals

1. *If the following relation holds:*

$$\lim_{\epsilon \to +0} \int\limits_\epsilon^t f(u)\, du = \int\limits_0^t f(u)du,$$

then also the relation holds:

$$\overline{\left|\int\limits_0^t f(u)\, du\right.} = \int\limits_0^t f(u)du.$$

Let us note that the condition of convergence as $\epsilon \to +0$ of the integral $\int\limits_\epsilon^t f(u)du$ implies the existence of the limit $\lim\limits_{\epsilon \to +0} \Phi(\epsilon)$; and from the structure of the function $\Phi(\epsilon)$ we see that this limit is equal to zero.

2. *If α is a number, then it can be taken out of the symbol of the finite part of the integral,* i.e.,

$$\overline{\left|\int\limits_0^t \alpha f(u)\, du\right.} = \alpha\,\overline{\left|\int\limits_0^t f(u)du.\right.}$$

3. *If there exist the finite parts* $\left|\overline{\int_0^t f(u)du}\right.$ *and* $\left|\overline{\int_0^t g(u)du}\right.$, *then there exists the finite part*

$$\left|\overline{\int_0^t (f(u) + g(u))\, du}\right.$$

and the following relation holds:

$$\left|\overline{\int_0^t f(u)\, du}\right. + \left|\overline{\int_0^t g(u)du}\right. = \left|\overline{\int_0^t (f(u) + g(u))du}\right. .$$

4. *The definite integral* $\int_\alpha^t f(u)du$ *is equal to the difference of the finite parts of the integrals* $\left|\overline{\int_0^t f(u)du}\right.$ *and* $\left|\overline{\int_0^\alpha f(u)du}\right.$, *i.e.,*

$$\int_\alpha^t f(u)du = \left|\overline{\int_0^t f(u)du}\right. - \left|\overline{\int_0^\alpha f(u)du}\right. .$$

This property follows as $\epsilon \to +0$ from the equation

$$\int_\alpha^t f(u)du = \Phi(\epsilon) + \int_\epsilon^t f(u)du - \Phi(\epsilon) - \int_\epsilon^\alpha f(u)du.$$

Corollary 2.3.32 *For $t > 0$ we have*

$$\frac{d}{dt}\left|\overline{\int_0^t f(u)\, du}\right. = f(t).$$

Let the function $f \in N_0$ be differentiable, and the derivative f' also belongs to the set N_0. In this case one can consider the finite part (see Property 4)

$$\left|\overline{\int_0^t f'(u)\, du}\right. = \left|\overline{\int_0^\delta f'(u)\, du}\right. + \int_\delta^t f'(u)\, du.$$

However (see (2.3.28)), for $0 < u \le \delta$ we have

$$f'(u) = \sum_{i,k} \beta_{ik} \frac{d}{du}(u^{\alpha_{ik}} \log^k u) + h'(u);$$

therefore,

$$\left|\overline{\int_0^\delta f'(u)du}\right. = \sum_{i,k} \beta_{ik} \left|\overline{\int_0^\delta \frac{d}{du}(u^{\alpha_{ik}} \log^k u)du}\right. + \int_0^\delta h'(u)du.$$

For all values of α and nonnegative integers k we have (see Property 4)

$$\overline{\left|\int_0^{\delta} \frac{d}{du}(u^{\alpha}\log^k u)du\right.} = \overline{\left|\int_0^{\delta} \alpha u^{\alpha-1}\log^k u\, du\right.} + \overline{\left|\int_0^{\delta} ku^{\alpha-1}\log^{k-1} u\, du\right.}$$

$$= \alpha\Phi_k(\delta;\alpha-1) + k\Phi_{k-1}(\delta;\alpha-1) = \delta^{\alpha}\log^k\delta;$$

therefore, the following formula holds:

$$\overline{\left|\int_0^{\delta} f'(u)\,du\right.} = f(\delta) - h(\delta) + \int_0^{\delta} h'(u)du,$$

hence also

$$\overline{\left|\int_0^{t} f'(u)\,du\right.} = f(\delta) - h(+0) + \int_{\delta}^{t} f'(u)du = f(t) - h(+0).$$

Thus, finally we have

$$f(t) - \overline{\left|\int_0^{t} f'(u)du\right.} = h(+0). \tag{2.3.32}$$

It is convenient to introduce for this difference a special notation

$$h(+0) =_{t=0} \left|\overline{f(t)},\right.$$

then

$$\overline{\left|\int_0^{t} f'(u)\,du\right.} = f(t) -_{t=0} \left|\overline{f(t)}.\right. \tag{2.3.33}$$

If $\alpha \in C^{\infty}[0,\infty)$, then we can write the equation

$$\alpha(t) = \sum_{k=0}^{n} \frac{\alpha^{(k)}(0)t^k}{k!} + \frac{t^{n+1}}{(n+1)!}\alpha^{(n+1)}(\Theta t), \qquad 0 < \Theta < 1,$$

where n may be taken arbitrarily large. Therefore, $\alpha f \in N_0$ if only $f \in N_0$. Now let f be a function such that $f' \in N_0$. In this case the following formula holds:

$$\overline{\left|\int_0^{t} [\alpha(u)f(u)]'du\right.} = \alpha(t)f(t) -_{t=0} \left|\overline{\alpha(t)f(t)}.\right.$$

On the other hand we have

$$\overline{\left|\int_0^{t} [\alpha(u)f(u)]'du\right.} = \overline{\left|\int_0^{t} \alpha'(u)f(u)du\right.} + \overline{\left|\int_0^{t} \alpha(u)f'(u)du\right.}$$

We obtain from here the following property of integration by parts of the finite part of an integral

5.

$$\left|\overline{\int_0^t \alpha(u)f'(u)du} = \alpha(t)f(t) \right.\left. -_{t=0} \overline{|\alpha(t)f(t)} - \left|\overline{\int_0^t \alpha'(u)f(u)du}\right.\right.\tag{2.3.34}$$

Let us compute the finite part of the integral

$$\frac{1}{(n-1)!}\int_0^t (t-u)^{n-1}u^\alpha \log^k u\,du.$$

Using the Properties 2 and 3 and the formula (2.3.31), we find the finite part of the integral:

$$\left|\overline{\frac{1}{(n-1)!}\int_0^t (t-u)^{n-1}u^\alpha \log^k u\,du}\right.$$

$$= \frac{1}{(n-1)!}\sum_{r=0}^{n-1}(-1)^r\binom{n-1}{r}t^{n-1-r}\left|\overline{\int_0^t u^{r+\alpha}\log^k u\,du}\right.$$

$$= \frac{1}{(n-1)!}\sum_{r=0}^{n-1}(-1)^r\binom{n-1}{r}t^{n-1-r}\Phi_k(t;r+\alpha).$$

Taking into account the relation

$$\Phi_k(t;r+\alpha) = \begin{cases} \frac{t^{r+\alpha+1}}{r+\alpha+1}\left(\log^k t - \frac{k}{r+\alpha+1}\log^{k-1}t + \cdots + \frac{(-1)^k k!}{(r+\alpha+1)^k}\right) \\ \qquad \text{for } r+\alpha+1 \neq 0, \\ \frac{\log^{k+1}t}{k+1} \qquad \text{for } r+\alpha+1=0, \end{cases}$$

we find

$$\left|\overline{\frac{1}{(n-1)!}\int_0^t (t-u)^{n-1}u^\alpha \log^k u\,du}\right.$$

$$= \sum_{r=0,r\neq r'}^{n-1}(-1)^r\frac{1}{(n-1)!}\binom{n-1}{r}\frac{t^{n+\alpha}}{r+\alpha+1}\left[\log^k t - \frac{k}{r+\alpha+1}\log^{k-1}t+\ldots\right.\tag{2.3.35}$$

$$\left.+\frac{(-1)^k k!}{(r+\alpha+1)^k}\right]+\frac{1}{(n-1)!}(-1)^{r'}\binom{n-1}{r'}t^{n+\alpha}\frac{\log^{k+1}t}{k+1},\qquad (r'=-\alpha-1).$$

If $r+\alpha+1\neq 0$ for $r=0,1,\ldots,n-1$, then the sum in the right-hand side of the last equation is taken over all r from 0 to $n-1$, and the summand

$$(-1)^{r'}\frac{1}{(n-1)!}\binom{n-1}{r'}t^{n+\alpha}\frac{\log^{k+1}t}{k+1}$$

is absent.

From the relation (2.3.35) we obtain the following lemma.

Lemma 2.3.36 *The functions*

$$\varphi_n(t) = \frac{1}{(n-1)!}\overline{\left|\int_0^t (t-u)^{n-1}u^\alpha \log^k u\, du,\right.} \qquad n = 1, 2, 3, \ldots$$

belong to the ring M for all $n > -\alpha$.

Indeed, if $n+\alpha > 0$, then the function $\varphi_n(t)$ is continuous for $0 \leq t < \infty$ and its derivative is integrable on any interval $0 < t < T$. Besides, obviously, $\varphi(0) = 0$.

Corollary 2.3.33 *If $f \in N_0$, then for all sufficiently large n the functions*

$$F_n(t) = \left|\frac{1}{(n-1)!}\overline{\int_0^t (t-u)^{n-1}f(u)du}\right| \tag{2.3.36}$$

belong to the set M and $F_n(0) = 0$.

The property 4 with $\alpha = \epsilon$ and the change of $f(u)$ by $\frac{1}{(n-1)!}(t-u)^{n-1}f(u)$ imply

$$\frac{1}{(n-1)!}\int_\epsilon^t (t-u)^{n-1}f(u)du = \left|\frac{1}{(n-1)!}\overline{\int_0^t (t-u)^{n-1}u^\alpha f(u)du}\right|$$

$$-\left|\frac{1}{(n-1)!}\overline{\int_0^\epsilon (t-u)^{n-1}u^\alpha f(u)du}.\right| \tag{2.3.37}$$

Obviously, the expression

$$P_{n-1}(t;\epsilon) = \left|\frac{1}{(n-1)!}\overline{\int_0^\epsilon (t-u)^{n-1}f(u)du}\right|$$

is a polynomial in t of degree $n-1$, whose coefficients depend on ϵ. In the conventional notation the expression (2.3.36) may be written in the form

$$F_n(t) = \frac{1}{(n-1)!}\int_\epsilon^t (t-u)^{n-1}f(u)du + P_{n-1}(t;\epsilon). \tag{2.3.38}$$

On noting that

$$\frac{d}{dt}P_{n-1}(t;\epsilon) = \frac{d}{dt}\sum_{r=0}^{n-1}(-1)^r\binom{n-1}{r}\frac{t^{n-l-r}}{(n-1)!}\left|\overline{\int_0^\epsilon u^r f(u)du}\right|$$

$$= \sum_{r=0}^{n-2}(-1)^r\binom{n-2}{r}\frac{t^{n-2-r}}{(n-2)!}\left|\overline{\int_0^\epsilon u^r f(u)du}\right| = P_{n-2}(t;\epsilon),$$

we obtain from (2.3.38) with $n > 1$

$$F'_n(t) = \frac{1}{(n-2)!} \int_0^t (t-u)^{n-2} f(u) du + P_{n-2}(t; \epsilon) = F_{n-1}(t).$$

For $n = 1$ it follows from (2.3.38) immediately

$$F'_1(t) = f(t) = F_0(t). \tag{2.3.39}$$

Thus, for all $n \geq 1$ we have

$$F'_n(t) = F_{n-1}(t), \tag{2.3.40}$$

and

$$\frac{d^n F_n(t)}{dt^n} = f(t). \tag{2.3.41}$$

Now let n_0 be such that $F_{n_0} \in M$. Then we have $F_n \in M$ for all $n \geq n_0$; hence, the operator $p^n F_n$ belongs to the field $\mathfrak{M}$.

Let us prove that the operator $p^n F_n$ does not depend on n. Let $m > n \geq n_0$ and $l = m - n$. It follows from (2.3.40) that

$$F'_m = F_{m-1}, \ldots, F_m^{(l)} = F_{m-1} = F_n,$$

and $F_n(0) = 0$ for all $n \geq n_0$, therefore,

$$F_n = F_m^{(l)} = p^l F_m = p^{m-n} F_m$$

or

$$p^n F_n = p^m F_m \quad \text{for any} \quad m \geq n \geq n_0.$$

Hence, the operator $p^n F_n$ depends only on the choice of the function f. Thus, one can put into correspondence to any function $f \in N_0$ the operator

$$a = p^n \left| \overline{\frac{1}{(n-1)!} \int_0^t (t-u)^{n-1} f(u) \, du} \in \mathfrak{M}.\right.$$

This correspondence has the following properties (see Properties 2 and 3 of finite parts):

1. *If the function f corresponds to the operator a, then the function λf, where λ is a number, corresponds to the operator λa.*

2. *If the function f corresponds to the operator a and the function $g \in N_0$ corresponds to the operator b, then the sum of the functions $f + g$ corresponds to the operator $a + b$.*

3. *If $f \in L$, then the following formula holds:*

$$p^n \left| \overline{\frac{1}{(n-1)!} \int_0^t (t-u)^{n-1} f(u) du} = p^n \int_0^t \frac{(t-u)^{n-1}}{(n-1)!} f(u) du = f(t);\right.$$

in this case the operator a coincides with the function f.

Let us denote the operator $p^n F_n$, see formula (2.3.36), where $f \in N_0$, by f:

$$f(t) = p^n \overline{\left| \int_0^t \frac{1}{(n-1)!}(t-u)^{n-1}f(u)du \right.} \qquad (2.3.42)$$

This notation is justified by the Properties 1, 2, 3, and the equation (2.3.41).

Suppose that the function f has the derivative $f' \in N_0$ on $\mathbb{R}_1$. Taking into account the Property 5, we find

$$\overline{\left| \frac{1}{(n-1)!} \int_0^t (t-u)^{n-1}f'(u)du \right.}$$

$$= \frac{1}{(n-1)!} \sum_{r=0}^{n-1} (-1)^r \binom{n-1}{r} t^{n-1-r} \overline{\left| \int_0^t u^r f'(u)du \right.}$$

$$= \frac{1}{(n-1)!} \sum_{r=0}^{n-1} (-1)^r \binom{n-1}{r} t^{n-1-r} \left[t^r f(t) -_{t=0}\overline{|t^r f(t)} - \overline{\left| r \int_0^t u^{r-1}f(u)\,du \right.} \right]$$

$$= -\frac{1}{(n-1)!} \sum_{r=0}^{n-1} (-1)^r \binom{n-1}{r} t^{n-1-r}\,_{t=0}\overline{|t^r f(t)} + \overline{\left| \int_0^t \frac{(t-u)^{n-2}}{(n-2)!}f(u)\,du \right.}$$

Let us introduce the notation

$$_{t=0}\overline{|t^r f(t)} = f_r, \quad r = 0, 1, 2, \ldots; \qquad (2.3.43)$$

then we have, see formula (2.3.36),

$$\overline{\left| \frac{1}{(n-1)!} \int_0^t (t-u)^{n-1}f'(u)\,du \right.} = -\sum_{r=0}^{n-1}(-1)^r \frac{f_r t^{n-1-r}}{r!(n-1-r)!} + F_{n-1}(t).$$

Taking into account that

$$t^{n-1-r} = \frac{(n-1-r)!}{p^{n-1-r}},$$

we obtain

$$p^n \overline{\left| \frac{1}{(n-1)!} \int_0^t (t-u)^{n-1}f'(u)\,du \right.} = -\sum_{r=0}^{n-1}(-1)^r \frac{f_r p^{r+1}}{r!} + p^n F_{n-1}(t),$$

or, taking into account (2.3.42),

$$f' = pf - \sum_{r=0}^{n-1} \frac{(-1)^r f_r p^{r+1}}{r!}, \qquad n > n_0.$$

It is convenient to represent the last equation in the form

$$f' = p\left[f - \sum_{r=0}^{\infty} \frac{(-1)^r f_r p^r}{r!} \right]. \qquad (2.3.44)$$

In fact, the series in this equation has only a finite number of nonzero terms, because, obviously (see formula 2.3.32) $f_r = 0$ for all sufficiently large r $(r \geq n_0)$.

If we suppose that $f' \in L$ in (2.3.44), then $f_r = 0$ for all $r > 0$ and $f_0 = f(0)$. Hence, the expression (2.3.44) takes the form

$$f' = p[f - f(0)],$$

i.e., it coincides with 2.3.2, (2.3.26).

The results above may be formulated as follows:

Theorem 2.3.84 *The set N_0 is contained in the field $\mathfrak{M}$.*

Let consider special cases. Suppose, that $f(t) = t^\alpha$, where α is not equal to a negative integer. In this case (see formula 2.3.32) we have

$$f_r =_{t=0} \overline{|t^r f(t)} =_{t=0} \overline{|t^{r+\alpha} f(t)} = 0$$

for all $r \geq 0$; therefore, (2.3.44) implies

$$\frac{d}{dt} t^\alpha = pt^\alpha. \tag{2.3.45}$$

Hence, if α is not equal to a negative integer, then the product of the operator p by t^α is to be computed by the rule of differentiating of the power function. If $\alpha = -m$, where m is a positive integer, then $f_r = 0$ for $r \neq m$ and $f_m = 1$ and from (2.3.44) we obtain

$$\frac{d}{dt}\left(\frac{1}{t^m}\right) = p\left[\frac{1}{t^m} - \frac{(-1)^m p^m}{m!}\right]. \tag{2.3.46}$$

Let find $F_n(t)$ when $f(t) = t^\alpha$. If α is not a negative integer, then we have from (2.3.35)

$$F_n(t) = \frac{t^{\alpha+n}}{(n-1)!} \sum_{r=0}^{n-1} (-1)^r \binom{n-1}{r} \frac{1}{\alpha+r+1}.$$

The sum in the right-hand side may be represented in terms of the Euler Gamma function. Indeed, if α is positive, then the following relation holds:

$$\sum_{r=0}^{n-1} (-1)^r \binom{n-1}{r} \frac{1}{\alpha+r+1} = \sum_{r=0}^{n-1} (-1)^r \binom{n-1}{r} \int_0^1 \xi^{\alpha+r} d\xi$$

$$= \int_0^1 \xi^\alpha (1-\xi)^{n-1} d\xi = \frac{\Gamma(1+\alpha)\Gamma(n)}{\Gamma(\alpha+n+1)}.$$

By virtue of the principle of analytical continuation the latter equation holds for all α for which the right- and left-hand sides have a meaning. Hence, we have

$$F_n(t) = \frac{t^{n+\alpha}\Gamma(1+\alpha)}{\Gamma(\alpha+n+1)};$$

therefore, the formula holds

$$t^\alpha = p^n F_n(t) = p^n \frac{t^{n+\alpha}\Gamma(1+\alpha)}{\Gamma(\alpha+n+1)}.$$

However, for $n+\alpha > 0$ we have

$$\frac{t^{n+\alpha}}{\Gamma(\alpha+n+1)} = \frac{1}{p^{n+\alpha}};$$

therefore, finally we obtain

$$\frac{t^\alpha}{\Gamma(1+\alpha)} = \frac{1}{p^\alpha}, \quad \alpha \neq -1, -2, \dots.$$

Thus, the formulae (2.3.23) and (2.3.24) in section 2.3.2 hold for all α, which are not equal to negative integers.

If α is equal to a negative integer $\alpha = -m$, then for $m \geq 1$ we have (see (2.3.35) when $n = m$:

$$F_m(t) = \frac{1}{(m-1)!}\left[\sum_{r=0}^{m-2}(-1)^r\binom{m-1}{r}\frac{1}{r+1-m} + (-1)^{m-1}\log t\right].$$

If $m = 1$, then $F_1(t) = \log t$.

Now we compute the sum

$$-\sum_{r=0}^{m-2}(-1)^r\binom{m-1}{r}\frac{1}{m-r-1} = -\sum_{r=0}^{m-2}(-1)^r\binom{m-1}{r}\int_0^1 \xi^{m-r-2}d\xi$$

$$= -\int_0^1\left[\sum_{r=0}^{m-2}(-1)^r\binom{m-1}{r}\xi^{m-1-r} - (-1)^{m-1}\right]\frac{d\xi}{\xi}$$

$$= -\int_0^1 \frac{(\xi-1)^{m-1} - (-1)^{m-1}}{\xi}d\xi = (-1)^{m-1}\int_0^1 \frac{1-(1-\xi)^{m-1}}{\xi}d\xi.$$

We introduce the notation

$$I_n = \int_0^1 \frac{1-(1-\xi)^{n-1}}{\xi}d\xi;$$

then we obtain

$$I_{n+1} - I_n = \int_0^1 \frac{(1-\xi)^{n-1}-(1-\xi)^n}{\xi}d\xi = \int_0^1(1-\xi)^{n-1}d\xi = \frac{1}{n}.$$

Thus, $I_{n+1} - I_n = \frac{1}{n}$; however, $I_1 = 0$, whence the following equation holds:

$$I_n = \int_0^1 \frac{1-(1-\xi)^{n-1}}{\xi}d\xi = \sum_{k=1}^{n-1}\frac{1}{k}$$

and

$$-\sum_{r=0}^{m-2}(-1)^r\binom{m-1}{r}\frac{1}{m-1-r}=(-1)^{m-1}\sum_{k=1}^{m-1}\frac{1}{k},$$

i.e., for $\alpha=-m$, $m=1,2,3,\ldots$ we have

$$F_1(t)=\log t,\quad F_m(t)=\frac{(-1)^{m-1}}{(m-1)!}\left[\log t+\sum_{k=0}^{m-1}\frac{1}{k}\right].\qquad(2.3.47)$$

Because of the lack of the space we note only that by the similar method it may be proven that nonintegrable functions with power or logarithmic singularities at a finite number of points of the region $0<t<\infty$ also belong to the field $\mathfrak{M}$.

2.3.4 Rational Operators

One of the main goals of operational calculus is the study of the operators of the form $R(p)$; $R(z)$ is a function of the variable z. In the simplest case, when $R(z)=\Sigma_k\alpha_k z^k$ is a polynomial, the operator $R(p)$ is equal to $\Sigma_k\alpha_k p^k$. The operations with such polynomials are executed in the same way, as in elementary algebra, for instance,

$$(p^2+2p+1)(p-1)=p^3+p^2-p-1=p^2(p+1)-(p+1)=(p^2-1)(p+1).$$

If two polynomials in the operator p are equal, i.e.,

$$P(p)=\sum_{k=0}^{n}\alpha_k p^k=Q(p)=\sum_{k=0}^{n}\beta_k p^k,$$

then the corresponding coefficients α_k and β_k are also equal. Indeed, if

$$\sum_{k=0}^{n}\alpha_k p^k=\sum_{k=0}^{n}\beta_k p^k,$$

then by multiplying both sides of this equation by the operator $\frac{1}{p^n}$ we obtain

$$\sum_{k=0}^{n}\frac{\alpha_k}{p^{n-k}}=\sum_{k=o}^{n}\frac{\beta_k}{p^{n-k}},$$

or

$$\sum_{k=0}^{n}\alpha_k\frac{t^{n-k}}{(n-k)!}=\sum_{k=0}^{n}\beta_k\frac{t^{n-k}}{(n-k)!},\quad 0\le t<\infty.$$

By virtue of the known theorem for (ordinary) polynomials it follows from this relation that

$$\alpha_k=\beta_k,\qquad k=0,1,2,\ldots,n.$$

Theorem 2.3.85 *If a polynomial*

$$\sum_{k=0}^{n}\alpha_k p^k$$

is reducible to a function, then $\alpha_1 = \alpha_2 = \cdots = \alpha_n = 0$.

This theorem implies that if the degree of a polynomial $\Sigma_{k=0}^n \alpha_k z^k$ is greater than or equal to one, then the corresponding operator $\Sigma_{k=0}^n \alpha_k p^k$ cannot be reduced to a function.

Proof. $\Sigma_{k=0}^n \alpha_k p^k = f(t)$, $n \geq 1$.   Multiplying this relation by the operator $\frac{1}{p^n}$, by virtue of 2.3.2, (2.3.25) we have

$$\sum_{k=0}^n \frac{\alpha_k}{p^{n-k}} = \frac{1}{p^n} f(t) = \int_0^t \frac{(t-u)^{n-1}}{(n-1)!} f(u) du,$$

or

$$\sum_{k=0}^n \frac{\alpha_k t^{n-k}}{(n-k)!} = \int_0^t \frac{(t-u)^{n-1}}{(n-1)!} f(u) du.$$

Putting $t = 0$ in this relation, we find $\alpha_n = 0$. Hence,

$$\sum_{k=0}^{n-1} \alpha_k p^k = f(t).$$

If $n - 1 \geq 1$, then multiplying the last relation by $\frac{1}{p^{n-1}}$, we find $\alpha_{n-1} = 0$; if $n - 2 \geq 1$, then similarly we obtain $\alpha_{n-2} = 0$ and so on, until we obtain $\alpha_1 = 0$.

Let $P_n(p) = \sum_{k=0}^n \alpha_k p^k$, and $Q_m(p) = \sum_{k=0}^m \beta_k p^k$. The collection of all operators of the form $\sum_{k=0}^n \alpha_k p^k$, $0 \leq n < \infty$, is a ring. This ring may be extended to the quotient field. The elements of this field are rational fractions of the operator p, i.e., operators of the form

$$\frac{\sum_{k=0}^n \alpha_k p^k}{\sum_{k=0}^m \beta_k p^k} = \frac{P_n(p)}{Q_m(p)} = R(p).$$

The operators $P_n(p)$ and $Q_m(p)$ belong to the field $\mathfrak{M}$. Hence, their ratio $R(p)$ also belongs to $\mathfrak{M}$. The operator $R(p)$ is called a *rational operator*. A rational operator is associated with every rational function $R(z) = \frac{P_n(z)}{Q_m(z)}$. This correspondence gives an isomorphism between the field of all rational functions and the field of rational operators. The field of rational operators is contained in the field $\mathfrak{M}$, which is a subfield of $\mathfrak{M}$.

Let us consider several examples. Suppose that $F(t) = e^{\mu t}$ in formula (2.3.26). Then we have $p e^{\mu t} = \mu e^{\mu t} + p$, or $p e^{\mu t} - \mu e^{\mu t} = p$, whence $(p - \mu) e^{\mu t} = p$. Therefore,

$$\boxed{\frac{p}{p - \mu} = e^{\mu t}}. \tag{2.3.48}$$

Multiplying this relation by the operator $\frac{1}{p} = t$, we find

$$\frac{1}{p - \mu} = t \star e^{\mu t} = \int_0^t e^{\mu u} du = \frac{e^{\mu t} - 1}{\mu}.$$

Thus,

$$\frac{\mu}{p-\mu} = e^{\mu t} - 1. \tag{2.3.49}$$

It follows from (2.3.48) and (2.3.49) that

$$\frac{p-\lambda}{p-\mu} = \frac{p}{p-\mu} - \frac{\lambda}{p-\mu} = e^{\mu t} - \frac{\lambda}{\mu}(e^{\mu t} - 1),$$

$$\frac{p-\lambda}{p\quad\mu} = \left(1 - \frac{\lambda}{\mu}\right)e^{\mu t} + \frac{\lambda}{\mu}. \tag{2.3.50}$$

Thus, (2.3.49) and (2.3.50) imply that any rational operators of the form $\frac{1}{p-\lambda}$ and $\frac{p-\lambda}{p-\mu}$ are functions belonging to the ring M.

Multiplying (2.3.48) by (2.3.49), we find

$$\frac{p}{(p-\mu)^2} = e^{\mu t} \star \frac{e^{\mu t} - 1}{\mu} = \frac{1}{\mu}(e^{\mu t} \star e^{\mu t} - e^{\mu t}).$$

However,

$$e^{\mu t} \star e^{\mu t} = \frac{d}{dt}\int_0^t e^{\mu(t-u)+\mu u}du = \frac{d}{dt}(te^{\mu t}) = e^{\mu t} + \mu t e^{\mu t}.$$

Hence, we have

$$\frac{p}{(p-\mu)^2} = \frac{1}{\mu}(\mu t e^{\mu t}) = t e^{\mu t}.$$

Let us prove that following the formula holds:

$$\frac{p}{(p-\mu)^{n+1}} = \frac{t^n e^{\mu t}}{n!}. \tag{2.3.51}$$

Indeed, if (2.3.51) holds for some n, then multiplying (2.3.51) by $\frac{1}{p-\mu}$ we obtain

$$\frac{p}{(p-\mu)^{n+2}} = \frac{t^n e^{\mu t}}{n!} \star \frac{e^{\mu t} - 1}{\mu} = \frac{1}{\mu n!}(t^n e^{\mu t} \star e^{\mu t} - t^n e^{\mu t}).$$

However,

$$t^n e^{\mu t} \star e^{\mu t} = \frac{d}{dt}\int_0^t e^{\mu(t-u)}u^n e^{\mu u}\,du = \frac{d}{dt}\left(\frac{t^{n+1}e^{\mu t}}{n+1}\right) = \frac{t^{n+1}\mu e^{\mu t}}{n+1} + t^n e^{\mu t};$$

therefore,

$$\frac{p}{(p-\mu)^{n+2}} = \frac{1}{\mu n!}\left(\frac{t^{n+1}\mu e^{\mu t}}{n+1} + t^n e^{\mu t} - t^n e^{\mu t}\right) = \frac{t^{n+1}e^{\mu t}}{(n+1)!}.$$

Thus, (2.3.51) holds for $n+1$ and for $n=0$. Hence, it is proved for all positive integers n. ∎

The question arises: for what rational functions $R(z)$ is the associated operator $R(p)$ reducible to a function? The answer to this question is given in the following theorem.

Theorem 2.3.86 *A rational operator $R(p) = \frac{P(p)}{Q(p)}$ is reducible to a function if and only if the degree of the polynomial $P(p)$ is less than or equal to the degree of the polynomial $Q(p)$.*

Proof. Suppose that the degree of $P(p)$ is n, the degree of $Q(p)$ is m and $n \le m$. Let us factorize the polynomials into linear factors:

$$P(p) = \alpha_n(p - \lambda_1)(p - \lambda_2)\ldots(p - \lambda_n);$$

$$Q(p) = \beta_m(p - \mu_1)(p - \mu_2)\ldots(p - \mu_m).$$

Here, $\lambda_1, \lambda_2, \ldots, \lambda_n$; $\mu_1, \mu_2, \ldots, \mu_m$ are the roots, perhaps multiple, of the polynomials $P(p)$ and $Q(p)$, respectively. Then for $n \le m$ we obtain

$$R(p) = \frac{P(p)}{Q(p)} = \frac{\alpha_n}{\beta_n} \frac{p - \lambda_1}{p - \mu_1} \frac{p - \lambda_2}{p - \mu_2} \cdots \frac{p - \lambda_n}{p - \mu_n} \frac{1}{(p - \mu_{n+1})\ldots(p - \mu_m)}.$$

We see from this factorization that the operator $R(p)$ is the product of a finite number of operators of the form $\frac{p-\lambda}{p-\mu}$ and $\frac{1}{p-\mu}$. However (see function 2.3.49 and 2.3.50) the operators of this form belong to the ring M, i.e., the operator $R(p)$ also belongs to M. Thus, $R(p)$ is reducible to a function belonging to M. Hence, the condition $n \le m$ is sufficient for the operator $R(p)$ to be reducible to a function.

Conversely: suppose that $R(p)$ is a function. Let us prove that $n \le m$. If $n < m$ then $R(p)$ may be represented in the form

$$R(p) = N(p) + R_1(p),$$

where $N(p)$ is a polynomial, whose degree is greater than zero, and $R_1(p)$ is a rational operator, whose nominator has a degree less than or equal to the degree of the denominator. According to the above reasoning, $R_1(p)$ is a function. Hence, the operator $N(p) = R(p) - R_1(p)$ is reducible to a function. However, in this case Theorem 2.3.85 implies that the degree of the polynomial $N(p)$ is equal to zero, i.e., $N(p)$ is a constant. This contradicts the supposition that $n > m$. Hence, $n \le m$.

Thus, if $n \le m$ then there exists a function $\varphi(t)$ such that $R(p) = \varphi(t)$, where $\varphi \in M$. The value of $R(p)f(t)$ for an arbitrary function $f \in L$ may be computed by the formula

$$R(p)f(t) = \frac{d}{dt} \int_0^t \varphi(t - r)f(r)dr, \qquad \varphi(t) \in M. \qquad (2.3.52)$$

If the roots $\mu_1, \mu_2, \ldots, \mu_m$ of the denominator of $Q(p)$ are simple and $Q(0) \ne 0$, then

$$\frac{P(p)}{pQ(p)} = \frac{1}{p}\frac{P(0)}{Q(0)} + \sum_{k=1}^{m} \frac{\alpha_k}{p - \mu_k},$$

where

$$\alpha_k = \lim_{p \to \mu_k} \frac{(p - \mu_k)P(p)}{pQ(p)} = \lim_{p \to \mu_k} \frac{P(p)}{p\frac{Q(p)-Q(\mu_k)}{p-\mu_k}} = \frac{P(\mu_k)}{\mu_k Q'(\mu_k)}.$$

Hence,

$$\frac{P(p)}{Q(p)} = \frac{P(0)}{Q(0)} + \sum_{k=1}^{m} \frac{P(\mu_k)}{\mu_k Q'(\mu_k)} \frac{p}{p - \mu_k} = \frac{P(0)}{Q(0)} + \sum_{k=1}^{m} \frac{P(\mu_k)}{\mu_k Q'(\mu_k)} \exp(\mu_k t).$$

Thus, we have

$$\varphi(t) = \frac{P(0)}{Q(0)} + \sum_{k=1}^{m} \frac{P(\mu_k)}{\mu_k Q'(\mu_k)} \exp(\mu_k t). \tag{2.3.53}$$

If $Q(0) = 0$, then it means that $Q(p) = pQ_1(p)$, where $Q_1(p) \neq 0$. Using (2.3.53) one can find $\frac{P(p)}{Q(p)} = \varphi_1(t)$, and then we obtain

$$\varphi(t) = \frac{1}{p} \star \varphi_1(t) = \int_0^t \varphi_1(u)du.$$

The case of multiple roots is more complicated. If, for instance, $\mu = \mu_1$ is a root of multiplicity r, then the partial fraction decomposition of $\frac{P(p)}{pQ(p)}$ contains fractions of the form $\frac{A_r}{(p-\mu_1)^r}$. Hence, $\varphi(t)$ contains terms of the form $A_r t^{r-1} e^{\mu_1 t}$. Thus, in the general case the function $\varphi(t)$ has the form

$$\varphi(t) = \sum_{k,r} A_{kr} t^{r-1} e^{\lambda_k t}.$$

This case will be investigated in detail when solving the differential equations in section 2.6.1. $\square$

2.3.5 Laplace Transformable Operators

Let S denote the set of all functions f, for which the Laplace integral

$$\mathcal{L}[f](z) = \int_0^\infty f(t)e^{-zt}dt \tag{2.3.54}$$

is absolutely convergent, while S^* denotes the set of all functions of the complex variable $z = x + iy$ representable by the integral (2.3.54), where $f \in S$.

The set S^* consists of functions analytical in half-planes H_γ (see 1.4.1). Obviously, S^* is a linear set. In addition, the convolution theorem (see 1.4.3, Theorem 1.4.11) implies that if two functions belong to S^*, then their product also belongs to S^*, i.e., S^* is a ring with respect to the ordinary operations of addition and multiplication.

Definition 2.3.41 *An operator $a \in \mathfrak{M}$ is called Laplace transformable if there exists a representative (F, G) such that $a = \frac{F}{G}$ and the Laplace integrals of the functions F and G are convergent, i.e., there exist the integrals*

$$\mathcal{L}[F](z) = \int\limits_0^\infty F(t)e^{-zt}dt;$$

$$\mathcal{L}[G](z) = \int\limits_0^\infty G(t)e^{-zt}dt. \tag{2.3.55}$$

It is known from the properties of Laplace integrals that if the Laplace integral of the function F is convergent, then the Laplace integral of the function $F_1(t) = \int\limits_0^t F(u)du = t \star F(t)$ is absolutely convergent. Besides, obviously, $a = \frac{F}{G} = \frac{t\star F(t)}{t\star G(t)}$. Therefore, if the operator $a = \frac{F}{G}$ is Laplace transformable, then without loss of generality we may always assume the integrals (2.3.55) absolutely convergent. This has to be taken into account below.

Theorem 2.3.87 *The set of all Laplace transformable operators is a field. This field will be denoted by $\mathfrak{M}(S)$.*

Proof. Let the operator $a = \frac{F}{G} \in \mathfrak{M}$ be Laplace transformable and $a \neq 0$. In this case, obviously, the operator $\frac{1}{a} = \frac{G}{F}$ is also Laplace transformable. Furthermore, if $a_1 = \frac{F_1}{G_1}$ and $a_2 = \frac{F_2}{G_2}$ are two Laplace transformable operators, then their sum $a_1 + a_2$ and product $a_1 \star a_2$ are also Laplace transformable operators. Indeed, we have

$$a_1 + a_2 = \frac{F_1 \star G_2 + F_2 \star G_1}{G_1 \star G_2},$$

or

$$a_1 + a_2 = \frac{\frac{d}{dt}\int\limits_0^t F_1(t-u)G_2(u)\,du + \frac{d}{dt}\int\limits_0^t F_2(t-u)G_1(u)du}{\frac{d}{dt}\int\limits_0^t G_1(t-u)G_2(u)\,du}$$

$$= \frac{\int\limits_0^t F_1(t-u)G_2(u)\,du + \int\limits_0^t F_2(t-u)G_1(u)\,du}{\int\limits_0^t G_1(t-u)G_2(u)du}.$$

The Laplace integrals of the functions F_1, F_2, G_1, G_2 are absolutely convergent. According to the convolution theorem for the Laplace transform the Laplace integral of the convolution of such functions is also absolutely convergent. Hence, Laplace integrals of the functions

$$H(t) = \int\limits_0^t F_1(t-u)G_2(u)\,du + \int\limits_0^t F_2(t-u)G_1(u)du,$$

$$R(t) = \int\limits_0^t G_1(t-u)G_2(u)du$$

are absolutely convergent; therefore, the operator $a_1 + a_2$ is Laplace transformable. □

Similarly, the relation

$$a_1 \star a_2 = \frac{\frac{d}{dt}\int\limits_0^t F_1(t-u)F_2(u)du}{\frac{d}{dt}\int\limits_0^t G_1(t-u)G_2(u)du} = \frac{\int\limits_0^t F_1(t-u)F_2(u)du}{\int\limits_0^t G_1(t-u)G_2(u)du}$$

implies Laplace transformability of the operator $a_1 \star a_2$. Obviously, $\mathfrak{M}(S)$ is a subfield of $\mathfrak{M}$.

Almost all problems of application of operational calculus are connected with the field $\mathfrak{M}(S)$. Therefore, it is sufficient for the reader interested in operational calculus as a tool for solving practical problems to restrict himself or herself to the investigation of the field $\mathfrak{M}$.

Definition 2.3.42 *Let* $a = \frac{F}{G} \in \mathfrak{M}(S)$. *The function of the complex variable* $z = x + iy$

$$A(z) = \frac{\mathcal{L}[F](z)}{\mathcal{L}[G](z)}, \tag{2.3.56}$$

is called the Laplace transform of the operator $a = \frac{F}{G}$.

Let us prove that the definition of the function $A(z)$ does not depend on the choice of the representative (F, G). Indeed, if the formulae

$$a = \frac{F}{G} = \frac{F_1}{G_1} \quad \text{and} \quad A_1(z) = \frac{\mathcal{L}[F_1](z)}{\mathcal{L}[G_1](z)}$$

hold, then the condition

$$F * G_1 = F_1 * G, \quad \text{or} \quad \int\limits_0^t F(t-u)G_1(u)du = \int\limits_0^t F_1(t-u)G(u)du$$

implies, by means of the convolution theorem of the Laplace transform, the relation

$$\mathcal{L}[F](z)\mathcal{L}[G_1](z) = \mathcal{L}[F_1](z)\mathcal{L}[G](z), \quad \text{or} \quad A_1(z) = A(z).$$

Hence, the function $A(z)$ is uniquely defined by the operator $a \in \mathfrak{M}(S)$.

Thus, the function $A(z)$, defined by (2.3.55), is associated with every operator $a \in \mathfrak{M}(S)$. This transformation of a into $A(z)$ is often denoted by the symbol

$$\boxed{a \doteq A(z)}. \tag{2.3.57}$$

We denote by $\bar{\mathfrak{M}}(S)$ the image of $\mathfrak{M}(S)$ under the transformation (2.3.57); the elements of the set $\bar{\mathfrak{M}}(S)$ are the functions $A(z) = \frac{\mathcal{L}[F](z)}{\mathcal{L}[G](z)}$, where $\mathcal{L}[F](z)$ and $\mathcal{L}[G](z)$ are absolutely convergent Laplace integrals.

Theorem 2.3.88 *The transformation (2.3.57) establishes a one-to-one correspondence between the fields $\mathfrak{M}(S)$ and $\bar{\mathfrak{M}}(S)$ such that the sum of operators $a_1 + a_2$ corresponds to the sum of functions $A_1(z) + A_2(z)$ and the product of operators $a_1 \star a_2$ corresponds to the ordinary product of functions $A_1(z)A_2(z)$. Zero and unity of the field $\mathfrak{M}(S)$ map onto zero and unity of the field $\bar{\mathfrak{M}}(S)$.*

Proof. Let $a_1 \in \mathfrak{M}(S)$, $a_2 \in \mathfrak{M}(S)$ and $a_1 \doteq A_1(z)$, $a_2 \doteq A_2(z)$. Let us prove that the following formulae hold:

$$a_1 + a_2 \doteq A_1(z) + A_2(z) \quad \text{and} \quad a_1 \star a_2 \doteq A_1(z)A_2(z).$$

We have

$$a_1 + a_2 = \frac{F_1 \star G_2 + F_2 \star G_1}{G_1 \star G_2} = \frac{\int\limits_0^t F_1(t-u)G_2(u)\,du + \int\limits_0^t F_2(t-u)G_1(u)du}{\int\limits_0^t G_1(t-u)G_2(u)du}.$$

As a representative of the operator $a_1 + a_2$, we take the pair

$$\left(\int\limits_0^t F_1(t-u)G_2(u)\,du + \int\limits_0^t F_2(t-u)G_1(u)du, \int\limits_0^t G_1(t-u)G_2(u)du \right),$$

which will be denoted briefly (H, R). The convolution theorem of the Laplace transform implies that

$$\mathcal{L}[H] = \mathcal{L}[F_1]\mathcal{L}[G_2] + \mathcal{L}[F_2]\mathcal{L}[G_1],$$

$$\mathcal{L}[R] = \mathcal{L}[G_1]\mathcal{L}[G_2].$$

Hence, the following relation holds:

$$a_1 + a_2 \doteq \frac{\mathcal{L}[F_1]}{\mathcal{L}[G_1]} + \frac{\mathcal{L}[F_2]}{\mathcal{L}[G_2]} = A_1(z) + A_2(z).$$

Similarly, we have

$$a_1 \star a_2 = \frac{F_1 \star F_2}{G_1 \star G_2} = \frac{\int\limits_0^t F_1(t-u)F_2(u)du}{\int\limits_0^t G_1(t-u)G_2(u)du};$$

therefore, we obtain

$$a_1 \star a_2 \doteq \frac{\mathcal{L}[F_1]\mathcal{L}[F_2]}{\mathcal{L}[G_1]G_2(z)} = A_1(z)A_2(z).$$

Obviously, the transformation (2.3.57) maps the zero element of $\mathfrak{M}(S)$ onto the zero element of the field $\bar{\mathfrak{M}}(S)$ and none of other operators of the field $\mathfrak{M}(S)$ maps onto the zero of the field $\bar{\mathfrak{M}}(S)$. Indeed, if some operator $a \doteq 0$, then $\int\limits_0^\infty F(t)e^{-zt}\,dt = 0$ and by virtue of Theorem 1.4.9, section 1.4.1, we have $F = 0$, i.e., $a = 0$. Thus, the bijectivity of the mapping (2.3.56) is proved.

Finally, if a is the unit operator $a = 1$, then

$$1 \div \frac{\int\limits_0^\infty e^{-zt}\,dt}{\int\limits_0^\infty e^{-zt}dt} = 1,$$

i.e., the unit of the field $\mathfrak{M}(S)$ maps onto the unit of the field $\bar{\mathfrak{M}}(S)$. ⧠

The proven theorem states an isomorphism of the fields $\mathfrak{M}(S)$ and $\bar{\mathfrak{M}}(S)$. The structure of the field $\bar{\mathfrak{M}}(S)$ is clear. Its elements are functions of a complex variable z. Every such function is a ratio of functions representable by absolutely convergent Laplace integrals.

Now we are going to investigate the properties of this isomorphism.

Properties of the Field Isomorphism $\mathfrak{M}(S) \div \bar{\mathfrak{M}}(S)$.

1. *Under the isomorphism $\mathfrak{M}(S) \div \bar{\mathfrak{M}}(S)$ the operator $p = \frac{1}{I}$ corresponds to the function* $I(z) = z$.

Proof. Indeed, the following formula holds:

$$p = \frac{1}{I} \div \frac{\int\limits_0^\infty e^{-zt}\,dt}{\int\limits_0^\infty te^{-zt}\,dt} = \frac{\frac{1}{z}}{\frac{1}{z^2}} = z;$$

thus,

$$\boxed{p \div z}. \qquad (2.3.58)$$

⧠

2. *If the operator a is reducible to a function belonging to S,*

$$a = \frac{F(t)}{I} = f \in S, \quad (F(0) = 0),$$

then the following formula holds:

$$f(t) \div z\mathcal{L}[f](z) = A(z). \qquad (2.3.59)$$

Proof. Indeed, on assuming $Re\,z > \gamma$, we have (see Definition 2.3.42) the operational correspondence

$$a = \frac{F(t)}{t} \div \frac{\int\limits_0^\infty F(t)e^{-zt}dt}{\int\limits_0^\infty te^{-zt}dt} = z^2 \int\limits_0^\infty F(t)d\left(\frac{e^{-zt}}{-z}\right)$$

$$= -zF(t)e^{-zt}\Big|_{t=0}^{t=\infty} + z\int\limits_0^\infty F'(t)e^{-zt}dt = z\int\limits_0^\infty f(t)e^{-zt}dt.$$

3. *If the formula $\frac{F}{G} \doteq \frac{\mathcal{L}[F](z)}{\mathcal{L}[G](z)} = A(z)$ holds, and $\alpha, \beta \in \mathbb{R}_+$, then we have*

$$\frac{F(\alpha t)}{G(\beta t)} \doteq \frac{\beta \mathcal{L}[F]\left(\frac{z}{\alpha}\right)}{\alpha \mathcal{L}[G]\left(\frac{z}{\beta}\right)}. \tag{2.3.60}$$

In particular, we have

$$\frac{F(\alpha t)}{G(\alpha t)} \doteq \frac{\mathcal{L}[F]\left(\frac{z}{\alpha}\right)}{\mathcal{L}[G]\left(\frac{z}{\alpha}\right)} = A\left(\frac{z}{\alpha}\right). \tag{2.3.61}$$

Proof. The proof is by straightforward calculation.

4. *The formula $\frac{F}{G} \doteq \frac{\mathcal{L}[F](z)}{\mathcal{L}[G](z)}$ implies the formula $\frac{e^{\alpha t} F(t)}{e^{\beta t} G(t)} \doteq \frac{\mathcal{L}[F](z-\alpha)}{\mathcal{L}[G](z-\beta)}$. Here, α and β are arbitrary numbers and $e^{\alpha t} F(t)$, $e^{\beta t} G(t)$ denote the ordinary product of functions. In particular,*

$$e^{\alpha t} f(t) \doteq \frac{z}{z-\alpha} \bar{f}(p - \alpha). \tag{2.3.62}$$

Proof. We have the relations

$$\int_0^\infty e^{\alpha t} F(t) e^{-zt} dt = \mathcal{L}[F](z - \alpha) \quad \text{and} \quad \int_0^\infty e^{\beta t} G(t) e^{-zt} dt = \mathcal{L}[G](z - \beta),$$

which imply (2.3.58).

5. *If $F(t) = 0$ for $t < 0$ and $G(t) = 0$ for $t < 0$, then for $\alpha \geq 0$ and $\beta \geq 0$ the formula $\frac{F}{G} \doteq \frac{\mathcal{L}[F](z)}{\mathcal{L}[G](z)} = a(z)$ implies the formula*

$$\frac{F(t - \alpha)}{G(t - \beta)} \doteq \frac{e^{-\alpha z} \mathcal{L}[F](z)}{e^{-\beta z} \mathcal{L}[G](z)} = e^{-(\alpha - \beta) z} A(z). \tag{2.3.63}$$

Proof. We have

$$\int_0^\infty F(t - \alpha) e^{-zt} dt = \int_\alpha^\infty F(t - \alpha) e^{-zt} dt, \qquad t - \alpha = u,$$

then

$$\int_0^\infty F(t - \alpha) e^{-zt} dt = \int_0^\infty F(u) e^{-z(\alpha + u)} du = e^{-z\alpha} \mathcal{L}[F](z),$$

and similarly

$$\int_0^\infty G(t - \beta) e^{-zt} dt = e^{-z\beta} \mathcal{L}[G](z).$$

For the operator $\frac{F(t+\alpha)}{G(t+\beta)}$, $\alpha \geq 0$, $\beta \geq 0$ we have

$$\frac{F(t + \alpha)}{G(t + \beta)} = \frac{e^{\alpha z} [\mathcal{L}[F](z) - \int_0^\alpha F(t) e^{-zt} dt]}{e^{\beta z} [\mathcal{L}[G](z) - \int_0^\beta G(t) e^{-zt} dt]}. \tag{2.3.64}$$

Indeed,

$$\int_0^\infty F(t+\alpha)e^{-zt}dt = \int_\alpha^\infty F(u)e^{-z(u-\alpha)}du = e^{z\alpha}\left(\mathcal{L}[F](z) - \int_0^\alpha F(u)e^{-zt}dt\right)$$

and analogously for the other integral. □

6. *The operational correspondence $\frac{F}{G} \doteq \frac{\mathcal{L}[F](z)}{\mathcal{L}[G](z)}$ implies the formula*

$$\frac{d}{dz}\left(\frac{\mathcal{L}[F](z)}{\mathcal{L}[G](z)}\right) \doteq \frac{-tF'(t)}{G(t)} + \frac{F(t)}{G(t)} \star \frac{tG(t)}{G(t)}. \tag{2.3.65}$$

Notice that $tF(t)$ and $tG(t)$ denote the ordinary product of functions, therefore, the operator $\frac{tG(t)}{G(t)}$ is not reducible to $\frac{t}{1}$.

Proof. We have

$$\frac{d}{dz}\left(\frac{\mathcal{L}[F](z)}{\mathcal{L}[G](z)}\right) = \frac{\frac{d}{dz}\mathcal{L}[F](z)}{\mathcal{L}[G](z)} - \frac{\mathcal{L}[F](z)}{\mathcal{L}[G](z)}\frac{\frac{d}{dz}\mathcal{L}[G](z)}{\mathcal{L}[G](z)};$$

however, $\frac{d}{dz}\mathcal{L}[F](z) = -\int_0^\infty tF(t)e^{-zt}dt$ (see 1.4.1, Theorem 1.4.8). Taking into account this property, we see that (2.3.65) holds. In particular, if the operator $\frac{F}{G}$ is reducible to a function (see formula 2.3.59) belonging to S, we have

$$z\frac{d}{dz}\left(\frac{\bar{f}(z)}{z}\right) = -tf(t). \tag{2.3.66}$$

□

7. *Suppose that F has the derivative $F' = f \in S$. In this case*

$$F'(t) \doteq z\bar{F}(z) - zF(0). \tag{2.3.67}$$

Proof. It follows from (2.3.59) that

$$F'(t) \doteq z\int_0^\infty f(t)e^{-zt}dt = z\int_0^\infty e^{-zt}dF(t) = -zF(0) + z^2\int_0^\infty F(t)e^{-zt}dt = z\bar{F}(z) - zF(0).$$

□

In the same way one can prove the more general formula

$$F^{(n)}(t) \doteq z^n\bar{F}(z) - z^nF(0) - z^{n-1}F'(0) - \cdots - zF^{(n-1)}(0). \tag{2.3.68}$$

The fields $\bar{\mathfrak{M}}(S)$ and $\mathfrak{M}(S)$ are isomorphic. This isomorphism put into correspondence to the operator p the function $a(z) = z$. Hence, an arbitrary polynomial of the operator p, namely $P(p) = \Sigma_k a_k p^k$, corresponds to an ordinary polynomial,

$$P(z) = \sum_k a_k z^k, \quad \text{i.e.,}$$

$$P(p) = \sum_k a_k p^k \doteq \sum_k a_k z^k = P(z).$$

(2.3.69)

Any rational operator $R(p) = \frac{P(p)}{Q(p)}$ corresponds to a rational fraction $R(z) = \frac{P(z)}{Q(z)}$, i.e.,

$$\boxed{R(p) \doteq R(z)}.$$

(2.3.70)

Comparing the formulae (2.3.68) and 2.3.2, (2.3.27) for the derivative $F^{(n)}(t)$, we obtain the formula

$$F^{(n)}(t) = p^n F(t) - p^n F(0) - \cdots - p F^{(n-1)}(0)$$
$$\doteq z^n \bar{F}(z) - z^n F(0) - z^{n-1} F'(0) - \cdots - z F^{(n-1)}(0).$$

(2.3.71)

Let $a = \frac{F}{G} \in \mathfrak{M}(S)$. In this case $a \doteq A(z)$ (see (2.3.60)). It is natural to introduce for the operator a the designation $a = A(p)$. Thus, we put into correspondence to every function $A(z)$ of the complex variable z belonging to the field $\overline{\mathfrak{M}}(S)$ the operator $A(p)$. It is just the operator a of the field $\mathfrak{M}$, which under the mapping $\mathfrak{M} \doteq \overline{\mathfrak{M}}(S)$ maps onto $A(z)$. Thus, we select in the field $\mathfrak{M}$ the subfield $\mathfrak{M}(S)$ of operators, which we can represent in the form of functions of the operator p:

$$a = A(p) \doteq A(z).$$

(2.3.72)

The formal difference between the right- and left-hand sides of formulae (2.3.70), (2.3.71) and (2.3.72) consists only of writing the letter z in the right-hand sides and the letter p in the left-hand sides. In fact, the letter p denotes the operator $\frac{1}{I}$, z denotes a complex variable, $A(p)$ is an operator, $A(z)$ is a function of the complex variable z. However, in view of the isomorphism of the fields $\mathfrak{M}(S)$ and $\overline{\mathfrak{M}}(S)$ the difference between $\mathfrak{M}(S)$ and $\overline{\mathfrak{M}}(S)$ in most cases is not essential. Therefore, the operator p and the complex number z may be denoted by the same letter. Below we shall write p instead of z. Sometimes the designation of the operator $\frac{1}{I}$ and the complex number z by the same letter makes the presentation more simple. Thus, the letter p denotes the operator $\frac{1}{I}$ in the field $\mathfrak{M}$, and p is the complex number $p = \sigma + i\tau$ in the field $\overline{\mathfrak{M}}(S)$. Hence, all Laplace transformable operators, i.e., the elements of $\mathfrak{M}(S)$ may be represented by the form

$$a = A(p) = \frac{\mathcal{L}[F](z)}{\mathcal{L}[G](z)}.$$

(2.3.73)

In particular, every function $f(t)$ of the set S, see formula (2.3.58), may be represented by the form

$$\boxed{f(t) = \bar{f}(p)}.$$

(2.3.74)

The expression $\bar{f}(p)$ will be called the *operational transform* of the function $f(t)$. Here

$$\bar{f}(p) = p \int_0^\infty f(t) e^{-pt} dt = p\mathcal{L}[f](p).$$

(2.3.75)

2.3.6 Examples

1. Now we are going to compute the operational transforms of functions. Let $f(t) = t^\alpha$. Then we find $f(p) = p\mathcal{L}[f](p)$ by means of formula (1.4.12):

$$t^\alpha = p^{-\alpha}\Gamma(1+\alpha), \quad \text{or} \quad \frac{1}{p^\alpha} = \frac{t^\alpha}{\Gamma(1+\alpha)}. \tag{2.3.76}$$

2. Let a function e be given by

$$e(\lambda) = \eta(t; \lambda) = \begin{cases} 0, & \text{if } t < \lambda, \\ 1, & \text{if } t \geq \lambda. \end{cases}$$

Obviously, we have

$$p\int_0^\infty \eta(t; \lambda) e^{-pt} dt = p\int_\lambda^\infty e^{-pt} dt = e^{-p\lambda};$$

hence,

$$\eta(t; \lambda) = e^{-p\lambda}. \tag{2.3.77}$$

This relation immediately implies that

$$\eta(t; \lambda) \star \eta(t; \mu) = \eta(t; \lambda + \mu).$$

This relation was proved in another way in section 2.2; see formula (2.2.2).

Let us find the product $e^{-p\lambda} f(t) = e^{-p\lambda} \star f(t)$, $\lambda \geq 0$. In order to find it we compute the convolution of the functions $e^{-p\lambda} = \eta(t; \lambda)$ and $f(t)$. We have

$$\int_0^t \eta(t-u; \lambda) f(u)\, du = \int_0^t f(t-u)\, \eta(u; \lambda)\, du = \begin{cases} 0, & \text{if } t < \lambda, \\ \int_\lambda^t f(t-u) du, & \text{if } t \geq \lambda \end{cases}.$$

We see from this that the convolution also belongs to the ring M. Hence, we have

$$e^{-p\lambda} f(t) = \frac{d}{dt}\int_0^t \eta(t-u; \lambda) f(u)\, du = \begin{cases} 0, & \text{if } t < \lambda, \\ f(t-\lambda), & \text{if } t \geq \lambda \end{cases}$$

and finally

$$e^{-p\lambda} f(t) = \begin{cases} 0, & \text{if } t < \lambda, \\ f(t-\lambda), & \text{if } t \geq \lambda. \end{cases} \tag{2.3.78}$$

3. Let $f(t) = \log t$.

In order to compute $\bar{f}(p)$ we use relation (2.3.76), which implies

$$\int_0^\infty t^\alpha e^{-pt} dt = \frac{\Gamma(\alpha+1)}{p^{\alpha+1}};$$

therefore,

$$\int_0^\infty \log t\, e^{-pt} dt = \frac{d}{d\alpha}\left[\frac{\Gamma(1+\alpha)}{p^{\alpha+1}}\right]_{\alpha=0} = \frac{\Gamma'(1)}{p} - \frac{\log p}{p},$$

or

$$p \int_0^\infty \log t e^{-pt} dt = -C - \log p,$$

where $\Gamma'(1) = -C$ is the Euler constant; finally we have

$$\boxed{\log t = -C - \log p}.$$ (2.3.79)

4. Let $f(t) = \log^2 t$.

Obviously,

$$\int_0^\infty \log^2 t e^{-pt} dt = \frac{d^2}{d\alpha^2}\left[\frac{\Gamma(1+\alpha)}{p^{\alpha+1}}\right]_{\alpha=0} = \frac{\Gamma''(1)}{p} - \frac{2\Gamma'(1)\log p}{p} + \frac{\log^2 p}{p};$$

however, $\Gamma'(1) = -C$ and $\Gamma''(1) = C^2 + \frac{\pi^2}{6}$; therefore,

$$\bar{f}(p) = \frac{\pi^2}{6} + (C + \log p)^2.$$ (2.3.80)

If instead of $\alpha = 0$ we put $\alpha = n$, where n is an integer, then we easily obtain

$$\frac{t^n}{n!}[\psi(n+1) - \log t] = \frac{\log p}{p^n},$$ (2.3.81)

$$\frac{t^n}{n!}[(\psi(n+1) - \log t)^2 - \psi'(n+1)] = \frac{\log^2 p}{p^n};$$ (2.3.82)

here

$$\psi(z) = \frac{\Gamma'(z)}{\Gamma(z)}.$$

5. Prove that

$$t^n \star \log^2 t = t^n \left[(\log t - \psi(n+1) - C)^2 + \frac{\pi^2}{6} - \psi'(n+1)\right].$$ (2.3.83)

We have

$$t^n \star \log^2 t = \frac{d}{dt} \int_0^t (t-\xi)^n \log^2 \xi d\xi.$$

Taking into account that

$$t^n = \frac{n!}{p^n}, \quad \log^2 t = \frac{\pi^2}{6} + (C + \log p)^2,$$

we obtain

$$t^n \star \log^2 t = \frac{n!}{p^n}\left(\frac{\pi^2}{6} + (C + \log p)^2\right) = \frac{n!\pi^2}{6p^n} + \frac{C^2 n!}{p^n} + \frac{2n!C\log p}{p^n} + \frac{n!\log^2 p}{p^n}.$$

Replacing the operational transforms of the functions by their inverse transforms (see formulas 2.3.82 and 2.3.81) we obtain (2.3.83).

6. Prove the identity

$$S_n = -2\sum_{k=1}^{n}(-1)^k\binom{n}{k}\frac{1}{k^2} = \sum_{k=1}^{n}\frac{1}{k^2} + \left(\sum_{k=1}^{n}\frac{1}{k}\right)^2.$$

We have $\frac{1}{k^2} = \int\limits_0^\infty te^{-kt}dt$; therefore,

$$S_n = 2\int\limits_0^\infty [1 - (1 - e^{-t})^n]dt,$$

on putting $1 - e^{-t} = \xi$, we find

$$S_n = -2\int\limits_0^1 (1 - \xi^n)\log(1 - \xi)\frac{d\xi}{1 - \xi} = n\int\limits_0^1 \xi^{n-1}\log^2(1 - \xi)d\xi,$$

or

$$S_n = n\int\limits_0^1 (1 - \xi)^{n-1}\log^2 \xi d\xi.$$

If we denote

$$f_n(t) = t^n \star \log^2 t = \frac{d}{dt}\int\limits_0^t (t - \xi)^n \log^2 \xi d\xi,$$

then, obviously, $f_n(1) = S_n$; therefore, see formula (2.3.83),

$$S_n = (\psi(n + 1) + C)^2 + \frac{\pi^2}{6} - \psi'(n + 1).$$

Taking into account the relations

$$\psi(n + 1) = -C + \sum_{k=1}^{n}\frac{1}{k}, \qquad \psi'(n + 1) = \frac{\pi^2}{6} - \sum_{k=1}^{n}\frac{1}{k^2},$$

we finally obtain

$$S_n = \left(\sum_{k=1}^{n}\frac{1}{k}\right)^2 + \sum_{k=1}^{n}\frac{1}{k^2}.$$

7. Let

$$\varphi_\sigma(t) = \begin{cases} \frac{\log^\sigma \frac{1}{t}}{\Gamma(1+\sigma)}, & \text{if } 0 < t \le 1, \\ 0, & \text{for } t > 1. \end{cases}$$

Prove that for the operational transform of the function $\varphi_\sigma(t)$ the following formula holds:

$$\bar{\varphi}_\sigma(p) = p\frac{d\bar{\varphi}_{\sigma+1}(p)}{dp}. \tag{2.3.84}$$

We have

$$\frac{\bar{\varphi}_\sigma(p)}{p} = \frac{1}{\Gamma(1 + \sigma)}\int\limits_0^1 \log^\sigma \frac{1}{t}e^{-pt}dt = -\frac{1}{\Gamma(2 + \sigma)}\int\limits_0^1 te^{-pt}d\left(\log^{\sigma+1}\frac{1}{t}\right)$$

$$= \frac{1}{\Gamma(2 + \sigma)}\int\limits_0^1 \log^{\sigma+1}\frac{1}{t}e^{-pt}dt - \frac{p}{\Gamma(2 + \sigma)}\int\limits_0^1 te^{-pt}\log^{\sigma+1}\frac{1}{t}dt,$$

or

$$\frac{\bar{\varphi}_\sigma(p)}{p} = \frac{\bar{\varphi}_{\sigma+1}(p)}{p} + p\frac{d}{dp}\frac{\bar{\varphi}_{\sigma+1}(p)}{p},$$

which implies formula (2.3.84).

8. Let

$$S_n(\sigma) = -\sum_{k=1}^{n}(-1)^k\binom{n}{k}\frac{1}{k^\sigma}, \qquad \sigma \geq 0, \ n = 1,2,\ldots$$

Prove that

$$S_n(\sigma+1) = \sum_{k=1}^{n}\frac{S_k(\sigma)}{k}. \tag{2.3.85}$$

We have

$$\frac{1}{k^\sigma} = \frac{1}{\Gamma(\sigma)}\int_0^\infty t^{\sigma-1}e^{-kt}dt;$$

therefore,

$$S_n(\sigma) = \frac{1}{\Gamma(\sigma)}\int_0^\infty [t - (1-e^{-t})^n]t^{\sigma-1}dt.$$

On putting $1 - e^{-t} = \xi$, we find $dt = \frac{d\xi}{1-\xi}$ and

$$S_n(\sigma) = \frac{1}{\Gamma(\sigma)}\int_0^\infty (1-\xi^n)\log^{\sigma-1}\left(\frac{1}{1-\xi}\right)\frac{d\xi}{1-\xi},$$

or

$$S_n(\sigma) = \frac{n}{\Gamma(1+\sigma)}\int_0^1 \xi^{n-1}\log^\sigma\left(\frac{1}{1-\xi}\right)d\xi;$$

therefore,

$$S_n(\sigma) = \frac{n}{\Gamma(1+\sigma)}\int_0^1 (1-\xi)^{n-1}\log^\sigma\frac{1}{\xi}d\xi.$$

Let us consider the function

$$\Phi_n(\sigma,t) = \frac{d}{dt}\int_0^t (t-\xi)^n\varphi_\sigma(\xi)d\xi, \tag{2.3.86}$$

where the function $\varphi_\sigma(\xi)$ was defined in Example 7.

Obviously,

$$S_n(\sigma) = \Phi_n(\sigma,1). \tag{2.3.87}$$

Furthermore, formula (2.3.86) implies

$$\Phi_n(\sigma,t) = \frac{n!}{p^n}\bar{\varphi}_\sigma(p); \tag{2.3.88}$$

therefore,

$$-t\Phi_n(\sigma+1,t) = p\frac{d}{dp}\left[\frac{n!}{p^{n+1}}\bar{\varphi}_{\sigma+1}(p)\right],$$

or

$$-t\Phi_n(\sigma+1,t) = -\frac{(n+1)!}{p^{n+1}}\bar{\varphi}_{\sigma+1}(p) + \frac{n!\bar{\varphi}'_{\sigma+1}(p)}{p^n}.$$

Taking into account (2.3.84) and (2.3.78) we find

$$-t\Phi_n(\sigma+1,t) = -\Phi_{n+1}(\sigma+1,t) + \frac{n!\varphi_\sigma(p)}{p^{n+1}},$$

or

$$\Phi_{n+1}(\sigma+1,t) - t\Phi_n(\sigma+1,t) = \frac{\Phi_{n+1}(\sigma,t)}{n+1}. \qquad (2.3.89)$$

It follows from (2.3.89) with $t=1$ and (2.3.87) that

$$S_{n+1}(\sigma+1) - S_n(\sigma+1) = \frac{S_{n+1}(\sigma)}{n+1},$$

or

$$S_k(\sigma+1) - S_{k-1}(\sigma+1) = \frac{S_k(\sigma)}{k}.$$

Taking into account that $S_1(\sigma)=1$, we find

$$\sum_{k=2}^{n}[S_k(\sigma+1) - S_{k-1}(\sigma+1)] = \sum_{k=2}^{n}\frac{S_k(\sigma)}{k},$$

or

$$S_n(\sigma+1) - 1 = \sum_{k=2}^{n}\frac{S_k(\sigma)}{k},$$

or

$$S_n(\sigma+1) = \sum_{k=1}^{n}\frac{S_k(\sigma)}{k}.$$

Note that $S_k(0)=1$; therefore,

$$S_n(1) = \sum_{k=1}^{n}\frac{1}{k};$$

therefore,

$$S_n(2) = \sum_{k=1}^{n}\sum_{r=1}^{k}\frac{1}{r} = \frac{1}{2}\left(\sum_{k=1}^{n}\frac{1}{k}\right)^2 + \frac{1}{2}\sum_{k=1}^{n}\frac{1}{k^2}$$

(compare with Example 6).

2.3.7 Periodic Functions

To conclude this section we find the operational transform of a periodic function f. Let $\omega > 0$ be the period of the function f, hence

$$f(t+n\omega) = f(t), \qquad 0 < t < \omega,\ n=1,2,3,\dots.$$

Let us consider the integral

$$\int_0^A f(t)e^{-pt}dt;$$

suppose the integer n is chosen such that $\omega n \leq A < (n+1)\omega$. In this case, we have

$$\int_0^A f(t)e^{-pt}dt = \int_0^{n\omega} f(t)e^{-pt}dt + \int_{n\omega}^A f(t)e^{-pt}dt$$

$$= \sum_{k=1}^n \int_{(k-1)\omega}^{k\omega} f(t)e^{-pt}dt + \int_{n\omega}^A f(t)e^{-pt}dt$$

$$= \sum_{k=1}^n \int_0^\omega f(t+(k-1)\omega)e^{-p(t+(k-1)\omega)}dt + \int_{n\omega}^A f(t)e^{-pt}dt$$

$$= \int_0^\omega f(t)e^{-pt}\left(\sum_{k=1}^n e^{(k-1)\omega p}\right)dt + \int_{n\omega}^A f(t)e^{-pt}dt$$

$$= \frac{1-e^{-n\omega p}}{1-e^{-\omega p}}\int_0^\omega f(t)e^{-pt}dt + \int_{n\omega}^A f(t)e^{-pt}dt.$$

Let us put $Re(p) \geq \epsilon > 0$ and $A \to \infty$; hence, $n \to \infty$. Then we obtain

$$\left|\int_\epsilon^A f(t)e^{-pt}dt\right| \leq \int_{n\omega}^{(n+1)\omega} |f(t)|e^{-Re\,pt}dt \leq \int_{n\omega}^{(n+1)\omega} |f(t)|e^{-\epsilon t}dt$$

$$= \int_0^\omega |f(t)|e^{-\epsilon(t+n\omega)}dt \leq e^{-\epsilon n\omega}\int_0^\omega |f(t)|dt.$$

This implies that for $Re(p) \geq \epsilon > 0$ we have

$$\lim_{A\to\infty}\int_{n\omega}^A f(t)e^{-pt}dt = 0.$$

However, $\lim_{A\to\infty} e^{-n\omega p} = 0$ when $Re(p) \geq \epsilon > 0$; therefore,

$$\lim_{A\to\infty}\int_0^A f(t)e^{-pt}dt = \int_0^\infty f(t)e^{-pt}dt = \frac{\int_0^\omega f(t)e^{-pt}dt}{1-e^{-\omega p}}.$$

Thus, any periodic function is Laplace transformable and its operational transform has the form

$$\bar{f}(p) = \frac{p\int_0^\omega f(t)e^{-pt}dt}{1-e^{-\omega p}} = f(t).$$

Conversely, if the operational transform of the function $f(t)$ has the form

$$\bar{f}(p) = \frac{p\int_0^\omega f(t)e^{-pt}dt}{1-e^{-\omega p}},$$

then f is a periodic function, whose period is equal to ω. Indeed, the last relation implies

$$(1 - e^{-\omega p})f(t) = p \int_0^\omega f(t)e^{-pt}dt = \begin{cases} f(t), & \text{if } t < \omega, \\ 0, & \text{if } t > \omega; \end{cases}$$

on the other hand, see formula (2.3.78).

$$(1 - e^{-\omega p})f(t) = \begin{cases} f(t), & \text{if } t < \omega, \\ f(t) - f(t - \omega), & \text{if } t > \omega. \end{cases}$$

When $t > \omega$, we have $f(t) - f(t - \omega) = 0$, and replacing t with $t + \omega > \omega$, we have $f(t + \omega) - f(t) = 0$ for all $t > 0$, where $f(t)$ is continuous. Hence, the function f is periodic and its period is equal to ω.

2.4 Bases of the Operator Analysis

2.4.1 Sequences and Series of Operators

Definition 2.4.43 *A sequence of operators $a_n \in \mathfrak{M}$ is called convergent to the operator $a = \frac{F}{G} \in \mathfrak{M}$, if there exist representatives (F_n, G_n) such that*

1) $a_n = \frac{F_n}{G_n}$;

2) *The sequences $(F_n(t))_{n\in\mathbb{N}}$ and $(G_n(t))_{n\in\mathbb{N}}$ converge to the limits $F(t)$ and $G(t)$, respectively, uniformly on any finite interval $[0, T]$:*

$$\lim_{n\to\infty} F_n(t) = F(t) \quad and \quad \lim_{n\to\infty} G_n(t) = G(t).$$

The operator $a = \frac{F}{G}$ is called the limit of the sequence of operators a_n and this limit is denoted

$$\lim_{n\to\infty} a_n = a. \tag{2.4.1}$$

Let us prove that the definition of the limit does not depend on the choice of representatives $(F_n(t), G_n(t))$. Indeed, let

$$a_n = \frac{F_n}{G_n} = \frac{\widetilde{F}_n}{\widetilde{G}_n}, \tag{2.4.2}$$

$$\lim_{n\to\infty} \widetilde{F}_n(t) = \widetilde{F}(t); \quad \lim_{n\to\infty} \widetilde{G}_n(t) = \widetilde{G}(t).$$

and the convergence be uniform on every finite segment $[0, T]$. It follows from (2.4.2) that

$$\frac{d}{dt} \int_0^t F_n(t - u)\widetilde{G}_n(u)du = \frac{d}{dt} \int_0^t \widetilde{F}_n(t - u)G_n(u)du;$$

therefore,

$$\int_0^t F_n(t-u)\widetilde{G}_n(u)du = \int_0^t \widetilde{F}_n(t-u)G_n(u)du. \qquad (2.4.3)$$

As $n \to \infty$, the uniform convergence on the segment $0 \le u \le t$ of the sequences

$$f_n(u) = F_n(t-u)\widetilde{G}_n(u) \quad \text{and} \quad \widetilde{f}_n(u) = \widetilde{F}_n(t-u)G_n(u)$$

implies

$$\int_0^t F(t-u)\widetilde{G}(u)du = \int_0^t \widetilde{F}(t-u)G(u)du.$$

Since the functions $F, \widetilde{F}, G$ and $\widetilde{G}$ belong to the set M, then

$$\frac{d}{dt}\int_0^t F(t-u)\widetilde{G}(u)du = \frac{d}{dt}\int_0^t \widetilde{F}(t-u)G(u)du,$$

and hence $a = \frac{F}{G} = \frac{\widetilde{F}}{\widetilde{G}}$; this completes the proof. Therefore, any convergent sequence has a unique limit.

Remark 2.4.98 *If a sequence of functions $f_n(t) \in L$ converges uniformly on every interval $0 \le t \le T$ to a function $f(t)$, then such a sequence is convergent in the sense of the above definition of convergence.*

Indeed, we have

$$a_n = \frac{F_n(t)}{t}, \quad \text{where} \quad F_n(t) = \int_0^t f_n(u)du;$$

$$a = \frac{F(t)}{t}, \quad \text{where} \quad F(t) = \int_0^t f(u)du.$$

Obviously, $\lim_{n\to\infty} F_n(t) = F(t)$, the convergence is uniform and the sequence of the operators a_n converges to the operator a, $\lim_{n\to\infty} a_n = a$.

Remark 2.4.99 *The ordinary convergence in classical calculus is a very special case of the operator convergence. Simple examples show that.*

Example 2.4.49 *The sequence of functions $f_n(t) = \cos nt$, $n \in \mathbb{N}$, is divergent in classical calculus, and in the operational sense it converges to zero.*

Indeed, the sequence

$$a_n = \frac{\frac{\sin nt}{n}}{t} = \cos nt$$

converges to zero, because $\lim_{n\to\infty} \frac{\sin nt}{n} = 0$ and the convergence is uniform on every interval $0 \le t \le T$.

Example 2.4.50 *The sequence of functions $n \sin nt$, $n \in \mathbb{N}$, is convergent.*

Indeed, we have

$$n \sin nt = \frac{1 - \cos nt}{t} = \frac{t - \frac{\sin nt}{n}}{t \star t} = a_n,$$

$$\lim_{n \to \infty} \left(t - \frac{\sin nt}{n} \right) = t$$

and the convergence is uniform. Hence,

$$\lim_{n \to \infty} a_n = \lim_{n \to \infty} n \sin nt = \frac{t}{t \star t} = \frac{1}{t} = p.$$

Example 2.4.51 *The sequence ne^{nt}, $n \in \mathbb{N}$, converges to the operator $-p$.*

Indeed, we have

$$ne^{nt} = \frac{np}{p - n} = \frac{n \cdot \frac{1}{t}}{\frac{1}{t} - n} = \frac{n}{1 - nt} = \frac{1}{\frac{1}{n} - t}.$$

All transformations given here, obviously, are made in the field $\mathfrak{M}$. For instance, on multiplying the nominator and the denominator of the fraction $\frac{n\frac{1}{t}}{\frac{1}{t}-n}$ by the function $F(t) = t$, we obtain

$$\frac{n\frac{1}{t} \star t}{\left(\frac{1}{t} - n\right) \star t} = \frac{n}{1 - nt}.$$

Obviously, the sequence $G_n(t) = \frac{1}{n} - t$ converges uniformly as $n \to \infty$ to the function $G(t) = -t$. Therefore, the sequence of the operators $a_n = \frac{1}{\frac{1}{n}-t}$ is convergent; obviously,

$$\lim_{n \to \infty} a_n = \lim_{n \to \infty} ne^{nt} = -\frac{1}{t} = -p.$$

Thus the sequence ne^{nt} converges to the operator $-p$.

Basic Properties of the Limit of a Sequence of Operators

1. *If a sequence of operators a_n, $n = 1, 2, 3, \ldots$, converges to a limit, then any of its subsequences converges to the same limit.*

Proof. Indeed, let $\lim_{n \to \infty} a_n = a$. This means that

$$\lim_{n \to \infty} F_n(t) = F(t), \quad \lim_{n \to \infty} G_n(t) = G(t), \quad a_n = \frac{F_n}{G_n}, \quad a = \frac{F}{G}.$$

If a_{n_k}, $k = 1, 2, 3, \ldots$, is a subsequence of a_n, then the appropriate subsequences of functions $F_{n_k}(t)$ and $G_{n_k}(t)$ converge to the functions $F(t)$ and $G(t)$, respectively. The convergence is uniform on every interval $[0, T]$. Hence, $\lim_{k \to \infty} a_{n_k} = a$. □

2. *If sequences of operators a_n and b_n, $n = 1, 2, 3, \ldots$, have limits,*

$$\lim_{n \to \infty} a_n = a \quad and \quad \lim_{n \to \infty} b_n = b,$$

then the sequences $(a_n + b_n)_{n \in \mathbb{N}}$ and $(a_n \star b_n)_{n \in \mathbb{N}}$ are convergent and

$$\lim_{n \to \infty} (a_n + b_n) = a + b, \tag{2.4.4}$$

$$\lim_{n\to\infty} (a_n \star b_n) = a \star b. \tag{2.4.5}$$

Proof. Let $a_n = \frac{F_n}{G_n}$, $b_n = \frac{\widetilde{F}_n}{\widetilde{G}_n}$, $a = \frac{F}{G}$ and $b = \frac{\widetilde{F}}{\widetilde{G}}$. By the assumption we have

$$\lim_{n\to\infty} F_n(t) = F(t), \qquad \lim_{n\to\infty} G_n(t) = G(t),$$

$$\lim_{n\to\infty} \widetilde{F}_n(t) = \widetilde{F}(t), \qquad \lim_{n\to\infty} \widetilde{G}_n(t) = \widetilde{G}(t)$$

and the convergence is uniform on every interval $0 \le t \le T$. We obtain

$$a_n + b_n = \frac{F_n}{G_n} + \frac{\widetilde{F}_n}{\widetilde{G}_n} = \frac{F_n \star \widetilde{G}_n + \widetilde{F}_n \star G_n}{G_n \star \widetilde{G}_n}.$$

Let us introduce the notation

$$H_n(t) = \int_0^t [F_n(t-u)\widetilde{G}_n(u) + \widetilde{F}_n(t-u)G_n(u)]\, du,$$

$$R_n(t) = \int_0^t G_n(t-u)\widetilde{G}_n(u)\, du;$$

obviously,

$$a_n + b_n = \frac{H_n(t)}{R_n(t)}.$$

The uniform convergence of the sequences $(F_n(t))_{n\in\mathbb{N}}$, $(G_n(t))_{n\in\mathbb{N}}$, $(\widetilde{F}_n(t))_{n\in\mathbb{N}}$ and $(\widetilde{G}_n(t))_{n\in\mathbb{N}}$ implies the uniform convergence of the sequences $(H_n(t))_{n\in\mathbb{N}}$, $(R_n(t))_{n\in\mathbb{N}}$ and

$$\lim_{n\to\infty} H_n(t) = \int_0^t [F(t-u)\widetilde{G}(u) + \widetilde{F}(t-u)G(u)]du = H(t),$$

$$\lim_{n\to\infty} R_n(t) = \int_0^t G(t-u)\widetilde{G}(u) = R(t).$$

Hence, the following relation holds:

$$\lim_{n\to\infty} (a_n + b_n) = \frac{H(t)}{R(t)} = \frac{\frac{d}{dt}H(t)}{\frac{d}{dt}R(t)} = \frac{F \star \widetilde{G} + \widetilde{F} \star G}{G \star \widetilde{G}} = \frac{F}{G} + \frac{\widetilde{F}}{\widetilde{G}} = a + b.$$

Similarly, for the product of operators $a_n \star b_n$ we have

$$a_n \star b_n = \frac{F_n \star \widetilde{F}_n}{G_n \star \widetilde{G}_n} = \frac{\int_0^t F_n(t-u)\widetilde{F}_n(u)\, du}{\int_0^t G_n(t-u)\widetilde{G}_n(u)\, du};$$

therefore,

$$\lim_{n \to \infty} (a_n \star b_n) = \frac{\lim_{n \to \infty} \int_0^t F_n(t-u)\widetilde{F}_n(u)du}{\lim_{n \to \infty} \int_0^t G_n(t-u)\widetilde{G}_n(u)du} = \frac{\int_0^t F(t-u)\widetilde{F}(u)du}{\int_0^t G(t-u)\widetilde{G}(u)du}$$

$$= \frac{\frac{d}{dt} \int_0^t F(t-u)\widetilde{F}(u)du}{\frac{d}{dt} \int_0^t G(t-u)\widetilde{G}(u)du} = \frac{F \star \widetilde{F}}{G \star \widetilde{G}} = a \star b.$$

□

3. *If there exist the limits*

$$\lim_{n \to \infty} a_n = a, \quad \lim_{n \to \infty} b_n = b \neq 0, \quad b_n \neq 0,$$

then there exists the limit

$$\lim_{n \to \infty} \left(\frac{a_n}{b_n} \right) = \frac{a}{b}.$$

Proof. Indeed, if the sequence of operators $b_n = \frac{\widetilde{F}_n}{\widetilde{G}_n}$ converges to the operator $b = \frac{\widetilde{F}}{\widetilde{G}}$ and $b \neq 0$ (hence, $\widetilde{F}(t) \neq 0$), then, obviously, the sequence of operators $\frac{1}{b_n} = \frac{\widetilde{G}_n(t)}{\widetilde{F}_n(t)}$ is convergent and

$$\lim_{n \to \infty} \frac{1}{b_n} = \frac{\widetilde{G}(t)}{\widetilde{F}(t)} = \frac{1}{b}.$$

Now the second property (see (2.4.2)) implies that $\lim_{n \to \infty} \frac{a_n}{b_n} = \frac{a}{b}$.

□

4. *If c is an arbitrary operator and*

$$\lim_{n \to \infty} a_n = a, \quad then \quad \lim_{n \to \infty} c\, a_n = c\, a.$$

This property follows from (2.4.5).

Along with sequences of operators operator series are also considered in operational calculus.

Definition 2.4.44 *Let $(a_n)_{n \in \mathbb{N}}$ be a sequence of operators, $a_n \in \mathfrak{M}$, $n \in \mathbb{N}$. An operator series is the sequence $(S_n)_{n \in \mathbb{N}}$ of partial sums*

$$S_n = a_1 + a_2 + \cdots + a_n. \tag{2.4.6}$$

An operator series is called convergent if there exists (in the operational sense) the limit of its partial sums (2.4.3).

The limit

$$\lim_{n \to \infty} S_n = \lim_{n \to \infty} (a_1 + a_2 + \cdots + a_n) = S \tag{2.4.7}$$

is called the sum of the series and we write

$$\sum_{n=0}^{\infty} a_n = a_1 + a_2 + \cdots + a_n + \cdots = S. \tag{2.4.8}$$

In section 2.3.6., Example 2, we considered the function $\eta(t; \lambda) = e^{-p\lambda}$. Recall that

$$e^{-p\lambda} = \begin{cases} 0, & \text{if } 0 \le t < \lambda, \\ 1, & \text{if } t \ge \lambda. \end{cases}$$

Let us consider an operator series

$$\sum_{k=0}^{\infty} a_k e^{-p\lambda_k}, \tag{2.4.9}$$

where α_k is an arbitrary sequence of numbers and λ_k are real positive numbers, which forms a monotone increasing sequence tending to infinity, i.e.,

$$0 = \lambda_0 < \lambda_1 < \cdots < \lambda_k < \ldots \quad \text{and} \quad \lim_{k \to \infty} \lambda_k = \infty.$$

Series (2.4.9) is in the operational sense always convergent.

Let us prove that the sequence of partial sums

$$\alpha_0 e^{-\lambda_0 p} + \alpha_1 e^{-\lambda_1 p} + \cdots + \alpha_n e^{-\lambda_n p} = S_n(p)$$

is a convergent sequence. We have

$$S_n(p) = \sum_{k=0}^{n} \alpha_k \frac{\eta_1(t; \lambda_k)}{t} = \frac{F_n(t)}{t},$$

where

$$\eta_1(t; \lambda) = \int_0^t \eta(u; \lambda) du = \begin{cases} 0, & \text{if } t < \lambda, \\ t - \lambda, & \text{if } t \ge \lambda, \end{cases}$$

and

$$F_n(t) = \sum_{k=0}^{n} \alpha_k \eta_1(t; \lambda_k).$$

The sequence of the functions $F_n(t) \in \mathfrak{M}$ converges as $n \to \infty$ uniformly on any interval $0 \le t \le T$. Indeed, let us choose n_0 sufficiently large, such that $\lambda_{n_0} > T$. Then the conditions $n \ge n_0$ and $\eta_1(t; \lambda_n) = 0$ for $0 \le t \le T < \lambda_{n_0}$ imply that $F_n(t) = F_{n_0}(t)$ for all $n \ge n_0$ and $0 \le t \le T$.

Now we calculate the sum of the series $\sum_{k=0}^{\infty} \alpha_k e^{-p\lambda_k}$. Let t be fixed. Taking into account that $e^{-p\lambda} = 0$ for $t < \lambda$ and $e^{-p\lambda} = 1$ for $t \ge \lambda$ we easily find

$$\sum_{k=0}^{\infty} \alpha_k e^{-p\lambda_k} = \sum_{\lambda_k \le t} \alpha_k, \tag{2.4.10}$$

where the sum in the right-hand side is taken over all indices k such that $\lambda_k \le t$. For instance, if $0 < t < \lambda_1$, then $\sum_{\lambda_k \le t} \alpha_k = \alpha_0$. The function defined by series (2.4.10) belongs to the class of step functions.

Definition 2.4.45 *A function f is said to be a step function on the interval* $[0\,\infty)$ *if the interval* $[0, \infty)$ *can be divided into a finite or denumerable number of nonoverlapping intervals, in each of which the function f has a constant value.*

Let us consider the operator $e^{-p\lambda} - e^{-p\mu}$, where $0 \le \lambda < \mu < \infty$. Obviously,

$$e^{-p\lambda} - e^{-p\mu} = \begin{cases} 0, & \text{if } 0 \le t < \lambda; \\ 1, & \text{if } \lambda \le t < \mu; \\ 0, & \text{if } \mu \le t. \end{cases} \qquad (2.4.11)$$

The graph of this step function is shown in Figure 8.

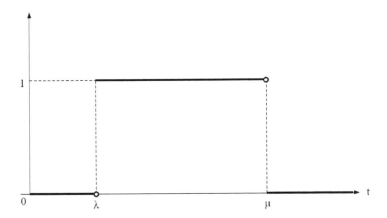

Figure 8

Now let $\varphi(t)$ be an arbitrary step function. In order to define $\varphi(t)$ we have to fix its values in all intervals

$$(\lambda_k, \lambda_{k+1}), \quad k = 0, 1, 2, \ldots; \quad \lambda_0 = 0 < \lambda_1 < \lambda_2 < \cdots < \lambda_n < \ldots, \quad \lim_{n \to \infty} \lambda_n = \infty.$$

Let us put $\varphi(t) = \beta_k$ for $\lambda_k < t < \lambda_{k+1}$; $k = 0, 1, 2, \ldots$. Using (2.4.8), it is easy to write the operational image of the function $\varphi(t)$:

$$\varphi(t) = \sum_{k=0}^{\infty} \beta_k (e^{-\lambda_k p} - e^{-\lambda_{k+1} p}). \qquad (2.4.12)$$

Hence, the set of all step functions coincides with the set of operator series of the form (2.4.12).

In applications series are often used where the numbers λ_k form an arithmetic progression $\lambda_k = kh$, $k = 0, 1, 2, \ldots$. In this case the function $\varphi(t)$ has the form

$$\varphi(t) = \sum_{k=0}^{\infty} \beta_k (e^{-khp} - e^{-(k+1)hp}) = (1 - e^{-hp}) \sum_{k=0}^{\infty} \beta_k e^{-khp};$$

thus,

$$\varphi(t) = (1 - e^{-hp}) \sum_{k=0}^{\infty} \beta_k e^{-khp}. \tag{2.4.13}$$

As an example we consider the function (Figure 9)

$$\varphi(t) = \sum_{k=0}^{\infty} e^{-khp} = \frac{1}{1 - e^{-hp}}.$$

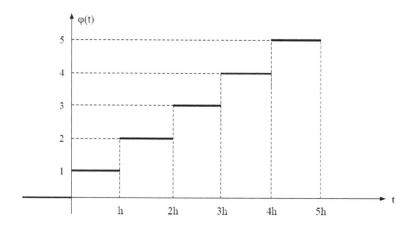

Figure 9

We have

$$(1 - e^{-hp}) \sum_{k=0}^{\infty} k e^{-khp} = \frac{(1 - e^{-hp})e^{-hp}}{(1 - e^{-hp})^2} = \frac{e^{-hp}}{1 - e^{-hp}};$$

therefore,

$$\frac{e^{-hp}}{1 - e^{-hp}} = k \quad \text{for} \quad hk \leq t < h(k+1).$$

2.4.2   Operator Functions

One can consider in the field $\mathfrak{M}$ operators depending on a parameter. Such operators are called *operator functions*. In this subsection the operators depending on one real parameter will be considered. If an operator $a = \frac{F}{G}$ depends on the parameter λ, $\alpha \leq \lambda \leq \beta$, then we shall write $a = a(\lambda)$. An operator function $a(\lambda)$ is defined by its representative (F, G). The functions F and G depend on the parameter λ, i.e., in the general case $F = F(t; \lambda)$ and $G = G(t; \lambda)$, the function $G(t; \lambda)$ does not vanish at any value of the parameter λ.

Example 2.4.52 $a(\lambda) = \frac{t^{\lambda}}{\Gamma(1+\lambda)}$, *if* $0 \leq \lambda < \infty$.

Example 2.4.53 $e(\lambda) = \eta(t; \lambda) = \begin{cases} 0, & \text{for } t < \lambda, \\ 1, & \text{for } t \geq \lambda, \end{cases}$ *where* $0 \leq \lambda < \infty$.

Example 2.4.54 $a(\lambda) = \frac{p}{p-\lambda} = \frac{\frac{1}{t}}{\frac{1}{t}-\lambda} = \frac{\frac{1}{t}\star t}{(\frac{1}{t}-\lambda)\star t} = \frac{1}{1-\lambda t} = e^{\lambda t}$, *where λ is an arbitrary complex number.*

If a is an arbitrary operator from the field $\mathfrak{M}$, then we always may indicate such a function $Q(t)$ belonging to the original ring M, that the product $a \star Q$ also belongs to this ring. If now $a = a(\lambda)$ depends on the parameter λ, then in the general case the function $Q(t)$ also depends on λ.

Definition 2.4.46 *An operator function $a(\lambda)$, $\alpha < \lambda < \beta$, belonging to $\mathfrak{M}(M)$ (see 2.3.2.) or $\mathfrak{M}(S)$ (see 2.3.5.), is called reducible on the interval (α, β) if there exists a function $Q(t) \in M$ (or $Q(t) \in S$), $Q(t) \not\equiv 0$ and $Q(t)$ does not depend on the parameter λ such that for all $\lambda, \alpha < \lambda < \beta$, the product*

$$Q(t) \star a(\lambda) = \varphi(t; \lambda) \tag{2.4.14}$$

belongs to the ring M (or S).

The sum and the product of functions reducible on a given interval (α, β) are also reducible functions. Let us prove this statement for the ring M. The proof for the ring S is analogous.

If $a_1(\lambda)$ and $a_2(\lambda)$ are functions reducible on an interval, then in M there exists functions $Q_1(t)$ and $Q_2(t)$ such that

$$Q_1(t) \star a_1(\lambda) = \varphi_1(t; \lambda) \in M$$

and $\alpha < \lambda < \beta$;

$$Q_2(t) \star a_2(\lambda) = \varphi_2(t; \lambda) \in M,$$

therefore,

$$Q_1 \star Q_2 \star [a_1(\lambda) + a_2(\lambda)] = (Q_2 \star \varphi_1) + (Q_1 \star \varphi_2) \in M$$

for all $\alpha < \lambda < \beta$. Hence, the sum $a_1(\lambda) + a_2(\lambda)$ is a reducible operator function.

Similarly, it follows from

$$Q_1 \star Q_2(a_1(\lambda) \star a_2(\lambda)) = (Q_2 \star \varphi_1) \star (Q_1 \star \varphi_2) \in M$$

that the product of two reducible functions is a reducible function.

However, if $a(\lambda)$ is a reducible function, then the function $\frac{1}{a(\lambda)}$ may be irreducible. Indeed, let us consider the operator function

$$e(\lambda) = \eta(t; \lambda) = \begin{cases} 0 & \text{for } t < \lambda, \\ 1 & \text{for } t \geq \lambda. \end{cases}$$

Obviously, $e(\lambda)$ in the interval $0 \leq \lambda < \infty$ is a reducible function, namely,

$$t \star e(\lambda) = \eta_1(t; \lambda) = \int_0^t e(\lambda)\, dt = \begin{cases} 0 & \text{for } t < \lambda, \\ t - \lambda & \text{for } \lambda \leq t. \end{cases}$$

Hence, in this case $Q(t) = t$; obviously, $e(\lambda) = \frac{\eta_1(t;\lambda)}{t}$.

The inverse function $\frac{1}{e(\lambda)} = \frac{t}{\eta_1(t;\lambda)}$ is not reducible for $0 \leq \lambda < \infty$. Indeed, if it is reducible, then there exists a function $Q(t) \in M$, $Q(t) \not\equiv 0$ such that

$$Q(t) \star \frac{1}{e(\lambda)} = \varphi(t;\lambda) \in M, \quad 0 \leq \lambda < \infty,$$

or

$$Q(t) = e(\lambda) \star \varphi(t;\lambda) = \frac{\frac{d}{dt}\int_0^t \eta_1(t-u;\lambda)\varphi(u;\lambda)\,du}{t},$$

or

$$Q(t) = \frac{d}{dt}\int_0^t \eta(t-u;\lambda)\varphi(u;\lambda)\,du, \quad 0 \leq \lambda < \infty.$$

Let an arbitrary $t = t_0$ be fixed and suppose that $\lambda > t_0$; then

$$Q(t_0) = \frac{d}{dt}\int_0^{t_0} \eta(t_0-u;\lambda)\varphi(u;\lambda)\,du = 0.$$

Thus, $Q(t) = 0$ for all $t \geq 0$. Hence, there exists no nonzero function $Q(t) \in M$, satisfying the condition $Q(t)\frac{1}{e(\lambda)} \in M$; therefore, the function $\frac{1}{e(\lambda)}$ is irreducible in the interval $(0,\infty)$.

The basic operations and notions of calculus may be easily extended to reducible operator functions. This can be done by a single general rule.

Definition 2.4.47 *A reducible operator function $a(\lambda)$ is called continuous in the region $\alpha < \lambda < \beta$ if there exists such a function $Q(t) \in M$, that $\varphi(t;\lambda) = Q(t) \star a(\lambda)$ is a continuous function in two variables t and λ in the region $0 \leq t < \infty$, $\alpha < \lambda < \beta$.*

Using the notion of the limit of a sequence of operators we can introduce the notion of the limit for an operator function.

Definition 2.4.48 *An operator function $a(\lambda)$ has a limit at the point $\lambda = \lambda_0$ if for any sequence $(\lambda_n)_{n\in\mathbb{N}}$ convergent to λ_0 there exists a $\lim_{n\to\infty} a(\lambda_n)$ and this limit does not depend on the choice of the sequence $(\lambda_n)_{n\in\mathbb{N}}$. In this case we write*

$$\lim_{\lambda\to\lambda_0} a(\lambda) = b.$$

Corollary 2.4.34 *If an operator function $a(\lambda)$ is continuous in the interval $a < \lambda < \beta$, then for any λ_0, $\alpha < \lambda_0 < \beta$, there exists*

$$\lim_{\lambda\to\lambda_0} a(\lambda) = a(\lambda_0).$$

This corollary follows immediately from the definition of a reducible continuous operator function.

2.4.3 The Derivative of an Operator Function

Definition 2.4.49 *A reducible operator function $a(\lambda)$ is called continuously differentiable in an interval (α, β), if the function $\varphi(t; \lambda)$ is differentiable with respect to the parameter λ and $\frac{\partial \varphi}{\partial \lambda} \in M$ is a continuous function of the variables t and λ in the region $t \geq 0$, $\alpha < \lambda < \beta$. The operator $\frac{1}{Q} \frac{\partial \varphi}{\partial \lambda}$ is called the continuous derivative of the function $a(\lambda)$ and is denoted $a'(\lambda)$ or $\frac{da(\lambda)}{d\lambda}$; thus,*

$$a'(\lambda) = \frac{da(\lambda)}{d\lambda} = \frac{1}{Q} \frac{\partial \varphi}{\partial \lambda} = \frac{1}{Q} \frac{\partial}{\partial \lambda}(Q \star a(\lambda)). \tag{2.4.15}$$

Let us prove that the definition of the derivative does not depend on the choice of the function $Q(t)$. Indeed, if $\varphi_1(t; \lambda) = Q_1(t) \star a(\lambda)$, then, obviously, the following relation holds:

$$Q \star \varphi_1 = Q_1 \star \varphi,$$

or

$$\frac{d}{dt} \int_0^t Q(t - u)\varphi_1(u; \lambda)du = \frac{d}{dt} \int_0^t Q_1(t - u)\varphi(u; \lambda)du.$$

Since $\frac{\partial \varphi_1}{\partial \lambda}$ and $\frac{\partial \varphi}{\partial \lambda}$ are continuous with respect to the variables t and λ, we have

$$\frac{\partial}{\partial \lambda} \frac{\partial}{\partial t} \int_0^t Q(t - u)\varphi_1(u; \lambda)du = \frac{d}{dt} \int_0^t Q(t - u)\frac{\partial \varphi_1(u; \lambda)}{\partial \lambda}du,$$

$$\frac{\partial}{\partial \lambda} \frac{\partial}{\partial t} \int_0^t Q_1(t - u)\varphi(u; \lambda)du = \frac{d}{dt} \int_0^t Q_1(t - u)\frac{\partial \varphi(u; \lambda)}{\partial \lambda}du.$$

Hence, the relations

$$Q \star \frac{\partial \varphi_1}{\partial \lambda} = Q_1 \star \frac{\partial \varphi}{\partial \lambda}, \quad \frac{1}{Q_1} \star \frac{\partial \varphi}{\partial \lambda} = \frac{1}{Q} \star \frac{\partial \varphi}{\partial \lambda}$$

hold. Thus, the definition of the derivative does not depend on the choice of the function Q.

The derivative of a reducible operator function has the properties of the ordinary derivative.

2.4.4 Properties of the Continuous Derivative of an Operator Function

1. *If operator functions $a(\lambda)$ and $b(\lambda)$ have in an interval (α, β) continuous derivatives, then their sum and product also have continuous derivatives in this interval and*

$$[a(\lambda) + b(\lambda)]' = a'(\lambda) + b'(\lambda), \tag{2.4.16}$$

$$[a(\lambda) \star b(\lambda)]' = a'(\lambda) \star b(\lambda) + a(\lambda) \star b'(\lambda). \tag{2.4.17}$$

Proof. By this assumption, there exist functions $Q_1(t) \in M$ and $Q_2(t) \in M$ such that

$$Q_1(t) \star a(\lambda) = \varphi_1(t; \lambda) \in M, \quad Q_2(t) \star b(\lambda) = \varphi_2(t; \lambda) \in M,$$

and there exist the derivatives $\frac{\partial \varphi_1}{\partial \lambda}$ and $\frac{\partial \varphi_2}{\partial \lambda}$. On multiplying these relations by Q_2 and Q_1, respectively, and denoting $Q = Q_1(t) \star Q_2(t)$, we have

$$Q \star a(\lambda) = Q_2(t) \star \varphi_1(t; \lambda) = \psi_1(t; \lambda) \in M,$$

$$Q \star b(\lambda) = Q_1(t) \star \varphi_2(t; \lambda) = \psi_2(t; \lambda) \in M.$$

By the assumption there exist the derivatives $\frac{\partial \varphi_1}{\partial \lambda}$ and $\frac{\partial \varphi_2}{\partial \lambda}$ belonging to M and continuous with respect to the variables $t, \lambda; 0 \leq t < \infty, \alpha < \lambda < \beta$. Therefore the functions $\psi_1(t; \lambda)$ and $\psi_2(t; \lambda)$ also have the derivatives $\frac{\partial \psi_1}{\partial \lambda}$ and $\frac{\partial \psi_2}{\partial \lambda}$ continuous in the region $0 \leq t < \infty, \alpha < \lambda < \beta$ and

$$\frac{\partial \psi_1}{\partial \lambda} = Q_2(t) \star \frac{\partial \varphi_1}{\partial \lambda} \quad \text{and} \quad \frac{\partial \psi_2}{\partial \lambda} = Q_1(t) \star \frac{\partial \varphi_2}{\partial \lambda}.$$

Hence, the following relation holds:

$$Q \star (a(\lambda) + b(\lambda)) = \psi_1(t; \lambda) + \psi_2(t; \lambda) = \psi(t; \lambda) \in M$$

and

$$[a(\lambda) + b(\lambda)]' = \frac{1}{Q} \frac{\partial \psi}{\partial \lambda} = \frac{1}{Q} \left[\frac{\partial \psi_1}{\partial \lambda} + \frac{\partial \psi_2}{\partial \lambda} \right] = \frac{1}{Q} \left[Q_2 \star \frac{\partial \varphi_1}{\partial \lambda} + Q_1 \star \frac{\partial \varphi_2}{\partial \lambda} \right]$$

$$= \frac{1}{Q_1} \frac{\partial \varphi_1}{\partial \lambda} + \frac{1}{Q_2} \frac{\partial \varphi_2}{\partial \lambda} = a'(\lambda) + b'(\lambda).$$

In order to prove (2.4.16) we take the product

$$Q \star Q \star [a(\lambda) \star b(\lambda)] = \psi_1(t; \lambda) \star \psi_2(t; \lambda).$$

On denoting $Q \star Q = Q^2$ we obtain

$$[a(\lambda) \star b(\lambda)]' = \frac{1}{Q^2} \frac{\partial}{\partial \lambda} [\psi_1 \star \psi_2] = \frac{1}{Q^2} \frac{d}{dt} \frac{\partial}{\partial \lambda} \int_0^t \psi_1(t - u; \lambda) \psi_2(u; \lambda) du$$

$$= \frac{1}{Q^2} \frac{d}{dt} \int_0^t \frac{\partial \psi_1(t - u; \lambda)}{\partial \lambda} \psi_2(u; \lambda) du$$

$$+ \frac{1}{Q^2} \frac{d}{dt} \int_0^t \psi_1(t - u; \lambda) \frac{\partial \psi_2(u; \lambda)}{\partial \lambda} du$$

$$= \frac{1}{Q^2} \left[\frac{\partial \psi_1}{\partial \lambda} \star \psi_2 + \psi_1 \star \frac{\partial \psi_2}{\partial \lambda} \right]$$

$$= \frac{1}{Q^2} \left[Q_1 \star \frac{\partial \varphi_1}{\partial \lambda} \star Q_2 \star \varphi_2 + Q_1 \star \varphi_1 \star Q_2 \star \frac{\partial \varphi_2}{\partial \lambda} \right]$$

$$= \frac{1}{Q_1} \star \frac{\partial \varphi_1}{\partial \lambda} \star \frac{1}{Q_2} \star \varphi_2 + \frac{1}{Q_1} \star \varphi_1 \star \frac{1}{Q_2} \star \frac{\partial \varphi_2}{\partial \lambda}$$

$$= a'(\lambda) \star b(\lambda) + a(\lambda) \star b'(\lambda).$$

2. *If a reducible operator function $a(\lambda)$ is constant in an interval (α, β), i.e., the same operator a is associated with every value of λ from this interval, then $a'(\lambda) = 0$. Conversely if $a'(\lambda) = 0$ for all $\alpha < \lambda < \beta$, then the operator function $a(\lambda)$ is constant in the interval $\alpha < \lambda < \beta$.*

Proof. If $a(\lambda)$ is constant in the interval (α, β), then we have $\frac{\partial \varphi}{\partial \lambda} = 0$. Hence, $a'(\lambda) = 0$. Conversely, if $a'(\lambda) = 0$, then $\frac{\partial \varphi}{\partial \lambda} = 0$ for $\alpha < \lambda < \beta$; therefore, φ does not depend on λ. Then the relation $a(\lambda) = \frac{1}{Q} \star \varphi$ implies that $a(\lambda)$ is constant in the interval (α, β).

3. *If c is an arbitrary operator that does not depend on λ and an operator function $a(\lambda)$ has a continuous derivative in an interval (α, β), then $[c \star a(\lambda)]' = c \star a'(\lambda)$.*

This property follows from Properties 1 and 2.

4. *If operator functions $a(\lambda)$ and $\frac{1}{a(\lambda)}$ have continuous derivatives, then*

$$\left(\frac{1}{a(\lambda)}\right)' = -\frac{a'(\lambda)}{a^2(\lambda)},$$

where $a^2(\lambda) = a(\lambda) \star a(\lambda)$.

Proof. Differentiating by λ the relation $1 = \frac{1}{a(\lambda)} \star a(\lambda)$, by virtue of Property 1, we have

$$0 = \left(\frac{1}{a(\lambda)}\right)' a(\lambda) + \frac{1}{a(\lambda)} \star a'(\lambda),$$

or

$$\left(\frac{1}{a(\lambda)}\right)' = \frac{-a'(\lambda)}{a(\lambda) \star a(\lambda)}.$$

5. *If operator functions $a(\lambda)$, $b(\lambda)$ and $\frac{1}{b(\lambda)}$ have in an interval (α, β) continuous derivatives, then*

$$\left(\frac{a(\lambda)}{b(\lambda)}\right)' = \frac{a'(\lambda)b(\lambda) - a(\lambda)b'(\lambda)}{b(\lambda) \star b(\lambda)}. \tag{2.4.18}$$

Proof. Properties 1 and 4 imply

$$\left(\frac{a(\lambda)}{b(\lambda)}\right)' = a'(\lambda)\frac{1}{b(\lambda)} + a(\lambda)\left(\frac{1}{b(\lambda)}\right)' = \frac{a'(\lambda)}{b(\lambda)} - \frac{a(\lambda)b'(\lambda)}{b^2(\lambda)} = \frac{a'(\lambda)b(\lambda) - a(\lambda)b'(\lambda)}{b^2(\lambda)}.$$

6. *If an operator function $f(\lambda)$ has a continuous derivative $f'(\lambda)$ in an interval $\alpha < \lambda < \beta$ and $\varphi(\lambda)$ is a continuously differentiable number function defined for $\mu < \lambda < \nu$, whose values belong to the interval (α, β), then the composite function $F(\lambda) = (f \circ \varphi)(\lambda)$ has a continuous derivative and the following formula holds*

$$F'(\lambda) = (f \circ \varphi)'(\lambda)\varphi'(\lambda).$$

Proof. There exists a function $Q(t) \in M$ such that $g(\lambda; t) = Q(t) \star f(\lambda)$ has the derivative $\frac{\partial g}{\partial \lambda}$. Obviously,

$$g[\varphi(\lambda); t] = Q(t) \star (f \circ \varphi)(\lambda) = Q(t) \star F(\lambda),$$

and therefore,

$$F'(\lambda) = \frac{1}{Q} \frac{\partial g[\varphi(\lambda); t]}{\partial \lambda} = \frac{1}{Q} \frac{\partial g}{\partial \varphi} \varphi'(\lambda) = (f \circ \varphi)'(\lambda)\varphi'(\lambda).$$

$\square$

Continuous derivatives of higher order are usually defined as:

$$f''(\lambda) = [f'(\lambda)]', \quad f^{(n)}(\lambda) = [f^{(n-1)}(\lambda)]'.$$

We also assume that the right-hand sides have a meaning.

2.4.5 The Integral of an Operator Function

A definite integral for continuous reducible operator functions may be introduced in the same way as the derivative was introduced.

Definition 2.4.50 *There exists always in the ring M a function $Q(t)$ such that $\int_\alpha^\beta Q \star a(\lambda) \, d\lambda = \varphi(t) \in M$; by the definition the integral $\int_\alpha^\beta a(\lambda) \, d\lambda$ is the operator $\frac{\varphi(t)}{Q(t)}$, i.e.,*

$$\int_\alpha^\beta a(\lambda) \, d\lambda = \frac{\varphi(t)}{Q(t)}.$$

This definition is correct. The integral $\int_\alpha^\beta a(\lambda) \, d\lambda$ does not depend on the choice of the function $Q(t)$. Indeed, if $P(t)$ is another operator such that

$$\int_\alpha^\beta P(t) \star a(\lambda) \, d\lambda = \int_\alpha^\beta \psi(t; \lambda) \, d\lambda = \psi(t),$$

then

$$\frac{\psi(t)}{P(t)} = \frac{\varphi(t)}{Q(t)}.$$

Indeed, on denoting

$$Q \star a(\lambda) = \varphi(t; \lambda) \quad \text{and} \quad P \star a(\lambda) = \psi(t; \lambda),$$

we have

$$Q(t) \star \psi(t; \lambda) = P(t) \star \varphi(t; \lambda),$$

or

$$\frac{d}{dt} \int_0^t Q(t-u)\psi(u; \lambda) \, du = \frac{d}{dt} \int_0^t P(t-u)\varphi(u; \lambda) \, du;$$

therefore,

$$\int_0^t Q(t-u)\psi(u;\lambda)du = \int_0^t P(t-u)\varphi(u;\lambda)du.$$

On integrating by λ from α to β, we find

$$\int_\alpha^\beta d\lambda \int_0^t Q(t-u)\psi(u;\lambda)du = \int_\alpha^\beta d\lambda \int_0^t P(t-u)\varphi(u;\lambda)du,$$

after the change of the order of integration

$$\int_0^t Q(t-u)\,du \int_\alpha^\beta \psi(u;\lambda)\,d\lambda = \int_0^t P(t-u)du \int_\alpha^\beta \varphi(u;\lambda)d\lambda,$$

or

$$\int_0^t Q(t-u)\psi(u)du = \int_0^t P(t-u)\varphi(u)du;$$

therefore, $Q \star \psi = P \star \varphi$.

In this way we have proven the relation

$$\int_\alpha^\beta Q(t) \star \psi(t;\lambda)\,d\lambda = Q(t) \star \int_\alpha^\beta \psi(t;\lambda)\,d\lambda. \qquad (2.4.19)$$

The integral of an operator function has all properties of the ordinary integral:

1) $\int_\alpha^\alpha a(\lambda)\,d\lambda = 0,$

2) $\int_\alpha^\beta c \star a(\lambda)\,d\lambda = c \star \int_\alpha^\beta a(\lambda)\,d\lambda$

(where c is an arbitrary operator that does not depend on the parameter λ);

3) $\int_\alpha^\beta a(\lambda)d\lambda = -\int_\beta^\alpha a(\lambda)d\lambda,$

4) $\int_\alpha^\beta a(\lambda)d\lambda + \int_\beta^\gamma a(\lambda)d\lambda = \int_\alpha^\gamma a(\lambda)d\lambda,$

5) $\int_\alpha^\beta [a(\lambda)+b(\lambda)]d\lambda = \int_\alpha^\beta a(\lambda)\,d\lambda + \int_\alpha^\beta b(\lambda)d\lambda.$

If operator functions have continuous derivatives, then

6) $\int_\alpha^\beta a'(\lambda) \star b(\lambda)d\lambda = a(\beta) \star b(\beta) - a(\alpha) \star b(\alpha) - \int_\alpha^\beta a(\lambda) \star b'(\lambda)d\lambda.$

If the values of a number function $\varphi(\lambda)$ defined in the interval $\mu < \lambda < \nu$ belong to the interval (α, β), $\varphi(\mu) = \alpha$, $\varphi(\nu) = \beta$ and $\varphi(\lambda)$ has continuous derivative $\varphi'(\lambda)$, then

7) $\int_\mu^\nu (f \circ \varphi)(\lambda)\,\varphi'(\lambda)d\lambda = \int_{\varphi(\mu)}^{\varphi(\nu)} f(\lambda)d\lambda.$

Remark 2.4.100 *Similarly, other notions from the theory of definite integrals may be extended to the integrals from operator functions. In particular, the improper integral $\int\limits_0^\infty f(\lambda)\,d\lambda$ may be defined as the limit when $n \to \infty$ of the sequence of operators $\int\limits_0^{A_n} f(\lambda)\,d\lambda$, and this definition does not depend on the choice of the sequence of numbers $A_n \to \infty$.*

Consider the function $e(\lambda) = \eta(t; \lambda)$. Let us find the derivative $e'(\lambda)$. We have

$$t^2 \star e(\lambda) = \eta_2(t; \lambda) = \begin{cases} 0 & \text{for } t < \lambda, \\ \frac{(t-\lambda)^2}{2} & \text{for } t \geq \lambda; \end{cases}$$

hence,

$$\frac{\partial \eta_2(t; \lambda)}{\partial \lambda} = -\eta_1(t; \lambda) = \begin{cases} 0 & \text{for } t < \lambda, \\ -(t - \lambda) & \text{for } t \geq \lambda. \end{cases}$$

Here $\eta_1(t; \lambda) = \int\limits_0^t \eta(u; \lambda)du$; therefore,

$$\frac{de(\lambda)}{d\lambda} = e'(\lambda) = -\frac{\eta_1(t; \lambda)}{t^2} = -\frac{1}{t}\frac{\eta_1(t; \lambda)}{t},$$

or, taking into account that $\frac{1}{t} = p$ and $\frac{\eta_1(t;\lambda)}{t} = \eta(t; \lambda) = e(\lambda)$, we obtain

$$e'(\lambda) = -pe(\lambda). \tag{2.4.20}$$

On the other hand, it was proven, see formula (2.3.14) that

$$e(\lambda) \star e(\mu) = e(\lambda + \mu). \tag{2.4.21}$$

Up to now the operator function $e(\lambda)$ was defined only for $\lambda \geq 0$. Let us define $e(\lambda)$ also for $\lambda < 0$ on setting

$$e(-\lambda) = \frac{1}{e(\lambda)}; \quad \lambda > 0.$$

Let us prove that in this case (2.4.21) holds for all real λ and μ.

Indeed, for $\lambda < 0$ and $\mu < 0$ we have

$$e(\lambda)e(\mu) = \frac{1}{e(-\lambda)} \star \frac{1}{e(-\mu)} = \frac{1}{e(-\lambda) \star e(-\mu)} = \frac{1}{e(-\lambda - \mu)} = e(\lambda + \mu).$$

If $\lambda > 0$ and $\mu < 0$, then we have to distinguish two cases. In the first one $\lambda + \mu \geq 0$, then, see formula (2.3.14),

$$e(\lambda)e(\mu) = \frac{e(\lambda)}{e(-\mu)} = \frac{\eta(t; \lambda)}{\eta(t; -\mu)} = \frac{\eta(t; \lambda + \mu)}{1} = e(\lambda + \mu).$$

If $\lambda + \mu < 0$, then

$$e(\lambda)e(\mu) = \frac{\eta(t; \lambda)}{\eta(t; -\mu)} = \frac{1}{\eta(t; -\lambda - \mu)} = e(\lambda + \mu).$$

Hence, the relation (2.4.21) holds for all real λ and μ. The function $\frac{1}{e(\lambda)}$ is not reducible, therefore (2.4.20) has no meaning for $\lambda < 0$. However, we can formally define a derivative of $\frac{1}{e(\lambda)}$, $\lambda > 0$, on setting $\left(\frac{1}{e(\lambda)}\right)' = -\frac{e'(\lambda)}{e(\lambda)\star e(\lambda)}$. In this case (2.4.20) implies

$$\left(\frac{1}{e(\lambda)}\right)' = \frac{pe(\lambda)}{e(\lambda)\star e(\lambda)} = p\frac{1}{e(\lambda)}. \tag{2.4.22}$$

The properties of $e(\lambda)$ for $\lambda < 0$ formulated above show that is convenient to denote

$$e(\lambda) = e^{-\lambda p} \tag{2.4.23}$$

for all real λ, $-\infty < \lambda < \infty$.

It was proven (see 2.3.6, (2.3.78)) that

$$e^{-p\lambda}f(t) = \begin{cases} 0 & \text{if } t < \lambda \\ f(t - \lambda) & \text{if } t \geq \lambda. \end{cases}$$

In the general case $e^{-p\lambda}f(t)$ for $\lambda < 0$ is an operator. Instead of $e^{-p\lambda}f(t)$ for $\lambda < 0$ it is more convenient to consider $e^{p\lambda}f(t)$ for $\lambda > 0$. The expression $e^{p\lambda}f(t)$ is reducible to a function if and only if $f(t)$ is equal to zero in the interval $(0, \lambda)$. More precisely, it must be $\int_0^t f(u)\,du = 0$ for all $0 \leq t \leq \lambda$. Indeed, if $e^{p\lambda}f(t) = \varphi(t)$ is a function, then $f(t) = e^{-p\lambda}\varphi(t)$ and therefore, $f(t) = 0$ (see formula 2.3.78) for $0 \leq t < \lambda$.

Conversely, if $f(t) = 0$ for $0 \leq t < \lambda$, then

$$e^{-p\lambda}f(t+\lambda) = \left\{ \begin{matrix} 0 & \text{for } 0 \leq t < \lambda, \\ f(t) & \text{for } t \geq \lambda \end{matrix} \right\} = f(t) \tag{2.4.24}$$

for all $t \geq 0$. Hence, $e^{p\lambda}f(t) = f(t+\lambda)$ is a function; thus,

$$e^{p\lambda}f(t) = f(t+\lambda) \tag{2.4.25}$$

under the condition that

$$\int_0^t f(u)\,du = 0 \quad \text{for all} \quad 0 \leq t \leq \lambda.$$

Let $f(\lambda) \in L$. The expression $a(\lambda) = e^{-p\lambda}f(\lambda)$ is, obviously, an operator function. Let us compute the integral $\int_0^A e^{-\lambda p}f(\lambda)\,d\lambda$. Since

$$t \star a(\lambda) = \begin{cases} 0 & \text{for } t < \lambda, \\ (t - \lambda)f(\lambda) & \text{for } t \geq \lambda, \end{cases}$$

then the function $t \star a(\lambda) = \varphi(t;\lambda)$ belongs to the ring M for all $\lambda \geq 0$. Using the definition of the integral of an operator function, we obtain

$$\int_0^A e^{-\lambda p}f(\lambda)\,d\lambda = \frac{1}{t} \star \int_0^A \varphi(t;\lambda)\,d\lambda = \begin{cases} p\int_0^t (t-\lambda)f(\lambda)\,d\lambda & \text{for } 0 < t < A \\ p\int_0^A (t-\lambda)f(\lambda)\,d\lambda & \text{for } t > A, \end{cases}$$

or

$$\int_0^A e^{-\lambda p} f(\lambda)\, d\lambda = \int_0^t f(\lambda) d\lambda \quad for \quad 0 < t < A.$$

Hence, the following relation holds:

$$p \int_0^A e^{-p\lambda} f(\lambda) d\lambda = f(t) \quad for \quad 0 < t < A.$$

Let us now

$$A_1 < A_2 < A_3 < \cdots < A_n < A_{n+1} < \ldots,$$

$$\lim_{n \to \infty} A_n = \infty \quad and \quad a_n = p \int_0^{A_n} e^{-\lambda p} f(\lambda)\, d\lambda.$$

Let us prove that the sequence a_n converges as $n \to \infty$. Indeed, let $(0, T)$ be an arbitrary fixed interval and n_0 be such that $A_{n_0} > T$. In this case for all $n > n_0$ we have

$$\frac{a_n}{p} = t \star a_n = \int_0^t f(\lambda)\, d\lambda \quad for \quad 0 < t < T.$$

This implies the convergence and the relation

$$\lim_{n \to \infty} a_n = f(t).$$

It is clear that the limit does not depend on the choice of the sequence $A_1 < A_2 < \cdots < A_n < \ldots$. Hence, there exists the improper integral

$$p \int_0^\infty e^{-p\lambda} f(\lambda) d\lambda = f(t). \qquad (2.4.26)$$

2.5   Operators Reducible to Functions

2.5.1   Regular Operators

An operator a belonging to the field $\mathfrak{M}(S)$ is called *regular* if the function $\bar{a}(p)$ associated with a in the field $\overline{\mathfrak{M}}(S)$ is analytic in a neighborhood of the point at infinity. The regular operators form a large and important class of operators for applications. Obviously, the sum of two regular operators is also a regular operator, and the product of two regular operators is a regular operator, too. Any regular operator is reducible to a function. This follows from the following theorem.

Theorem 2.5.89 *Let $\bar{a}(p)$ be a regular operator. Hence, in a neighborhood of the point at infinity $|p| > R$ we have the expansion $\bar{a}(p) = \sum_{k=0}^{\infty} \frac{a_k}{p^k}$; then the following relation holds*

$$\bar{a}(p) = \sum_{k=0}^{\infty} \frac{a_k}{p^k} = \sum_{k=0}^{\infty} \frac{a_k t^k}{k!} = a(t), \qquad (2.5.1)$$

and the radius of convergence is equal to infinity.

Proof. The function $\sum_{k=1}^{\infty} \frac{a_k}{p^k}$ may be represented by a Laplace integral, see section 1.4.2, formula (1.4.13),

$$\sum_{k=1}^{\infty} \frac{a_k}{p^k} = \int_0^{\infty} \left(\sum_{k=1}^{\infty} \frac{a_k t^{k-1}}{(k-1)!} \right) e^{-zt} dt;$$

therefore

$$\sum_{k=1}^{\infty} \frac{a_k}{p^{k-1}} = \sum_{k=1}^{\infty} \frac{a_k t^{k-1}}{(k-1)!},$$

the series converges for all values of t, and multiplying it by the operator $\frac{1}{p}$, we find

$$\sum_{k=1}^{\infty} \frac{a_k}{p^k} = \sum_{k=1}^{\infty} \frac{a_k t^k}{k!};$$

hence,

$$a = \bar{a}(p) = \sum_{k=0}^{\infty} \frac{a_k}{p^k} = \sum_{k=0}^{\infty} \frac{a_k t^k}{k!}.$$

□

Theorem 2.5.90 *If for a sequence of regular operators $\left(\bar{F}_n(p) \right)_{n \in \mathbb{N}}$, the series*

$$\sum_{n=0}^{\infty} \bar{F}_n(p) = \bar{F}(p) \qquad (2.5.2)$$

converges uniformly in the region $|p| > R$, then $\bar{F}(p)$ is a regular operator and

$$\bar{F}(p) = \sum_{n=0}^{\infty} F_n(t), \qquad (2.5.3)$$

where

$$F_n(t) = \bar{F}_n(p).$$

Series (2.5.3) converges uniformly on every segment $0 \le t \le A$.

Proof. Every function $\bar{F}_n(p)$ is analytic in the region $|p| > R$. The uniform convergence implies that the function $F(p)$ is also analytic in the region $|p| > R$. If C_{R_1} is a circle with the center at the point $p = 0$, then it is known that the following formula holds:

$$\frac{t^k}{k!} = \frac{1}{2\pi i} \int_{C_{R_1}} \frac{e^{pt}}{p^{k+1}} dp.$$

Replacing $\frac{t^k}{k!}$ in (2.5.1) with $\frac{1}{2\pi i}\int_{C_{R_1}}\frac{e^{pt}}{p^{k+1}}dp$ and supposing the radius R_1 of the circle C_{R_1} to be larger than R, we find

$$\frac{1}{2\pi i}\int_{C_{R_1}}\frac{\bar{a}(p)}{p}e^{pt}dp = a(t).\tag{2.5.4}$$

Applying this formula to $\bar{F}_n(p)$ we have

$$F_n(t) = \frac{1}{2\pi i}\int_{C_{R_1}}\frac{\bar{F}_n(p)}{p}e^{pt}dp\tag{2.5.5}$$

and by virtue of the uniform convergence of series (2.5.2) we have

$$\sum_{n=0}^{\infty}F_n(t) = \frac{1}{2\pi i}\int_{C_{R_1}}\sum_{n=0}^{\infty}\frac{\bar{F}_n(p)}{p}e^{pt}dp$$

or

$$\sum_{n=0}^{\infty}F_n(t) = \frac{1}{2\pi i}\int_{C_{R_1}}\frac{\bar{F}_n(p)}{p}e^{pt}dp = F(t).$$

It remains now to prove the uniform convergence of series (2.5.3) in the segment $0 \le t \le A$. Let $\epsilon > 0$. Then there exists N such that

$$\left|\sum_{n=N}^{\infty}\bar{F}_n(p)\right| < \epsilon e^{-R_1 A}\quad\text{in the region}\quad |p| \ge R_1 > R.$$

Therefore, (2.5.4) and the last inequality imply (setting $p = R_1 e^{i\varphi}, 0 \le \varphi \le 2\pi$)

$$\left|\sum_{n=N}^{\infty}F_n(p)\right| \le \frac{1}{2\pi}\int_{C_{R_1}}\frac{e^{-R_1 A_\epsilon}|e^{pt}|}{R_1}d\varphi \le \frac{1}{2\pi R_1}2\pi R_1\epsilon = \epsilon.$$

Thus,

$$\left|\sum_{n=N}^{\infty}F_n(t)\right| < \epsilon\quad\text{for all}\quad 0 \le t \le A.$$

The uniform convergence is proven.   □

Example 2.5.55 *The operator $e^{\frac{\lambda}{p}}$ is, obviously, regular. We have the expansion*

$$e^{-\frac{\lambda}{p}} = \sum_{n=0}^{\infty}\frac{(-\lambda)^n}{n!p^n}.$$

The series converges for $|p| > 0$; therefore,

$$e^{-\frac{\lambda}{p}} = \sum_{n=0}^{\infty}(-1)^n\frac{\lambda^n t^n}{(n!)^2} = J_o(2\sqrt{\lambda t}).$$

Similarly, we have

$$e^{\frac{\lambda}{p}} = J_o(2\sqrt{-\lambda t}) = J_o(i2\sqrt{\lambda t}) = I_o(2\sqrt{\lambda t}).$$

In the same way one can easily obtain

$$\frac{1}{p^\nu}e^{-\frac{\lambda}{p}} = \left(\frac{t}{\lambda}\right)^{\frac{\nu}{2}} J_\nu(2\sqrt{\lambda t}), \tag{2.5.6}$$

$$\frac{1}{p^\nu}e^{\frac{\lambda}{p}} = \left(\frac{t}{\lambda}\right)^{\frac{\nu}{2}} I_\nu(2\sqrt{\lambda t}) \tag{2.5.7}$$

Example 2.5.56 *The operator $\frac{p}{\sqrt{p^2+\lambda^2}}$ is, obviously, regular, and for $|p| > \lambda > 0$ we have*

$$\frac{p}{\sqrt{p^2+\lambda^2}} = \frac{1}{\sqrt{1+\frac{\lambda^2}{p^2}}} = \left(1+\frac{\lambda^2}{p^2}\right)^{-\frac{1}{2}} = \sum_{k=0}^{\infty}\binom{-\frac{1}{2}}{k}\left(\frac{\lambda^2}{p^2}\right)^k = \sum_{k=0}^{\infty}\binom{-\frac{1}{2}}{k}\frac{(\lambda t)^{2k}}{(2k)!},$$

or

$$\frac{p}{\sqrt{p^2+\lambda^2}} = \sum_{k=0}^{\infty}\frac{-\frac{1}{2}(-\frac{1}{2}-1)\cdots(-\frac{1}{2}-k+1)(\lambda t)^{2k}}{k!\cdot 1\cdot 2\cdot 3\cdots k(k+1)\cdots(2k-1)2k}$$

$$= \sum_{k=0}^{\infty}\frac{(-1)^k 1\cdot 3\cdot 5\cdots(2k-1)(\lambda t)^{2k}}{2^k\cdot k!\cdot 1\cdot 3\cdots(2k-1)\cdot 2^k k!}$$

$$= \sum_{k=0}^{\infty}\frac{(-1)^k(\frac{\lambda t}{2})^{2k}}{(k!)^2} = J_0(\lambda t);$$

Thus

$$\frac{p}{\sqrt{p^2+\lambda^2}} = J_0(\lambda t).$$

We have obtained the operator transform of the Bessel function of order zero.

2.5.2   The Realization of Some Operators

In operational calculus, one deals, as a rule, with operators represented by the form $\bar{a}(p)$. Two problems arise here:

1) some criteria must be formulated, which allow us to decide if a given operator $\bar{a}(p)$ is reducible to a function;

2) if an operator is reducible to a function, then we need to find this function.

This problem often may be solved only approximately, i.e., one can calculate only individual values of the function.

An operator $\bar{a}(p)$ is reducible to a function $\varphi(t) \in S$ if and only if $\bar{a}(p)$ and $\varphi(t)$ satisfy the relation

$$\bar{a}(p) = p\int_0^\infty \varphi(t)e^{-pt}dt = p\mathcal{L}[\varphi](p), \quad p \in H_\gamma. \tag{2.5.8}$$

Hence, an operator $\bar{a}(p)$ is reducible to a function $\varphi(t)$ if the function $\frac{\bar{a}(p)}{p}$ of a complex variable is representable by an absolutely convergent Laplace integral. In section 1.4.4 some sufficient conditions for an analytical function in the half-plane H_γ to be representable by the Laplace integral were given. If it is shown that the operator $\bar{a}(p)$ is reducible to a function, then in order to find this function we have to use the inversion theorem for the Laplace integral (see Theorem 1.4.13 and formula 2.5.8) and we obtain

$$\varphi(t) = \frac{1}{2\pi i} \int\limits_{(\gamma)} \frac{\bar{a}(p)}{p} e^{pt} dp. \tag{2.5.9}$$

In order to obtain a convenient expression for the function $\varphi(t)$ we often must deform the path of integration in formulae (2.5.9). Sometimes, applying Jordan's lemma and the Cauchy theorem on the residues, one can obtain an explicit series expansion of $\varphi(t)$; see 1.4.5.

Theorem 2.5.91 *Let*

1) *$\bar{a}(p)$ be a regular function in any finite part of the plane of the complex variable p except the set of points $p_1, p_2, \cdots, p_n, \cdots$ ($|p_1| < |p_2| < \cdots < |p_n| < \cdots$) which are the poles of the function $\frac{\bar{a}(p)}{p}$ such that $Re(p_n) < \gamma_0$ for all n;*

2) *the following limit exists*

$$\lim_{\omega \to \infty} \frac{1}{2\pi i} \int\limits_{\gamma - i\omega}^{\gamma + i\omega} \frac{\bar{a}(p)}{p} e^{pt} dp = \frac{1}{2\pi i} \int\limits_{(\gamma)} \frac{\bar{a}(p)}{p} e^{pt} dp, \quad \gamma > \gamma_0;$$

3) *there exist a sequences of simple contours C_n relying on the line $Re(p) = \gamma$ at the points $\gamma + i\beta_n$, $\gamma - i\beta_n$. (These contours lie in the half-plane $Re(p) < \gamma$ and do not go through the poles p_n.) Each contour C_n contains the origin and n first poles $p_1, p_2, \cdots, p_n$ (see Figure 10).*

4) *for all $t > 0$*

$$\lim_{n \to \infty} \int\limits_{C_n} \frac{\bar{a}(p)}{p} e^{pt} dp = 0;$$

then the value of the integral is equal to the convergent series

$$\frac{1}{2\pi i} \int\limits_{(\gamma)} \frac{\bar{a}(p)}{p} e^{pt} dp = \sum_{n=0}^{\infty} r_n(t),$$

where $r_n(t)$ is the residue of the function $\frac{\bar{a}(p)}{p} e^{pt}$ at the point $p = p_n (n = 1, 2, \cdots)$ and $r_0(t)$ is the residue at zero.

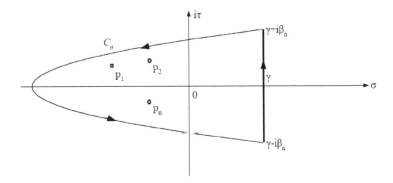

Figure 10

NOTE. If the function $\frac{\bar{a}(p)}{p}$ satisfies the conditions of Jordan's lemma, then it is natural to choose as C_n arcs of circles.

If there exist sequences of positive numbers β_n and δ_n and a number $Q > 0$ such that

1) $\quad \lim_{n \to \infty} \beta_n = \infty, \quad \lim_{n \to \infty} \delta_n = 0;$

2) $\quad \left| \dfrac{\bar{a}(\sigma \pm i\beta_n)}{\sigma \pm i\beta_n} \right| < \delta_n, \quad \left| \dfrac{\bar{a}(-\beta_n + ir)}{-\beta_n + ir} \right| < Q, \quad -\beta_n \le \sigma \le \gamma, \quad |r| \le \beta_n,$

then one can take as the contours C_n the contour shown in Figure 11.

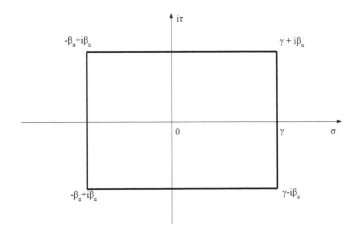

Figure 11

Example 2.5.57 *Let us consider the operator* $\sqrt{p}\,e^{-\lambda\sqrt{p}}, \lambda > 0$. *The function* $\bar{f}(p) = \frac{\sqrt{p}}{p}e^{-\lambda\sqrt{p}}$ *is, obviously, bounded on the half-plane* $H_{\gamma_o}, \gamma_o > 0$ *and*

$$\int_{-\infty}^{\infty} \left| \frac{1}{\sqrt{\sigma + ir}} e^{-\lambda\sqrt{\sigma + ir}} \right| dr < \infty,$$

therefore (see section 1.4.5, Theorem 1.4.15) the function $\bar{f}(p)$ is representable by a Laplace integral. Hence, the operator $\sqrt{p}e^{-\lambda\sqrt{p}}$ is reducible to a function. The value of this function is equal to the integral

$$\frac{1}{2\pi i}\int\limits_{(\gamma)}\frac{e^{-\lambda\sqrt{p}+tp}}{\sqrt{p}}dp, \quad \gamma>0.$$

In order to calculate this integral we use the formula $\sqrt{tp}-\frac{\lambda}{2\sqrt{t}}=\omega$ to change the variable of integration. We have

$$\frac{1}{2\pi i}\int\limits_{(\gamma)}\frac{e^{-\lambda\sqrt{p}+tp}}{\sqrt{p}}dp=\frac{1}{\pi i\sqrt{t}}\int\limits_{(\gamma)}e^{(\sqrt{tp}-\frac{\lambda}{2\sqrt{p}})^2-\frac{\lambda^2}{4t}}d(\sqrt{tp})=\frac{e^{-\frac{\lambda^2}{4t}}}{\pi i\sqrt{t}}\int\limits_{L}e^{\omega^2}d\omega;$$

in this case the line $Re(p)=\gamma$ transforms into the hyperbola L; thus,

$$\sqrt{p}e^{-\lambda\sqrt{p}}=\frac{e^{-\frac{\lambda^2}{4t}}}{\pi i\sqrt{t}}C,$$

where $C=\int\limits_{L}e^{\omega^2}d\omega$ is constant. In order to find C, let us set $\lambda=0$. Then $\sqrt{p}=\frac{C}{\pi i\sqrt{t}}$, and from 2.3.2, (2.3.24) and the remarks on this formula in 2.3.3, we have

$$\sqrt{p}=p^{\frac{1}{2}}=\frac{t^{-\frac{1}{2}}}{\Gamma(1-\frac{1}{2})}=\frac{1}{\sqrt{\pi t}},$$

hence, we have $\frac{1}{\sqrt{\pi t}}=\frac{C}{\pi i\sqrt{t}}$. or $C=i\sqrt{\pi}$; thus,

$$\sqrt{p}e^{-\lambda\sqrt{p}}=\frac{1}{\sqrt{\pi t}}e^{-\frac{\lambda^2}{4t}}. \tag{2.5.10}$$

2.5.3   Efros Transforms

The following theorem is useful for solving the question about the reducibility of a given operator $\bar{a}(p)$ to a function and computation of this function.

Theorem 2.5.92 *If an operator $\bar{a}(p)$ can be represented by the form*

$$\bar{a}(p)=pH\left[\frac{\bar{h}(p)}{p}\right], \tag{2.5.11}$$

where the operator $\bar{h}(p)$ is reducible to a function $h(p)\in S$ and $H(z)$ is an analytic function in the disk $|z|<\rho$ and $H(0)=0$, then $\bar{a}(p)$ is reducible to the function $a(t)$ belonging to S.

Proof. Indeed, the formulated assumptions imply that the function $\frac{\bar{h}(p)}{p}$ is reducible by an absolutely convergent Laplace integral. This implies the representability of the function $H\left[\frac{\bar{h}(p)}{p}\right]$ by an absolutely convergent Laplace integral, i.e.,

$$\bar{a}(p)=pH\left[\frac{\bar{h}(p)}{p}\right]=p\int\limits_{0}^{\infty}a(t)e^{-pt}dt,$$

and this integral is absolutely convergent for $Re(p) > \gamma$; hence, the operator $\bar{a}(p)$ is reducible to a function.

Consider the function $G(z) = \frac{1}{\zeta-z} - \frac{1}{\zeta}$, where ζ is a complex parameter. Obviously, the function $G(z)$ is analytical in the region $|z| < |\zeta| = \rho$; therefore, the operator $\frac{p}{\zeta - \frac{\bar{h}(p)}{p}} - \frac{p}{\zeta}$ is reducible to a function. Let us introduce the notation

$$\frac{p}{\zeta - \frac{\bar{h}(p)}{p}} - \frac{p}{\zeta} = \frac{p\bar{h}(p)}{\zeta(\zeta p - \bar{h}(p))} = K(t;\zeta). \tag{2.5.12}$$

Multiplying both sides of this relation by $\frac{1}{2\pi i}H(\zeta)$ and integrating along the circle $|\zeta| = \rho$, we have

$$\frac{1}{2\pi i}\int\limits_{|\zeta|=\rho} \frac{pH(\zeta)}{\zeta - \frac{\bar{h}(p)}{p}}d\zeta - \frac{p}{2\pi i}\int\limits_{|\zeta|=\rho}\frac{H(\zeta)}{\zeta}d\zeta = \frac{1}{2\pi i}\int\limits_{|\zeta|=\rho}K(t;\zeta)H(\zeta)d\zeta.$$

But $H(0) = 0$; therefore, from the Cauchy integral formula we have

$$pH\left[\frac{\bar{h}(p)}{p}\right] = \frac{1}{2\pi i}\int\limits_{|\zeta|=\rho}K(t;\zeta)H(\zeta)d\zeta = a(t). \tag{2.5.13}$$

☐

This formula may be applied to compute the function $a(t)$. The simplest example is the case of $\bar{h}(p) = 1$; then the operator $pH\left(\frac{1}{p}\right)$ is regular. For $\bar{h}(p) = 1$ we have

$$K(t;\zeta) = \frac{p}{\zeta(p\zeta - 1)} = \frac{p}{\zeta^2(p - \frac{1}{\zeta})} = \frac{1}{\zeta^2}\exp\left(\frac{t}{\zeta}\right)$$

and

$$pH\left(\frac{1}{p}\right) = \frac{1}{2\pi i}\int\limits_{|\zeta|=\rho}\exp\left(\frac{t}{\zeta}\right)H(\zeta)\frac{d\zeta}{\zeta^2}.$$

On setting $\frac{1}{\zeta} = z$, we have

$$\bar{a}(p) = pH\left(\frac{1}{p}\right) = \frac{1}{2\pi i}\int\limits_{|\xi|=\frac{1}{\rho}}e^{tz}H\left(\frac{1}{z}\right)dz.$$

This equation coincides with (2.5.4), section 2.5.1.

Consider now another method of computing the function by its operator transform.

Theorem 2.5.93 *If an operator $\bar{a}(p)$ may be represented by the form* $\bar{a}(p) = \frac{\bar{\Phi}(q(p))}{pq(p)}$, *and*

a) $\bar{\Phi}(p) = p\int\limits_0^\infty \Phi(t)\exp(-pt)dt$ *and* $\int\limits_0^\infty |\Phi(t)|\exp(-\gamma_o t)dt < \infty$, *where* $\gamma_o > 0$;

b) $q(p)$ *is analytic in the half-plane* $\bar{H}_{\gamma_o}$, $\gamma_o \geq 0$, *satisfying in this half-plane the condition* $Re(q(p)) \geq \gamma_o$, *then the operator $\bar{a}(p)$ is reducible to a function belonging to S:*

$$\bar{a}(p) = \int_0^\infty L(t;\xi)\Phi(\xi)d\xi, \qquad (2.5.14)$$

where

$$L(t;\xi) = \frac{1}{p}e^{-\xi q(p)}. \qquad (2.5.15)$$

Proof. The conditions a) and b) imply the absolute and uniform convergence in the half-plane $\bar{H}_{\gamma_o}$ of the integral

$$\frac{1}{q(p)}\bar{\Phi}(q(p)) = \int_0^\infty e^{-\xi q(p)}\Phi(\xi)d\xi. \qquad (2.5.16)$$

Indeed,

$$\left|\int_0^\infty e^{-\xi q(p)}\Phi(\xi)d\xi\right| \le \int_0^\infty e^{-\xi Re(q(p))}|\Phi(\xi)|d\xi \le \int_0^\infty e^{-\xi\gamma_o}|\Phi(\xi)|d\xi < \infty; \qquad (2.5.17)$$

therefore, the integral (2.5.16) represents in the half-plane $\bar{H}_{\gamma_o}$ an analytic function. The inequality

$$\left|\frac{\bar{\Phi}(q(p))}{q(p)}\right| \le \int_0^\infty e^{-\xi\gamma_o}|\Phi(\xi)|d\xi$$

implies the uniform boundedness of the function $\frac{\bar{\Phi}(q(p))}{q(p)}$ in the half-plane $\bar{H}_{\gamma_o}$; therefore, there exists the integral

$$\frac{1}{2\pi i}\int_{(\gamma)}\frac{\bar{\Phi}(q(p))}{p^2 q(p)}e^{pt}dp = a(t),$$

and

$$p\int_0^\infty a(t)e^{-pt}dt = \frac{\bar{\Phi}(q(p))}{pq(p)} = \bar{a}(p). \qquad (2.5.18)$$

Hence, the operator $\bar{a}(p)$ is reducible to the function $a(t)$. In order to compute $a(t)$ we note that for any ξ, $0 \le \xi < \infty$, the operator $\frac{1}{p}e^{-\xi q(p)}$ is reducible to the function

$$L(t;\xi) = \frac{1}{2\pi i}\int_{\gamma-i\infty}^{\gamma+i\infty}\frac{e^{-\xi q(p)+pt}}{p^2}dp \qquad (2.5.19)$$

because the function $\frac{e^{\xi q(p)}}{p^2}$, $0 \le \xi < \infty$, satisfies all the conditions of Theorem 1.4.13, section 1.4.4, and hence it may be represented by an absolutely convergent Laplace integral. In particular, integral (2.5.19) is absolutely convergent, and taking into account (2.5.15) and

(2.5.17) we find

$$\int\limits_0^\infty L(t;\xi)\Phi(\xi)d\xi = \lim_{A\to\infty} \frac{1}{2\pi i}\int\limits_{(\gamma)} \frac{e^{pt}}{p^2}dp \int\limits_0^A e^{-\xi q(p)}\Phi(\xi)d\xi$$

$$= \frac{1}{2\pi i}\int\limits_{(\gamma)} \frac{e^{pt}}{p^2}dp \int\limits_0^\infty e^{-\xi q(p)}\Phi(\xi)d\xi$$

$$= \frac{1}{2\pi i}\int\limits_{(\gamma)} \frac{e^{pt}\bar\Phi(q(p))}{q(p)}\frac{dp}{p^2} = a(t).$$

Comparing this relation with (2.5.18), we obtain the formula

$$\frac{\bar\Phi(q(p))}{pq(p)} = \int\limits_0^\infty L(t;\xi)\Phi(\xi)d\xi,$$

which implies

$$\bar\Phi(q(p)) = pq(p)\int\limits_0^\infty L(t;\xi)\Phi(\xi)d\xi. \qquad (2.5.20)$$

On multiplying both sides by the operator $\bar b(p)$, we obtain

$$\bar b(p)\bar\Phi(q(p)) = p\bar b(p)q(p)\int\limits_0^\infty L(t;\xi)\Phi(\xi)d\xi.$$

After denoting $p\bar b(p)q(p)L(t;\xi) = \Psi(t;\xi)$ we obtain

$$\bar b(p)\bar\Phi(q(p)) = \int\limits_0^\infty \Psi(t;\xi)\Phi(\xi)d\xi.$$

Hence, the integral, generally speaking, is convergent in the operational sense (see section 2.4). Formula (2.5.20) is called the *Efros transform*. ⬚

Example 2.5.58 *Let* $q(p) = \frac{1}{p}$. *Obviously,* $Re(\frac{1}{p}) > 0$, *if* $Re(p) > 0$. *Hence,* $\gamma_o = 0$. *In accordance with (2.5.6), section 2.5.1, we have*

$$L(t;\xi) = \frac{1}{p}e^{-\frac{\xi}{p}} = \sqrt{\frac{t}{\xi}}J_1(2\sqrt{t\xi});$$

hence,

$$\bar\Phi\left(\frac{1}{p}\right) = \int\limits_0^\infty \Phi(\xi)J_1(2\sqrt{t\xi})\sqrt{\frac{t}{\xi}}d\xi.$$

Applying the operator p^{1-n} *to both sides of this relation and taking into account that* $p^{1-n}L(t;\xi) = \frac{1}{p^n}e^{-\frac{\xi}{t}} = \left(\frac{t}{\xi}\right)^{\frac{n}{2}}J_n(2\sqrt{t\xi})$, *we find*

$$p^{1-n}\Phi\left(\frac{1}{p}\right) = \int\limits_0^\infty \Phi(\xi)J_1(2\sqrt{t\xi})\left(\frac{t}{\xi}\right)^{\frac{n}{2}}d\xi.$$

Example 2.5.59 *Let $q(p) = \sqrt{p}$. From (2.5.10), section 2.5.2, we have*

$$\sqrt{p}e^{-\xi\sqrt{p}} = \frac{1}{\sqrt{\pi t}}e^{-\frac{\xi^2}{4t}};$$

hence,

$$\frac{1}{p}e^{-\xi\sqrt{p}} = L(t;\xi) = \frac{1}{p\sqrt{p}}\frac{1}{\sqrt{\pi t}}e^{-\frac{\xi^2}{4t}},$$

and therefore,

$$\frac{\bar{\Phi}(\sqrt{p})}{p\sqrt{p}} = \frac{1}{p\sqrt{p}}\int_0^\infty \frac{1}{\sqrt{\pi t}}e^{-\frac{\xi^2}{4t}}\Phi(\xi)d\xi,$$

or

$$\bar{\Phi}(\sqrt{p}) = \int_0^\infty \frac{1}{\sqrt{\pi t}}e^{-\frac{\xi^2}{4t}}\Phi(\xi)d\xi. \qquad (2.5.21)$$

Example 2.5.60 *Let $q(p) = \frac{1}{\sqrt{p}}$. The simplest way to compute the operator $\Phi\left(\frac{1}{\sqrt{p}}\right)$ is the following: we compute $\bar{\Phi}\left(\frac{1}{p}\right)$ from the first example, then, using the second example, we compute $\bar{\Phi}\left(\sqrt{\frac{1}{p}}\right) = \bar{\Phi}\left(\frac{1}{\sqrt{p}}\right)$. So we obtain*

$$\bar{\Phi}\left(\frac{1}{\sqrt{p}}\right) = \frac{1}{\sqrt{\pi t}}\int_0^\infty e^{-\frac{\xi^2}{4t}}d\xi \int_0^\infty \Phi(u)J_1(2\sqrt{\xi u})\sqrt{\frac{\xi}{u}}du.$$

Example 2.5.61 *Let $q(p) = p + \frac{1}{p}$. We have*

$$L(t;\xi) = \frac{1}{p}e^{-\xi(p+\frac{1}{p})} = \frac{e^{-\xi p}}{p}e^{-\frac{\xi}{p}} = \frac{1}{p}e^{-\xi p}J_0(2\sqrt{t\xi}),$$

or, see formula (2.3.78),

$$L(t;\xi) = \begin{cases} 0 & \text{for } t < \xi, \\ \frac{1}{p}J_0(2\sqrt{\xi(t-\xi)}) & \text{for } t \geq \xi; \end{cases}$$

hence, we obtain

$$\frac{\Phi(p+\frac{1}{p})}{p(p+\frac{1}{p})} = \frac{1}{p}\int_0^t J_0(2\sqrt{\xi(t-\xi)})\Phi(\xi)d\xi,$$

or

$$\frac{\Phi(p+\frac{1}{p})}{(p+\frac{1}{p})} = \int_0^t J_0(2\sqrt{\xi(t-\xi)})\Phi(\xi)d\xi. \qquad (2.5.22)$$

2.6 Application of Operational Calculus

2.6.1 Ordinary Differential Equations

Problem 2.6.1. Consider a linear ordinary differential equation of nth order with constant coefficients

$$x^{(n)}(t) + a_{n-1}x^{(n-1)}(t) + \cdots + a_1x'(t) + a_0x(t) = f(t), \quad 0 \le t < \infty \qquad (2.6.1)$$

with the initial conditions

$$x(0) = x_0, \quad x'(0) = x_1, \quad \ldots, \quad x^{(n-1)}(0) = x_{n-1}. \qquad (2.6.2)$$

Applying the formula

$$x^{(k)}(t) = p^k x(t) - p^k x(0) - p^{k-1}x'(0) - \cdots - px^{(k-1)}(0), \qquad (2.6.3)$$

we can rewrite (2.6.1) in the form

$$L(p)\left[x(t) - x_0 - \frac{x_1}{p} - \cdots - \frac{x_{n-1}}{p^{n-1}}\right] = f(t) - b_0 - \frac{b_1}{p} - \cdots - \frac{b_{n-1}}{p^{n-1}},$$

where

$$L(p) = p^n + a_{n-1}p^{n-1} + \cdots + a_0;$$

$$b_k = \sum_{s=k}^{n-1} x_s a_{s-k} \quad (k = 0, 1, \ldots, n-1);$$

therefore, we obtain

$$x(t) = \frac{1}{L(p)}f(t) - \frac{1}{L(p)}\sum_{k=0}^{n-1}\frac{b_k}{p^k} + x_0 + \frac{x_1}{p} + \cdots + \frac{x_{n-1}}{p^{n-1}}. \qquad (2.6.4)$$

This formula gives us the solution of equation (2.6.1). We see immediately that the right-hand side of (2.6.4) is an n times differentiable function, satisfying the initial conditions.

The first part of solution (2.6.4), namely,

$$x_1(t) = \frac{f(t)}{L(p)}, \qquad (2.6.5)$$

is a solution of the nonhomogeneous equation (2.6.1) with zero initial conditions, and the second part

$$x_2(t) = -\frac{1}{L(p)}\sum_{k=0}^{n-1}\frac{b_k}{p^k} + x_0 + \frac{x_1}{p} + \cdots + \frac{x_{n-1}}{p^{n-1}} \qquad (2.6.6)$$

is the solution of the associated homogeneous equation with arbitrary initial conditions (2.6.2). In the case when $\lambda_1, \lambda_2, \ldots, \lambda_n$ are simple roots of $L(p)$, we have

$$L(p) = (p - \lambda_1)(p - \lambda_2)\ldots(p - \lambda_n) = \prod_{\rho=1}^{n}(p - \lambda_\rho),$$

and the function $\bar{z}(p) = \frac{1}{L(p)}$ may be decomposed into partial fractions:

$$\bar{z}(p) = \frac{1}{L(p)} = \sum_{\nu=1}^{n} \frac{c_\nu}{p - \lambda_\nu}. \tag{2.6.7}$$

Multiplying (2.6.7) by $(p - \lambda_\mu)$, we obtain

$$\frac{p - \lambda_\mu}{L(p)} = c_\mu + (p - \lambda_\mu) \sum_{\nu=1}^{n}{}' \frac{c_\nu}{p - \lambda_\nu},$$

where the prime sign by the sum means that in the sum the term with $\mu = \nu$ is omitted. After passing to the limit as $p \to \lambda_\mu$, we obtain

$$c_\mu = \lim_{p \to \lambda_\mu} \frac{p - \lambda_\mu}{L(p)} = \lim_{p \to \lambda_\mu} \frac{1}{\frac{L(p) - L(\lambda_\mu)}{p - \lambda_\mu}} = \frac{1}{L'(\lambda_\mu)}.$$

Therefore, the decomposition of $\bar{z}(p)$ into partial fractions has the form

$$\bar{z}(p) = \frac{1}{L(p)} = \sum_{\nu=1}^{n} \frac{1}{(p - \lambda_\nu)L'(\lambda_\nu)}.$$

According to formula (2.3.49) we have

$$z(t) = \sum_{\nu=1}^{n} \frac{e^{\lambda_\nu t} - 1}{\lambda_\nu L'(\lambda_\nu)},$$

therefore, by formula (2.3.52) and (2.3.4), we have

$$x_1(t) = \bar{z}(p)f(t) = \frac{d}{dt} \int_0^t z(t - \tau)f(\tau)d\tau = \sum_{\nu=1}^{n} \frac{e^{\lambda_\nu t}}{L'(\lambda_\nu)} \int_0^t e^{-\lambda_\nu \tau} f(\tau)d\tau. \tag{2.6.8}$$

In the case where $L(p)$ has multiple roots

$$\lambda_1 = \lambda_2 = \cdots = \lambda_r, \lambda_{r+1}, \lambda_{r+2}, \ldots, \lambda_n,$$

we have

$$L(p) = (p - \lambda_1)^r (p - \lambda_{r+1})(p - \lambda_{r+2}) \ldots (p - \lambda_n) = (p - \lambda_1)^r L_r(p),$$

where

$$L_r(p) = (p - \lambda_{r+1})(p - \lambda_{r+2}) \ldots (p - \lambda_n),$$

and the function $\bar{z}(p)$ may be represented by the following sum of partial fractions

$$\bar{z}(p) = \frac{1}{L(p)} = \frac{c_{11}}{p - \lambda_1} + \frac{c_{12}}{(p - \lambda_1)^2} + \cdots + \frac{c_{1r}}{(p - \lambda_1)^r} + \frac{c_{r+1}}{p - \lambda_{r+1}} + \cdots + \frac{c_n}{p - \lambda_n},$$

or

$$\bar{z}(p) = \frac{1}{L(p)} = \sum_{\nu=1}^{r} \frac{c_{1\nu}}{(p - \lambda_1)^\nu} + \sum_{\nu=r+1}^{n} \frac{c_\nu}{p - \lambda_\nu}. \tag{2.6.9}$$

Multiplying (2.6.9) by $(p - \lambda_1)^r$, we obtain

$$\frac{(p - \lambda_1)^r}{L(p)} = \frac{1}{L_r(p)} = c_{1r} + \sum_{\nu=1}^{r-1} c_{1\nu}(p - \lambda_1)^{r-\nu} + (p - \lambda_1)^r \sum_{\nu=r+1}^{n} \frac{c_\nu}{p - \lambda_\nu}. \quad (2.6.10)$$

Passing to the limit as $p - \lambda_1$, we find

$$c_{1r} = s(\lambda_1) = \frac{1}{L_r(\lambda_1)}. \quad (2.6.11)$$

Similarly, differentiating (2.6.10) $(r - \mu)$ times ($\mu = 1, 2, \ldots, r - 1$) and passing to the limit as $p \to \lambda_1$, we obtain

$$c_{1\mu} = \frac{s_r^{(r-\mu)}(\lambda_1)}{(r - \mu)!} \quad (\mu = 1, 2, \ldots, r - 1), \quad (2.6.12)$$

where

$$s_r(p) = \frac{1}{L_r(p)}.$$

It follows from (2.6.11) and (2.6.12) that

$$c_{1\nu} = \frac{s_r^{(r-\nu)}(\lambda_1)}{(r - \nu)!} \quad (\nu = 1, 2, \ldots, r). \quad (2.6.13)$$

In the same way the coefficients c_ν may be found. After multiplying (2.6.9) by $p - \lambda_\mu$ ($\mu = r + 1, r + 2, \ldots, n$), we have

$$\frac{p - \lambda_\mu}{L(p)} = (p - \lambda_\mu) \sum_{\nu=1}^{r} \frac{c_{1\nu}}{(p - \lambda_1)^\nu} + c_\mu + (p - \lambda_\mu) \sum_{\nu=r+1}^{n}{}' \frac{c_\nu}{p - \lambda_\nu},$$

where the prime sign by the sum denotes again that the term with $\mu = \nu$ is omitted in the sum. On setting $p = \lambda_\mu$, we find

$$c_\mu = \lim_{p \to \lambda_\mu} \frac{p - \lambda_\mu}{L(p)} = \lim_{p \to \lambda_\mu} \frac{1}{\frac{L(p)-L(\lambda_\mu)}{p-\lambda_\mu}} = \frac{1}{L'(\lambda_\mu)};$$

hence,

$$c_\nu = \frac{1}{L'(\lambda_\nu)} = \frac{1}{(\lambda_\nu - \lambda_1)^r L_r'(\lambda_\nu)}, \quad \nu = r + 1, r + 2, \ldots, n. \quad (2.6.14)$$

According to formulas (2.3.48) and (2.3.51) we have

$$\bar{z}(p) = z(t) = \int_0^t e^{\lambda_1 \tau} \sum_{\nu=1}^{r} \frac{c_{1\nu}\tau^{\nu-1}}{(\nu - 1)!} d\tau + \int_0^t \sum_{\nu=r+1}^{n} \frac{e^{\lambda_\nu \tau}}{L'(\lambda_\nu)} d\tau. \quad (2.6.15)$$

Finally, using formula (2.3.52) we find

$$x_1(t) = \bar{z}(p)f(t) = \frac{d}{dt} \int_0^t z(t - \tau)f(\tau)d\tau$$

$$= e^{\lambda_1 t} \sum_{\nu=1}^{r} \frac{c_{1\nu}}{(\nu - 1)!} \int_0^t (t - \tau)^{\nu-1} e^{-\lambda_1 \tau} f(\tau)d\tau \quad (2.6.16)$$

$$+ \sum_{\nu=r+1}^{n} \frac{e^{\lambda_\nu t}}{L'(\lambda_\nu)} \int_0^t e^{-\lambda_\nu \tau} f(\tau)d\tau.$$

In particular, if λ_1 is a simple root of $L(p)$, i.e., $r = 1$, then $c_{11} = \frac{1}{L'(\lambda_1)}$ and (2.6.16) implies (2.6.8).

Now we are going to consider the homogeneous differential equation associated with (2.6.1) with initial conditions (2.6.2). Setting in formula (2.6.2)

$$x_0 = x_1 = x_2 = \cdots = x_{n-2} = 0, \quad x_{n-1} = 1,$$

from (2.6.6) we find

$$x_2(t) = \frac{p}{L(p)} = \bar{\Psi}(p). \tag{2.6.17}$$

In order to find the function $\bar{\Psi}(p) = \Psi(t)$, according to the formulas (2.3.48), (2.6.7), (2.3.51), and (2.6.9) we have

$$\Psi(t) = \sum_{\nu=1}^{n} \frac{e^{\lambda_\nu t}}{L'(\lambda_\nu)} \tag{2.6.18}$$

or

$$\Psi(t) = e^{\lambda_1 t} \sum_{\nu=1}^{r} \frac{c_{1\nu} t^{\nu-1}}{(\nu-1)!} + \sum_{\nu=r+1}^{n} \frac{e^{\lambda_\nu t}}{L'(\lambda_\nu)}. \tag{2.6.19}$$

Obviously, the function $\Phi(t)$ satisfies the conditions

$$\Psi(0) = \Psi'(0) = \Psi''(0) = \cdots = \Psi^{(n-2)}(0) = 0, \quad \Psi^{(n-1)}(0) = 1. \tag{2.6.20}$$

Representing the expression (2.6.6) in the form

$$x_2(t) = \frac{1}{L(p)} \sum_{k=0}^{n-1} x_k (p^{n-k} + a_1 p^{n-k-1} + a_2 p^{n-k-2} + \cdots + a_{n-k-1} p), \quad (a_0 = 0), \tag{2.6.21}$$

and using (2.6.3), (2.6.20), we find the solution of the homogeneous equation, satisfying arbitrary initial conditions

$$x_2(t) = \sum_{k=0}^{n-1} x_k [\Psi^{(n-k-1)}(t) + a_1 \Psi^{(n-k-2)}(t) + \cdots + a_{n-k-1} \Psi(t)]. \tag{2.6.22}$$

Problem 2.6.2. Let us introduce into consideration a system of linear differential equations of the first order with constant coefficients (a_{ik}), solved with respect to the first derivatives:

$$\frac{dx_1}{dt} = a_{11} x_1 + a_{12} x_2 + \cdots + a_{1n} x_n,$$

$$\frac{dx_2}{dt} = a_{21} x_1 + a_{22} x_2 + \cdots + a_{2n} x_n,$$

$$\cdots\cdots\cdots\cdots\cdots\cdots\cdots\cdots\cdots\cdots\cdots\cdots \tag{2.6.23}$$

$$\frac{dx_n}{dt} = a_{n1} x_1 + a_{n2} x_2 + \cdots + a_{nn} x_n,$$

$$0 < t < \infty.$$

It is known that any system of differential equations solvable with respect to the highest derivatives of unknown functions may be reduced to a system of the form (2.6.23). We shall derive a solution of the system (2.6.23), satisfying the initial conditions

$$x_1|_{t=0} = x_1^0, \quad x_2|_{t=0} = x_2^0, \quad ,\ldots, x_n|_{t=0} = x_n^0. \tag{2.6.24}$$

The system of equations (2.6.23) may be rewritten in the form

$$\frac{dx_k}{dt} = \sum_{s=1}^{n} a_{ks} x_s, \quad k = 1, 2, \ldots, n. \tag{2.6.25}$$

Using formulae (2.6.3), we reduce the system (2.6.25) to a system of algebraic equations

$$p\bar{x}_k(p) = \sum_{s=1}^{n} a_{ks} x_s(p) + px_k^0, \quad k = 1, 2, \ldots, n. \tag{2.6.26}$$

The extended form of the system (2.6.26) is

$$
\begin{array}{llll}
(a_{11} - p)\bar{x}_1(p) + & a_{12}\bar{x}_2(p) & +\ldots + & a_{1n}\bar{x}_n(p) & = -px_1^0, \\
a_{21}\bar{x}_1(p) & +(a_{22} - p)\bar{x}_2(p) +\ldots + & a_{2n}\bar{x}_n(p) & = -px_2^0, \\
\ldots & +\ldots & +\ldots + & \ldots & = \ldots, \\
a_{n1}\bar{x}_1(p) & + \quad a_{n2}\bar{x}_2(p) +\ldots + & (a_{nn} - p)\bar{x}_n(p) & = -px_n^0.
\end{array}
\tag{2.6.27}
$$

Let

$$\Delta(p) = \begin{vmatrix} a_{11} - p & a_{12} & \ldots & a_{1n} \\ a_{21} & a_{22} - p & \ldots & a_{2n} \\ \ldots & \ldots & \ldots & \ldots \\ a_{n1} & a_{n2} & \ldots & a_{nn} - p \end{vmatrix} \tag{2.6.28}$$

be the determinant of the system (2.6.27), $\Delta_{ks}(p)$ the minor of the entry in the kth row and sth column of this determinant, i.e., the determinant obtained after omitting the kth row and the sth column in the determinant $\Delta(p)$ and multiplied by $(-1)^{k+s}$.

The solution of the system (2.6.27) may be expressed in the form

$$\bar{x}_s(p) = -p \sum_{k=1}^{n} x_k^0 \frac{\Delta_{ks}(p)}{\Delta(p)}, \quad s = 1, 2, \ldots, n. \tag{2.6.29}$$

In order to find $x_s(t)$ we have to find the functions

$$\psi_{ks}(t) = \bar{\psi}_{ks}(p) = -p\frac{\Delta_{ks}(p)}{\Delta(p)}. \tag{2.6.30}$$

The functions ψ_{ks} are easy to determine after breaking down $\bar{\psi}_{ks}(p)$ into partial fractions. In order to break down $\bar{\psi}_{ks}(p)$ into partial fractions we need to know the roots of the characteristic equation

$$\Delta(p) = 0. \tag{2.6.31}$$

After determining the functions $\psi_{ks}(t)$ and finding $x_s(t)$ we have

$$x_s(t) = \sum_{k=1}^{n} x_k^0 \psi_{ks}(t), \quad s = 1, 2, \ldots, n. \tag{2.6.32}$$

The formulated method may be applied to the integration of a nonhomogeneous system of linear differential equations of the first order with constant coefficients of the form

$$\frac{dx_k}{dt} = \sum_{k=1}^{n} a_{ks}x_s + f_k(t), \quad k = 1, 2, \ldots, n. \tag{2.6.33}$$

We seek the solution of system (2.6.33), satisfying the initial conditions (2.6.24). The operator transform has the form

$$p\bar{x}_k(p) = px_0^k + \sum_{s=1}^{n} a_{ks}\bar{x}_s(p) + \bar{f}_k(p). \tag{2.6.34}$$

Similar to the previous case, we find the solution of the linear system (2.6.34):

$$\bar{x}_s(p) = -p\sum_{k=1}^{n}\left(x_o^k + \frac{\bar{f}_k(p)}{p}\right)\frac{\Delta_{ks}(p)}{\Delta(p)}, \tag{2.6.35}$$

or

$$\bar{x}_s(p) = -p\sum_{k=1}^{n}x_o^k\frac{\Delta_{ks}(p)}{\Delta(p)} - \sum_{k=1}^{n}\bar{f}_k(p)\frac{\Delta_{ks}(p)}{\Delta(p)}. \tag{2.6.36}$$

Taking into account (2.6.32) we have

$$x_s(t) = \sum_{k=1}^{n}\left[x_k^0\psi_{ks}(t) + \int_0^t f_k(\tau)\psi_{ks}(t-\tau)d\tau\right]. \tag{2.6.37}$$

Similarly, one can consider a more general system of linear ordinary differential equations of the form

$$\sum_{k=1}^{n}\left(a_{\nu k}\frac{d^2x_k}{dt^2} + b_{\nu k}\frac{dx_k}{dt} + c_{\nu k}x_k\right) = f_\nu(t), \quad \nu = 1, 2, \ldots, n, \tag{2.6.38}$$

with the initial conditions

$$x_k(0) = \alpha_k, \quad x_k'(0) = \beta_k, \quad k = 1, 2, \ldots, n. \tag{2.6.39}$$

The operator transform of the system (2.6.38) with conditions (2.6.39) is the system of algebraic equations with respect to unknown functions $\bar{x}_k(p)$:

$$\sum_{k=1}^{n}(a_{\nu k}p^2 + b_{\nu k}p + c_{\nu k})\bar{x}_k(p) = \bar{f}_\nu(p) + \sum_{k=1}^{n}(a_{\nu k}p^2 + b_{\nu k}p)\alpha_k + a_{\nu k}p\beta_k, \tag{2.6.40}$$
$$\nu = 1, 2, \ldots, n,$$

After finding from this system the functions $\bar{x}_k(p)$ and passing to the originals, one obtains the desired solution.

Example 2.6.62 *Find the solution of the system of two linear ordinary differential equations*

$$\frac{dx}{dt} = ay + f(t),$$
$$\frac{dy}{dt} = -ax + g(t) \tag{2.6.41}$$

with initial conditions

$$x(0) = 0, \quad y(0) = 0. \tag{2.6.42}$$

The operator transform of the system (2.6.41) *with conditions* (2.6.42) *has the form*

$$p\bar{x}(p) - a\bar{y}(p) = \bar{f}(p),$$
$$a\bar{x}(p) + p\bar{y}(p) = \bar{g}(p).$$

Let us find the solution of these equations:

$$\bar{x}(p) = \frac{p\bar{f}(p) + a\bar{g}(p)}{p^2 + a^2}, \quad \bar{y}(p) = \frac{-a\bar{f}(p) + p\bar{g}(p)}{p^2 + a^2}.$$

Using the operational formulae

$$\frac{ap}{p^2 + a^2} = \sin at, \quad \frac{p^2}{p^2 + a^2} = \cos at,$$

we obtain the desired solution

$$x(t) = \int_0^t [f(\tau) \cos a(t - \tau) + g(\tau) \sin a(t - \tau)] d\tau,$$

$$y(t) = \int_0^t [-f(\tau) \sin a(t - \tau) + g(\tau) \cos a(t - \tau)] d\tau.$$

Example 2.6.63 *Find the solution of the system of three linear ordinary differential equations*

$$\frac{dx}{dt} = -x + y + z,$$
$$\frac{dy}{dt} = x - y + z, \tag{2.6.43}$$
$$\frac{dz}{dt} = x + y + z,$$

with initial conditions

$$x(0) = 0, \quad y(0) = 0 \quad z(0) = 0. \tag{2.6.44}$$

The operator transform of the system (2.6.43) *with conditions* (2.6.44) *has the form*

$$(p + 1)\bar{x}(p) - \bar{y}(p) - \bar{z}(p) = p,$$
$$- \bar{x}(p) + (p + 1)\bar{y}(p) - \bar{z}(p) = 0,$$
$$- \bar{x}(p) - \bar{y}(p) + (p - 1)\bar{z}(p) = 0.$$

From the last system we find:

$$\bar{x}(p) = \frac{1}{3}\frac{p}{p + 1} + \frac{1}{2}\frac{p}{p + 2} + \frac{1}{6}\frac{p}{p - 2},$$
$$\bar{y}(p) = \frac{1}{3}\frac{p}{p + 1} - \frac{1}{2}\frac{p}{p + 2} + \frac{1}{6}\frac{p}{p - 2},$$
$$\bar{z}(p) = -\frac{1}{3}\frac{p}{p + 1} + \frac{1}{3}\frac{p}{p - 2}.$$

Using the formula (2.3.49), we obtain the desired solution:

$$x(t) = \frac{1}{3}e^{-t} + \frac{1}{2}e^{-2t} + \frac{1}{6}e^{2t},$$

$$y(t) = \frac{1}{3}e^{-t} - \frac{1}{2}e^{-2t} + \frac{1}{6}e^{2t},$$

$$z(t) = -\frac{1}{3}e^{-t} + \frac{1}{3}e^{2t}.$$

Problem 2.6.3. In some cases operational calculus may be applied in order to solve linear ordinary differential equations with variable coefficients

$$x^{(n)}(t) + a_1(t)x^{(n-1)}(t) + a_2(t)x^{(n-2)}(t) + \cdots + a_n(t)x(t) = f(t). \tag{2.6.45}$$

We consider only the case when all functions $a_i(t)$ are polynomials. In order to obtain the operator transform of the equation in this case, it suffices, obviously, to know how to write the operator transform of the function $t^k x^{(n)}(t)$. As it is known,

$$tx(t) = -p\frac{d}{dp}\left[\frac{\bar{x}(p)}{p}\right]. \tag{2.6.46}$$

Applying this formula to the function $tx(t)$ we obtain

$$t^2 x(t) = (-1)^2 p\frac{d^2}{dp^2}\left[\frac{\bar{x}(p)}{p}\right]. \tag{2.6.47}$$

In a similar way one can easily obtain the formulae

$$t^3 x(t) = (-1)^3 p\frac{d^3}{dp^3}\left[\frac{\bar{x}(p)}{p}\right], \tag{2.6.48}$$

$$\cdots\cdots\cdots\cdots\cdots\cdots\cdots\cdots$$

$$t^k x(t) = (-1)^k p\frac{d^k}{dp^k}\left[\frac{\bar{x}(p)}{p}\right]. \tag{2.6.49}$$

After expanding the operator $\frac{d^k}{dp^k}$ in the last formula, we have

$$t^k x(t) = \sum_{\nu=0}^{k}(-1)^\nu \binom{k}{\nu}\frac{(k-\nu)!}{p^{k-\nu}}\bar{x}^{(\nu)}(p) = \sum_{\nu=0}^{k}(-1)^\nu \frac{k!}{\nu!}\frac{1}{p^{k-\nu}}\bar{x}^{(\nu)}(p). \tag{2.6.50}$$

In order to obtain the operator transform of the function $t^k x^{(r)}(t)$ we need to substitute in (2.6.50) instead of $\bar{x}(p)$ the operator transform of the function $x^{(r)}(t) = p^r\bar{x}(p) - p^r x(0) - p^{r-1}x'(0) - \cdots - px^{(r-1)}(0)$:

$$t^k x^{(r)}(t) = \sum_{\nu=0}^{k}\frac{(-1)^\nu k!}{\nu!}\frac{[p^r\bar{x}(p) - p^r x(0) - p^{r-1}x'(0) - \cdots - px^{(r-1)}(0)]^{(\nu)}}{p^{k-\nu}}. \tag{2.6.51}$$

For zero initial conditions

$$x(0) = x'(0) = \cdots = x^{(r-1)}(0) = 0. \tag{2.6.52}$$

The formula (2.6.51) takes the form

$$t^k x^{(r)}(t) = \sum_{\nu=0}^{n} \frac{(-1)^\nu k!}{\nu!} \frac{[p^r \bar{x}(p)]^\nu}{p^k}. \qquad (2.6.53)$$

As an example, we consider Hermite's equation

$$x''(t) - tx'(t) + nx(t) = 0 \qquad (n \in \mathbb{N}_o).$$

We shall seek the solution for the following initial conditions:

$$x(0) = 1, \quad x'(0) = 0, \quad \text{if} \quad n = 2k \text{ is an even integer;}$$
$$x(0) = 0, \quad x'(0) = 1, \quad \text{if} \quad n = 2k+1 \text{ is an odd integer.}$$

On putting $k = 1$ and $r = 1$ in (2.6.51), we obtain

$$tx'(t) = -p\frac{d\bar{x}(p)}{dp}.$$

Hence, the operator transform for Hermite's equation has the form

$$p^2\left[\bar{x}(p) - x(0) - \frac{1}{p}x'(0)\right] + p\frac{d\bar{x}(p)}{dp} + n\bar{x}(p) = 0,$$

or

$$\frac{d\bar{x}(p)}{dp} = -\left(p + \frac{n}{p}\right)\bar{x}(p) + px(0) + \bar{x}'(0);$$

therefore, we have the solution:

$$\bar{x}(p) = cp^{-n}e^{-\frac{p^2}{2}} + p^{-n}e^{-\frac{p^2}{2}}\int [px(0) + x'(0)]p^n e^{\frac{p^2}{2}}\,dp. \qquad (2.6.54)$$

Consider the first case, when $n = 2k$ is an even integer, $x(0) = 1$, $x'(0) = 0$:

$$\bar{x}(p) = cp^{-2k}e^{-\frac{p^2}{2}} + p^{-2k}e^{-\frac{p^2}{2}}\int p^{2k+1}e^{\frac{p^2}{2}}\,dp.$$

Let us denote

$$I_k = \int p^{2k+1}e^{\frac{p^2}{2}}\,dp.$$

After integrating by parts, we have

$$I_k = \int p^{2k+1}e^{\frac{p^2}{2}}\,dp = \int p^{2k}d\left(e^{\frac{p^2}{2}}\right) = p^{2k}e^{\frac{p^2}{2}} - \int 2kp^{2k-1}e^{\frac{p^2}{2}}\,dp,$$

i.e.,

$$I_k = p^{2k}e^{\frac{p^2}{2}} - 2kI_{k-1}.$$

Thus,

$$I_k = \frac{e^{p^2}}{2}\sum_{s=0}^{k}(-1)^s 2^s s!\binom{k}{s}p^{2k-2s}.$$

Hence, from (2.6.54) we have

$$\bar{x}(p) = cp^{-2k}e^{-\frac{p^2}{2}} + \sum_{s=0}^{k}(-1)^s 2^s s!\binom{k}{s}p^{-2s}.$$

The function $f(p) = f(\sigma + i\tau) = cp^{-2k}e^{-\frac{p^2}{2}}$ is not bounded for a fixed σ and $\tau \to \infty$. Indeed,

$$\frac{c}{p^{2k}}e^{-\frac{p^2}{2}} = \frac{c}{p^{2k}}e^{-\frac{(\sigma^2-\tau^2+2i\sigma\tau)}{2}} = \frac{c}{p^{2k}}e^{\frac{\tau^2-\sigma^2}{2}+i\sigma\tau};$$

therefore,

$$\left|\frac{c}{p^{2k}}e^{-\frac{p^2}{2}}\right| = \frac{c}{|p|^{2k}}e^{\frac{\tau^2-\sigma^2}{2}} \to \infty \quad \text{as} \quad \tau \to \infty.$$

It follows from here, that for $c \neq 0$ the function $f(p)$ is not representable by a Laplace integral; therefore, $c = 0$ and the solution is

$$x(t) = \sum_{s=0}^{k}(-1)^s 2^s s!\binom{k}{s}\frac{t^{2s}}{(2s)!}.$$

One can show that $x(t)$ up to a constant factor is the Hermite polynomial.

The second case, when $x(0) = 0$, $x'(0) = 1$ and $n = 2k+1$ is an odd integer, may be solved analogously. From (2.6.54) we have

$$\bar{x}(p) = cp^{-n}e^{-\frac{\tau^2}{2}} + p^{-n}e^{-\frac{p^2}{2}}\int p^n e^{\frac{p^2}{2}}\,dp,$$

where $n = 2k+1$. Previously we investigated the integral $\int p^{2k+1}e^{\frac{p^2}{2}}\,dp$ and found that it is equal to I_k; therefore, $c = 0$ and the solution has the form

$$x(t) = \sum_{s=0}^{k}(-1)^s 2^s s!\binom{k}{s}\left(\frac{t^{2s+1}}{(2s+1)!}\right).$$

Problem 2.6.4. Consider a differential equation with delayed argument and constant coefficients

$$x^{(n)}(t) = \sum_{k=0}^{n-1}a_k x^{(k)}(t-h_k) + g(t), \quad 0 \leq t < \infty, \quad h_k \geq 0. \tag{2.6.55}$$

For the sake of simplicity we shall assume the initial conditions to be equal to zero. Hence, we have to find the solution of equation (2.6.55) under the condition that

$$x(0) = x'(0) = \cdots = x^{(n-1)}(0) = 0. \tag{2.6.56}$$

In addition, we suppose that

$$x(t) = x'(t) = \cdots = x^{(n-1)}(t) = 0 \quad \text{for} \quad t < 0.$$

Taking into account that

$$x^{(k)}(t-h_k) = p^k e^{-h_k p}x(t),$$

we find the operator transform of equation (2.6.57):

$$p^n x(t) = \sum_{k=0}^{n-1}a_k p^k e^{-h_k p}x(t) + g(t);$$

therefore,

$$x(t) = \left(p^n - \sum_{k=0}^{n-1} a_k p^k e^{-h_k p} \right)^{-1} g(t).$$

Let us introduce the notation

$$\omega(p) = \frac{1}{p^n} \sum_{k=0}^{n-1} a_k p^k e^{-h_k p}.$$

Then we obtain

$$x(t) = \frac{g(t)}{p^n} \frac{1}{1 - \omega(p)}. \qquad (2.6.57)$$

In order to show that (2.6.57) is the solution of equation (2.6.55), satisfying initial conditions (2.6.57), it is sufficient to show that the operator $\frac{1}{1-\omega(p)}$ is reducible to a function. Obviously, there exists a constant Q such that for all p in the half-plane $\bar{H}_{\sigma_o}, \sigma_0 > 0$ the following inequalities hold:

$$|\omega(p)| < \frac{Q}{|p|} < 1.$$

Let us represent the operator $\frac{1}{1-\omega(p)}$ in the form

$$\frac{1}{1 - \omega(p)} = \left[1 + \omega(p) + \frac{[\omega(p)]^2}{1 - \omega(p)} \right]. \qquad (2.6.58)$$

Since the operator $[1 + \omega(p)]$, obviously, is reducible to a function, it is sufficient to show that the operator $\frac{[\omega(p)]^2}{1-\omega(p)}$ is reducible to a function, or, the equivalent, that the function of the complex variable $p = \sigma + i\tau$

$$\frac{[\omega(p)]^2}{p[1 - \omega(p)]} \qquad (2.6.59)$$

may be represented by the Laplace integral. Indeed, in the half-plane $\bar{H}_{\sigma_o}, \sigma_o > 0$ the function $\frac{[\omega(p)]^2}{p[1-\omega(p)]}$, which is analytical in this half-plane, satisfies the inequality

$$\left| \frac{[\omega(p)]^2}{p(1 - \omega(p))} \right| < \frac{\left(\frac{Q}{|p|}\right)^2}{|p|\left(1 - \frac{Q}{|p|}\right)} = \frac{Q^2}{|p|^3 \left(1 - \frac{Q}{|p|}\right)};$$

therefore, in the half-plane $\Re(p) \geq \sigma_0$ the function (2.6.59) is uniform with respect to arg p tending to zero as $|p| \to \infty$, and the integral

$$\int_{-\infty}^{\infty} \left| \frac{[\omega(p)]^2}{p(1 - \omega(p))} \right| d\tau, \quad (p = \sigma + i\tau),$$

is convergent.

Then from section 1.4.4, Theorem 1.4.13 it follows that the function $\frac{[\omega(p)]^2}{p(1-\omega(p))}$ is representable by an absolutely convergent Laplace integral. Thus, the operator $\frac{1}{1-\omega(p)}$ is reducible to a function, and the solution, in fact, is given by the formula

$$x(t) = \frac{g(t)}{p^n} \sum_{m=0}^{\infty} [\omega(p)]^m,$$

where

$$\omega(p) = \frac{1}{p^n} \sum_{k=0}^{n-1} a_k p^k e^{-h_k p}.$$

The function $x(t)$ is n times differentiable and

$$x(0) = x'(0) = \cdots = x^{(n-1)}(0) = 0.$$

2.6.2 Partial Differential Equations

Consider a partial differential equation whose coefficients $a_{\mu\nu}(x)$ are numerical functions of the variable x:

$$\sum_{\mu=0}^{m} \sum_{\nu=0}^{n} a_{\mu\nu}(x) \frac{\partial^{\mu+\nu} u(x,t)}{\partial x^\mu \partial t^\nu} = f(x,t). \qquad (2.6.60)$$

On applying the formula

$$\frac{\partial^{\mu+\nu} u(x,t)}{\partial x^\mu \partial t^\nu} = p^\nu \frac{\partial^\mu u(x,t)}{\partial x^\mu} - p^\nu \frac{\partial^\mu u(x,0)}{\partial x^\mu} - p^{\nu-1} \frac{\partial^{\mu+1} u(x,0)}{\partial x^\mu \partial t} - \cdots - p \frac{\partial^{\mu+\nu-1} u(x,0)}{\partial x^\mu \partial t^{\nu-1}},$$

we reduce equation (2.6.60) to the form

$$\sum_{\mu=0}^{m} a_\mu(x,p) \frac{\partial^\mu u(x,t)}{\partial x^\mu} = f(x,t) + \sum_{\mu=0}^{m} \sum_{\nu=1}^{n} \sum_{k=0}^{\nu-1} p^{\nu-k} \frac{\partial^{\mu+k} u(x,0)}{\partial x^\mu \partial t^k},$$

where

$$a_\mu = a_\mu(x,p) = \sum_{\nu=0}^{n} a_{\mu\nu}(x) p^\nu.$$

On denoting the right-hand side of this equation by $\Phi(x,p)$ and considering $u(x,t)$ as an operator function depending on the parameter $x, u(x,t) = \bar{u}(x,p) = \bar{u}(x)$, we have

$$a_m \bar{u}^{(m)}(x) + a_{m-1} \bar{u}^{(m-1)}(x) + \cdots + a_0 \bar{u}(x) = \Phi(x,p). \qquad (2.6.61)$$

Here the coefficients a_k are also operator functions depending on x. Thus, the problem of integrating equation (2.6.60) is reduced to integrating a linear operator differential equation. Equation (2.6.61), which is the operator transform equation (2.6.60), is called the *operator* or *transformed* equation.

When solving equation (2.6.61), the isomorphism of fields $\mathfrak{M}(S)$ and $\overline{\mathfrak{M}}(S)$ must be used. In the field $\overline{\mathfrak{M}}(S)$ the transformed equation (2.6.61) becomes an ordinary linear differential equation of the nth order, whose coefficients and right-hand side depend on the parameter (complex number) p. Such equations have been fully investigated. Let $\bar{u}(x,p)$ be a solution of the equation. If it turns out that $\bar{u}(x,p)$ belongs to the field $\overline{\mathfrak{M}}(S)$ for the given values of $x, \alpha < x < \beta$, this will imply that (2.6.60) has the solution $\bar{u}(x,p)$ in the field $\mathfrak{M}$, where p is regarded as the operator $p = \frac{1}{t}$.

The application of operational calculus to the solution of partial differential equations is performed as follows:

1) Replacement of the original equation by the transformed equation. Similarly, the boundary conditions of the problem are replaced by transformed boundary conditions, which will be the boundary conditions for the solution $\bar{u}(x,p)$ of the transformed equation (2.6.61).

2) Finding the solution $\bar{u}(x,p)$ of the transformed equation with the given transformed boundary conditions.

3) Investigation of the solution obtained for the purpose of proving that the solution $\bar{u}(x,p)$ belongs to the field $\overline{\mathfrak{M}}(S)$. In the case when $\bar{u}(x,p)$ belongs to $\overline{\mathfrak{M}}(S)$, auxiliary investigations have to be made to establish whether the solution $u(x,t) = \bar{u}(x,p)$ is a generalized solution or whether it can be reduced to a function having partial derivatives with respect to the variables x and t up to and including the derivative $\frac{\partial^{m+n}u(x,t)}{\partial x^m \partial t^n}$. This latter case will imply that $u(x,t)$ satisfies the initial partial differential equation in the ordinary classical sense.

4) Realization of the operator $\bar{u}(x,p)$, i.e. determination of the function $u(x,t) = \bar{u}(x,p)$.

The investigation of point 3) can often be considerably simplified if point 4) is carried out.

5) Proof of the fact that the solution $u(x,t)$ satisfies the initial and boundary conditions of the problem.

Let us take as an example the equations

$$\rho(x)u_t = \rho_0(x)u_{xx} + \rho_1(x)u_x + \rho_2(x)u; \qquad (2.6.62)$$

$$\rho(x)u_{tt} = \rho_0(x)u_{xx} + \rho_1(x)u_x + \rho_2(x)u; \qquad (2.6.63)$$

in the domain $0 \leq x < l$, $t > 0$. Here, $\rho(x)$, $\rho_0(x)$, $\rho_1(x)$, $\rho_2(x)$ are given continuous functions in the interval $0 < x \leq l$ and $\rho(x) > 0$. The solution $u(x,t)$ must have continuous partial derivatives up to and including the second order in the domain $(0 < x \leq l, \ t > 0)$ and must satisfy the initial conditions

$$\lim_{t \to +0} u(x,t) = \varphi(x), \quad 0 < x \leq l \qquad (2.6.64)$$

for equation (2.6.62) and

$$\lim_{t \to +0} u(x,t) = \varphi(x), \quad \lim_{t \to +0} u_t(x,t) = \Psi(x), \quad 0 < x \leq l \qquad (2.6.65)$$

for equation (2.6.63), as well as the boundary conditions

$$\lim_{t \to +0} u(x,t) = f(x), \quad au_x(l,t) + bu_t(l,t) = cu(l,t) \qquad (2.6.66)$$

for $t > 0$, where $\varphi(x)$, $\Psi(x)$ are given piecewise continuous functions; $f(t) \in S$ and is continuous for $t > 0$; and a,b,c are given constants.

We shall seek the solution of the equations in the form $u(x,t) = \bar{u}(x,p)$. The transformed equations for (2.6.62) and (2.6.63) will be

$$\rho_0(x)\frac{d^2\bar{u}}{dx^2} + \rho_1(x)\frac{d\bar{u}}{dx} + [\rho_2(x) - p\rho(x)]\bar{u} = -\rho(x)p\varphi(x); \qquad (2.6.67)$$

$$\rho_0(x)\frac{d^2\bar{u}}{dx^2} + \rho_1(x)\frac{d\bar{u}}{dx} + [\rho_2(x) - p^2\rho(x)]\bar{u} = -p^2\rho(x)p\varphi(x) - p\rho(x)p\Psi(x). \quad (2.6.68)$$

We obtain from the boundary conditions of the problem the boundary conditions for the solution

$$\begin{cases} \bar{u}(+0,p) = \bar{f}(p), \quad \text{where} \quad \bar{f}(p) = f(t), \\ \bar{u}_x(l,p) + bp[\bar{u}(l,p) - \varphi(l)] = c\bar{u}(l,p). \end{cases} \quad (2.6.69)$$

Theorem 2.6.94 *Let $\bar{u}(x,p)$ be a solution of equation (2.6.67) or (2.6.68) with conditions (2.6.69). Furthermore, let*

1) *the operators $\bar{u}(x,p)$, $\bar{u}_x(x,p)$ and $\bar{u}_{xx}(x,p)$ reduce to functions for $0 < x \leq l$;*

2) *there exist a number σ_0 such that, as $t \to \infty$,*

$$\bar{u}(x,p) = 0(e^{\sigma_0 t}), \quad \bar{u}_x(x;p) = 0(e^{\sigma_0 t}), \quad \bar{u}_{xx}(x;p) = 0(e^{\sigma_0 t}),$$

uniformly with respect to x in any segment $\epsilon \leq x \leq l$;

3) *there exist an integer $k \geq 0$ such that $|p^{-k}\bar{u}(x,p)| < Q=$ const in the field $\overline{\mathfrak{M}}(S)$ for all $0 \leq x \leq \varepsilon < l$, $Re(p) > \sigma_1 > \sigma_0$;*

4) *there exist the limit $\lim_{t\to +0} \bar{u}(x,p) = g(t)$, $t > 0$, where $g(t)$ is a continuous function for $t > 0$ and is bounded as $t \to 0$.*

Then $u(x,t) = \bar{u}(x,p)$ is the solution of equation (2.6.62) or (2.6.63), satisfying the given boundary and initial conditions.

Proof. First, we prove that the conditions of the theorem imply the existence of the derivatives $u_x(x,t)$ and $u_{xx}(x,t)$ for $0 < x \leq l$. Indeed, let $\bar{u}_x(x,p) = v(x,t)$, then we have

$$\bar{u}(x,p) = p\int_0^\infty u(x,t)e^{-pt}dt; \quad (2.6.70)$$

$$\bar{u}_x(x,p) = p\int_0^\infty v(x,t)e^{-pt}dt, \quad (2.6.71)$$

and according to condition *2)* the integrals for $Re(p) > \sigma_0$ are absolutely and uniformly convergent with respect to $\varepsilon \leq x \leq l$. Therefore, the second integral may be integrated by the variable x from ε to l:

$$\bar{u}(x;p) - \bar{u}(\varepsilon,p) = p\int_0^\infty \left(\int_\varepsilon^y v(y,t)dy\right)e^{-pt}dt, \quad Re(p) > \sigma_0,$$

or

$$\bar{u}(x,p) = p\int_0^\infty \left[u(\varepsilon,t) + \int_\varepsilon^x v(y,t)dy\right]e^{-pt}dt, \quad Re(p) > \sigma_0.$$

Comparing the latter integral with (2.6.70), we conclude that

$$u(x,t) = u(\varepsilon,t) + \int_\varepsilon^x v(y,t)dy.$$

It follows from here that the solution $u(x,t)$ is differentiable with respect to x and the following equation holds:

$$u_x(x,t) = v(x,t) = \bar{u}_x(x;t), \quad 0 < x \leq l. \tag{2.6.72}$$

If we introduce the notation $\bar{u}_{xx}(x;p) = w(x;t)$, taking into account condition *2)* of the theorem, equation (2.6.71) and equation (2.6.72), in which $u(x,t)$ must be replaced by $u_x(x,t)$ and $v(x,t)$ by $w(x,t)$, then we obtain

$$u_{xx}(x,t) = w(x,t) = \bar{u}_{xx}(x,p), \quad 0 < x \leq l. \tag{2.6.73}$$

Thus, the existence of the derivatives $u_x(x,t)$ and $u_{xx}(x,t)$ is proved.

Now, (2.6.72) and (2.6.73) imply

$$\rho_0(x)u_{xx}(x,t) + \rho_1(x)u_x(x,t) + \rho_2(x)u(x,t) = \rho_0(x)\bar{u}_{xx}(x,p) + \rho_1(x)\bar{u}_x(x,p) + \rho_2(x)\bar{u}(x,p),$$

or, taking into account the transformed equation (2.6.68), we have

$$\begin{aligned} \rho_0(x)u_{xx}(x,t) &+ \rho_1(x)u_x(x,t) + \rho_2(x)u(x,t) \\ &= \rho(x)p^2\left[\bar{u}(x,p) - \varphi(x) - \frac{1}{p}\Psi(x)\right] \\ &= \rho(x)p^2[u(x,t) - \varphi(x) - t\Psi(x)]. \end{aligned} \tag{2.6.74}$$

It follows from (2.6.73), (2.6.72) and the second condition of the theorem that the sum

$$\rho_0(x)u_{xx}(x,t) + \rho_1(x)u_x(x,t) + \rho_2(x)u(x,t)$$

belongs for $0 < x \leq l$ to the set S, therefore, the function

$$\rho(x)p^2[u(x,t) - \varphi(x) - t\Psi(x)]$$

also belongs to the set S; however, $\rho(x) > 0$ for $0 < x \leq l$, and hence, for $0 < x \leq l$ the operator

$$p^2[u(x,t) - \varphi(x) - t\Psi(x)]$$

is reducible to a function belonging to S. Let us introduce the notation

$$p^2[u(x,t) - \varphi(x) - t\Psi(x)] = q(x,t) \in S;$$

then we obtain

$$u(x,t) - \varphi(x) - t\Psi(x) = \frac{1}{p^2}q(x,t) = \int_0^t (t-\xi)q(x,\xi)d\xi.$$

Hence, the function $u(x,t)$ for $0 < x \leq l$ is twice differentiable with respect to the variable t and

$$u(x,0) = \varphi(0); \quad u_t(x,0) = \Psi(0), \quad 0 < x \leq l.$$

It follows from (2.6.74) that in the domain $t > 0$, $0 < x \leq l$, we have

$$\rho_0(x)u_{xx}(x,t) + \rho_1(x)u_x(x,t) + \rho_2(x)u(x,t) = \rho(x)u_{tt}.$$

Thus, it has been proved that

$$u(x,t) = \bar{u}(x,p) \tag{2.6.75}$$

is the solution of equation (2.6.68), satisfying the initial conditions (2.6.65). The condition

$$a\bar{u}_x(l,p) + bp[\bar{u}(l,p) - \varphi(l)] = c\bar{u}(l,p)$$

and (2.6.72) and (2.6.75) for $x = l$ imply that

$$au_x(l,t) + bu_t(l,t) = cu(l,t),$$

i.e., the boundary conditions hold for $x = l$.

It remains now to consider the behavior of the solution as $x \to 0$. Let us introduce the notation

$$\lim_{x \to +0} u(x,t) = g(t).$$

According to the fourth condition of the theorem, this limit exists and is a continuous function for $t > 0$ and $g(t)$ is bounded as $t \to 0$. Hence, it remains only to prove that $g(t) = f(t)$, $t > 0$. However, this immediately follows from the condition $\bar{u}(+0,p) = \bar{f}(p)$, see (2.6.69) and condition (2.6.62) of the theorem. Indeed, condition (2.6.62) implies the continuity of the operator function $\bar{u}(x,p)$ in the domain $0 \leq x \leq l$. Therefore, we have

$$\lim_{x \to 0} \bar{u}(x,p) = \bar{u}(o,p) = \bar{f}(p) = f(t), \quad \text{i.e.,} \quad g(t) = f(t).$$

□

Remark 2.6.101 *Instead of using the notion of an operator function one can in the proof of the latter fact explicity prove, starting from conditions 3) and 4) of* Theorem 2.6.94, *the relation $f(t) = g(t)$. Indeed, for sufficiently large n we have*

$$\frac{1}{2\pi i} \int_{\sigma - i\infty}^{\sigma + i\infty} \frac{\bar{u}(x,p)}{p^{n+1}} e^{pt} dp = \frac{1}{(n-1)!} \int_0^t (t-\xi)^{n-1} u(x,\xi) d\xi, \tag{2.6.76}$$

and the integral in the left-hand side converges absolutely and uniformly with respect to $x \to 0$. Therefore, we can pass in (2.6.76) to the limit as $x \to 0$, then we have

$$\frac{1}{2\pi i} \int_{\sigma - i\infty}^{\sigma + i\infty} \frac{\bar{f}(p)}{p^{n+1}} e^{pt} dp = \frac{1}{(n-1)!} \int_0^t (t-\xi)^{n-1} g(\xi) d\xi,$$

or

$$\int_0^t (t-\xi)^{n-1} f(\xi) d\xi = \int_0^t (t-\xi)^{n-1} g(\xi) d\xi;$$

therefore, $f(t) = g(t)$ for all $t > 0$.

Let us consider some problems of mathematical physics.

Problem 2.6.5. Find the distribution of the temperature in a semiinfinite line $0 < x < \infty$, if its left end has the constant temperature equal to zero, and the initial temperature of the line is equal to one.

We need to find the solution of the heat equation

$$\frac{\partial u}{\partial t} = \frac{\partial^2 u}{\partial x^2} \quad (x > 0, \ t > 0) \tag{2.6.77}$$

under the conditions

$$u = 0 \quad \text{for} \quad x = 0, \ t > 0; \tag{2.6.78}$$

$$u = 1 \quad \text{for} \quad x > 0, \ t = 0. \tag{2.6.79}$$

The operator transform of equation (2.6.77) has the form

$$\frac{\partial^2 \bar{u}}{\partial x^2} = p\bar{u} - p, \quad (x > 0) \tag{2.6.80}$$

under the condition

$$\bar{u}(x, p) = 0 \quad \text{for} \quad x = 0. \tag{2.6.81}$$

The general solution of equation (2.6.80) is

$$\bar{u}(x, p) = 1 + Ae^{x\sqrt{p}} + Be^{-x\sqrt{p}}, \tag{2.6.82}$$

where "the constants" A and B in the general case depend on p and are defined by the boundary conditions. The condition of boundedness of the solution as $x \to \infty$ implies that $A = 0$. From (2.6.81) we find $0 = 1 + B$. Thus, we have

$$\bar{u}(x, p) = 1 - e^{-x\sqrt{p}}.$$

In this case by the inversion theorem of the Laplace transform (see section 1.4.4, Theorem 1.4.12) we have

$$u(x, t) = \lim_{\tau \to \infty} \frac{1}{2\pi i} \int_{\sigma - i\tau}^{\sigma + i\tau} \left\{ \frac{1}{p} - \frac{e^{-x\sqrt{p}}}{p} \right\} e^{pt} dp, \quad \sigma > 0. \tag{2.6.83}$$

As it is known (see formula 1.4.8) the inversion of $\frac{1}{p}$ gives the function $f(t) \equiv 1$ for $t > 0$.

Now we begin the inversion of the function $\frac{1}{p} e^{-x\sqrt{p}}$, i.e., the computation of the integral

$$J = \frac{1}{2\pi i} \int_{\sigma - i\tau}^{\sigma + i\tau} \frac{e^{-x\sqrt{p}}}{p} e^{pt} dp, \quad \sigma > 0.$$

The function $\frac{1}{p}e^{-x\sqrt{p}}$ is analytical on the whole plane of the complex variable p except the origin, therefore, $\frac{1}{p}e^{-x\sqrt{p}}$ is univalent and analytical in the plane with the cut along the negative part of the real axis. According to the Cauchy theorem, the integration along the line $(\sigma - i\tau, \sigma + i\tau)$ may be replaced by the integration along any curve with the ends at the points $\sigma \pm i\tau$, which does not intersect the cut.

In particular, it is convenient to use the contour shown in Figure 12; we have

$$\int\limits_{\sigma-i\tau}^{\sigma+i\tau} = \int\limits_{AC} + \int\limits_{CD} + \int\limits_{DE} + \int\limits_{EF} + \int\limits_{FB} . \qquad (2.6.84)$$

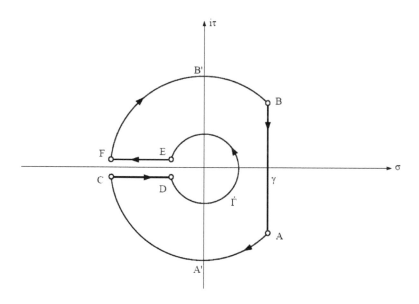

Figure 12

Let us show that the integrals $\int\limits_{AC}$ and $\int\limits_{FB}$ approach zero as $\tau \to \infty$. We have

$$\left| \frac{e^{-x\sqrt{p}}}{p} \right| = \frac{e^{-x\sqrt{p}}}{|p|}, \qquad (-\pi < \arg p < \pi);$$

hence, $-\frac{\pi}{2} < \arg\sqrt{p} < \frac{\pi}{2}$, and therefore, $Re(\sqrt{p}) \ge 0$ and for $x \ge 0$

$$\left| \frac{e^{-x\sqrt{p}}}{p} \right| \le \frac{1}{|p|}.$$

According to Jordan's lemma (see section 1.4.5., Lemma 1.4.1) for $t > 0$ and $R \to \infty$ the integral of the function $\frac{e^{-x\sqrt{p}+pt}}{p}$ along the arcs AC and FB approaches zero.

Now we begin the computation of the integrals along the lines CD and EF. On these lines $p^{\frac{1}{2}}$ is equal to $i|p|^{\frac{1}{2}}$ and $-i|p|^{\frac{1}{2}}$, respectively. On setting $\rho = |p|$, we have

$$\int_{CD} + \int_{EF} = -2i \int_r^R \rho^{-1} \sin\left(x\rho^{\frac{1}{2}}\right)e^{-t\rho}d\rho = -4i \int_{r^{1/2}}^{R^{1/2}} \frac{\sin x\xi}{\xi} e^{-t\xi^2} d\xi; \tag{2.6.85}$$

hence, there exists the limit

$$\lim_{\substack{r \to 0 \\ R \to \infty}} \left\{ \int_{CD} + \int_{EF} \right\} \left(\frac{1}{p}e^{-x\sqrt{p}}e^{pt}\right) dp = -4i \int_0^\infty \frac{\sin x\xi}{\xi} e^{-t\xi^2} d\xi. \tag{2.6.86}$$

Finally, we have

$$\int_{DE} = \int_{DE} \frac{e^{-x\sqrt{p}+pt}}{p} dp = \int_{-\pi}^\pi \frac{e^{-x\sqrt{\varepsilon}e^{\frac{i\varphi}{2}}+\varepsilon e^{i\varphi}}\varepsilon e^{i\varphi}id\varphi}{\varepsilon e^{i\varphi}} = \int_{-\pi}^\pi e^{-x\sqrt{\varepsilon}e^{\frac{i\varphi}{2}}+\varepsilon e^{i\varphi}}id\varphi.$$

Hence,

$$\lim_{r\to 0}\int_{DE} = \lim_{\varepsilon\to 0}\int_{-\pi}^\pi e^{-x\sqrt{\varepsilon}e^{\frac{i\varphi}{2}}+\varepsilon e^{i\varphi}}id\varphi = 2\pi i. \tag{2.6.87}$$

Thus, combining (2.6.84) through (2.6.87), we find

$$\lim_{r\to\infty}\frac{1}{2\pi i}\int_{\sigma-i\tau}^{\sigma+i\tau}\frac{1}{p}\exp\left(-xp^{\frac{1}{2}}+tp\right)dp = 1 - \frac{2}{\pi}\int_0^\infty \frac{\sin x\xi}{\xi}e^{-t\xi^2}d\xi.$$

According to (2.6.83), the final solution of our problem has the form

$$u(x,t) = \frac{2}{\pi}\int_0^\infty \frac{\sin x\xi}{\xi}e^{-t\xi^2}d\xi. \tag{2.6.88}$$

Let us reduce this solution to another form. On differentiating (2.6.88) with respect to x, we obtain

$$\frac{\partial u}{\partial x} = \frac{2}{\pi}\int_0^\infty e^{-t\xi^2}\cos x\xi d\xi. \tag{2.6.89}$$

The computation of this integral may be made with the help of the theory of residues. Let us introduce into consideration the function

$$f(z) = e^{-tz^2},$$

whose integral along the real axis may be computed with the help of Poisson's integral, known from calculus:

$$\int_0^\infty e^{-u^2} du = \frac{\sqrt{\pi}}{2}. \tag{2.6.90}$$

On the line $\tau = h$ we have

$$e^{-t(\sigma+ih)^2} = e^{th^2} e^{-t\sigma^2} (\cos 2\tanh\sigma - i\sin 2\tanh\sigma).$$

The real part of this expression at $h = \frac{x}{2t}$ differs from the integrand by a constant factor. In accordance with this, we use the contour of integration, shown in Figure 13.

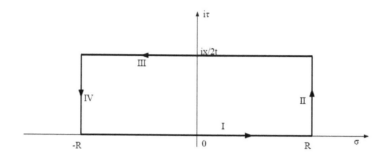

Figure 13

By Cauchy's theorem, we have

$$\int_{I} + \int_{II} + \int_{III} + \int_{IV} = 0, \tag{2.6.91}$$

here,

$$\int_{I} = \int_{-R}^{R} e^{-t\xi^2} d\xi = \frac{2}{\sqrt{t}} \int_{0}^{\frac{R}{\sqrt{t}}} e^{-\xi^2} d\xi;$$

$$\int_{II} = -e^{\frac{x^2}{4t}} \int_{-R}^{R} e^{-t\xi^2} e^{-ix\xi} d\xi.$$

On the segments II and IV, where $x = \pm R$, we have

$$|e^{-tz^2}| = e^{-t(R^2-\tau^2)} \le e^{\frac{x^2}{4t}} e^{-tR^2};$$

therefore, assuming $t > 0$, we find $\int_{II,IV} \to 0$ as $R \to \infty$. Performing in (2.6.91) the limit as $R \to \infty$, and using (2.6.90), we obtain

$$\frac{\sqrt{\pi}}{\sqrt{t}} - e^{-\frac{x^2}{4t}} \int_{-\infty}^{\infty} e^{-t\xi^2} e^{-ix\xi} d\xi = 0,$$

and comparing the real parts, we find

$$\int_{0}^{\infty} e^{-t\xi^2} \cos x\xi \, d\xi = \frac{1}{2}\sqrt{\frac{\pi}{t}} e^{-\frac{x^2}{4t}}, \quad t > 0. \tag{2.6.92}$$

Thus, according to (2.6.92), formula (2.6.89) takes the form

$$\frac{\partial u}{\partial x} = (\pi t)^{-\frac{1}{2}} e^{-\frac{x^2}{4t}}. \tag{2.6.93}$$

Taking into account that $u(0,t) = 0$, and integrating equation (2.6.93), we obtain

$$u(x,t) = (\pi t)^{-\frac{1}{2}} \int_0^x e^{-\frac{y^2}{4t}} dy.$$

On changing the variable by the formula $\xi = \frac{y}{\sqrt{2t}}$, we reduce this equation to the form

$$u(x,t) = 1 - e^{-x\sqrt{p}} = \sqrt{\frac{2}{\pi}} \int_0^{\frac{x}{\sqrt{2t}}} -e^{\frac{\xi^2}{2}} d\xi = \frac{2}{\sqrt{\pi}} \int_0^{\frac{x}{2\sqrt{t}}} e^{-\xi^2} d\xi = erf\left(\frac{x}{2\sqrt{t}}\right). \tag{2.6.94}$$

It is obvious from here that $u(0,t) = 0$, $u(x,0) = 1$.

Note that equation (2.6.94) may be obtained easily with the help of operational calculus in the following way. By formula (2.5.10) we have

$$\sqrt{p}e^{-\lambda\sqrt{p}} = \frac{1}{\sqrt{\pi t}} e^{-\frac{\lambda^2}{4t}}.$$

However,

$$\int_\lambda^\infty e^{-\xi\sqrt{p}} d\xi = \frac{1}{\sqrt{p}} e^{-\lambda\sqrt{p}};$$

hence,

$$e^{-\lambda\sqrt{p}} = \int_\lambda^\infty \sqrt{p} e^{-\xi\sqrt{p}} d\xi = \frac{1}{\sqrt{\pi t}} \int_\lambda^\infty e^{-\frac{\xi^2}{4t}} d\xi = \frac{2\sqrt{t}}{\sqrt{\pi t}} \int_{\frac{\lambda}{2\sqrt{t}}}^\infty e^{-u^2} du = \frac{2}{\sqrt{\pi}} \int_{\frac{\lambda}{2\sqrt{t}}}^\infty e^{-u^2} du \tag{2.6.95}$$

$$= erfc\left(\frac{\lambda}{2\sqrt{t}}\right) = 1 - erf\left(\frac{\lambda}{2\sqrt{t}}\right).$$

Problem 2.6.6. On an infinite cylinder of the radius a the constant temperature $u_0 = 1$ is maintained. Determine the temperature at any point of the external space at an instant t, if at the initial instant the temperature in the space was equal to zero.

The problem is reducible to the solution of the differential equation

$$\frac{\partial u}{\partial t} = k\left(\frac{\partial^2 u}{\partial r^2} + \frac{1}{r}\frac{\partial u}{\partial r}\right) \tag{2.6.96}$$

under the initial condition

$$u = 0 \quad \text{for} \quad t = 0 \quad \text{and} \quad a < r < \infty \tag{2.6.97}$$

and the boundary conditions

$$u = 1 \quad \text{for} \quad r = a, \, t > 0 \quad \text{and} \quad \lim_{r\to\infty} u(r,t) = 0, \quad t > 0. \tag{2.6.98}$$

The transformed equation has the form

$$\frac{d^2\bar{u}}{dr^2} + \frac{1}{r}\frac{du}{dr} - \frac{p}{k}\bar{u} = 0.$$

The solution must satisfy the conditions

$$\bar{u} = 1 \quad \text{for} \quad r = a, \quad \text{and} \quad \lim_{r\to\infty} u(r,p) = 0.$$

The equation has two linearly independent solutions $K_0(\nu r)$ and $I_0(\nu r)$, where

$$\nu = \sqrt{\frac{p}{k}}. \tag{2.6.99}$$

Hence, the general solution is $c_1 K_0(\nu r) + c_2 I_0(\nu r)$. Taking into account the boundary conditions of the problem, we find the desired solution:

$$\bar{u}(r,p) = \frac{K_0(\nu r)}{K_0(\nu a)};$$

we have chosen the branch of root (2.6.99) for which $\lim_{r\to+\infty} K_0(\nu r) = 0$, i.e., $\sqrt{p} > 0$, if p is a real positive number and arg $p = 0$. Thus, if the operator $\frac{K_0(\nu r)}{K_0(\nu a)}$, where ν is the same, as in (2.6.99), satisfies the conditions of Theorem 2.6.94 (see 2.6.2), then the solution of the problem is the function

$$u(r,t) = \frac{1}{2\pi i}\int_{(\gamma)} \frac{K_0(\nu r)}{K_0(\nu a)}\frac{e^{pt}}{p}dp, \quad \gamma \in \mathbb{R}_+. \tag{2.6.100}$$

The function $\frac{K_0(\nu r)}{K_0(\nu a)}$ is analytical in the complex plane with a cut along the negative part of the real axis. For large values of ν the asymptotic representation, see [E.1], vol. 2, section 7.13.1, formula (7), holds:

$$\frac{K_0(\nu r)}{K_0(\nu a)} \sim \sqrt{\frac{a}{r}}e^{-\nu(r-a)},$$

where ν is the same as in (2.6.99). The integral (2.6.100) converges uniformly in the domain $r - a \geq \varepsilon > 0$ and $0 \leq t \leq T$; hence, the function $u(r,t)$ is continuous in this domain. Let us transform expression (2.6.100) to the form convenient for computation. For this purpose we consider the integral

$$\frac{1}{2\pi i}\int_L \frac{K_0(\nu r)}{K_0(\nu a)}e^{pt}\frac{dp}{p}, \tag{2.6.101}$$

where L is the contour shown in Figure 12.

Taking into account that the integrand satisfies the conditions of Jordan's lemma, we deduce from (2.6.101) as $R \to \infty$

$$u(r,t) = \frac{1}{2\pi i}\int_\Gamma \frac{K_0(\nu r)}{K_0(\nu a)}\frac{e^{pt}}{p}dp,$$

where Γ is the contour, shown in Figure 14.

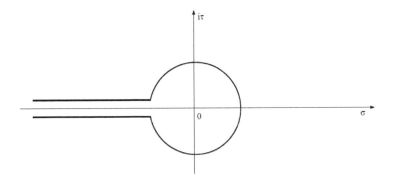

Figure 14

Contracting the contour Γ to the edges of the cut and taking into account the singularity at the point $p = 0$, we find

$$u(r,t) = 1 - \frac{1}{\pi} \int\limits_0^\infty \frac{J_0(a\nu)Y_0(r\nu) - J_0(r\nu)Y_0(a\nu)}{J_0^2(a\nu) + Y_0^2(a\nu)} \frac{e^{-pt}}{p} \, dp, \quad \nu = \sqrt{\frac{p}{k}}.$$

In these transformations we used the equations

$$K_0(ze^{i\frac{\pi}{2}}) = -\pi i[J_0(z) - iY_0(z)],$$

$$K_0(ze^{-i\frac{\pi}{2}}) = \pi i[J_0(z) + iY_0(z)],$$

(see [E.1], vol. 2, section 7.2.1, formulas 16, 17, 5, and 6). The function $u(r,t)$ may be, obviously, represented by the form

$$u(r,t) = 1 + \frac{2}{\pi} \int\limits_0^\infty e^{-tk\alpha^2} \frac{J_0(r\alpha)Y_0(a\alpha) - J_0(a\alpha)Y_0(r\alpha)}{J_0^2(a\alpha) + Y_0^2(a\alpha)} \frac{d\alpha}{\alpha}. \tag{2.6.102}$$

It follows immediately from the expression obtained for $u(r,t)$ the continuity of $u(r,t)$ when approaching any point (a,t), $t > 0$. Indeed, for $t > 0$ we have

$$\lim_{r \to a} u(r,t) = 1.$$

Chapter 3

Generalized Functions

3.1 Introduction

Distributions are a generalization of locally integrable functions on the real line, or more generally, a generalization of functions that are defined on an arbitrary open set in the Euclidean space.

Distributions were introduced as a result of difficulties with solving some problems of mathematical physics, quantum mechanics, electrotechnics and so forth. In these domains there are many theoretical and practical problems where the notion of function is not sufficient in this or that sense.

In 1926 the English physicist Paul Dirac introduced a new element of mathematical formalism into quantum mechanics, [Dir]. He named it the *Delta function* and denoted it by $\delta(t)$. Dirac assumed that the delta function is defined on the real line and fulfills the following conditions:

$$\delta(t) = \begin{cases} 0 & \text{for } t \neq 0, \\ +\infty & \text{for } t = 0, \end{cases} \tag{3.1.1}$$

and

$$\int_{-\infty}^{+\infty} \delta(t)dt = 1. \tag{3.1.2}$$

In the theory of real functions the two conditions (3.1.1) and (3.1.2) are contradictory. No real function exists to fulfill both conditions at the same time. On the other hand, both conditions give incorrect but highly convincing evidence of a physical intuition: $\delta(t)$ represents an infinitely large growth of electric tension in the infinitely short time where a unit of electricity loads. Nevertheless, the existence of mathematical models for which the search for mathematical description would lead naturally to *Dirac's deltas* should not provide an excuse to use the imprecise mathematical notion that hides under $\delta(t)$, treat it as a function and at the same time assume that it fulfills conditions (3.1.1) and (3.1.2). Despite all formal objections many important results were achieved with the help of *Dirac's delta*.

In the 1930s it became obvious that *Dirac's delta* has a fixed place in theoretical physics. As a result scientists sought a new mathematical theory that would help define *Dirac's delta* as it was previously used in the precise definition of real numbers, and so forth. The mathematical theory known as the theory of distributions, which enabled introducing *Dirac's delta* without any logical objections, was developed in the 1940s. This theory allowed the generalization of the notion of function as it was once done for real numbers.

There are many ways to define distributions (as generalizations of functions), i.e., I. Halperin [Ha], II. König [Kö.1], J. Korevar [Ko.1], J. Mikusiński and R. Sikorski [MiS.1], W. Słowikowski [Sło], S. Soboleff [So], L. Schwartz [S.1], and G. Temple [Te.1], [Te.2]. The two most important in theory and practice are the functional approach of S. Soboleff (1936) and L. Schwartz (1945), where distributions are defined as linear continuous functionals in certain linear spaces; and the sequential approach of J. Mikusiński (1948), where distributions are defined as classes of equivalent sequences.

The functional theory is more general but more complicated. It uses difficult notions of functional analysis and the theory of linear spaces. L. Schwartz developed and presented his theory together with applications in a two volume manual [S.2].

The sequential approach is easier because it is based on fundamental notions of mathematical analysis. It has a geometrical and physical interpretation that relies on an intuitive understanding of *Dirac's delta*, which is common in physics. The sequential approach is easier to understand and easier to apply. A formal definition of distribution was given by J. Mikusiński in [Mi.2]. Based on that definition J. Mikusiński together with R. Sikorski developed the sequential theory of distributions and published it in [MiS.1] and [MiS.2].

The theory was developed and together with P. Antosik a monograph *"Theory of Distribution, The Sequential Approach,"* [AMS], was published.

The starting point for the definition of distributions in the sequential approach are sequences of continuous or smooth functions (i.e., of class $\mathcal{C}^\infty$) in a certain fixed interval (a, b), $(-\infty \le a < b \le +\infty)$. The definition given by J. Mikusiński is analogous to the definition of real numbers in Cantor's theory. The aim of introducing real numbers was the performance of certain mathematical operations within this set. Similarly, the introduction of distributions enabled differentiation that cannot always be performed in the set of functions.

3.2 Generalized Functions — Functional Approach

3.2.1 Introduction

In the following we will assume that the functions to be considered can have complex values. Let f and φ be functions of class $\mathcal{C}^\infty(\mathbb{R})$ and the function φ vanishes outside a finite interval. Integrating by parts we get

$$\int\limits_{-\infty}^{+\infty} f(x)\varphi'(x)dx = -\int\limits_{-\infty}^{+\infty} f'(x)\varphi(x)dx. \tag{3.2.1}$$

Note that the left-hand side of (3.2.1) makes sense also for much weaker assumptions concerning the function f. Namely, it is sufficient to assume that f is integrable on every bounded interval. Therefore, we can use (3.2.1) to define a generalized derivative Df. Namely, we assume that a generalized derivative is a function $g \in L_1^{loc}(\mathbb{R})$ such that for an arbitrary function φ fulfilling the above conditions we have:

$$\int\limits_{-\infty}^{+\infty} f(x)\varphi'(x)dx = -\int\limits_{-\infty}^{+\infty} g(x)\varphi(x)dx. \tag{3.2.2}$$

If we want to apply the same reasoning to define the generalized second derivative $D^2 f$ for a function $f \in L_1^{loc}$ we would have to strengthen the regularity conditions and assume that φ belongs to $C^2(\mathbb{R})$. Therefore, it is better to assume at once that φ has derivatives of all orders, i.e., $\varphi \in C^\infty$.

We define the *support of a continuous function* φ as a closure of a set $\{x : \varphi(x) \neq 0\}$ and denote it by *suppφ*. By C_o^∞ we denote a set of all continuous functions on $\mathbb{R}$ as well as all their derivatives of any order with compact support.

Let $[a, b] = supp\varphi$. The regularity conditions imply that the x axis is a tangent of an "infinite order" to the graph of the function φ at the points $x = a$ and $x = b$. The fact that such a function exists is not so obvious. We will, however, show that it is easy to construct such a function. Let, for example,

$$h(t) = \begin{cases} e^{-\frac{1}{t}} & \text{for } t > 0, \\ 0 & \text{for } t \leq 0. \end{cases} \tag{3.2.3}$$

The function h is of the class $C^\infty(\mathbb{R})$ and its support is a half line $[a, \infty)$. Taking a constant $a > 0$ we define a function

$$\varphi_a(x) = h\left(1 - \frac{x^2}{a^2}\right), \tag{3.2.4}$$

i.e.,

$$\varphi_a(x) = \begin{cases} e^{\frac{a^2}{(x^2 - a^2)}} & \text{for } |x| \leq a, \\ 0 & \text{for } |x| > a. \end{cases} \tag{3.2.5}$$

It follows from the equation (3.2.4) that φ_a, being a superposition of two infinitely many times differentiable functions, has derivatives of all orders. On the other hand, (3.2.5) implies that $supp\varphi = [-a, a]$.

Let f be a locally integrable function on $\mathbb{R}$. The function $g \in L_1^{loc}(\mathbb{R})$ is called a generalized derivative of f if (3.2.2) holds true for every function $\varphi \in C_o^\infty(\mathbb{R})$. In this chapter we denote the generalized derivative by $g = Df$.

Example 3.2.64 *Let $f(x) = x \cdot 1_+(x)$, where $1_+(x)$ is the Heaviside function:*

$$1_+(x) = \begin{cases} 1 & \text{for } x > 0, \\ 0 & \text{for } x \le 0. \end{cases}$$

Integrating by parts gives

$$\int_{-\infty}^{+\infty} 1_+(x) x\varphi'(x)dx = \int_0^\infty x\varphi'(x)dx = -\int_0^\infty \varphi(x)dx,$$

which means that $Df = 1_+(x)$. Because Df is a locally integrable function we will try to find its generalized derivative, i.e., $D^2 f$. Integrating, we have

$$\int_{-\infty}^{+\infty} 1_+(x)\varphi'(x)dx = -\varphi(0). \tag{3.2.6}$$

It is, therefore, sufficient to represent the right-hand side of (3.2.6) in the form of the integral appearing on the right-hand side of (3.2.2). It is easy to show that such a representation is not possible. Namely, we may assume that such a function exists. In particular, we may take $g \in L_1^{loc}(\mathbb{R})$ and let

$$\int_{-\infty}^{+\infty} g(x)\varphi(x)dx = \varphi(0) \quad \text{for any} \quad \varphi \in C_o^\infty(\mathbb{R}). \tag{3.2.7}$$

In particular, we take $\varphi = \varphi_a$ and let $a \to 0$. Then the right-hand side of (3.2.7) is constant and equal to e^{-1} but the left-hand side tends to 0, which follows from the estimation

$$\left| \int_{-a}^a g(x)\varphi_a(x)dx \right| \le e^{-1} \int_{-a}^a |g(x)|dx.$$

The example given above shows that not every locally integrable function has a generalized derivative that also belongs to $L_1^{loc}(\mathbb{R})$. In other words, repeating the operation of generalized differentiation, i.e., obtaining generalized derivatives of higher orders is in general not possible. This is a serious drawback of the definition introduced above. What makes the formula (3.2.1) incorrect is the assumption made about its result, namely the assumption that a generalized derivative is a function. In the next chapter we will show that it is possible to find a more general class of objects for which generalized derivatives can be obtained without restrictions.

3.2.2   Distributions of One Variable

Let us understand the equality (3.2.1), section 3.2.1 in a different way. The class $C_o^\infty(\mathbb{R})$ is a linear space over the field of complex numbers and the integral on the right-hand side of (3.2.8) defines a linear functional Λ_g, namely,

$$\Lambda_g(\varphi) := \int\limits_{-\infty}^{+\infty} g(x)\varphi(x)dx. \tag{3.2.8}$$

Using the previously introduced notation we can write (3.2.1), section 3.2.1 in a different way:

$$\Lambda_f(\varphi') = -\Lambda_g(\varphi), \quad \varphi \in \mathcal{C}_o^\infty(\mathbb{R}).$$

Let us return to Example 3.2.64, section 3.2.1. Differentiating twice the function f in the generalized sense we obtain a linear functional on

$$\mathcal{C}_o^\infty(\mathbb{R}) \ni \varphi \to \varphi(0).$$

This idea comes from contemporary mathematicians L. Soboleff [So] and L. Schwartz [S.1], who are regarded as the creators of the theory of distributions. The theory of distributions proved to be very useful in the theory of linear partial differential equations.

The Space $\mathcal{D}(\mathbb{R})$

As we have mentioned before the space $\mathcal{C}_o^\infty(\mathbb{R})$ is a linear space.

We will denote by $\mathcal{D}(\mathbb{R})$ the set of all infinitely differentiable functions with compact supports contained in $\mathbb{R}$. Such *functions* will be called *smooth*. In order to define a class of convergent sequences on that space it is sufficient to introduce the notion of a sequence convergent to zero, i.e., to a function equal to zero identically. A sequence $(\varphi_\nu)_{\nu\in\mathbb{N}}$ will be called convergent to a function φ if and only if the difference $(\varphi_\nu - \varphi)_{\nu\in\mathbb{N}}$ is convergent to zero.

Definition 3.2.51 *We say that a sequence* $(\varphi_\nu)_{\nu\in\mathbb{N}}$, $\varphi_\nu \in \mathcal{D}(\mathbb{R})$, *converges to zero if*

(i) *there exists a bounded interval* $[a,b] \subset \mathbb{R}$ *such that* $\mathrm{supp}\varphi_\nu \subset [a,b]$ *for each* $\nu \in \mathbb{N}$;

and

(ii) *the sequence* $(\varphi_\nu^{(j)})_{\nu\in\mathbb{N}}$ *converges to zero uniformly on* $[a,b]$ *for each* $j \in \mathbb{N}$.

The sequence $(\varphi_\nu)_{\nu\in\mathbb{N}}$ *converges to* φ *in* $\mathcal{D}(\mathbb{R})$ *if the sequence* $(\varphi_\nu - \varphi)_{\nu\in\mathbb{N}}$ *converges to zero in* $\mathcal{D}(\mathbb{R})$.

Definition and Examples of Distributions

Definition 3.2.52 *A functional* $\Lambda : \mathcal{D}(\mathbb{R}) \to \mathbb{C}$ *fulfilling the conditions*

(iii) $\Lambda(c_1\varphi_1 + c_2\varphi_2) = c_1\Lambda(\varphi_1) + c_2\Lambda(\varphi_2)$, *where* $c_j \in \mathbb{C}, \varphi_j \in \mathcal{D}(\mathbb{R}), j = 1, 2;$

(iv) *every sequence* $(\varphi_\nu)_{\nu\in\mathbb{N}} \subset \mathcal{D}(\mathbb{R})$ *that converges to zero* $\Lambda(\varphi_\nu) \to 0$

is called a distribution.

The condition (iii) means that functional Λ is linear while the condition (iv) means that the functional is continuous with respect to the convergence introduced in the space $\mathcal{D}(\mathbb{R})$. The set of all distributions on the real line, i.e., the set of all linear and continuous functionals on $\mathcal{D}(\mathbb{R})$ will be denoted by $\mathcal{D}'(\mathbb{R})$.

The symbol $< \Lambda, \varphi >$ or $\Lambda(\varphi)$ will denote the value of a functional $\Lambda \in \mathcal{D}(\mathbb{R})$ on the function $\varphi \in \mathcal{D}(\mathbb{R})$.

Theorem 3.2.95 *A linear form* Λ *defined on* $\mathcal{D}(\mathbb{R})$ *is continuous iff for every bounded closed interval* $I \subset \mathbb{R}$ *there exist a constant* $C > 0$ *and an integer* $m \in \mathbb{N}$ *such that*

$$|\Lambda(\varphi)| \leq C\|\varphi\|_{m,I}, \tag{3.2.9}$$

where

$$\|\varphi\|_{m,I} = \max_{|s|\leq m} \sup_{x\in I} |\varphi^{(s)}(x)| \quad and \quad supp\varphi \subset I.$$

Proof. That it is sufficient is evident; we now prove it is necessary. Assume that $\Lambda \in \mathcal{D}'(\mathbb{R})$ and (3.2.9) does not hold for some compact (bounded, closed) interval $I \subset \mathbb{R}$. Therefore, for each $\nu \in \mathbb{N}$ there exists a function

$$\varphi_\nu \in \mathcal{D}(\mathbb{R}) \quad such \ that \quad supp\varphi_\nu \subset I \quad and \quad |\Lambda(\varphi_\nu)| > \nu\|\varphi_\nu\|_{\nu,I}.$$

Obviously, one can choose φ_ν such that $\Lambda(\varphi_\nu) = 1$ for each $\nu \in \mathbb{N}$. From this we obtain

$$1 = \Lambda(\varphi_\nu) > \nu\|\varphi_\nu\|_{\nu,I},$$

consequently, $\|\varphi_\nu\|_{\nu,I} < \frac{1}{\nu}$. Hence,

$$\|\varphi_{\nu+\mu}\|_{\mu,I} \leq \|\varphi_{\nu+\mu}\|_{\nu+\mu,I} \leq \frac{1}{\nu+\mu} \leq \frac{1}{\nu}.$$

From this it follows that $\|\varphi_{\nu+\mu}\|_{\mu,I} \to 0$ as $\nu \to \infty$ for fixed μ. Therefore, $(\varphi_\nu)_{\nu\in\mathbb{N}}$ converges to zero as $\nu \to \infty$ in $\mathcal{D}(\mathbb{R})$. This contradicts the continuity of Λ. Thus, the proof is complete.

We give some examples of distributions.

Example 3.2.65 *Let $f \in L_1^{loc}$. We take*

$$\Lambda_f(\varphi) := \int_{-\infty}^{+\infty} f(x)\varphi(x)dx. \tag{3.2.10}$$

It is easy to see that

$$|\Lambda_f(\varphi)| \leq \|f\|_{L_1(I)}\|\varphi\|_{0,I}, \quad if \quad supp\varphi \subset I.$$

Therefore, Λ_f belongs to $\mathcal{D}'$.

Property 3.2.1 The mapping $f \rightarrow \Lambda_f$ is an injection from $\mathcal{C}_o(\mathbb{R})$ into $\mathcal{D}'(\mathbb{R})$.

Proof. The mapping is linear so it is sufficient to show that the condition

$$< \Lambda_f, \varphi >= 0, \quad \varphi \in \mathcal{D}(\mathbb{R}) \tag{3.2.11}$$

implies that the function f vanishes identically. Suppose, on the contrary, that it does not vanish. Therefore, for a certain $x_o \in \mathbb{R}$, we have $f(x_o) \neq 0$. Let $f(x_o) > 0$. Because f is continuous, there exists a number $\eta > 0$ such that $f(x) > 0$ for $|x - x_o| < \eta$. Let $\varphi(x) = \varphi_\eta(x - x_o)$. Then we have

$$< \Lambda_f, \varphi >= \int_{|x-x_o|<\eta} f(x)\varphi_\eta(x - x_o)dx > 0,$$

which contradicts (3.2.11). This completes the proof. ▯

The Property 3.2.1 can be generalized to the case when f is an arbitrary locally integrable function. Namely, we have:

Property 3.2.2 The mapping $f \rightarrow \Lambda_f$ is an injection from $L_1^{loc}(\mathbb{R})$ into $\mathcal{D}'(\mathbb{R})$.

This is an immediate consequence of the following.

Lemma 3.2.37 (du Bois Reymond) *If f is in $L_1^{loc}(\mathbb{R})$ and*

$$\int_{-\infty}^{+\infty} f(x)\varphi(x)dx = 0 \quad for \quad \varphi \in \mathcal{D}(\mathbb{R})$$

then $f(x) = 0$ a.e. in $\mathbb{R}$.

For a proof see [Vl.1], p. 18.

Distributions that can be represented by the form (3.2.10) are called *regular distributions*.

In the following we will identify the distribution Λ_f with the function f. In that sense, all continuous functions, and more generally all locally integrable functions, can be regarded as special cases of distributions. Therefore, in addition to the term "distribution," one finds the term "generalized functions," which has, however, a slightly broader meaning.

Example 3.2.66 *The distribution*

$$< \delta, \varphi >= \varphi(0), \quad \varphi \in \mathcal{D}(\mathbb{R}) \tag{3.2.12}$$

is called the Dirac delta distribution. We have already shown in Example 3.2.64, section 3.2.1, that it is not a regular distribution. Therefore, it cannot be identified with any locally integrable function. Nevertheless, in the physical and technical literature it is called a "delta function." The equality (3.2.12) can be written in the "integral" form:

$$\int\limits_{-\infty}^{+\infty} \delta(x)\varphi(x)dx = \varphi(0).$$

This representation is, however, purely formal.

Example 3.2.67 *The function $\frac{1}{x}$ is not locally integrable, because its integral over every interval of the form $[0, a]$ is divergent. Nevertheless, we can assign to it a distribution defined by the identity*

$$\Lambda(\varphi) = \frac{1}{(\cdot)}(\varphi) := \lim_{\epsilon \to 0} \int\limits_{|x|>\epsilon} \frac{\varphi(x)}{x}dx, \tag{3.2.13}$$

where $\varphi \in \mathcal{D}$ and $supp\varphi \subset [-a, a]$.

Let us take a new function

$$\psi(x) = \begin{cases} \frac{\varphi(x)-\varphi(0)}{x} & \text{for } x \neq 0, \\ \varphi'(0) & \text{for } x = 0. \end{cases}$$

Clearly, ψ is continuous and $\varphi(x) = \varphi(0) + x\psi(x)$ for $x \in \mathbb{R}$. Of course, $\psi(x) = \varphi'(\Theta_x \cdot x)$, $0 < \Theta_x < 1$ for $x \neq 0$. This implies that

$$|\psi(x)| \leq \max_{x \in [-a,a]} |\varphi'(x)| \leq \|\varphi\|_{1,[-a,a]}$$

for $x \in \mathbb{R}$. Note that

$$\int\limits_{|x|>\epsilon} \frac{\varphi(x)}{x}dx = \int\limits_{\epsilon<|x|\leq a} \frac{\varphi(x)}{x}dx = \int\limits_{\epsilon<|x|\leq a} \psi(x)dx.$$

By the Lebesgue dominated convergence theorem we have

$$\Lambda(\varphi) = \lim_{\epsilon \to 0} \int\limits_{|x|>\epsilon} \frac{\varphi(x)}{x}dx = \int\limits_{|x|<a} \psi(x)dx.$$

Hence, we obtain the following inequality:

$$|\Lambda(\varphi)| \le 2a\|\varphi\|_{1,[-a,a]},$$

if supp$\varphi \subset [-a, a]$. Thus, we have shown that Λ is a distribution. The distribution $\frac{1}{(\cdot)}$ is called the Cauchy finite part of the integral $\int\limits_{\mathbb{R}} \frac{1}{(\cdot)}$.

3.2.3 Distributional Convergence

Let Λ, Λ_α be in $\mathcal{D}'(\mathbb{R})$, $\alpha \in \mathbb{R}$.

Definition 3.2.53 *We say that Λ_α distributionally converges to Λ as $\alpha \to \alpha_o$ if*

$$\lim_{\alpha \to \alpha_o} \Lambda_\alpha(\varphi) = \Lambda(\varphi) \quad \text{for each} \quad \varphi \in \mathcal{D}(\mathbb{R}).$$

In particular, if $n \in \mathbb{N}$ then we take

$$\lim_{n \to \infty} \Lambda_n = \lim_{\frac{1}{n} \to 0} \Lambda_n = \Lambda \quad \text{if} \quad \lim_{n \to \infty} \Lambda_n(\varphi) = \Lambda(\varphi) \quad \text{for each} \quad \varphi \in \mathcal{D}(\mathbb{R}).$$

Example 3.2.68 *The following sequences*

$$\left(\frac{1}{2} \sqrt{\frac{n}{2\pi}} e^{-nx^2/2} \right)_{n \in \mathbb{N}} \quad \text{(Picard)},$$

$$\left(\frac{2}{\pi} \frac{n}{e^{nx} + e^{-nx}} \right)_{n \in \mathbb{N}} \quad \text{(Stieltjes)},$$

and

$$\left(\frac{1}{\pi} \frac{n}{1 + (xn)^2} \right)_{n \in \mathbb{N}} \quad \text{(Cauchy)}$$

distributionally converge to δ (see also section 3.5.3).

Example 3.2.69 *For each φ in $\mathcal{D}$ the Sochozki formulas*

$$\lim_{\epsilon \to 0^+} \int\limits_{\mathbb{R}} \frac{\varphi(x)}{x + i\epsilon} \, dx = -i\pi\delta(\varphi) + \frac{1}{(\cdot)}(\varphi) \tag{3.2.14}$$

and

$$\lim_{\epsilon \to 0^+} \int\limits_{\mathbb{R}} \frac{\varphi(x)}{x - i\epsilon} \, dx = i\pi\delta(\varphi) + \frac{1}{(\cdot)}(\varphi), \tag{3.2.15}$$

where

$$\frac{1}{(\cdot)}(\varphi) = \lim_{\epsilon \to 0} \int\limits_{|x| \ge \epsilon} \frac{\varphi(x)}{x} dx \quad \text{(see Example 3.2.67, section 3.2.2)},$$

hold.

We shall show the equality (3.2.14). Let φ be in $\mathcal{D}$ and supp$\varphi \subset [-a, a]$. We know that φ may be written as follows:

$$\varphi(x) = \varphi(0) + x\psi(x), \quad \text{where} \quad \psi \quad \text{is a continuous function.}$$

Moreover,

$$\frac{1}{(\cdot)}(\varphi) = \int\limits_{-a}^{a} \psi(x)dx \quad (see\ Example\ 3.2.67,\ section\ 3.2.2).$$

Hence, we have

$$\int\limits_{\mathbb{R}} \frac{\varphi(x)}{x + i\epsilon}dx = \int\limits_{-a}^{a} \frac{\varphi(0)}{x + i\epsilon}dx + \int\limits_{-a}^{a} \frac{x\psi(x)}{x + i\epsilon}dx.$$

Note that

$$\int\limits_{-a}^{a} \frac{\varphi(0)}{x + i\epsilon}dx = \varphi(0)\int\limits_{-a}^{a} \frac{x}{x^2 + \epsilon^2}dx - \pi i\varphi(0)\frac{1}{\pi}\int\limits_{-a}^{a} \frac{\epsilon}{x^2 + \epsilon^2}dx = -\pi i\varphi(0)\frac{1}{\pi}\int\limits_{\frac{-a}{\epsilon}}^{\frac{a}{\epsilon}} \frac{1}{t^2 + 1}dt.$$

Finally, we have

$$\lim\limits_{\epsilon \to 0^+} \int\limits_{\mathbb{R}} \frac{\varphi(0)}{x + i\epsilon}dx = -\pi i\varphi(0) = -\pi i\delta(\varphi).$$

We see that

$$\lim\limits_{\epsilon \to 0} \int\limits_{-a}^{a} \frac{x\psi(x)}{x + i\epsilon}dx = \int\limits_{-a}^{a} \psi(x)dx;$$

thus, the formula (3.2.14) is proved. The proof of formula (3.2.15) is similar.

Example 3.2.70 *Let $f, f_n \in L_1^{loc}(\mathbb{R})$ for $n \in \mathbb{N}$ and the sequence $(f_n)_{n \in \mathbb{N}}$ converges to f as $n \to \infty$ in the sense of $L_1^{loc}(\mathbb{R})$. It is easy to see that for each $\varphi \in \mathcal{D}(\mathbb{R})$ the sequence*

$$\left(\int\limits_{\mathbb{R}} f_n(x)\varphi(x) \right)_{n \in \mathbb{N}} \quad converges\ to \quad \int\limits_{\mathbb{R}} f(x)\varphi(x)dx \quad as \quad n \to \infty.$$

This means that the sequence $(f_n)_{n \in \mathbb{N}}$ distributionally converges to f.

3.2.4   Algebraic Operations on Distributions

As we have mentioned before, all continuous functions, hence, also all differentiable functions can be regarded as distributions. This indicates that one could generalize onto the class $\mathcal{D}'(\mathbb{R})$ all operations that can be done on functions having certain properties. In this chapter we provide definitions of operations on distributions and examine the properties of those operations.

Definition 3.2.54 *Two distributions S and T are regarded to be equal if they are identical as functionals on $\mathcal{D}(\mathbb{R})$, i.e., if*

$$< S, \varphi >=< T, \varphi >, \quad for\ all\ functions \quad \varphi \in \mathcal{D}(\mathbb{R}).$$

Addition and Multiplication by a Constant

The operations of addition of two distributions and multiplication by a constant can be defined in a manner similar to the case of any space of linear functionals on an arbitrary space over the field of complex numbers.

Definition 3.2.55 *Let S and T be in $\mathcal{D}'(\mathbb{R})$.*

The expression $S + T$ defined in the following way

$$(S + T)(\varphi) = S(\varphi) + T(\varphi), \quad \varphi \in \mathcal{D}(\mathbb{R})$$

is called the sum of the distribution S and T.

The expression λS is defined in the following way

$$(\lambda S)(\varphi) = \lambda S(\varphi), \quad \varphi \in \mathcal{D}(\mathbb{R})$$

and is called the product of arbitrary distribution S by an arbitrary complex number λ.

Obviously, $S + T$ is also a continuous linear form on $\mathcal{D}(\mathbb{R})$. Moreover, if S_f and T_g are regular distributions corresponding to the functions f and g, then $S_f + T_g$ is also a regular distribution and it corresponds to the function $f + g$.

Multiplication of Distribution by a Smooth Function

There is no a natural way to define the product of two arbitrary distributions. Nevertheless, it is possible to define the product of any distribution Λ by an infinitely differentiable function ω. Note that the product of an infinitely differentiable function ω and function φ from $\mathcal{D}(\mathbb{R})$ belongs to $\mathcal{D}(\mathbb{R})$. Moreover, if the sequence $(\varphi_\nu)_{\nu \in \mathbb{N}}, \varphi_\nu \in \mathcal{D}(\mathbb{R})$ converges to zero in $\mathcal{D}(\mathbb{R})$ then the sequence $(\omega \varphi_\nu)_{\nu \in \mathbb{N}}$ also converges to zero in $\mathcal{D}(\mathbb{R})$.

Definition 3.2.56 *Let Λ be a distribution, $\omega \in \mathcal{C}^\infty(\mathbb{R})$. The mapping $\varphi \to \Lambda(\omega \varphi)$ is said to be the product of the distribution Λ and the function ω, i.e., the product $\omega \Lambda$ can be defined as a functional:*

$$< \omega \Lambda, \varphi > = < \Lambda, \omega \varphi >, \quad \varphi \in \mathcal{D}(\mathbb{R}). \tag{3.2.16}$$

For the regular distribution Λ_f corresponding to the locally integrable function f, multiplication by ω corresponds to multiplication of f and ω in the usual sense.

Differentiation of Distributions

Let φ be in $\mathcal{D}(\mathbb{R})$ and $f \in \mathcal{C}^m(\mathbb{R})$. Integrating by parts and recalling that φ is in $\mathcal{D}(\mathbb{R})$ we arrive at

$$\int\limits_{-\infty}^{+\infty} f'(x)\varphi(x)dx = - \int\limits_{-\infty}^{+\infty} f(x)\varphi'(x)dx.$$

By induction we have

$$\int\limits_{-\infty}^{+\infty} f^{(m)}(x)\varphi(x)dx = (-1)^m \int\limits_{-\infty}^{+\infty} f(x)\varphi^{(m)}(x)dx. \qquad (3.2.17)$$

Of course, the mapping

$$\varphi \to (-1)^m \int\limits_{-\infty}^{+\infty} f(x)\varphi^{(m)}(x)dx$$

is a linear continuous form on $\mathcal{D}(\mathbb{R})$. From (3.2.17) it follows that the mapping above is the regular distribution corresponding to the function $f^{(m)}$, where $f^{(m)}$ denotes the m-th derivatives of f. We shall use the equality (3.2.17) to define the derivative of a distribution.

Definition 3.2.57 *Let $\Lambda \in \mathcal{D}'(\mathbb{R})$ and let φ be in $\mathcal{D}(\mathbb{R})$. The linear form defined by*

$$\varphi \to (-1)^m \Lambda(\varphi^{(m)})$$

will be called the m-derivative of Λ. This form will be denoted by $D^m\Lambda$. In other words, the derivative $D^m\Lambda$ of distribution Λ can be defined as the functional

$$< D^m\Lambda, \varphi >:= (-1)^m < \Lambda, \varphi^{(m)} >, \quad \varphi \in \mathcal{D}(\mathbb{R}). \qquad (3.2.18)$$

From now on, for simplicity of notation, we write f instead of Λ_f if Λ_f is the regular distribution corresponding to the function f. Note that if f is in $\mathcal{C}^m(\mathbb{R})$ then $D^m f = f^{(m)}$.

In particular, for a regular distribution $f \in \mathcal{C}^1(\mathbb{R})$ the identity

$$< f', \varphi >= - < f, \varphi' > \quad \text{for} \quad \varphi \in \mathcal{D}(\mathbb{R}) \qquad (3.2.19)$$

becomes a rule of intergrating by parts (3.2.1), section 3.2.1, in which f' denotes the derivative in the classical sense. As a result, the operation of differentiating in the distributional sense and classical sense coincide in the class $\mathcal{C}^1(\mathbb{R})$. It is noteworthy that the theorem is not valid for weaker assumptions; see Example 3.2.71.

Example 3.2.71 *Let us consider a function f:*

$$f(x) = \begin{cases} g(x) & \text{for } x < 0 \\ h(x) & \text{for } x \geq 0, \end{cases}$$

where functions $g, h \in \mathcal{C}^1(\mathbb{R})$.

The function f is continuous in every point except for zero. Its derivative in the classical sense, $\frac{df}{dx}$, is a function defined and continuous for $x \neq 0$ and has right and left finite limits at zero. Thus, it is a locally integrable function and may be considered as a regular distribution. *Now, we calculate the distributional derivative of f. We can write the right-hand side of (3.2.19) in the integral form:*

$$< f', \varphi > = - \int_{-\infty}^{0} g(x)\varphi'(x)dx - \int_{0}^{\infty} h(x)\varphi'(x)dx, \qquad (3.2.20)$$

hence, after differentiating the right-hand side of (3.2.20) by parts, we get

$$< f', \varphi > = \int_{-\infty}^{0} \frac{dg}{dx}\varphi(x)dx + \int_{0}^{\infty} \frac{dh}{dx}\varphi(x)dx + [h(0) - g(0)]\varphi(0). \qquad (3.2.21)$$

We can write the right-hand side of (3.2.21) in the functional form

$$< f', \varphi > = \left\langle \frac{df}{dx}, \varphi \right\rangle + [h(0) - g(0)] < \delta, \varphi >,$$

in which

$$f' = \frac{df}{dx} + \sigma_o \delta, \qquad (3.2.22)$$

where

$$\sigma_o = \lim_{x \to 0^+} f(x) - \lim_{x \to 0^-} f(x) \qquad (3.2.23)$$

is the jump of the function f in the origin. From (3.2.23) we see that the derivatives $\frac{df}{dx}$ and f' are equal, if f is continuous at 0.

Remark 3.2.102 *For the generalization of Example 3.2.71 see Theorem 3.7.161, section 3.7.2.*

The differentiation of distributions is a linear continuous operation in $\mathcal{D}'(\mathbb{R})$. Namely, we have:

Theorem 3.2.96 *If Λ and Λ_n ($n \in \mathbb{N}$) are in $\mathcal{D}'(\mathbb{R})$ and sequence $(\Lambda_n)_{n\in\mathbb{N}}$ tends to Λ as $n \to \infty$ in $\mathcal{D}'(\mathbb{R})$ then*

$$sequence \quad (D^m \Lambda_n)_{n\in\mathbb{N}} \quad tends \ to \quad D^m \Lambda \quad as \quad n \to \infty.$$

Proof. The theorem is an immediate consequence of the definition of the derivative of distributions. □

Theorem 3.2.97 *If $\omega, \omega_1, \omega_2 \in \mathcal{C}^\infty(\mathbb{R})$ and $S, T \in \mathcal{D}'(\mathbb{R})$, then the following equalities hold:*

$$(\omega_1 + \omega_2)S = \omega_1 S + \omega_2 S,$$
$$\omega(S + T) = \omega S + \omega T,$$
$$D(\omega S) = \omega' S + \omega DS \quad (Leibniz's \ formula)$$
$$D(S + T) = DS + DT.$$

Proof. We shall show Leibniz's formula. The remaining follows from definition. For any function $\varphi \in \mathcal{D}(\mathbb{R})$ we have

$$< D(\omega S), \varphi > = - < S, \omega \varphi' > .$$

According to the formula of differentiating the product of smooth functions we can write

$$\omega \varphi' = (\omega \varphi)' - \omega' \varphi,$$

hence,

$$< D(\omega S), \varphi > = - < S, (\omega \varphi)' > + < S, \omega' \varphi >,$$

and finally we obtain

$$< D(\omega S), \varphi > = < \omega DS, \varphi > + < \omega' S, \varphi >,$$

which proves our assertion.   ☐

Example 3.2.72 *We shall show that*

$$D\ln |\cdot| = \frac{1}{(\cdot)}.$$

Let $\varphi \in \mathcal{D}$. Note that

$$D\ln |\cdot|(\varphi) = - \int_{-\infty}^{+\infty} \ln |x| \varphi'(x) dx = - \int_{-\infty}^{0} \ln (-x)\, \varphi'(x) dx - \int_{0}^{+\infty} \ln x\, \varphi'(x) dx.$$

We have

$$\int_{0}^{+\infty} \ln x \varphi'(x) dx = \lim_{\epsilon \to 0^+} \int_{\epsilon}^{+\infty} \ln x \varphi'(x) dx,$$

$$- \int_{-\infty}^{-\epsilon} \ln (-x)\, \varphi'(x) dx = -\varphi(-\epsilon)\ln \epsilon + \int_{-\infty}^{-\epsilon} \frac{\varphi(x)}{x} dx$$

and

$$- \int_{\epsilon}^{+\infty} \ln x\, \varphi'(x) dx = \varphi(\epsilon)\ln \epsilon + \int_{\epsilon}^{\infty} \frac{\varphi(x)}{x} dx.$$

It is easy to verify that

$$\lim_{\epsilon \to 0^+} [\varphi(\epsilon) - \varphi(-\epsilon)]\ln \epsilon = 0.$$

Hence,

$$- \int_{-\infty}^{+\infty} \ln |x| \varphi'(x) dx = \lim_{\epsilon \to 0^+} \int_{|x| \geq \epsilon} \frac{\varphi(x)}{x} dx$$

and finally we obtain

$$D\ln |\cdot| = \frac{1}{|\cdot|}$$

which completes the proof.

Linear Transformations of an Independent Variable

Let f be a continuous function defined on the real axis. Then the function composed of $f(ax + b)$, where $a \neq 0$, may be considered a regular distribution.

We make a substitution in the integral

$$y = ax + b, \quad a \neq 0, \quad \text{for} \quad \varphi \in \mathcal{D}(\mathbb{R}), \tag{3.2.24}$$

and we get

$$\int\limits_{-\infty}^{+\infty} f(ax + b)\varphi(x)dx = \frac{1}{|a|} \int\limits_{-\infty}^{+\infty} f(y)\varphi\left(\frac{y - b}{a}\right) dy,$$

i.e., in the functional form

$$< f \circ d, \varphi >= \frac{1}{|a|} < f, \varphi \circ d^{-1} >, \tag{3.2.25}$$

where d is a transformation defined by the formula (3.2.24).

According to (3.2.25) we define a linear substitution in an arbitrary distribution $\Lambda \in \mathcal{D}'(\mathbb{R})$. We use the notation for functions

$$< \Lambda(ax + b), \varphi(x) >= \frac{1}{|a|}\left\langle \Lambda(y), \varphi\left(\frac{y - b}{a}\right)\right\rangle,$$

We have to remember that the notation $\Lambda(ax + b)$ or $\Lambda(y)$ should be treated as purely formal.

Translation is a special and important case of linear substitution:

$$y = x + b. \tag{3.2.26}$$

We use the notation for functions:

$$(r_b\varphi)(x) := \varphi(x + b).$$

According to (3.2.25) we define a *translation of a distribution* by putting

$$< r_b\Lambda, \varphi >:= < \Lambda, r_{-b}\varphi > . \tag{3.2.27}$$

The second example of a linear substitution is a reflection from the origin:

$$y = -x. \tag{3.2.28}$$

The reflection for a function is defined by

$$\varphi^\vee(x) = \varphi(-x)$$

and for a distribution $\Lambda \in \mathcal{D}'(\mathbb{R})$ by

$$< \Lambda^\vee, \varphi >:= < \Lambda, \varphi^\vee > . \tag{3.2.29}$$

Example 3.2.73 *We will show the result of applying the distribution $r_{-b}\delta$ on a function $\varphi \in \mathcal{D}(\mathbb{R})$. According to (3.2.27) we have*

$$< r_{-b}\delta, \varphi > = \varphi(b). \tag{3.2.30}$$

In the technical and physical texts the translated distribution $r_{-b}\delta$ is frequently denoted by $\delta(x-b)$ and the equality (3.2.30) is written in the form

$$\int_{-\infty}^{+\infty} \delta(x-b)\varphi(x)dx = \varphi(b).$$

The symbol of an integral is purely formal here because δ is not a regular distribution and it cannot be written in the integral form.

The Antiderivative of a Distribution

By an antiderivative of a continuous $(-\infty, +\infty)$ function f we mean a differentiable function g fulfilling the condition $g'(x) = f(x)$ for $x \in \mathbb{R}$. It is well known that every continuous function has an antiderivative, and two arbitrary antiderivatives of the same function differ by a constant. Thus, the theorem can be generalized to distribution.

Definition 3.2.58 *By an antiderivative of a distribution $\Lambda \in \mathcal{D}'(\mathbb{R})$ we mean a distribution $S \in \mathcal{D}'(\mathbb{R})$ such that*

$$DS = \Lambda.$$

Theorem 3.2.98 *Every distribution $\Lambda \in \mathcal{D}'(\mathbb{R})$ has infinitely many antiderivatives that differ by a constant.*

Proof. Let $\mathcal{H} = \{\varphi \in \mathcal{D}(\mathbb{R}) : \quad \varphi = \psi', \quad \psi \in \mathcal{D}(\mathbb{R})\}$. It is easy to verify that

$$\varphi \in \mathcal{H} \quad \text{iff} \quad \int_{-\infty}^{+\infty} \varphi(x)dx = 0;$$

moreover $\psi(x) = \int_{-\infty}^{x} \varphi(t)\, dt$. From the definition it follows that the antiderivative S is a functional defined on the set by the formula

$$< S, \varphi > = - < \Lambda, \psi >, \quad \varphi \in \mathcal{H}. \tag{3.2.31}$$

This functional has to be generalized to the whole class $\mathcal{D}(\mathbb{R})$.

Let us consider a function $\varphi^{o} \in \mathcal{D}(\mathbb{R})$ such that

$$\int_{-\infty}^{+\infty} \varphi^{o}(x)dx = 1,$$

we can, for example, take $\varphi^o = \varphi_a \left[\int\limits_{-\infty}^{+\infty} \varphi_a(x)dx \right]^{-1}$, where φ_a is a function given by (3.2.5), section 3.2.1. An arbitrary function φ can be uniquely represented by the form

$$\varphi = \lambda\varphi^o + \varphi^1 \qquad (3.2.32)$$

where $\varphi^1 \in \mathcal{H}$ and $\lambda = \int\limits_{-\infty}^{+\infty} \varphi(x)dx$. By the above, according to (3.2.31) and (3.2.32) we take

$$< S, \varphi > = \lambda < \overset{.}{S}, \varphi^v > - < \Lambda, \psi^1 >, \qquad (3.2.33)$$

where $\psi^1(x) = \int\limits_{-\infty}^{x} \varphi^1(t)\,dt$, and the number $< S, \varphi^o >$ is arbitrary. An easy computation shows that the above-defined functional is a distribution, and $DS = \Lambda$. If S_1 and S_2 are the antiderivatives of the same distribution then $D(S_1 - S_2) = 0$. To complete the proof it is enough to show that every distribution $S \in \mathcal{D}'(\mathbb{R})$ that fulfills the condition $DS = 0$ is a regular distribution equal to a constant function. The above property results from the decomposition (3.2.32) which gives

$$< S, \varphi > = \int\limits_{-\infty}^{+\infty} c\varphi(x)dx,$$

where $c = < S, \varphi^o >$ is an integration constant. ⬚

3.3 Generalized Functions — Sequential Approach

3.3.1 The Identification Principle

The identification principle relies on the identification of objects that have a common property. It is often applied in mathematics to construct new concepts. In this chapter the identification principle will be defined and then explained by means of examples.

Definition 3.3.59 *We say that the relation $\rho \subset X \times X$ is a equivalence relation in the set X, if ρ satisfies the condition of reflexivity, symmetry and transitivity, i.e., if*

(i) $(x\rho x)$ *for all $x \in X$*   *(reflexivity),*

(ii) $[(x\rho y) \Longrightarrow (y\rho x)]$ *for all $x, y \in X$*   *(symmetry),*

(iii) $[(x\rho y) \wedge (y\rho z) \Longrightarrow (x\rho z)]$ *for all $x, y, z \in X$ (transitivity).*

We denote the relation of equivalence by $\sim$.

Let $\sim$ be an arbitrary equivalence relation in $X \neq \emptyset$. For each element $x \in X$, let us denote by $[x]$ the class of all elements $y \in X$ satisfying the relation $x \sim y$. By definition it follows that

$$[x] = \{y \in X : x \sim y\}, \tag{3.3.1}$$

$$\left[(y \in [x]) \Leftrightarrow (x \sim y)\right], \quad \text{for all} \quad x, y \in X. \tag{3.3.2}$$

The class $[y]$ thus obtained will be called an equivalence class in X.

Theorem 3.3.99 *If $\sim$ is an arbitrary equivalence relation in the set $X \neq \emptyset$, then for each $x, x_1, x_2 \in X$ the following conditions are satisfied:*

$$x \in [x], \tag{3.3.3}$$

if $x_1 \sim x_2$, then $[x_1] = [x_2]$, i.e., the classes $[x_1]$ $\qquad$ (3.3.4)

and $[x_2]$ have the same elements,

if the relation $x_1 \sim x_2$ does not hold, then the classes $\qquad$ (3.3.5)

$[x_1]$ and $[x_2]$ have no common element.

Proof. Property (3.3.3) follows from (i). To prove (3.3.4) suppose that $x_1 \sim x_2$. If $x \in [x_1]$, then $x \sim x_1$. Hence, by (iii), $x \sim x_2$, i.e., $x \in [x_2]$. Thus, $[x_1] \subset [x_2]$. On the other hand, it follows from (ii) that $x_2 \sim x_1$. Therefore, if $x \in [x_2]$, i.e., $x \sim x_2$, then also $x \sim x_1$, by (iii), i.e., x belongs to $[x_1]$. Thus, $[x_2] \subset [x_1]$. Hence, by previously proven inclusion, we get $[x_1] = [x_2]$.

To prove (3.3.5) suppose that the relation $x_1 \sim x_2$ does not hold and that there exists an element x belonging to $[x_1]$ and $[x_2]$. Then $x \sim x_1$ and $x \sim x_2$, and $x_1 \sim x_2$, by (ii) and (iii), contrary to the hypothesis. $\qquad\qquad$ □

Theorem 3.3.100 *The set $X \neq \emptyset$ with an equivalence relation $\sim$ in it, can be decomposed into equivalence classes without common elements so that two elements $x, y \in X$ are in the same equivalence class if and only if $x \sim y$, i.e., they are equivalent.*

Example 3.3.74 *Let X will be a set of directed segments on the plane. We say that two directed segments x and y are equivalent if they are parallel, have the same length, and the same direction. It is easy to check that the relation defined above is an equivalence relation. By identifying equivalent objects we can obtain the notion of a free vector. By means of the equivalence relation we obtain a decomposition of the set of all directed segments into disjoint classes such that segments in the same class are equivalent. Thus, each free vector is a class of equivalent segments.*

Example 3.3.75 *Let X be the set of all fundamental sequences of rational numbers. By a fundamental sequence we mean a sequence $(a_n)_{n \in \mathbb{N}}$ satisfying the Cauchy condition: for every $\epsilon > 0$ there exists a number $n_o(\epsilon)$ such that*

$$|a_n - a_m| < \epsilon, \qquad n, m > n_o(\epsilon), \quad m, n \in \mathbb{N}.$$

We say that two fundamental sequences (a_n) and (b_n) are equivalent if the sequence $(a_n - b_n)_{n \in \mathbb{N}}$ converges to 0, i.e.,

$$\lim_{n \to \infty} (a_n - b_n) = 0.$$

In this case we write $(a_n)_{n \in \mathbb{N}} \sim (b_n)_{n \in \mathbb{N}}$. Thus,

$$(a_n)_{n \in \mathbb{N}} \sim (b_n)_{n \in \mathbb{N}} \iff \lim_{n \to \infty} (a_n - b_n) = 0.$$

It is easy to verify that the relation of equivalence defined above is an equivalence relation, i.e., it has the properties (i), (ii), and (iii) of Definition 3.3.59.

Identifying equivalent fundamental sequences we obtain the notion of real numbers. Thus, in the Cantor theory, a real number is a class of equivalent fundamental sequences of rational numbers.

3.3.2 Fundamental Sequences

We recall:

Definition 3.3.60 *We say that a sequence $(f_n)_{n \in \mathbb{N}}$ of functions is convergent to a function f uniformly in the interval I, open or closed, if the function f is defined on I and, for any given number $\epsilon > 0$, there is an index n_o such that for every $n > n_o$ the function f_n is defined on the interval I and for every $x \in I$ the inequality*

$$|f_n(x) - f(x)| < \epsilon, \quad n > n_o,$$

is true.

The symbol
$$f_n(x) \rightrightarrows f(x), \quad x \in I; \quad f_n(x) \rightrightarrows \quad x \in I,$$
will denote the *uniform convergence in the interval I.*

We write
$$f_n(x) \rightrightarrows\leftleftarrows g_n(x), \quad x \in I$$
if both sequences (f_n) and (g_n) converge uniformly on I to the same limit.

Definition 3.3.61 *We say that a sequence of functions $(f_n)_{n \in \mathbb{N}}$ is convergent to function f almost uniformly in the open interval $(a; b)$ if it converges uniformly to f on every interval I inside $(a; b)$.*

The symbol
$$f_n(x) \overset{\text{a.u.c.}}{\rightrightarrows} f(x), \quad x \in (a; b)$$

will denote *almost uniform convergence in the interval* $(a; b)$.

It is easy to see that every uniformly convergent sequence is also convergent almost uniformly. The limit of an almost uniformly convergent sequence of continuous functions is itself a continuous function.

Lemma 3.3.38 *The sequence of functions* $\left(f_n(x)\right)_{n\in\mathbb{N}}$ *is uniformly convergent to the function* $f(x)$ *in the interval* $(a; b)$ *if*

$$\lim_{n\to\infty}\left(\sup_{a<x<b} h_n(x)\right) = 0, \quad \text{where} \quad h_n(x) = |f(x) - f_n(x)|.$$

The proof of Lemma 3.3.38 is based on Weierstrass's theorem.

Example 3.3.76 *Show that the sequence*

$$F_n(x) = \frac{x}{2} + \frac{x}{\pi}\text{arctg}\,(nx) - \frac{1}{2\pi n}\ln\left(1 + n^2 x^2\right), \quad n \in \mathbb{N},$$

is uniformly convergent in the interval $(-\infty; +\infty)$.

Let us discuss the pointwise convergence of the sequence $\left(F_n(x)\right)_{n\in\mathbb{N}}$.

Applying l'Hospital's rule (with respect to n) we obtain:

$$\lim_{n\to\infty}\frac{\ln\left(1 + n^2 x^2\right)}{2\pi n} = \lim_{n\to\infty}\frac{nx^2}{\pi(1 + n^2 x^2)} = \lim_{n\to\infty}\frac{x^2}{2\pi n x^2}$$

$$= \lim_{n\to\infty}\frac{1}{2\pi n} = 0, \quad x \in \mathbb{R}.$$

On the other hand,

$$\lim_{n\to\infty}\left(\frac{x}{2} + \frac{x}{\pi}\text{arctg}\,(nx)\right) = \begin{cases} x, & \text{for } x > 0, \\ 0, & \text{for } x \leq 0. \end{cases}$$

Hence, we obtain

$$\lim_{n\to\infty} F_n(x) = F(x) = \begin{cases} x, & \text{for } x > 0, \\ 0, & \text{for } x \leq 0. \end{cases}$$

In order to investigate the uniform convergence we apply Lemma 3.3.38.

We note that

$$\sup_{x\in\mathbb{R}}|F_n(x) - F(x)| = \sup_{x\in\mathbb{R}}\left|\frac{x}{2} + \frac{x}{\pi}\text{arctg}\,(nx) - F(x) - \frac{1}{2\pi n}\ln\left(1 + n^2 x^2\right)\right|$$

$$\leq \sup_{x\in\mathbb{R}}\left|\frac{x}{2} + \frac{x}{\pi}\text{arctg}\,(nx) - F(x)\right| + \sup_{x\in\mathbb{R}}\left|\frac{1}{2\pi n}\ln\left(1 + n^2 x^2\right)\right|.$$

We have

a) $\quad \sup_{x\in\mathbb{R}}\left|\frac{x}{2} + \frac{x}{\pi}\text{arctg}\,(nx) - F(x)\right| =$

$$= \begin{cases} \sup_{x\in\mathbb{R}^+}\left|\frac{x}{\pi}\text{arctg}\,(nx) - \frac{x}{2}\right| = \sup_{x\in\mathbb{R}^+}\left|\frac{x}{\pi}\text{arctg}\,\left(\frac{1}{nx}\right)\right| \leq \sup_{x\in\mathbb{R}^+}\left|\frac{x}{\pi nx}\right| < \frac{1}{n}, \\ \sup_{x\in\mathbb{R}^-}\left|\frac{x}{\pi}\text{arctg}\,(nx) + \frac{x}{2}\right| < \sup_{x\in\mathbb{R}^-}\left|\frac{x}{\pi}\text{arctg}\,\left(\frac{-1}{nx}\right)\right| < \frac{1}{n}. \end{cases}$$

b)

$$\sup_{x \in \mathbb{R}} \left| \frac{1}{2\pi n} \ln \left(1 + n^2 x^2 \right) \right| \le \frac{1}{2\pi} \sup_{x \in \mathbb{R}} \left| \ln \left(1 + n^2 x^2 \right)^{\frac{1}{n}} \right|$$

$$\le \frac{1}{2\pi} \sup_{x \in \mathbb{R}} \ln \left(2 n^2 x^2 \right)^{\frac{1}{n}} \to 0, \quad \text{if} \quad n \to \infty,$$

because

$$\lim_{n \to \infty} \sqrt[n]{n^2} \cdot \sqrt[n]{x^2} = 1, \quad x \in \mathbb{R}.$$

From (a), (b) and Lemma 3.3.38 *it follows that*

$$F_n(x) \overset{a.u.c.}{\rightrightarrows} F(x).$$

We recall: *We say that a function defined in $\mathbb{R}$ is smooth if it is continuous in $\mathbb{R}$ as well as its derivatives of any order and is denoted by $\mathcal{C}^\infty(\mathbb{R}) = \mathcal{C}^\infty$.*

Definition 3.3.62 *We say that a sequence $(f_n)_{n \in \mathbb{N}}$ of smooth functions, defined in the interval (a, b), $(-\infty \le a < b \le +\infty)$, is fundamental in (a, b) if for every interval I inside (a, b) there exists a number $k \in \mathbb{N}_o$ and a sequence of smooth functions $(F_n)_{n \in \mathbb{N}}$ such that*

$$(E_1) \quad F_n^{(k)}(x) = f_n(x), \quad x \in I$$

and

$$(E_2) \qquad F_n(x) \rightrightarrows, \quad x \in I,$$

i.e., the sequence $F_n(x)$ is uniformly convergent in I.

Obviously constant functions are smooth. On this basis, identifying numbers with constant functions we can assume that every number is a smooth function on $\mathbb{R}$ and a convergent sequence of numbers is fundamental.

By Definition 3.3.60 it follows:

Theorem 3.3.101 *Each almost uniformly convergent sequence of smooth functions in (a, b) is fundamental, in (a, b).*

Theorem 3.3.102 *If $(f_n)_{n \in \mathbb{N}}$ is a fundamental sequence of smmoth functions, then the sequence $(f_n^{(m)})_{n \in \mathbb{N}}$, $m \in \mathbb{N}$, is fundamental, too.*

Proof. In fact, if the sequence $(F_n)_{n \in \mathbb{N}}$ satisfies conditions (E_1) and (E_2) then $(F_n^{(k+m)})_{n \in \mathbb{N}} = f_n^{(m)}$ and condition (E_2) holds. This proves that the sequence $(f_n^{(m)})_{n \in \mathbb{N}}$ is fundamental.

□

Theorem 3.3.103 *If the sequence* $(F_n)_{n \in \mathbb{N}}$ *of smooth functions satisfies conditions* (E_1) *and* (E_2) *and if* $m \geq k$, $m \in \mathbb{N}$, *then the sequence of smooth functions*

$$\widetilde{F}_n(x) = \int_{x_o}^{x} F_n(t)dt^{m-k} \quad (x_o \in I)$$

also satisfies conditions (E_1) *and* (E_2), *with* k *replaced by* $k + m$. *Moreover, if* $F_n \rightrightarrows F$, *then* $\widetilde{F}_n \rightrightarrows \widetilde{F}$, *where*

$$\widetilde{F}(x) = \int_{x_o}^{x} F(t)dt^{m-k},$$

and the integral is defined in the following way:

$$\int_{x_o}^{x} f(t)\, dt^k = \int_{x_o}^{x} dt_k \int_{x_o}^{t_k} dt_{k-1} \cdots \int_{x_o}^{t_2} f(t_1)dt_1.$$

Theorem 3.3.104 *If a sequence* $(f_n)_{n \in \mathbb{N}}$ *of smooth functions is bounded and* $f_n \overset{a.u.c.}{\rightrightarrows} f$ *in the intervals* (a, x_o) *and* (x_o, b), *then*

$$\int_{x_o}^{x} f_n(t)dt \overset{a.u.c.}{\rightrightarrows} \int_{x_o}^{x} f(t)dt \quad \text{in the interval} \quad a < x < b,$$

i.e., the sequence $(f_n)_{n \in \mathbb{N}}$ *is fundamental in* (a, b).

Proof. The sequence $(f_n)_{n \in \mathbb{N}}$ is bounded; thus, there is a number $M > 0$, such that

$$|f_n(x)| < M, \quad \text{for} \quad n \in \mathbb{N}.$$

From almost uniform convergence of the sequence $(f_n)_{n \in \mathbb{N}}$ it follows that for any $\epsilon > 0$ and an interval $\bar{a} \leq x \leq \bar{b}$ $(a < \bar{a} < x_o < \bar{b} < b)$, there exists an index n_o such that

$$|f_n(x) - f(x)| < \frac{\epsilon}{2(\bar{b} - \bar{a})} \quad \text{for} \quad n > n_o$$

in the intervals $\bar{a} \leq x \leq x_o - \frac{\epsilon}{4M}$ and $x_o + \frac{\epsilon}{4M} \leq x \leq \bar{b}$.
  Hence,

$$\left| \int_{x_o}^{x} f_n(t)dt - \int_{x_o}^{x} f(t)dt \right| < \epsilon \quad \text{for} \quad n > n_o,$$

in the interval $\bar{a} \leq x \leq \bar{b}$. □

Theorem 3.3.105 *A sequence* $(W_n(x))_{n \in \mathbb{N}}$ *of polynomials of degree less than* m *is fundamental iff it converges almost uniformly.*

To prove Theorem 3.3.105 the following two lemmas will be helpful.

Lemma 3.3.39 *If a sequence of polynomials $(W_n(x))_{n \in \mathbb{N}}$ of degree less than k, where*

$$W_n(x) = a_{no} + a_{n1}x + \cdots + a_{n(k-1)}x^{k-1}, \tag{3.3.6}$$

converges at k points, then the limits

$$a_j = \lim_{n \to \infty} a_{nj} \quad (j = 0, 1, \cdots, k-1) \tag{3.3.7}$$

exist. Conversely, if the limits (3.3.7) exist, then $W_n(x) \overset{a.u.c.}{\rightrightarrows} W(x)$, where $W(x) = a_o + a_1 x + \cdots + a_{k-1}x^{k-1}$.

Proof. **a)** Suppose that the sequence of polynomials $(W_n(x))_{n \in \mathbb{N}}$ (3.3.6), is convergent at k different points: $x_1, x_2, \cdots, x_k$. We get the system of equations

$$\begin{cases} W_n(x_1) = a_{no} + a_{n1}x_1 + \cdots + a_{n(k-1)}x_1^{k-1} \\ \cdots\cdots\cdots\cdots\cdots\cdots\cdots\cdots\cdots\cdots\cdots \\ \cdots\cdots\cdots\cdots\cdots\cdots\cdots\cdots\cdots\cdots\cdots \\ W_n(x_k) = a_{no} + a_{n1}x_k + \cdots + a_{n(k-1)}x_k^{k-1}. \end{cases} \tag{3.3.8}$$

We shall show that the sequence of coefficients $(a_{nj})_{n \in \mathbb{N}}$ $(j = 0, 1, \cdots, k-1)$ is convergent. Let us assume that the coefficients are given. The determinant for the system (3.3.8) is equal

$$A = \det \begin{pmatrix} 1 & x_1 & \dots & x_1^{k-1} \\ \cdots & \cdots & \cdots & \cdots \\ \cdots & \cdots & \cdots & \cdots \\ \vdots & \vdots & \ddots & \vdots \\ 1 & x_k & \dots & x_k^{k-1} \end{pmatrix} = \prod_{\substack{l > m \\ m,l=1}}^{k} (x_l - x_m) \neq 0.$$

By Cramer's rule, the system has only one solution of the form

$$a_{n,j} = \frac{W_n(x_1)A_{1j} + \cdots + W_n(x_k)A_{kj}}{A}, \quad j = 0, 1, \cdots, k-1,$$

where A_{jk} are minors of $A(i = 1, ..., k)$ and are independent of n. Hence, the sequence of polynomials $(W_n(x))_{n \in \mathbb{N}}$ is convergent at k points which means that

$$\lim_{n \to \infty} W_n(x_1) = g_1, \cdots, \lim_{n \to \infty} W_n(x_k) = g_k.$$

Hence,

$$\lim_{n \to \infty} a_{nj} = \lim_{n \to \infty} \frac{1}{A}[W_n(x_1)A_{1j} + \cdots + W_n(x_k)A_{kj}] = \frac{1}{A}[g_1 A_{1j} + \cdots + g_k A_{kj}] = a_j.$$

This means that the sequences of coefficients $(a_{nj})_{n \in \mathbb{N}}, (j = 0, \cdots, k-1)$ are convergent

$$\lim_{n \to \infty} a_{nj} = a_j, \quad j = 0, 1, \cdots, k-1.$$

b) Let us assume that the limits (3.3.7) exist and let $W(x) = a_o + a_1 x + \cdots + a_{k-1}x^{k-1}$.

The uniform convergence of the sequence $(W_n(x))_{n\in\mathbb{N}}$ in each interval $<-b,b>$ results from the estimation

$$|W_n(x) - W(x)| \leq |a_{no} - a_o| + |a_{n1} - a_1||x| + \cdots + |a_{n(k-1)} - a_{k-1}||x| \leq \sum_{j=0}^{k-1} |a_{nj} - a_j|b^j.$$

Hence, it is easy to see that $W_n(x) \overset{\text{a.u.c.}}{\rightrightarrows} W(x)$. $\qquad\qquad\qquad$ □

Lemma 3.3.40 *Let a sequence $(f_n)_{n\in\mathbb{N}}$ of continuous functions be given. If $f_n(x) \rightrightarrows f(x)$, for $x \in I$ then the function f is continuous on I.*

Examples of Fundamental Sequences

1. Consider the sequence of smooth functions on $\mathbb{R}$ defined by

$$f_n(x) = \frac{1}{1 + e^{-nx}}, \quad n \in \mathbb{N}.$$

The graphs of f_1, f_2, and f_3 are sketched in Figure 15. We shall show that the sequence is fundamental. In fact, this sequence is bounded by the number 1 and $\frac{1}{1+e^{-nx}} \overset{\text{a.u.c.}}{\rightrightarrows}$ in the interval $(0, \infty)$. By Theorem 3.3.104 this sequence is fundamental.

2. Consider the sequence of the Picard functions on $\mathbb{R}$ defined by

$$f_n(x) = \sqrt{\frac{n}{2\pi}} \cdot e^{-nx^2/2}, \quad n \in \mathbb{N}.$$

The graphs of f_1, f_2 and f_3 are sketched in Figure 16. We shall show that the sequence is fundamental. In fact, the sequence $(G_n)_{n\in\mathbb{N}}$, where

$$G_n(x) = \int_{-\infty}^{x} f_n(t)dt$$

is bounded by the number 1 and

$$G_n(x) \overset{\text{a.u.c.}}{\rightrightarrows} \begin{cases} 0 & \text{in the interval } -\infty < x < 0 \\ 1 & \text{in the interval } 0 < x < \infty. \end{cases}$$

In view of Theorem 3.3.104, the sequence $(G_n)_{n\in\mathbb{N}}$ is fundamental. In view of Theorem 3.3.102, the sequence $(f_n)_{n\in\mathbb{N}}$ is also fundamental.

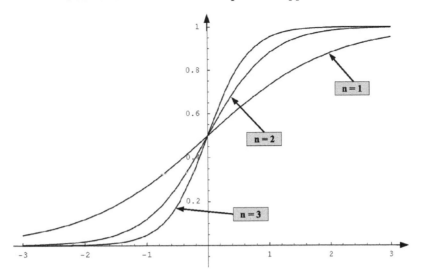

Figure 15. The graphs of functions: $f_n(x) = \frac{1}{1+e^{-nx}}; n = 1, 2, 3$

3. Consider the sequences of functions defined by:

$$f_n(x) = \frac{1}{\pi}\arctan(nx) + \frac{1}{2}, \qquad n = 1, 2, \cdots;$$

$$g_n(x) = \frac{1}{\pi}\frac{n}{n^2x^2 + 1}, \qquad n = 1, 2, \cdots \text{(see Figure 17)};$$

$$h_n(x) = -\frac{2}{\pi}\frac{n^3x}{(n^2x^2 + 1)^2}, \qquad n = 1, 2, \cdots \text{(see Figure 18)};$$

$$v_n(x) = \frac{2n^3}{\pi}\frac{3n^2x^2 - 1}{(n^2x^2 + 1)^3}, \qquad n = 1, 2, \cdots.$$

(3.3.9)

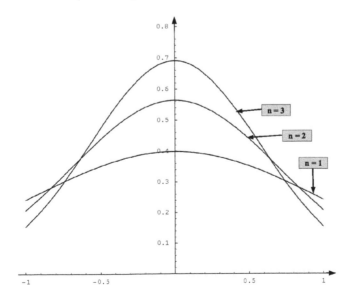

Figure 16. The graphs of the Picard functions: $f_n(x) = \sqrt{\frac{n}{2\pi}} \cdot e^{-nx^2/2}; n = 1, 2, 3$

Sequences (3.3.9) are fundamental, since in the interval $(-\infty, +\infty)$, we have

$$f_n = F_n^{(1)}, \quad g_n = F_n^{(2)},$$
$$h_n = F_n^{(3)}, \quad v_n = F_n^{(4)},$$

where

$$F_n(x) = \frac{x}{2} + \frac{x}{\pi}\operatorname{arctg}(nx) - \frac{1}{2\pi n}\ln(1 + n^2 x^2), \quad n \in \mathbb{N},$$

and

$$F_n \overset{\text{a.u.c.}}{\rightrightarrows} \quad \text{in the interval} \quad (-\infty, \infty) \quad (\text{see Example 3.3.76}).$$

4. The sequence of functions

$$f_n(x) = \cos nx, \quad n \in \mathbb{N},$$

is fundamental because if $F_n(x) = \frac{1}{n}\sin nx, n \in \mathbb{N}$ and $k = 1$, then (E_1) and (E_2) are satisfied.

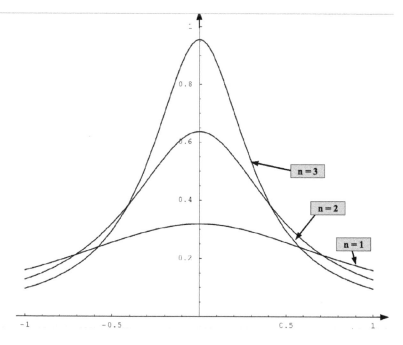

Figure 17. The graphs of the Cauchy functions: $g_n(x) = \frac{1}{\pi}\frac{n}{n^2 x^2 + 1}$, $n = 1, 2, 3$

5. The sequence of functions

$$f_n(x) = n\cos nx, \quad n \in \mathbb{N},$$

is fundamental since if $F_n(x) = -\frac{1}{n}\cos nx$, $n \in \mathbb{N}$ and $k = 2$, then (E_1) and (E_2) are satisfied.

6. The interlaced sequence of sequences **4** and **5**

$$\cos x, \ \cos x, \ \cos 2x, \ 2\cos 2x, \ \cos 3x, \ 3\cos 3x, \ \cdots$$

is fundamental because if $(F_n)_{n \in \mathbb{N}}$ is the sequence

$$- \cos x, \; - \cos x, \; -\frac{1}{4} \cos 2x, \; -\frac{1}{2} \cos 2x, \; -\frac{1}{9} \cos 3x, \; -\frac{1}{3} \cos 3x, \; \cdots$$

and $k = 2$, then the conditions (E_1) and (E_2) are satisfied.

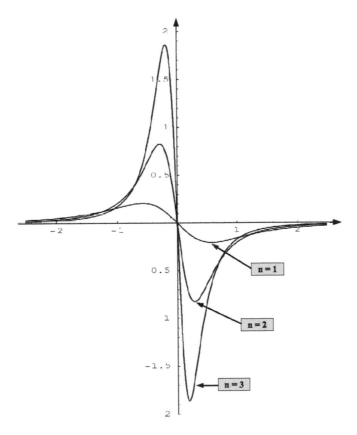

Figure 18. The graphs of the functions: $h_n(x) = -\frac{2}{\pi} \frac{n^3 x}{(n^2 x^2 + 1)^2}$, $n = 1, 2, 3$

3.3.3 Definition of Distributions

Definition 3.3.63 *We say that two sequences $(f_n)_{n \in \mathbb{N}}$ and $(g_n)_{n \in \mathbb{N}}$ fundamental in (a, b) are equivalent in (a, b) and we write*

$$(f_n)_{n \in \mathbb{N}} \sim (g_n)_{n \in \mathbb{N}} \tag{3.3.10}$$

if the interlaced sequence

$$f_1, \; g_1, \; f_2, \; g_2, \; \cdots \tag{3.3.11}$$

is fundamental in (a, b).

Theorem 3.3.106 *Two sequences $(f_n)_{n\in\mathbb{N}}$ and $(g_n)_{n\in\mathbb{N}}$ are equivalent in (a,b) if for each interval I inside (a,b) there exist sequences $(F_n)_{n\in\mathbb{N}}$ and $(G_n)_{n\in\mathbb{N}}$ of smooth functions and an integer $k \in \mathbb{N}_o$ such that*

$$(F_1) \quad F_n^{(k)}(x) = f_n(x) \quad and \quad G_n^{(k)}(x) = g_n(x), \quad x \in I$$

$$(F_2) \quad F_n(x) \rightrightarrows\leftleftarrows G_n(x), \quad x \in I.$$

Proof. Suppose that sequences $(f_n)_{n\in\mathbb{N}}$ and $(g_n)_{n\in\mathbb{N}}$ are equivalent. This means that the interlaced sequence (3.3.11) $f_1, g_1, f_2, g_2, \cdots$ is fundamental. Then there exist an integer $k \in \mathbb{N}_o$ and smooth functions F_n and G_n such that $F_n^{(k)} = f_n$ and $G_n^{(k)} = g_n$, and the sequence

$$F_1, G_1, F_2, G_2, \cdots \tag{3.3.12}$$

converges almost uniformly in $(a;b)$. Consequently the conditions (F_1) and (F_2) are satisfied.

Conversely, suppose that conditions (F_1) and (F_2) are satisfied. Then the sequence (3.3.12) converges almost uniformly, i.e., the sequence (3.3.11) satisfies conditions (E_1) and (E_2) with order k_1, k_2, respectively. Then by the Theorem 3.3.103, sequences of smooth functions $\widetilde{F}_n$ and $\widetilde{G}_n$, such that

$$\widetilde{F}_n(x) = \int_{x_o}^{x} F_n(t)dt^{m-k_1} \quad (x_o \in I), \quad n \in \mathbb{N},$$

$$\widetilde{G}_n(x) = \int_{x_o}^{x} G_n(t)dt^{m-k_2} \quad (x_o \in I), \quad n \in \mathbb{N},$$

satisfy the condition (F_1) with order m. Moreover

$$\widetilde{F}_n^{(m)} = F_n^{(k_1)} = f_n, \quad \widetilde{G}_n^{(m)} = G_n^{(k_2)} = g_n,$$

$$\widetilde{F}_n(x) \rightrightarrows\leftleftarrows \widetilde{G}_n(x), \quad x \in I.$$

Thus, the sequence

$$\widetilde{F}_1, \widetilde{G}_1, \widetilde{F}_2, \widetilde{G}_2, \cdots$$

is almost uniformly convergent in (a,b); therefore the sequence (3.3.11) $f_1, g_1, f_2, g_2, \cdots$ satisfies conditions (E_1) and (E_2), i.e., is fundamental. Sequences $(F_n)_{n\in\mathbb{N}}$ and $(G_n)_{n\in\mathbb{N}}$ and the order k depend, in general, on the choice of the interval I. $\quad\square$

Corollary 3.3.35 *The integer k, appearing in the condition (F_1) of equivalent sequences, can, if necessary, be replaced by any greater integer.*

Proposition 3.3.63 *The relation (3.3.10) is an equivalence relation.*

Proof. In fact. It is easy to observe that the relation (3.3.10) satisfies conditions (i) and (ii), i.e., it is reflexive and symmetric.

We shall show that the condition (iii) is also satisfied. If $(f_n)_{n \in \mathbb{N}} \sim (g_n)_{n \in \mathbb{N}}$ and $(g_n)_{n \in \mathbb{N}} \sim (h_n)_{n \in \mathbb{N}}$ in (a, b), then for each bounded open interval I inside (a, b) there exists an integer $k \geq 0$ and sequences $(F_n)_{n \in \mathbb{N}}$ and $(G_n)_{n \in \mathbb{N}}$ satisfying conditions (F_1) and (F_2) and there exists an integer $m \geq 0$ and sequences $(\widetilde{G}_n)_{n \in \mathbb{N}}$ and $(H_n)_{n \in \mathbb{N}}$ satisfying conditions

$$\widetilde{G}_n^{(m)} = g_n, \quad H_n^{(m)} = h_n,$$

$$\widetilde{G}_n(x) \rightrightarrows \leftleftarrows H_n(x), \quad x \in I$$

By Theorem 3.3.106 we may assume that $k = m$. Then, writing

$$\widetilde{\widetilde{H}}_n = G_n - \widetilde{G}_n + H_n, \quad \text{we get}$$

$$F_n^{(k)} = f_n \quad \text{and} \quad \widetilde{\widetilde{H}}_n^{(k)} = h_n,$$

$$F_n(x) \rightrightarrows \leftleftarrows \widetilde{\widetilde{H}}_n(x), \quad x \in I.$$

which implies that $(f_n)_{n \in \mathbb{N}} \sim (h_n)_{n \in \mathbb{N}}$ in (a, b). $\qquad\square$

Since the relation (3.3.10) $\sim$ is a relation of equivalence, the set of all sequences fundamental in (a, b) can be decomposed into disjoint classes (equivalence classes of the relation $\sim$) such that two fundamental sequences are in the same class if they are equivalent.

These equivalence classes are called **distributions** in the sense of *Mikusiński* (defined on the interval (a, b)). An analogous definition can be formulated for distributions in an arbitrary open set in $\mathbb{R}^n$.

All fundamental sequences that are equivalent define the same distribution. The distribution defined by fundamental sequence $(f_n(x))_{n \in \mathbb{N}}$ will be denoted by the symbol $[f_n]$.

Two fundamental sequences $(f_n)_{n \in \mathbb{N}}$ and $(g_n)_{n \in \mathbb{N}}$ define the same distribution if they are equivalent. The equality

$$[f_n] = [g_n] \quad \text{holds iff} \quad (f_n)_{n \in \mathbb{N}} \sim (g_n)_{n \in \mathbb{N}}.$$

Thus

$$f = [f_n] = \Big\{ (g_n)_{n \in \mathbb{N}} : (f_n)_{n \in \mathbb{N}} \sim (g_n)_{n \in \mathbb{N}} \Big\}.$$

Distributions will be denoted by f, g, etc. in the same way as functions. The set of all distributions is denoted by $\mathcal{D}'$.

The sequences from the examples 2 (Figure 16) and 3 (Figure 17) are equivalent. Therefore, they define the same distribution. This distribution is called the **delta distribution** or the **Dirac delta distribution** and is denoted by δ:

$$\left[\sqrt{\frac{n}{2\pi}} \cdot e^{-nx^2/2} \right] = \left[\frac{1}{\pi} \frac{n}{n^2 x^2 + 1} \right] = \delta(x).$$

We will discuss the δ Dirac distribution in section 3.4.

The sequences from examples 4 and 5 are equivalent. They define the same distribution:

$$[\cos nx] = [n \cos nx] = [0] = 0.$$

Obviously, an arbitrary smooth function ϕ can be identified with a class of equivalence of the sequence $\phi, \phi, \phi, \cdots$.

Remark 3.3.103 *Smooth, continuous, and locally integrable functions are easily identified with respective distributions.*

Definition 3.3.64 *If a fundamental sequence $(f_n)_{n \in \mathbb{N}}$ which defines the distribution $f = [f_n]$ is such a sequence that there exists an integer $k \in \mathbb{N}_o$ and a sequence of smooth functions $(F_n)_{n \in \mathbb{N}}$ such that*

$$F_n^{(k)}(x) = f_n(x), \quad F_n(x) \rightrightarrows$$

for all bounded intervals I included in (a, b), then we say that the distribution

$$f = [f_n]$$

is of a finite order in (a, b).

In the opposite case we say that the distribution is of an infinite order.

Examples of Distributions of Finite Order in $\mathbb{R}$

(1°) $\left[\frac{x}{2} + \frac{x}{\pi} \operatorname{arctg}(nx) - \frac{1}{2\pi n} \ln(1 + n^2 x^2) \right]$;

(2°) $\left[\frac{1}{\pi} \operatorname{arctg}(nx) + \frac{1}{2} \right]$;

(3°) $\left[\frac{1}{\pi} \frac{n}{n^2 x^2 + 1} \right]$;

(4°) $\left[-\frac{2}{\pi} \frac{n^3 x}{(n^2 x^2 + 1)^2} \right]$;

(5°) $\left[\frac{2n^3}{\pi} \frac{3n^2 x^2 - 1}{(n^2 x^2 + 1)^3} \right]$.

Example of a Distribution of an Infinite Order

The series

$$\delta(x) + \delta'(x - 1) + \delta^{(2)}(x - 2) + \cdots,$$

where derivatives are defined in the distributional sense, represents a distribution of infinite order.

3.3.4 Operations with Distributions

Multiplication by a Number. The operation $\lambda f(x)$ of multiplication of a function $f(x)$ by a number λ has the following property:

(i) If $(f_n(x))_{n \in \mathbb{N}}$ is a fundamental sequence, so is $\left(\lambda f_n(x) \right)_{n \in \mathbb{N}}$.

This property enables us to extend the operation to arbitrary distribution $f(x) = [f_n(x)]$.

Definition 3.3.65 *By the product* λf *of a distribution* $f = [f_n]$ *with a number* λ *we mean the distribution* $[\lambda f_n]$.

It is easy to see that

(ii) The product $\lambda f(x)$ does not depend on the choice of the fundamental sequence $(f_n(x))_{n \in \mathbb{N}}$.

Addition. The operation $f + g$ of addition of two functions f and g has the following property:

1. if $(f_n)_{n \in \mathbb{N}}$ and $(g_n)_{n \in \mathbb{N}}$ are fundamental sequences, so is the sequence $(f_n + g_n)_{n \in \mathbb{N}}$.

Proof. **1.** Suppose that there exist integers $k, k_1 \geq 0$ and sequences $(F_n)_{n \in \mathbb{N}}$ and $(G_n)_{n \in \mathbb{N}}$ of smooth functions such that

$$F_n^{(k)} = f_n \quad \text{and} \quad F_n \rightrightarrows,$$
$$G_n^{(k_1)} = g_n \quad \text{and} \quad G_n \rightrightarrows.$$

By Theorem 3.3.106 we can assume that $k = k_1$. Since

$$(F_n + G_n)^{(k)} = f_n + g_n \quad \text{and} \quad F_n + G_n \rightrightarrows,$$

the sequence $(f_n + g_n)_{n \in \mathbb{N}}$ is fundamental. ☐

Property 1 enables us to extend addition to arbitrary distributions.

Definition 3.3.66 *By the sum* $f + g$ *of the distributions* $f = [f_n]$ *and* $g = [g_n]$ *we mean the distribution* $[f_n + g_n]$.

To verify the consistency of this definition we must prove that:

2. the distribution $[f_n + g_n]$ does not depend on the choice of sequences $(f_n)_{n \in \mathbb{N}}$ and $(g_n)_{n \in \mathbb{N}}$ representing the distributions f and g.

Proof. **2.** We must show that if

$$(f_n)_{n \in \mathbb{N}} \sim (\widetilde{f}_n)_{n \in \mathbb{N}} \quad \text{and} \quad (g_n)_{n \in \mathbb{N}} \sim (\widetilde{g}_n)_{n \in \mathbb{N}}$$

then

$$(f_n + g_n)_{n \in \mathbb{N}} \sim (\widetilde{f}_n + \widetilde{g}_n)_{n \in \mathbb{N}}.$$

By definition of equivalence of fundamental sequences it follows that

$$f_1, \widetilde{f}_1, f_2, \widetilde{f}_2, \ldots \quad \text{and} \quad g_1, \widetilde{g}_1, g_2, \widetilde{g}_2, \ldots$$

are fundamental. By the property 1 the sequence

$$f_1 + g_1, \widetilde{f}_1 + \widetilde{g}_1, f_2 + g_2, \widetilde{f}_2 + \widetilde{g}_2, \ldots$$

is fundamental. Thus, by (3.3.10), we have

$$(f_n + g_n)_{n \in \mathbb{N}} \sim (\tilde{f}_n + \tilde{g}_n)_{n \in \mathbb{N}},$$

which implies the assertion. □

Subtraction. The operation $f - g$ of difference of two functions f and g has the property

1) if $(f_n)_{n \in \mathbb{N}}$ and $(g_n)_{n \in \mathbb{N}}$ are fundamental sequences, so is the sequence $(f_n - g_n)_{n \in \mathbb{N}}$.

Property 1) enables us to extend subtraction to arbitrary distributions.

Definition 3.3.67 *By the difference $f - g$ of the distribution $f = [f_n]$ and $g = [g_n]$ we mean the distribution $[f_n - g_n]$.*

The consistency of this definition can be checked by a procedure similar to that already used for the sum.

It is easy to see from the definitions of the operation introduced above that the following properties are true:

Properties of Algebraic Operations on Distributions

1. $f + g = g + f$;

2. $(f + g) + h = f + (g + h)$;

3. the difference $g = h - f$ is the only solution of the equation $f + g = h$;

4. $\lambda(f + g) = \lambda f + \lambda g$, $\lambda \in \mathbb{R}$;

5. $(\lambda + \mu)f = \lambda f + \mu f$, $\lambda, \mu \in \mathbb{R}$;

6. $\lambda(\mu f) = (\lambda \mu)f$, $\lambda, \mu \in \mathbb{R}$;

7. $1f = f$.

Denoting by 0 the *zero distribution*, i.e., the distribution that is defined by the class $[f_n]$, where f_n are functions identically equal to zero, we have

$$0 + f = f \quad \text{and} \quad 0f = f.$$

In the last formula the symbol 0 has two different meanings: on the left-hand side it denotes the zero distribution, on the right-hand side it denotes the number zero. This ambiguity in practice does not lead to any confusion.

3.3.5 Regular Operations

An advantage of the sequential approach in the theory of distributions is the ease of extending to distributions many operations that are defined for smooth functions.

Definition 3.3.68 *We say that an operation A, which to every system $(\varphi_1, \ldots, \varphi_k)$ of smooth functions in $\mathbb{R}$ assigns a smooth function in $\mathbb{R}$ (or a number), is regular if for arbitrary fundamental sequences $(\varphi_{1n})_{n \in \mathbb{N}}, \ldots, (\varphi_{kn})_{n \in \mathbb{N}}$ of smooth functions in $\mathbb{R}$ the sequence $\Big(A(\varphi_{1n}, \ldots, \varphi_{kn})\Big)_{n \in \mathbb{N}}$ is fundamental.*

Every regular operation A defined on smooth functions can be extended automatically to distributions in the following way.

Definition 3.3.69 *If $f_1, \ldots, f_k$ are arbitrary distributions in $\mathbb{R}$ and $(\varphi_{1n})_{n \in \mathbb{N}}, \ldots, (\varphi_{kn})_{n \in \mathbb{N}}$ the corresponding fundamental sequences, i.e., $f_1 = [\varphi_{1n}], \ldots, f_k = [\varphi_{kn}]$, then the operation A on $f_1, \quad, f_k$ is defined by the formula*

$$A(f_1, \ldots, f_k) = \Big[A(\varphi_{1n}, \ldots, \varphi_{kn})\Big].$$

Remark 3.3.104 *This extension is always unique, i.e., it does not depend on the choice of the fundamental sequences $(\varphi_{1n})_{n \in \mathbb{N}}, \cdots (\varphi_{kn})_{n \in \mathbb{N}}$ representing the distributions $f_1, \cdots, f_k$. In other words, if*

$$(\varphi_{1n})_{n \in \mathbb{N}} \sim (\widetilde{\varphi}_{1n})_{n \in \mathbb{N}}, \cdots, (\varphi_{kn})_{n \in \mathbb{N}} \sim (\widetilde{\varphi}_{kn})_{n \in \mathbb{N}}$$

then

$$\Big(A(\varphi_{1n}, \cdots, \varphi_{kn})\Big)_{n \in \mathbb{N}} \sim \Big(A(\widetilde{\varphi}_{1n}, \cdots, \widetilde{\varphi}_{kn})\Big)_{n \in \mathbb{N}}.$$

Indeed, by hypothesis, the sequences

$$\varphi_{11}, \widetilde{\varphi}_{11}, \varphi_{12}, \widetilde{\varphi}_{12}, \ldots$$
$$\varphi_{21}, \widetilde{\varphi}_{21}, \varphi_{22}, \widetilde{\varphi}_{22}, \ldots$$
$$\cdots\cdots\cdots\cdots\cdots$$

are fundamental. By the definition of regular operation the sequence

$$A(\varphi_{11}, \varphi_{21}, \ldots), A(\widetilde{\varphi}_{11}, \widetilde{\varphi}_{21}, \ldots), A(\varphi_{12}, \varphi_{22}, \ldots), \ldots$$

is also fundamental, which proves the assertion.

Remark 3.3.105 *Multiplication by a number, addition and difference are regular operators, as we have seen in the previous section.*

The Following Operations Are Regular:

(1) addition of smooth functions;

(2) difference of smooth functions;

(3) mulitplication of a smooth function by a fixed number λ: $\lambda\varphi$;

(4) translation of the argument of a smooth function $\varphi(x+h)$;

(5) derivation of a smooth function of a fixed order m: $\varphi^{(m)}$;

(6) multiplication of a smooth function by a fixed smooth function ω: $\omega\varphi$;

(7) substitution a fixed smooth function $\omega \neq 0$;

(8) product of smooth functions with separated variables: $\varphi_1(x)\varphi_2(y)$;

(9) convolution of a smooth function with a fixed function ω from the space $\mathcal{D}$ (of smooth functions whose supports are bounded)

$$(f * \omega)(x) = \int_{\mathbb{R}} \varphi(x-t)\omega(t)dt,$$

(10) inner product of a smooth function with a fixed function from the space $\mathcal{D}$

$$(\varphi, \omega) = \int_{\mathbb{R}} \varphi(x)\omega(x)dx.$$

Remark 3.3.106 *The support of a distribution is the smallest closed set outside which the distribution vanishes.*

It is easy to check that:

Theorem 3.3.107 *Substitution of regular operations is regular, too.*

All formulae involving regular operations, which hold true for smooth functions, are extended automatically to distributions.

Translation. The translation $\varphi(x) \to \varphi(x+h)$ is a regular operation.
Moreover, if $\left(\varphi_n(x)\right)_{n\in\mathbb{N}}$ is a fundamental sequence in the open interval (c,d), then $\left(\varphi_n(x+h)\right)_{n\in\mathbb{N}}$ is a fundamental sequence in the translated inteval (c_h, d_h), where

$$(c_h, d_h) = \{x : x+h \in (c,d)\}.$$

Thus, if $f(x) = [\varphi_n(x)]$ is a distribution defined in (c,d), then

$$f(x+h) = [\varphi_n(x+h)] \quad \text{is a distribution defined in} \quad (c_h, d_h).$$

Derivation. One of the most important benefits of extending the notion of a function to the notion of a distribution is the fact that every distribution has all derivatives which are again distributions.

The derivation $\varphi^{(m)}$ of an arbitrary order m is a regular operation. In fact, if $(\varphi_n)_{n \in \mathbb{N}}$ is a fundamental sequence, then so is sequence $(\varphi_n^{(m)})_{n \in \mathbb{N}}$ (see 3.3.2, Theorem 3.3.102). Thus, we can define the derivative of order m of distribution f in the following way:

Definition 3.3.70 *We define the derivative of order m, $m \in \mathbb{N}$, of any distribution $f = [\varphi_n]$ by setting*

$$f^{(m)} = [\varphi_n^{(m)}].$$

It is easy to see that the following theorem is true:

Theorem 3.3.108 *Each distribution has derivatives of all orders.*

The following formulas occurring in the ordinary differential calculus follow immediately for distributions from the definition:

$$f^{(o)} = f \tag{3.3.13}$$

$$(f + g)^{(m)} = f^{(m)} + g^{(m)} \tag{3.3.14}$$

$$(\lambda f)^{(m)} = \lambda f^{(m)}, \quad \lambda \in \mathbb{R} \tag{3.3.15}$$

$$(f^{(m)})^{(k)} = f^{(m+k)}. \tag{3.3.16}$$

Multiplication of a Distribution by a Smooth Function. The multiplication $\varphi \omega$, considered as an operation on two functions φ and ω, is not regular. Namely, if the sequences $(\varphi_n)_{n \in \mathbb{N}}$ and $(\omega_n)_{n \in \mathbb{N}}$ are fundamental, their product $(\varphi_n)_{n \in \mathbb{N}}$ and $(\omega_n)_{n \in \mathbb{N}}$ need not be fundamental.

However, multiplication may also be thought of as an operation on a single function, the other factor being kept fixed.

Definition 3.3.71 *We define the product of an arbitrary distribution $f = [\varphi_n]$ by a smooth function ω by means of the formula*

$$\omega f = [\omega \varphi_n].$$

To verify the consistency of this definition we must prove that the multiplication $\omega \varphi$ by a smooth function ω is a regular operation, i.e., we must prove the theorem:

Theorem 3.3.109 *If a sequence $(\varphi_n)_{n \in \mathbb{N}}$ is fundamental, $\omega \in \mathcal{C}^\infty$, then the sequence $(\omega \varphi_n)_{n \in \mathbb{N}}$ is fundamental, too.*

Proof. Since $(\varphi_n)_{n\in\mathbb{N}}$ is fundamental for every interval I inside $(a; b)$ there exist a number k and smooth functions G_n such that

$$G_n^{(k)}(x) = (\varphi_n), \quad x \in I \quad \text{and}$$

$$G_n \rightrightarrows \quad x \in I.$$

We shall show that for every order m and for every smooth function ω, the sequence

$$(\omega G_n^{(m)})_{n\in\mathbb{N}} \tag{3.3.17}$$

is fundamental. The proof follows by induction.

The case $m = 0$ follows from Theorem 3.3.101 in section 3.3.2. If the sequence is fundamental for some m, then the sequence is also fundamental for $m + 1$, since

$$\omega G_n^{(m+1)} = (\omega G_n^{(m)})' - \omega' G_n^{(m)}$$

and the right-hand side in the difference of two sequences that are fundamental by Theorem 3.3.101 (see section 3.3.2) and the induction hypothesis. Since the interval I is arbitrary, the sequence (3.3.17) is fundamental in (a, b). □

The following usual properties of multiplication follow directly from the definition:

$$\begin{array}{llll}
\omega_1(\omega_2 f) & = (\omega_1\omega_2)f, & \omega_1, \omega_2 \in C^\infty, & f \in \mathcal{D}'; \\
(\omega_1 + \omega_2)f & = \omega_1 f + \omega_2 f, & \omega_1, \omega_2 \in C^\infty, & f \in \mathcal{D}'; \\
\omega(f + g) & = \omega f + \omega g, & \omega \in C^\infty, & f, g \in \mathcal{D}',
\end{array}$$

where $\mathcal{D}'$ is the space of distributions.

Remark 3.3.107 *We note that if f is a function, then the product defined above is the ordinary product of functions. Moreover, if ω is a constant function and f is an arbitrary distribution, this product coincides with the product in* Definition 3.3.65; *see section 3.3.4.*

It is easy to prove the formula

$$(\omega f)' = \omega' f + \omega f', \quad \omega \in C^\infty, \quad f \in \mathcal{D}'. \tag{3.3.18}$$

This formula may be considered a particular case (when $k = 1$) of the formula

$$\omega f^{(k)} = \sum_{j=o}^{k} (-1)^j \binom{k}{j} \left(\omega^{(j)} f\right)^{(k-j)}, \quad \omega \in C^\infty, \quad f \in \mathcal{D}', \tag{3.3.19}$$

which can be proved by induction in the same way as for functions. The following Leibniz-Schwarz formulas hold true:

$$(\omega f)^{(k)} = \sum_{0 \le j \le k} \binom{k}{j} \omega^{(j)} f^{(k-j)}, \quad \omega \in C^\infty, \quad f \in \mathcal{D}'; \tag{3.3.20}$$

$$\omega^{(k)} f = \sum_{0 \le j \le k} (-1)^j \binom{k}{j} \left(\omega f^{(j)}\right)^{(k-j)}, \quad \omega \in C^\infty, \quad f \in \mathcal{D}'. \tag{3.3.21}$$

Substitution. Let ψ be a fixed smooth function defined in the interval (a, b) such that $\psi'(x) \neq 0$ for $x \in (a, b)$ and suppose that the values of the function ψ belong to the open interval (c, d). Composition $(\varphi \circ \psi)(x)$ defined by means of

$$(\varphi \circ \psi)(x) = \varphi(\psi(x))$$

is a regular operation on $\varphi(y)$ (ψ being fixed). We shall show the following theorem:

Theorem 3.3.110 *If $(\varphi_n(y))_{n \in \mathbb{N}}$ is fundamental in open interval (c, d), the values of smooth function ψ belong to (c, d) and $\psi'(x) \neq 0$ for $x \in (a, b)$, then $\left(\varphi_n(\psi(x)) \right)_{n \in \mathbb{N}}$ is fundamental in (a, b).*

Proof. Observe that if the sequence $\left(\varphi_n(\psi(x)) \right)_{n \in \mathbb{N}}$ is fundamental then the sequence $\left(\varphi_n'(\psi(x)) \right)_{n \in \mathbb{N}}$ is fundamental, too. This follows from the equality

$$\left(\varphi_n'(\psi(x)) \right) = \frac{1}{\psi'(x)} \left(\varphi_n(\psi(x)) \right)'; \tag{3.3.22}$$

see Theorem 3.3.102 (section 3.3.2) and Theorem 3.3.109.

Let I be an open interval inside (a, b). The function ψ maps I into an interval $I' \subset (c, d)$. Let $(F_n)_{n \in \mathbb{N}}$ be a sequence of smooth functions such that

$$F_n(y) \rightrightarrows \quad y \in I',$$
$$F_n^{(k)}(y) = \varphi_n(y), \quad y \in I'.$$

The sequence $\left(F_n(\psi(x)) \right)_{n \in \mathbb{N}}$ of smooth functions converges uniformly for $x \in I$. Since the interval I is arbitrary, the sequence

$$\left(F_n(\psi(x)) \right)_{n \in \mathbb{N}}$$

is fundamental in (a, b). Consequently, the sequences

$$\left(F_n'(\psi(x)) \right)_{n \in \mathbb{N}}, \cdots, \left(F_n^{(k)}(\psi(x)) \right)_{n \in \mathbb{N}}$$

are also fundamental. The last of these sequences coincides with the sequence $\left(\varphi_n(\psi(x)) \right)_{n \in \mathbb{N}}$. This completes the proof of this theorem. $\square$

Calculation with substitution of distributions can be carried out in the same way as those with substitution of functions. In particular, we have the formula

$$\left(f(\psi(x)) \right)' = f'\left(\psi(x) \right) \psi'(x), \quad \psi \in C^\infty, \ \psi' \neq 0, \ f \in \mathcal{D}', \tag{3.3.23}$$

since

$$\left(f(\psi(x)) \right)' = \left[\varphi_n(\psi(x)) \right]' = \left[\left(\varphi_n(\psi(x)) \right)' \right] = \left[\varphi_n'(\psi(x)) \psi'(x) \right] = f'\left(\psi(x) \right) \psi'(x),$$

where $(\varphi_n)_{n\in\mathbb{N}}$ is a fundamental sequence of f.

We have thus proved that the substitution of a given smooth function ϕ, satisfying condition $\phi'(x) \neq 0$, is a regular operation. Following the general method we define the substitution.

Definition 3.3.72 *We define the substitution of a fixed function $\phi \in C^\infty$ ($\phi'(x) \neq 0$ for $x \in (a,b)$, with values in (c,d)), into an arbitrary distribution $f(y) = [\varphi_n(y)]$ in (c,d) by the formula*

$$f(\phi(x)) = [\varphi_n(\phi(x))].$$

Theorem 3.3.111 *For every distribution f and every integer $k \geq 0$ we have the equality*

$$\left(f(\alpha x + \beta)\right)^{(k)} = \alpha^k f^{(k)}(\alpha x + \beta), \quad \alpha \neq 0.$$

Proof. Let $(\varphi_n)_{n\in\mathbb{N}}$ be a fundamental sequence of the distribution f. Then

$$\left(f(\alpha x + \beta)\right)^{(k)} = \left[\varphi_n(\alpha x + \beta)\right]^{(k)} = \left[\alpha^k \varphi_n^{(k)}(\alpha x + \beta)\right]$$
$$= \alpha^k f^{(k)}(\alpha x + \beta).$$

From this theorem we have the following corollary:

Corollary 3.3.36 *If a distribution $f(x)$ is the k-th derivative of a continuous function $F(x)$, then the distribution $f(\alpha x + \beta)$ is the k-th derivative of the function*

$$\frac{1}{\alpha^k} F(\alpha x + \beta).$$

Convolution with a Fixed Smooth Function. The convolution $f * \omega$ of a distribution f with a fixed smooth function ω of bounded support is meant here as a regular operation

$$A(f) = f * \omega,$$

(the proof will be given later), which for the smooth function φ is defined in the known manner

$$A(\varphi)(x) = (\varphi * \omega)(x) = \int_{-\infty}^{+\infty} \varphi(x - t)\omega(t)dt;$$

see also section 1.3.3, formula (1.3.27).

Definition 3.3.73 *We define the convolution of any distribution $f = [\varphi_n]$ with the smooth function ω of bounded support by setting*

$$f * \omega = [\varphi_n * \omega].$$

To verify the consistency of this definition we must prove that the convolution $w * f$ by a function $w \in \mathcal{D}$ is a regular operation. To obtain this we need three preparatory lemmas.

Lemma 3.3.41 *If f is a continuous function (or local integrable) in $\mathbb{R}$ and $w \in \mathcal{D}$, i.e., is a smooth function of bounded support in $\mathbb{R}$, then the convolution $w * f$ is a smooth function and the equality holds:*

$$(f * w)^{(m)} = f * w^{(m)} \quad \text{for all} \quad m \in \mathbb{N}. \tag{3.3.24}$$

Proof. To prove that $f * w$ is a smooth function it suffices to show that for any fixed $m \in \mathbb{N}$ the convolution $(f * w)^{(m)}$ is a smooth function.

Let $k = f * w$. For any fixed x the product $f(t)w(x - t)$ is an integrable function, since $f(x)$ is integrable on the set where $w(x - t) \neq 0$. Thus, convolution

$$k(x) = (f * w)(x)$$

exists everywhere. Then

$$|k^{(m)}(x) - k^{(m)}(x_o)| \leq \int_{\mathbb{R}} |f(t)||w^{(m)}(x - t) - w^{(m)}(x_o - t)|dt.$$

Since the function w has a bounded support there exists integer $r > 0$ such that $w(t) = 0$ for $|t| > r$. If x_o is fixed and $|x - x_o| < 1$ then the difference $w(x - t) - w(x_o - t)$ vanishes for t satisfying inequality $|t - x_o| > r - 1$. Hence,

$$|k^{(m)}(x) - k^{(m)}(x_o)| \leq M \int_{\mathbb{R}} |w^{(m)}(x - t) - w^{(m)}(x_o - t)|dt$$

$$= M \int_{\mathbb{R}} |w^{(m)}(t + (x - x_o)) - w^{(m)}(t)|dt$$

for $|x - x_o| < 1$.

By Lebesgue's theorem the last integral tends to 0, if $x \to x_o$, and this shows the continuity of considered convolution. The equality (3.3.24) can be obtained by simple transformations. □

From Lemma 3.3.41 it follows:

Lemma 3.3.42 *If f is a smooth function and $w \in \mathcal{D}$ then*

$$(f * w)^{(m)} = f^{(m)} * w = f * w^{(m)} \quad \text{for all} \quad m \in \mathbb{N}.$$

Lemma 3.3.43 *If $(\varphi_n)_{n \in \mathbb{N}}$ is a fundamental sequence and $w \in \mathcal{D}$, then the sequence of convolutions $(\varphi_n * w)_{n \in \mathbb{N}}$ is fundamental, too.*

Proof. Since ω is a function with bounded support there exists an integer $\alpha \in \mathbb{R}_+$ such that

$$\omega(x) = 0 \quad \text{for} \quad |x| > \alpha.$$

Let I be any bounded open interval in $\mathbb{R}$ and I' be a bounded open interval too such that $I \subset I'_{-\alpha}$, where $I'_{-\alpha}$ denotes the set of all points $x \in I'$ whose distance from the boundary of I' is greater than α. Since $(\varphi_n)_{n \in \mathbb{N}}$ is a fundamental sequence there are integer $k \in \mathbb{N}$, continuous functions F_n, $(n \in \mathbb{N})$ and F such that

$$F_n^{(k)} = \varphi_n \quad \text{and} \quad F_n \rightrightarrows F, \quad x \in I'.$$

Note that

$$|(F_n * \omega)(x) - (F * \omega)(x)| \leq \int_{\mathbb{R}} |F_n(t) - F(t)||\omega(x - t)|dt \leq \epsilon \int_{\mathbb{R}} |\omega(t)| \leq \epsilon \int_{-\alpha}^{+\alpha} |\omega(t)|dt.$$

This means that

$$F_n * \omega \rightrightarrows F * \omega \quad \text{on} \quad I.$$

By Lemma 3.3.42 we have

$$(F_n * \omega)^{(k)} = F_n^{(k)} * \omega = \varphi_n * \omega \quad \text{on} \quad I.$$

This means that $(\varphi_n * \omega)_{n \in \mathbb{N}}$ is fundamental. This proves the lemma. $\quad\square$

Corollary 3.3.37 *If a sequence $(\varphi_n)_{n \in \mathbb{N}}$ is fundamental and $\omega \in \mathcal{D}$, then the sequence of convolutions $(\varphi_n * \omega)_{n \in \mathbb{N}}$ converges almost uniformly.*

Remark 3.3.108 *It follows from* Lemma 3.3.43 *and* Theorem 3.3.101, *section 3.3.2, that convolution with a smooth function with bounded support is a regular operation.*

Since the convolution with function $\omega \in \mathcal{D}$ is a regular operation all formulae that hold true for smooth functions can be extended to distributions. Thus, we have:

Proposition 3.3.64 *If $\omega \in \mathcal{D}$ and f_1, f_2, f_3 are any distributions then the following formulae are true:*

$$f * \omega = \omega * f;$$

$$(\lambda f) * \omega = f * (\lambda \omega) = \lambda(f * \omega), \quad \lambda \in \mathbb{R};$$

$$(f_1 + f_2) * \omega = f_1 * \omega + f_2 * \omega;$$

$$f * (\omega_1 + \omega_2) = f * \omega_1 + f * \omega_2, \quad \omega_1, \omega_2 \in \mathcal{D}.$$

3.4 Delta Sequences

In section 3.3.3 we mentioned that an arbitrary smooth function ϕ can be identified with a distribution that is an equivalence class of a constant sequence $\phi, \phi, \phi, \dots$. In order to identify an arbitrary continuous function and more generally a locally integrable function with a distribution we have to approximate these functions by smooth functions, e.g., by delta sequences. We will present one definition of a delta sequence.

3.4.1 Definition and Properties

Definition 3.4.74 *By a delta sequence in $\mathbb{R}$ we mean any sequence of smooth functions $\left(\delta_n(x)\right)_{n\in\mathbb{N}}$, $x \in \mathbb{R}$, with the following properties:*

(Δ_1) *There is a sequence of positive numbers $\alpha_n \to 0$ such that*

$$\delta_n(x) = 0 \quad for \quad |x| \geq \alpha_n, \quad n \in \mathbb{N};$$

(Δ_2) $\int_{\mathbb{R}} \delta_n(x)dx = 1$, *for $n \in \mathbb{N}$;*

(Δ_3) *For every $k \in \mathbb{N}_o$ there is a positive integer M_k such that*

$$\alpha_n^k \int_{\mathbb{R}} |\delta_n^{(k)}(x)|dx < M_k \quad for \quad n \in \mathbb{N}.$$

Various definitions of delta sequences can be found in the literature (see [AMS], [MiS.2]).

Example 3.4.77 *As an example of a delta sequence we can take*

$$\delta_n(x) = \alpha_n^{-1}\Omega(\alpha_n^{-1}x) \quad for \quad n \in \mathbb{N},$$

where $\Omega \in \mathcal{D}$ (i.e., Ω is any smooth function of bounded support) and such that

$$\int_{\mathbb{R}} \Omega(x)dx = 1,$$

where $(\alpha_n)_{n\in\mathbb{N}}$ is an arbitrary sequence, different from 0 and tending to 0.

The delta sequences (δ sequences) have the following properties:

Property 3.4.3 The convolution of two delta sequences is a delta sequence.

Proof. Let $(\delta_{1n})_{n\in\mathbb{N}}$ and $(\delta_{2n})_{n\in\mathbb{N}}$ be two delta sequences. We have to prove that $(\delta_n)_{n\in\mathbb{N}} = (\delta_{1n})_{n\in\mathbb{N}} * (\delta_{2n})_{n\in\mathbb{N}}$ is another delta sequence. In fact, δ_n are smooth functions. Moreover, if

$$\delta_{1n}(x) = 0 \quad for \quad |x| \geq \alpha_{1n} \quad and \quad \delta_{2n}(x) = 0 \quad for \quad |x| \geq \alpha_{2n}, \quad n \in \mathbb{N},$$

$$then \quad \delta_n(x) = 0 \quad for \quad |x| \geq \alpha_{1n} + \alpha_{2n}, \quad n \in \mathbb{N}.$$

This implies that δ_n satisfies condition Δ_1. Since

$$\int_{\mathbb{R}} \delta_n(x)dx = \int_{\mathbb{R}} dx \int_{\mathbb{R}} \delta_{1n}(x-t)\delta_{2n}(t)dt = \int_{\mathbb{R}} \delta_{2n}(t)dt \int_{\mathbb{R}} \delta_{1n}(x-t)dx = 1\cdot 1 = 1,$$

so that condition Δ_2 is satisfied. Because

$$\int_{\mathbb{R}} |\delta_n^{(k)}(x)|dx = \int_{\mathbb{R}} dx \int_{\mathbb{R}} |\delta_{1n}^{(k)}(x-t)||\delta_{2n}(t)|dt = \int_{\mathbb{R}} |\delta_{2n}(t)|dt \int_{\mathbb{R}} |\delta_{1n}^{(k)}(x-t)|dx,$$

thus

$$\alpha_{1n}^k \int_{\mathbb{R}} |\delta_n^{(k)}(x)|dx \le M_{1k}M_{2o}.$$

Similarly, we obtain

$$\alpha_{2n}^k \int_{\mathbb{R}} |\delta_n^{(k)}(x)|dx \le M_{1o}M_{2k}.$$

Since

$$(\alpha_{1n}+\alpha_{2n})^k \le 2^k(\alpha_{1n}^k + \alpha_{2n}^k),$$

by the last two inequalities, we get

$$(\alpha_{1n}+\alpha_{2n})^k \int_{\mathbb{R}} |\delta_n^k(x)|dx \le M_k, \quad \text{where} \quad M_k = 2^k(M_{1k}M_{2o} + M_{1o}M_{2k}),$$

which proves condition Δ_3. $\square$

Property 3.4.4 Every delta sequence is fundamental.

Proof. Let $(\delta_n)_{n\in\mathbb{N}}$ be an arbitrary delta sequence. Let us consider a sequence $(\gamma_n)_{n\in\mathbb{N}}$ such that

$$\gamma_n(x) = \int_{-\infty}^{x} dt \int_{-\infty}^{t} \delta_n(t_1)dt_1.$$

It is easy to see that $\gamma_n, n \in \mathbb{N}$, are smooth functions and $\gamma_n \rightrightarrows$ on $\mathbb{R}$ and $\gamma_n^{(2)} = \delta_n$ for $n \in \mathbb{N}$. This means that the sequence $(\delta_n)_{n\in\mathbb{N}}$ is fundamental. $\square$

It is easy to see that if $(\delta_{1n})_{n\in\mathbb{N}}$ and $(\delta_{2n})_{n\in\mathbb{N}}$ are two delta sequences, then the interlaced sequence

$$\delta_{11}, \delta_{21}, \delta_{12}, \delta_{22}, \delta_{13}, \delta_{23}, \cdots$$

is also a delta sequence. Thus, we have the following property:

Property 3.4.5 Every two delta sequences are equivalent.

Since all delta sequences are equivalent, they represent the same distribution, which is called **Dirac's delta distribution**:

$$\boxed{\delta(x) = [\delta_n(x)].}$$

By Theorem 3.3.109, we have

Property 3.4.6 The product of a smooth function with a delta sequence $(\delta_n)_{n\in\mathbb{N}}$ is fundamental.

Property 3.4.7 If ω is a smooth function, $(\delta_n)_{n\in\mathbb{N}}$ is a delta sequence, then the fundamental sequence $\left(\omega(x)\delta_n(x)\right)_{n\in\mathbb{N}}$ is equivalent to a sequence $\left(\omega(0)\delta_n(x)\right)_{n\in\mathbb{N}}$.

Proof. For an arbitrary positive integer ϵ, there exists an index n_o such that for $n > n_o$

$$|\omega(x) - \omega(0)| < \epsilon \quad \text{for} \quad -\alpha_n < x < \alpha_n.$$

Hence

$$\left| \int_{-\infty}^{x} \left(\omega(t) - \omega(0)\right)\delta_n(t)dt \right| \leq \epsilon \int_{-\infty}^{\infty} |\delta_n(x)|dx = \epsilon \cdot M_o,$$

which proves that the integral converges uniformly to 0. Hence

$$\left(\omega(x)\delta_n(x) - \omega(0)\delta_n(x)\right)_{n\in\mathbb{N}} \sim (0)_{n\in\mathbb{N}},$$

i.e.,

$$\left(\omega(x)\delta_n(x)\right)_{n\in\mathbb{N}} \sim \left(\omega(0)\delta_n(x)\right)_{n\in\mathbb{N}}. \tag{3.4.1}$$

□

Sequences appearing on the left and right sides of the equality (3.4.1) are fundamental for the products $\omega(x)\delta(x)$ and $\omega(0)\delta(x)$, respectively. This means that the considered products represent the same distribution. Thus,

$$\boxed{\omega(x)\delta(x) = \omega(0)\delta(x), \quad \omega \in C^\infty.} \tag{3.4.2}$$

Similarly, we obtain

$$\boxed{\omega(x)\delta(x - x_o) = \omega(x_o)\delta(x - x_o), \quad \omega \in C^\infty.} \tag{3.4.3}$$

In particular, if $\omega(x) = x^n$, $n \in \mathbb{N}$, we obtain

$$\boxed{x^n\delta(x) = 0 \quad \text{for} \quad n \in \mathbb{N}.} \tag{3.4.4}$$

By induction we can prove that

$$\boxed{x^n\delta^{(n-k)}(x) = 0 \quad \text{for} \quad k = 1,\cdots,n; \quad n \in \mathbb{N},} \tag{3.4.5}$$

where $\delta^{(o)} = \delta$.

Property 3.4.8 If f is a continuous function in (a,b) and $(\delta_n)_{n\in\mathbb{N}}$ is a delta sequence, then the sequence of smooth functions

$$(f * \delta_n)_{n\in\mathbb{N}}$$

converges to f, almost uniformly in (a, b).

Proof. Let I be any bounded interval inside (a, b). For every positive number ϵ there is an index n_o such that for $n > n_o$ the inequality

$$|f(x - t) - f(x)| < \epsilon \quad \text{holds for} \quad x \in I \quad \text{and} \quad t \in (-\alpha_n, \alpha_n),$$

where $(\alpha_n)_{n \in \mathbb{N}}$ is a sequence of positive numbers such that $\alpha_n \to 0$ and $\delta_n(t) = 0$ for $|t| \geq \alpha_n, n \in \mathbb{N}$. Hence,

$$|(f * \delta_n)(x) - f(x)| \leq \int_{-\infty}^{\infty} |f(x - t) - f(x)||\delta_n(t)|dt = \int_{-\alpha_n}^{\alpha_n} |f(x - t) - f(x)||\delta_n(t)|dt$$

$$\leq \epsilon \int_{-\alpha_n}^{\alpha_n} |\delta_n(t)|dt \leq \epsilon M_o$$

for $n \geq n_o$ and $x \in I$.

This proves that $f * \delta_n$ converges to f almost uniformly in (a, b). □

The generalization of the Property 3.4.8 is the following:

Property 3.4.9 If $(f_n)_{n \in \mathbb{N}}$ is a sequence of continuous functions convergent to f, almost uniformly in (a, b) and $(\delta_n)_{n \in \mathbb{N}}$ is a delta sequence, then the sequence of smooth functions $(f_n * \delta_n)_{n \in \mathbb{N}}$ converges to f, almost uniformly in (a, b).

Proof. Note that

$$f_n * \delta_n = f * \delta_n + (f_n - f) * \delta_n.$$

By Property 3.4.8 we have

$$f * \delta_n \overset{\text{a.u.c.}}{\rightrightarrows} f \quad \text{in} \quad (a, b).$$

It suffices to show that

$$(f_n - f) * \delta_n \overset{\text{a.u.c.}}{\rightrightarrows} 0 \quad \text{in} \quad (a, b).$$

In fact, given any bounded interval I inside (a, b) and any positive number ϵ, we have, for sufficiently large n,

$$|(f_n - f) * \delta_n| \leq |f_n - f| * |\delta_n| \leq \epsilon * |\delta_n| \leq \epsilon \cdot M_o \quad \text{in} \quad I.$$

This proves the property. □

Property 3.4.10 If f is a smooth function and $(\delta_n)_{n \in \mathbb{N}}$ is a delta sequence, then

$$(f * \delta_n)^{(k)} \overset{\text{a.u.c.}}{\rightrightarrows} f^{(k)} \quad \text{for} \quad k \in \mathbb{N}_o.$$

Proof. It is easy to see that

$$(f * \delta_n)^{(k)} = f^{(k)} * \delta_n.$$

By Property 3.4.8 it follows immediately that

$$f^{(k)} * \delta_n \overset{\text{a.u.c.}}{\Rightarrow} f^{(k)},$$

which completes the proof. □

Property 3.4.11 If f and f_n $(n \in \mathbb{N})$ are smooth functions and

$$f_n^{(k)} \overset{\text{a.u.c.}}{\Rightarrow} f^{(k)}, \quad \text{for every} \quad k \in \mathbb{N}_o,$$

then

$$(f_n * \delta_n)^{(k)} \overset{\text{a.u.c.}}{\Rightarrow} f^{(k)}.$$

Proof. It suffices to see that

$$(f_n * \delta_n)^{(k)} = f_n^{(k)} * \delta_n$$

and to use Property 3.4.8. □

Property 3.4.12 If f is a locally integrable function in $\mathbb{R}$, then $(f * \delta_n)_{n \in \mathbb{N}}$ converges in norm to f, i.e.,

$$\int_{\mathbb{R}} |(f * \delta_n)(x) - f(x)| dx \to 0, \quad n \to \infty.$$

Proof. To prove this property, we first observe that

$$\int_{\mathbb{R}} |(f * \delta_n - f)(x)| dx \leq \int_{\mathbb{R}} |\int_{\mathbb{R}} \left(f(x-t) - f(x) \right) \delta_n(t) dt | dx$$

$$\leq \int_{\mathbb{R}} \left(|\delta_n(t)| \int_{\mathbb{R}} |f(x-t) - f(x)| \, dx \right) dt.$$

We now apply a well-known Lebesgue theorem to get

$$\int_{\mathbb{R}} |f(x-t) - f(x)| \, dx \to 0, \quad \text{as} \quad t \to 0.$$

Hence, for any positive number ϵ there is an index n_o such that for $n > n_o$

$$\int_{\mathbb{R}} |f(x-t) - f(x)| \, dx < \epsilon, \quad \text{for} \quad |t| \leq \alpha_n,$$

where $(\alpha_n)_{n \in \mathbb{N}}$ is a sequence of positive number such that $\alpha_n \to 0$ and $\delta_n(t) = 0$ for $|t| \geq \alpha_n$. We conclude that

$$\int_{\mathbb{R}} |(f * \delta_n)(x) - f(x)| dx \leq \int_{\mathbb{R}} |\delta_n(t)| \, dt \cdot \epsilon = M_o \cdot \epsilon \to 0,$$

which proves our property.   ⬚

Property 3.4.13 If f is a locally integrable function in (a, b), then the sequence $(f * \delta_n)_{n \in \mathbb{N}}$ converges locally in norm to f, i.e., given any interval I inside (a, b), we have

$$\int_I |(f * \delta_n)(x) - f(x)| dx \to 0.$$

Property 3.4.14 If a sequence of function $(f_n)_{n \in \mathbb{N}}$ integrable in $\mathbb{R}$ converges in norm to f, then the sequence $(f_n * \delta_n)_{n \in \mathbb{N}}$ also converges to f with respect to the norm in L_1.

Proof. Note that

$$\int_{\mathbb{R}} |((f_n - f) * \delta_n)(x)| dx \le \int_{\mathbb{R}} dx \int_{\mathbb{R}} |f_n(t) - f(t)||\delta_n(x - t)| dt$$

$$= \int_{\mathbb{R}} |f_n(t) - f(t)| dt \int_{\mathbb{R}} |\delta_n(x - t)| dx$$

$$\le M_o \int_{\mathbb{R}} |f_n(t) - f(t)| dt \to 0, \quad \text{as} \quad n \to \infty.$$

From this it follows that $\left((f_n - f) * \delta_n\right)_{n \in \mathbb{N}}$ converges to 0 with respect to the norm in L_1. Hence, by Property 3.4.12 and equality $f_n - \delta_n = f * \delta_n + (f_n - f) * \delta_n$, we get the assertion.
⬚

Property 3.4.15 If $(f_n)_{n \in \mathbb{N}}$ is a sequence of locally integrable functions that converges in (a, b), to f locally in L_1 norm, then the sequence $(f_n * \delta_n)_{n \in \mathbb{N}}$ converges locally in norm to f in (a, b).

Proof. In order to prove this property note that

$$f_n * \delta_n = f * \delta_n + (f_n - f) * \delta_n.$$

Next, it suffices to show, using Property 3.4.13, that the sequence

$$\left((f_n - f) * \delta_n\right)_{n \in \mathbb{N}}$$

converges locally in norm to 0.   ⬚

Remark 3.4.109 *It is known that, if f and g are integrable functions in $\mathbb{R}$, then the convolution $f * g$ is an integrable function in $\mathbb{R}$. In particular, if f is an integrable in $\mathbb{R}$, then $f * \delta_n$, for all $n \in \mathbb{N}$, are integrable, too.*

Property 3.4.16 If f and g are locally integrable functions in $\mathbb{R}$ such that the convolution of their moduli exists a.e. and represents a locally integrable function in $\mathbb{R}$, then for every delta sequence $(\delta_n)_{n \in \mathbb{N}}$ the sequence

$$\Big((f * \delta_n) * (g * \delta_n) \Big)_{n \in \mathbb{N}},$$

converges locally in norm to $f * g$.

3.4.2 Distributions as a Generalization of Continuous Functions

Every continuous function can be treated as a distribution. In order to obtain the identification rule for continuous functions we will prove the following two lemmas.

Lemma 3.4.44 *If a sequence of smooth functions $(\varphi_n)_{n \in \mathbb{N}}$ is such that for every $k \in \mathbb{N}$*

$$\varphi_n \rightrightarrows 0 \quad and \quad \varphi_n^{(k)} \rightrightarrows \quad in \quad (a, b),$$

then

$$\varphi_n^{(k)} \rightrightarrows 0 \quad in \quad (a, b).$$

Proof. This lemma is true for $k = 0$. We now argue by induction. Suppose that the assertion holds for an order k, and that

$$\varphi_n^{(k)}(x + \eta) - \varphi_n^{(k)}(x) = \int_0^\eta \varphi_n^{(k+1)}(x + t)dt \rightrightarrows \int_0^\eta f(x + t)dt$$

in the interval $a + |\eta| < x < a - |\eta|$. By the induction hypothesis, the last integral vanishes. Because the number η is arbitrary, we obtain $f(x) = 0$. ☐

Lemma 3.4.45 *Almost uniformly convergent sequences of smooth functions are equivalent if they converge to the same continuous function.*

Proof. Let the sequences $(\varphi_n)_{n \in \mathbb{N}}$ and $(\psi_n)_{n \in \mathbb{N}}$ converge almost uniformly to a function f. Then they satisfy condition (F_1) and (F_2) with k=0 (see 3.3.2, Theorem 3.3.101). Thus,

$$(\varphi_n)_{n \in \mathbb{N}} \sim (\psi_n)_{n \in \mathbb{N}}.$$

Conversely, if $(\varphi_n)_{n \in \mathbb{N}} \sim (\psi_n)_{n \in \mathbb{N}}$, then for every open and bounded interval I there exist smooth functions Φ_n and Ψ_n and order k such that conditions (E_1) and (E_2) are satisfied. Hence, $\Phi_n(x) - \Psi_n(x) \rightrightarrows 0$, $x \in I$. By Lemma 3.4.44

$$\varphi_n(x) - \psi_n(x) \rightrightarrows 0, \quad x \in I.$$

This means that the limits of sequences $(\varphi_n)_{n \in \mathbb{N}}$ and $(\psi_n)_{n \in \mathbb{N}}$ are the same. ☐

Now we are able to establish the correspondence between continuous functions and certain distributions.

It follows from Property 3.4.8, section 3.4.1, that for every continuous function there exists a sequence of smooth functions $(\varphi_n)_{n\in\mathbb{N}}$ which converges almost uniformly to $f(x)$. By Theorem 3.3.101, section 3.3.2, this sequence is fundamental. Thus to every continuous function $f(x)$ there corresponds a distribution $[\varphi_n]$. By Lemma 3.4.45 the correspondence is one to one.

In the sequel we will identify a continuous function f with a distribution $[\varphi_n]$, i.e., $f = [\varphi_n]$, as $\varphi_n \overset{\text{a.u.c.}}{\rightrightarrows} f$.

In particular, by Property 3.4.8, section 3.4.1, we can have the equality

$$\boxed{f = [f * \delta_n],}$$ (3.4.6)

for every continuous function and for every delta sequence.

In that way we have proved what can be stated as the following theorem:

Theorem 3.4.112 *Every continuous function f can be identified with the distribution $[f * \delta_n]$, where $(\delta_n)_{n\in\mathbb{N}}$ is a delta sequence.*

Of course, smooth functions φ are distributions. For them we have the simpler identity

$$\varphi = [\varphi].$$

In particular, the zero distribution, i.e., the distribution identified with the function that vanishes everywhere, is denoted by 0.

The identification presented above shows that distributions are a generalization of continuous functions. This justifies using the same symbols for functions and distributions.

Theorem 3.4.113 *The convolution $f * \omega$ of a distribution f with a smooth function $\omega \in \mathcal{D}$ is a smooth function.*

Proof. Let $f = [\varphi_n]$. The sequence $(\varphi_n)_{n\in\mathbb{N}}$ is fundamental. By Lemma 3.3.43, section 3.3.5, the sequence $(\varphi_n * \omega)_{n\in\mathbb{N}}$ is fundamental, too, and converges almost uniformly to a continuous function g. Moreover, for any order $k \in \mathbb{N}$

$$(\varphi_n * \omega)^{(k)} = \varphi_n^{(k)} * \omega = \varphi_n * \omega^{(k)}, \quad \text{for} \quad k \in \mathbb{N},$$

and, by Lemma 3.3.42, section 3.3.5, the sequence $(\varphi_n * \omega)_{n\in\mathbb{N}}^{(k)}$ converges almost uniformly. By a classical theorem this sequence of convolutions converges to $g^{(k)}$. Thus, g is a smooth function. On the other hand,

$$f * \omega = [\varphi_n * \omega] = g,$$

by the definition of convolution and the identification of continuous function with distribution. Therefore, $f * \omega \in \mathcal{C}^\infty$.   □

Corollary 3.4.38 *For every smooth function the identity*

$$\boxed{\varphi = \varphi * \delta, \quad \varphi \in \mathcal{C}^{\infty}} \tag{3.4.7}$$

is true.

Proof. In fact, replacing f by φ in (3.4.6), we get

$$\varphi = [\varphi * \delta_n] - \varphi * [\delta_n] = \varphi * \delta.$$

$\Box$

Corollary 3.4.39 *The distribution δ considered in $\mathbb{R}$ is not equal to the zero distribution, but $\delta(x) = 0$ for $x \neq 0$.*

Proof. Since $0 = \varphi * 0$ for any smooth function φ, it follows by (3.4.7) that distribution δ is not equal to the zero distribution when considered in the whole space. Since every delta sequence $(\delta_n)_{n \in \mathbb{N}}$ converges almost uniformly to 0 for $x \neq 0$, thus $\delta(x) = 0$ for $x \neq 0$. $\Box$

Theorem 3.4.114 *Every distribution in (a, b) is, in every interval I inside (a, b), a derivative of some order of a continuous function.*

Proof. Let $f = [\varphi_n]$. By properties (F_1) and (F_2) there exist an order $k \in \mathbb{N}$, smooth functions Φ_n and continuous function F such that

$$\Phi_n^{(k)} = \varphi_n \quad \text{and} \quad \Phi_n \rightrightarrows F \quad x \in I.$$

Hence, $F(x) = [\Phi_n(x)]$ in I and

$$f(x) = [\Phi_n^{(k)}(x)] = [\Phi_n(x)]^{(k)} = F^{(k)}(x) \quad \text{in} \quad I.$$

$\Box$

Theorem 3.4.115 *If $(\delta_n)_{n \in \mathbb{N}}$ is a delta sequence and f is any distribution, then*

$$f = [f * \delta_n]. \tag{3.4.8}$$

Proof. In fact, by Theorem 3.4.114, for every interval I inside (a, b), when f is defined, there exist an order k and a continuous function F such that $F^{(k)} = f$ in I. By (3.4.6)

$$F = [F * \delta_n] \quad \text{in} \quad I.$$

Hence, differentiating k times, we obtain

$$f(x) = F^{(k)}(x) = [(F * \delta_n)(x)]^{(k)} = [(F^{(k)} * \delta_n)(x)] = [(f * \delta_n)(x)], \quad \text{for} \quad x \in I.$$

Hence, by Lemma 3.4.44, formula (3.4.8) holds true in (a, b). $\Box$

3.4.3   Distributions as a Generalization of Locally Integrable Functions

In the previous section we have shown that distributions are a generalization of continuous functions. We will prove that distributions embrace a wider class of functions, namely locally integrable functions.

We recall:

Definition 3.4.75 *We say that a function f defined in (a, b) is locally integrable in (a, b), $f \in L_1^{loc}$, if the integral*

$$\int_I f(x)dx$$

exists for every open interval I inside (a, b).

Note that if f is a continuous function in an interval I, then in I

$$\left(\int_{x_o}^x f(t)dt \right)' = f(x), \quad x_o \in I. \tag{3.4.9}$$

If the function f is not continuous but it is locally integrable, then

$$\int_{x_o}^x f(t)dt$$

is a continuous function. In this case equality (3.4.9) holds almost everywhere, where the derivative on the left-hand side is defined in the usual way as the limit of the expression

$$f_h(x) = \frac{1}{h} \int_x^{x+h} f(t)dt,$$

as $h \to 0, h > 0$. The left-hand side of (3.4.9) can be interpreted as a distribution that is a distributional derivative of order 1 of the continuous function

$$\int_{x_o}^x f(t)dt. \tag{3.4.10}$$

One can easily see that this distribution does not depend on the choice of x_o in I. This remark suggests the following identification:

Definition 3.4.76 *We say that a distribution g is equal to a locally integrable function in the interval (a, b) if for every bounded, open interval I included in (a, b) this distribution is a distributional derivative of the function (3.4.10), i.e.,*

$$g(x) = \left(\int_{x_o}^x f(t)dt \right)', \quad x_o \in I, \quad x \in I.$$

Such a derivative, if it exists, is uniquely defined by a locally integrable function f. We will prove that the distribution always exists. Namely, the following theorem holds true:

Theorem 3.4.116 *Every locally integrable function f can be identified with the distribution $[f * \delta_n]$, where $(\delta_n)_{n \in \mathbb{N}}$ is a delta sequence.*

Proof. Let I be any given bounded interval inside (a, b) and let

$$F(x) = \int\limits_{x_o}^{x} f(t)dt, \quad x_o \in I.$$

By Property 3.4.8 of delta sequences, we have

$$F * \delta_n \overset{\text{a.u.c.}}{\rightrightarrows} F \quad in \quad I.$$

Hence, by the identification at continuous functions with distributions, we obtain

$$[(F * \delta_n)(x)] = F(x) \quad \text{for} \quad x \in I,$$

and hence,

$$[F' * \delta_n] = F',$$

i.e.,

$$[f * \delta_n] = f \quad in \quad I.$$

In that way we have proved that every locally integrable function f can be identifed with the distribution $[f * \delta_n]$. $\quad\Box$

If f is a continuous function then by Property 3.4.8 the sequence is convergent almost uniformly to f. Therefore, the identification of integrable functions with distributions is consistent with the identification in section 3.4.2.

In applications we often come across the so-called *Heaviside function:*

$$1_+(x) = \begin{cases} 0 & \text{for } x < 0, \\ 1 & \text{for } x \geq 0. \end{cases}$$

Its integral

$$G(x) = \int\limits_{0}^{x} 1_+(x)dt = \begin{cases} 0 & \text{for } x < 0, \\ x & \text{for } x \geq 0 \end{cases}$$

is continuous. The Heaviside function $1_+(x)$ is the distributional derivative of $G(x)$; it is also its ordinary derivative except at the point $x = 0$. Since G is the limit of the integrals

$$\int\limits_{-\infty}^{x} g_n(x)dt, \quad n \in \mathbb{N},$$

where

$$g_n(x) = \frac{1}{1 + e^{-nx}}$$

are the functions from example 1° of section 3.3.2, we obtain

$$[g_n] = G' = 1_+,$$

i.e., the fundamental sequence $(g_n(x))_{n \in \mathbb{N}}$ represents the Heaviside function.

Similarly, the sequence

$$G_n(x) = \int\limits_{-\infty}^{x} f_n(t)dt, \quad n \in \mathbb{N},$$

from Example 2, section 3.3.2, where $(f_n)_{n \in \mathbb{N}}$ is a sequence of Picard functions, represents the Heaviside function. Hence, for the Picard functions f_n from Example 2, section 3.3.2, we have $\delta = [f_n] = [G'] = (1_+)'$, thus

$$\boxed{\delta = (1_+)'} \tag{3.4.11}$$

i.e., the Dirac distribution $\delta(x)$ is the distributional derivative of the Heaviside function.

The Dirac distribution $\delta(x)$ in the interval $-\infty < x < \infty$ is an example of a distribution that is not a locally integrable function. Indeed, if the Dirac delta distribution was a locally integrable function then from the identity it would follow that 1_+ is continuous, which is not true.

We note that it can happen that both derivatives ordinary and distributional of a locally integrable function exist but are different. For example, the ordinary derivative of the Heaviside function is equal to 0 everywhere except for the point x where it does not exist. The distributional derivative of this function is a Dirac delta distribution.

Remark 3.4.110 *In the theory of distributions the ordinary derivative plays a marginal role. Therefore, if not otherwise stated, a derivative of a function in the sequel will be understood as a distributional derivative.*

3.4.4   Remarks about Distributional Derivatives

In section 3.3.5 we introduced the notion of the derivative of order m of a distribution. Let us recall that definition.

The derivative of order m of a distribution f given as $f = [\varphi_n]$ is a distribution $[\varphi^{(m)}]$ denoted by $f^{(m)}$ or $[\varphi_n]^{(m)}$, i.e.,

$$f^{(m)} = [\varphi_n^{(m)}].$$

It is easy to see that the following theorem is true:

Theorem 3.4.117 *If a distribution is a function with a continuous mth derivative, then its mth derivative in the distributional sense coincides with its derivative in the ordinary sense.*

The notion of the derivative of a distribution is a generalization of the notion of derivative in the domain of continuous differentiable functions.

Theorem 3.4.118 *The equality $f^{(m)}(x) = 0$ holds iff the distribution f is a polynomial of degree less than m.*

Proof. The sufficiency is obvious. To prove the necessity, suppose that

$$f^{(m)} = 0. \tag{3.4.12}$$

Let $f = [\varphi_n]$, where $(\varphi_n)_{n\in\mathbb{N}}$ is a fundamental sequence. By the definition of distributional derivative, we have

$$f^{(m)} = [\varphi_n^{(m)}]. \tag{3.4.13}$$

By (3.4.12) and (3.4.13) it follows that $(\varphi_n^{(m)}(x))_{n\in\mathbb{N}} \sim (0)_{n\in\mathbb{N}}$. There exists, an integer $k \leq m$ and sequences $(F_n)_{n\in\mathbb{N}}$, $(G_n)_{n\in\mathbb{N}}$ such that

$$F_n^{(k)} = \varphi_n^{(m)}, \quad F_n \rightrightarrows P: \tag{3.4.14}$$

$$G_n^{(k)} = 0, \quad G_n \rightrightarrows P. \tag{3.4.15}$$

By (3.4.15) it follows that the functions G_n are polynomials of degree less than k. Hence, by Lemma 3.3.39, section 3.3.2, it follows that P is a polynomial of degree less than k, too.

By (3.4.14) it follows that $(\varphi_n - F_n^{(k-m)})_{n\in\mathbb{N}}$ is a fundamental sequence of polynomial of degree less than m. By Theorem 3.3.102, section 3.3.2, this sequence converges almost uniformly to some polynomial p. Thus,

$$[\varphi_n - F_n^{(k-m)}] = p$$

and consequently

$$f = [\varphi_n] = [F_n^{(k-m)}] + p. \tag{3.4.16}$$

Since $P = [F_n]$, where P is a polynomial of degree $< k$, the distribution

$$[F_n^{(k-m)}] = P^{(k-m)}$$

is a polynomial of degree less than m. Hence, by (3.4.16), f is a polynomial of degree less than m. □

From Theorem 3.4.117 it follows

Corollary 3.4.40 *The equality $f' = 0$ holds if the distribution f is a constant function.*

Replacing f by $f - g$, in Corollary 3.4.40, we get:

Corollary 3.4.41 *The equality $f' = g'$ holds if the distributions f and g differ from each other by a constant function.*

Theorem 3.4.119 *If a derivative $f^{(m)}$ of a distribution f is a continuous function, then f is a continuous function and $f^{(m)}$ is its ordinary derivative.*

Proof. Let

$$g(x) = \int_{x_o}^{x} dt_1 \int_{x_o}^{t_1} dt_2 \cdots \int_{x_o}^{t_{m-1}} f(t_m) dt_m.$$

By Theorem 3.4.118 it follows that $p = f - g$ is a polynomial of degree less than m. Thus, f is the function of the form $g + p$ and by Theorem 3.4.117, $f^{(m)}$ is its ordinary derivative.
□

The fact that we can include locally integrable functions to the set of distributions allows us to state Theorem 3.4.119 in a stronger form:

Theorem 3.4.120 *If a derivative $f^{(m)}$, $m \in \mathbb{N}$, of a distribution is a locally integrable function, then f is a continuous function and $f^{(m)}$ is its ordinary mth derivative.*

In particular, from Theorem 3.3.108, section 3.3.5, it follows that every continuous function has a distributional derivative. This derivative in general is not a continuous function but a distribution. For example, a nowhere differentiable Weierstrass function is differentiable in the distributional sense but its derivative is not a function.

Theorem 3.4.121 *If $f(x) = 0$ for $x \neq x_o$, then the distribution f is of the form*

$$f(x) = \alpha_o \delta(x - x_o) + \alpha_1 \delta'(x - x_o) + \cdots + \alpha_k \delta^{(k)}(x - x_o). \tag{3.4.17}$$

Proof. By Theorem 3.4.114, section 3.4.2, it follows that there exist an order k and a continuous function f such that $F^{(k)} = f$. From Theorem 3.4.118, by assertion it follows that F is a polynomial of degree $< k$ in each of the interval $-\infty < x < x_o$ and $x_0 < x < \infty$, i.e.,

$$F(x) = \begin{cases} F(x_o) + \alpha_1(x - x_o) + \cdots + \alpha_{k-1}(x - x_o)^{k-1}, & \text{for } -\infty < x < x_o, \\ F(x_o) + \beta_1(x - x_o) + \cdots + \beta_{k-1}(x - x_o)^{k-1}, & \text{for } x_o < x < \infty. \end{cases}$$

The function F can be written in the form

$$F(x) = F(x_o) + \phi_1(x) + \cdots + \phi_{k-1}(x),$$

where

$$\phi_\mu = \begin{cases} \alpha_\mu (x - x_o)^\mu, & \text{for} \quad -\infty < x < x_o, \\ \beta_\mu (x - x_o)^\mu, & \text{for} \quad x_o < x < \infty. \end{cases}$$

It is easy to check that

$$\phi_\mu^{(\mu)}(x) = \mu!(\alpha_\mu + (\beta_\mu - \alpha_\mu) 1_+(x - x_o))$$

$$\phi_\mu^{(k)}(x) = \mu!(\beta_\mu - \alpha_\mu)\delta^{(k-\mu-1)}(x - x_o), \quad \mu = 1, \ldots, (k-1),$$

which proves the theorem.

If $f(x) = 0$ for $x \neq 0$, then the representation of the distribution f in the form (3.4.17) is unique. This follows from:

Theorem 3.4.122 *If g is a function and*

$$g(x) + \alpha_o \delta(x - x_o) + \cdots + \alpha_k \delta^{(k)}(x - x_o) = 0$$

on the whole real line, then $g(x) = 0$ and $\alpha_o = \cdots = \alpha_k = 0$.

The proof follows by induction.

3.4.5 Functions with Poles

Let us consider a function $f(x) = \frac{1}{x}$. This function is not locally integrable because it is not integrable in any neighborhood of zero. Excluding an arbitrary neighborhood of zero we obtain an integrable function. This means that the function is a distribution in intervals $(-\infty, 0) \cup (0, \infty)$. It cannot be identified with a distribution on the real line (since it is not integrable in any neighborhood of $x = 0$).

However, there exists a distribution f defined in $\mathbb{R}$ such that

$$f(x) = \frac{1}{x} \quad \text{for} \quad x \neq 0. \tag{3.4.18}$$

For instance,

$$(\ln |x|)' = \frac{1}{x} \quad \text{for} \quad x \neq 0, \tag{3.4.19}$$

where the derivative is understood in the distributional sense. If we add an arbitrary linear combination of $\delta(x)$ and its derivatives to the left-hand side of (3.4.19), then the equality (3.4.19) will be true.

The equality

$$(\ln |x|)' = \frac{1}{x} \quad \text{for} \quad x \neq 0$$

can be considered the identification of the function $\frac{1}{x}$ with the distribution $(\ln|x|)'$.

Such an identification can be extended to a wider class of functions which have poles at some points and are locally integrable elsewhere. We shall consider functions f in some interval (a, b) which, in a neighborhood of any point x_o are of the form

$$f(x) = f_o(x) + \sum_{\nu=1}^{k} \frac{c_\nu}{(x - x_o)^\nu}, \tag{3.4.20}$$

where f_o is an integrable function. The decomposition into the integrable function f_o and the remaining singular part is unique. The singular part can vanish.

The point x_o, for which at least one of the coefficients c_ν differs from zero, is called a *pole* of f.

In every finite closed subinterval with end point α and β included in (a, b), there is at most a finite number of poles. The function f can be written in the form

$$f(x) = f_1(x) + \sum_{\mu=1}^{m} \sum_{\nu=1}^{k} \frac{c_{\mu\nu}}{(x - x_\mu)^\mu}, \tag{3.4.21}$$

where f_1 is an integrable function and $x_1, ..., x_m$ are points of the interval (α, β).

If α, β are not poles, we define the integral from α to β by the formula:

$$\int_\alpha^\beta f(t)dt = \int_\alpha^\beta f_1(t)dt + \sum_{\mu=1}^{m} c_\mu \ln|x - x_\mu| \Big|_\alpha^\beta + \sum_{\mu=1}^{m} \sum_{\nu=2}^{k} \frac{-c_{\mu\nu}}{(\nu-1)(x-x_\mu)^{\nu-1}} \Big|_\alpha^\beta,$$

which is obtained from (3.4.21) by formal integration.

By these methods we have included in the space of distributions all rational functions, all rational expressions of sine and cosine. In particular, we have the formulas

$$\frac{1}{(x - x_o)^k} = \left(\frac{(-1)^{k-1}}{(k-1)!} \ln|x - x_o| \right)^{(k)},$$

$$\operatorname{tg}(x) = (-\ln|\cos(x)|)', \quad \operatorname{ctg}(x) = (\ln|\sin(x)|)',$$

where the derivatives are understood in the distributional sense.

Similarly, we have included in the calculus of distributions, for example, the elliptic functions and the Euler function, etc.

3.4.6   Applications

Problem 3.4.1 Prove that for every smooth function ω we have

$$\boxed{\omega(x)\delta'(x) = \omega(0)\delta'(x) - \omega'(0)\delta(x), \quad \omega \in C^\infty.} \tag{3.4.22}$$

Proof. By (3.3.18), section 3.3.5, we have

$$\omega(x)\delta'(x) = \Big(\omega(x)\delta(x)\Big)' - \omega'(x)\delta(x).$$

Hence, by (3.4.2), section 3.4.1, we get

$$\omega(x)\delta'(x) = \omega(0)\delta'(x) - \omega'(0)\delta(x).$$

☐

By means of induction one solves:

Problem 3.4.2 Prove that for every smooth function ω and $n \in \mathbb{N}$ the formula

$$\omega(x)\delta^{(n)}(x) = \sum_{k=0}^{n}(-1)^k \binom{n}{k}\omega^{(k)}(0)\delta^{(n-k)}(x), \quad \omega \in \mathcal{C}^\infty. \tag{3.4.23}$$

is true.

Problem 3.4.3 Prove that

$$x\delta'(x) = -\delta(x), \tag{3.4.24}$$

$$x^k\delta'(x) = 0 \quad \text{for} \quad k = 2, 3, ... \tag{3.4.25}$$

Proof. By equality (3.4.22),

(i) for $\omega(x) = x$, we have formula (3.4.24)

(ii) for $\omega(x) = x^k$, we get formula (3.4.25).

☐

Problem 3.4.4 Find all derivatives of the function

$$f(x) = |x|.$$

Proof. Let ϕ be any function of class $\mathcal{D}$ and let 1_+ be Heaviside's function. We shall show that

$$\int_{-\infty}^{\infty} |x|'\phi(x)dx = \int_{-\infty}^{\infty} (2 \cdot 1_+(x) - 1)\phi(x)dx, \quad \phi \in \mathcal{D}. \tag{3.4.26}$$

Indeed, integrating by part, and making simple calculations, we obtain

$$\int_{-\infty}^{\infty} |x|'\phi(x)dx = -\int_{-\infty}^{\infty} |x|\phi'(x)dx = \int_{-\infty}^{0} x\phi'(x)dx - \int_{0}^{\infty} x\phi'(x)dx$$

$$= x\phi(x)\Big|_{-\infty}^{0} - \int_{-\infty}^{0} \phi(x)dx - x\phi(x)\Big|_{0}^{\infty} + \int_{0}^{\infty} \phi(x)dx$$

$$= -\int_{-\infty}^{0} \phi(x)dx + \int_{0}^{\infty} \phi(x)dx = 2\int_{0}^{\infty} \phi(x)dx - \int_{-\infty}^{+\infty} \phi(x)dx$$

$$= 2\int_{-\infty}^{+\infty} 1_+(x)\phi(x)dx - \int_{-\infty}^{+\infty} \phi(x)dx = \int_{-\infty}^{+\infty} (2 \cdot 1_+(x) - 1)\phi(x)dx.$$

Thus,

$$\int\limits_{-\infty}^{+\infty} |x|'\phi(x)dx = \int\limits_{-\infty}^{+\infty} (2 \cdot 1_+(x) - 1)\phi(x)dx, \quad \phi \in \mathcal{D}.$$

Hence, we obtain the distributional derivatives of function $|x|$:

$$|x|' - 2 \cdot 1_+(x) - 1,$$

i.e.,

$$\boxed{|x|' = sgn x.} \tag{3.4.27}$$

□

Since $(1_+)'(x) = \delta(x)$, thus

$$|x|^{(2)} = 2\delta(x)$$

and generally

$$\boxed{|x|^{(k)} = 2\delta^{(k-2)}(x) \quad \text{for} \quad k = 2, 3, \cdots.} \tag{3.4.28}$$

We have shown that the function $f(x) = |x|$ is, in a distributional sense, infinitely many times differentiable at any point x. It is obvious that this function has no derivative at $x = 0$ in the usual sense.

Problem 3.4.5 Prove that

$$\boxed{x^n \delta^{(n)}(x) = (-1)^n \, n! \, \delta(x), \quad n \in \mathbb{N};} \tag{3.4.29}$$

$$\boxed{x^n \delta^{(n+1)}(x) = (-1)^n \, (n+1)! \, \delta'(x), \quad n \in \mathbb{N}.} \tag{3.4.30}$$

Proof. We will prove the above formulas by induction.

1. For $n = 1$ the formula (3.4.29) takes the form

$$x \, \delta^{(1)}(x) = (-1)\delta(x). \tag{3.4.31}$$

which is the formula (3.4.22), in Problem 3.4.1, for $\omega(x) = x$.

Let us assume that the formula (3.4.29) holds true for a certain $k \in \mathbb{N}$

$$x^k \delta^{(k)}(x) = (-1)^k \, k! \, \delta(x). \tag{3.4.32}$$

We will prove that the formula holds true for $k + 1$, i.e.,

$$x^{k+1}\delta^{(k+1)}(x) = (-1)^{k+1} \, (k+1)! \, \delta(x).$$

Indeed, multiplying equality (3.4.32) by $k + 1$ we get

$$x^k \, (k+1) \, \delta^{(k)}(x) = (-1)^k \, k! \, (k+1)\delta(x). \tag{3.4.33}$$

By the formula (3.4.23), in Problem 3.4.2, for $n = k+1$, $\omega(x) = x$, it follows

$$x\,\delta^{(k+1)}(x) = -(k+1)\,\delta^{(k)}(x).$$

Hence, and from (3.4.32), we get

$$x^{k+1}\delta^{(k+1)}(x) = (-1)^{k+1}(k+1)!\,\delta(x),$$

which was to be shown.

2^0 For $n = 1$ the formula (3.4.30) takes the form

$$x\delta^{(2)}(x) = (-1)2!\delta'(x).$$

Moreover, for $n = 2$, $\omega(x) = x$, we get the formula (3.4.23), in Problem 3.4.2.

Let us assume that the formula (3.4.30) holds true for a certain $k > 1, k \in \mathbb{N}$, i.e.,

$$x^{k-1}\delta^{(k)}(x) = (-1)^{k-1}k!\delta'(x). \tag{3.4.34}$$

We will prove that the formula holds true for $k+1$, i.e.,

$$x^k\delta^{(k+1)}(x) = (-1)^k(k+1)!\delta'(x).$$

Indeed, multiplying equality (3.4.34) by $k+1$, we obtain

$$x^{k-1}(k+1)\delta^{(k)}(x) = (-1)^{k-1}k!(k+1)\delta'(x). \tag{3.4.35}$$

By formula (3.4.23), Problem 3.4.2, for $n = k+1$, $\omega(x) = x$, we get

$$(k+1)\delta^{(k)}(x) = -x\,\delta^{(k+1)}(x);$$

thus,

$$x^k\delta^{(k+1)}(x) = (-1)^k(k+1)!\,\delta'(x),$$

what was to be shown.

Problem 3.4.6 Prove that the following formula is valid:

$$x^n\delta^{(n+k)}(x) = (-1)^n\frac{(n+k)!}{k!}\delta^{(k)}(x), \quad k,n \in \mathbb{N}_o. \tag{3.4.36}$$

Applying mathematical induction we obtain formula (3.4.36).

Problem 3.4.7 Prove that for any smooth function ω the following formula is valid:

$$\omega(x)\delta^{(m)}(x-x_o) = \sum_{i=0}^{m}(-1)^i\binom{m}{i}\omega^{(i)}(x_o)\delta^{(m-i)}(x-x_o), \quad \omega \in C^\infty, \tag{3.4.37}$$

$m \in \mathbb{N}, x, x_o \in \mathbb{R}$.

Remark 3.4.111 *From the formula (3.4.37) it is easy to obtain other formulas, e.g., (3.4.29), (3.4.30) and (3.4.36). In the case when $w(x) = \sin x$, we have*

$$\sin x \cdot \delta^{(m)}(x) = -\binom{m}{1}\delta^{(m-1)} + \binom{m}{3}\delta^{(m-3)}(x) - \cdots +$$

$$+ (-1)^p\binom{m}{2p-1}\delta^{(m-2p+1)}(x) + \binom{m}{m}\sin\frac{m\pi}{2}\delta(x).$$

Problem 3.4.8 Prove that for the δ distribution the following formulas are valid:

$$\delta(\alpha x + \beta) = \frac{1}{|\alpha|}\delta\left(x + \frac{\alpha}{\beta}\right), \quad \alpha \neq 0, \tag{3.4.38}$$

$$\delta^{(m)}(\alpha x + \beta) = \frac{1}{|\alpha|\alpha^m}\delta^{(m)}\left(x + \frac{\alpha}{\beta}\right), \quad \alpha \neq 0. \tag{3.4.39}$$

Proof. By (3.4.11), section 3.4.3, and (3.3.23), section 3.3.5, it follows

$$\delta(\varphi(x)) = 1'_+(\varphi(x)) = \frac{\left(1_+(\varphi(x))\right)'}{\varphi'(x)}, \quad \varphi \in C^\infty, \varphi' \neq 0.$$

If $\varphi > 0$, then the function $1_+(\varphi(x))$ is equal to 1 everywhere,

if $\varphi < 0$, then function $1_+(\varphi(x))$ is equal to 0 everywhere.

Hence,

$$\delta(\varphi) = \begin{cases} \frac{(1)'}{\varphi'} = 0, & \text{for } \varphi' > 0, \\ \frac{(0)'}{\varphi'} = 0, & \text{for } \varphi' < 0, \end{cases}$$

If there exists only one point x_o such that $\varphi(x_o) = 0$ (since $\varphi' \neq 0$ everywhere), then

$$1_+(\varphi(x)) = \begin{cases} 1_+(x - x_o), & \text{for } \varphi' > 0, \\ -1_+(x - x_o), & \text{for } \varphi' < 0. \end{cases}$$

Hence,

$$\delta(\varphi) = \begin{cases} \frac{1'_+(x-x_o)}{\varphi'(x_o)} = \frac{\delta(x-x_o)}{\varphi'(x_o)}, & \text{for } \varphi(x_o) = 0, \varphi' > 0, \\ -\frac{1'_+(x-x_o)}{\varphi'(x_o)} = -\frac{\delta(x-x_o)}{\varphi'(x_o)}, & \text{for } \varphi(x_o) = 0, \varphi' < 0 \end{cases}$$

Finally, if $\varphi \in C^\infty$, $\varphi' \neq 0$, then

$$\delta(\varphi(x)) = \begin{cases} 0, & \text{for } \varphi \neq 0, \\ \frac{1}{|\varphi'(x_o)|}\delta(x - x_o), & \varphi(x_o) = 0. \end{cases}$$

☐

In the case if $\varphi(x) = \alpha x + \beta$, $\alpha \neq 0$, we obtain formula (3.4.38). Differentiating (3.4.38) further we get formula (3.4.39).

Problem 3.4.9 Calculate f' and $f^{(2)}$ of the following function:

$$f(x) = x\,1_+(x) - (x - 1)\,1_+(x - 1) + 1_+(x - 1).$$

Solution. Applying formula (3.3.18), section 3.3.5, we obtain

$$f'(x) = 1_+(x) + x\delta(x) - 1_+(x-1) - (x-1)\delta(x-1) + \delta(x-1).$$

Since $w(x)\delta(x) = w(0)\delta(x)$ (see formula 3.4.2, section 3.4.1) thus,

$$f'(x) = 1_+(x) - 1_+(x-1) + \delta(x-1),$$

and next

$$f^{(2)}(x) = \delta(x) - \delta(x-1) + \delta'(x-1).$$

Problem 3.4.10 Calculate the second derivative of the function

$$g(x) = 1_+(x)\exp(x),$$

where 1_+ is a Heaviside function.

Solution. By formulas (3.3.18), section 3.3.5, (3.4.2), section 3.4.1, and by the du Bois-Reymond Lemma (see [Vl.1], p. 26), in turn we get

$$g'(x) = \delta(x)\exp(x) + 1_+(x)\exp(x),$$
$$g'(x) = \delta(x)\exp(0) + 1_+(x)\exp(x).$$

Thus,

$$g'(x) = \delta(x) + 1_+(x)\exp(x).$$

Analogously, we have

$$g^{(2)}(x) = \delta'(x) + \delta(x)\exp(x) + 1_+(x)\exp(x).$$

Since

$$\delta(x)\exp(x) = \delta(x)\exp(0).$$

thus,

$$g^{(2)}(x) = \delta'(x) + \delta(x) + 1_+(x)\exp(x).$$

Problem 3.4.11 Calculate derivatives of the following distributions:

1. $f(x) = \delta\exp(-x)$

2. $g(x) = \delta(x)\sin(x).$

Answer. 1. $f' = \delta'$; 2. $g' = 0$.

Problem 3.4.12 Prove that

$$\left(\frac{d}{dx} + \lambda\right)1_+(x)\exp(-\lambda x) = \delta(x).$$

Proof. By formulas (3.3.18), section 3.3.5 and (3.4.2), section 3.4.1, we obtain in turn

$$\left(\frac{d}{dx} + \lambda\right)1_+(x)\exp(-\lambda x)$$
$$= (1_+(x)\exp(-\lambda x))' + \lambda 1_+(x)\exp(-\lambda x)$$
$$= \delta(x)\exp(-\lambda x) - \lambda 1_+(x)\exp(-\lambda x) + \lambda 1_+(x)\exp(-\lambda x)$$
$$= \delta(x)\exp(0) = \delta(x).$$

$\Box$

3.5 Convergent Sequences

3.5.1 Sequences of Distributions

Definition 3.5.77 *We say that a sequence of distributions $\left(f_n(x)\right)_{n\in\mathbb{N}}$ converges in (a,b) to a distribution f if the distribution f is defined in (a,b) and for every open bounded interval I inside (a,b), there exist an order $k \in \mathbb{N}_o$, a sequence of continuous functions $(F_n)_{n\in\mathbb{N}}$ and a continuous function F such that in I*

$$F_n^{(k)}(x) = f_n(x) \quad for \quad n > n_o$$
$$F^{(k)}(x) = f(x) \quad and \quad F_n \rightrightarrows F \quad in \quad I.$$
(3.5.1)

The notation

$$f_n \xrightarrow{\text{d}} f \quad in \quad (a,b) \quad or \quad \lim_{n\to\infty} f_n(x) \stackrel{\text{d}}{=} f(x) \quad in \quad (a,b)$$

will be used to denote the convergence of a sequence of distributions $(f_n)_{n\in\mathbb{N}}$ in (a,b) to a distribution f.

Definition 3.5.78 *We say that a sequence of distributions $\left(f_n(x)\right)_{n\in\mathbb{N}}$ converges in (a,b) if for every open bounded interval I inside (a,b), there exist an order $k \in \mathbb{N}_o$, a sequence of continuous functions $(F_n)_{n\in\mathbb{N}}$ such that in I*

$$F_n^{(k)}(x) = f_n(x) \quad for \quad n > n_o \quad and \quad F_n \rightrightarrows \quad in \quad I.$$

The notation $f_n \xrightarrow{\text{d}}$ in (a,b) will be used to denote the convergence of a sequence of distributions $(f_n)_{n\in\mathbb{N}}$ in (a,b).

Lemma 3.5.46 *The order $k \in \mathbb{N}_o$, which occurs in (3.5.1), can be replaced, if necessary, by any order $l \geq k$.*

Proof. It suffices to observe that if conditions (3.5.1) hold, then also

$$\tilde{F}_n^{(m)} = f_n, \quad \tilde{F}^{(m)} = f, \quad \tilde{F}_n(x) \rightrightarrows \tilde{F}(x), \quad x \in I,$$

and

$$\tilde{F}_n(x) = \int_{x_o}^x F_n(t)dt^{m-k}, \quad \tilde{F}(x) = \int_{x_o}^x F(t)dt^{m-k}, \quad x_o \in I,$$

where

$$\int_{x_o}^x F(t)dt^l = \int_{x_o}^x dt_1 \int_{x_o}^{t_1} dt_2 \cdots \int_{x_o}^{t_{l-1}} F(t_{l-1})dt_l.$$

☐

Lemma 3.5.47 *If a sequence of continuous functions $(f_n)_{n\in\mathbb{N}}$ converges almost uniformly to f in (a,b) and if $f_n^{(k)}(x) = 0$ in (a,b) for $n \in \mathbb{N}$, then $f^{(k)}(x) = 0$ in (a,b).*

Proof. Let $(\delta_n)_{n\in\mathbb{N}}$ be a delta sequence. By Property 3.4.9, section 3.4.1, sequence

$$(\varphi_n)_{n\in\mathbb{N}} = (f_n * \delta_n)_{n\in\mathbb{N}}$$

converges almost uniformly to f in (a,b), such that

$$f = [f_n * \delta_n].$$

Hence,

$$f^{(k)} = [f_n * \delta_n]^{(k)} = [f_n^{(k)} * \delta_n] = [0 * \delta_n] = 0.$$

☐

Theorem 3.5.123 *The limit of a sequence of distributions, if it exists, is unique.*

Proof. Let I be an arbitrary open interval inside (a,b). If $(f_n)_{n\in\mathbb{N}}$ is such that

$$f_n \xrightarrow{d} f \quad \text{in} \quad (a,b) \quad \text{and} \quad f_n \xrightarrow{d} g \quad \text{in} \quad (a,b)$$

then there exist orders $k, m \in \mathbb{N}_o$, sequences of continuous functions $(F_n)_{n\in\mathbb{N}}$, $(G_n)_{n\in\mathbb{N}}$ and continuous functions f, G such that

$$F_n(x) \rightrightarrows F(x), \quad G_n(x) \rightrightarrows G(x), \quad x \in I$$

where

$$F_n^{(k)}(x) = f_n(x), \quad G_n^{(m)}(x) = f_n(x), \quad x \in I \quad \text{and}$$
$$F^{(k)}(x) = f(x), \quad G^{(m)}(x) = g(x), \quad x \in I.$$

We may assume that $k = m$, for otherwise we could replace both orders by a greater order. Since

$$\left(F_n(x) - G_n(x)\right)^{(k)} = 0 \quad \text{and} \quad F_n(x) - G_n(x) \rightrightarrows F(x) - G(x), \quad x \in I,$$

according to Lemma 3.5.47, we have

$$\left(F(x) - G(x)\right)^{(k)} = 0,$$

which implies that $f(x) = g(x)$ in I. Since I is arbitrary, it follows that the limit is unique.
$\square$

Directly from the definition of the limit the following theorems follow:

Theorem 3.5.124 *If a sequence of continuous functions converges almost uniformly, then it also converges distributionally to the same limit.*

Theorem 3.5.125 *If a sequence $(f_n)_{n \in \mathbb{N}}$ of locally integrable functions converges almost everywhere to a function f and it is bounded by a locally integrable function, then it also converges distributionally.*

In the proof of the theorem the following property should be used:
    If a sequence $(f_n)_{n \in \mathbb{N}}$ converges almost everywhere to f and it is bounded, then

$$\int_a^x f_n(t)dt \overset{\text{a.u.c.}}{\rightrightarrows} \int_a^x f(t)dt.$$

Theorem 3.5.126 *If a sequence converges distributionally, then its every subsequence converges distributionally to the same limit, i.e., if $f_n \overset{d}{\longrightarrow} f$, then $f_{r_n} \overset{d}{\longrightarrow} f$, for every sequence $(r_n)_{n \in \mathbb{N}}$ of positive integers such that $r_n \to \infty$.*

Theorem 3.5.127 *If $f_n \overset{d}{\longrightarrow} f$ and $g_n \overset{d}{\longrightarrow} f$, then the interlaced sequence*

$$f_1, g_1, f_2, g_2, \cdots,$$

also converges to f.

It follows from the definition of distributional convergence and Lemma 3.5.46 that the arithmetic operations on limits of the sequences of distributions can be done in the same way as for limits of functions:

Theorem 3.5.128 *If $(f_n)_{n \in \mathbb{N}}$, $(g_n)_{n \in \mathbb{N}}$ are sequences of distributions such that*

$$f_n \overset{d}{\longrightarrow} f, \quad g_n \overset{d}{\longrightarrow} g, \quad \text{then} \quad f_n + g_n \overset{d}{\longrightarrow} f + g; \quad f_n - g_n \overset{d}{\longrightarrow} f - g;$$

$$\lambda_n f_n \overset{d}{\longrightarrow} \lambda f \quad \text{if} \quad \lambda_n \to \lambda \quad (\lambda_n, \lambda \in \mathbb{R}).$$

Theorem 3.5.129 *If $f_n \xrightarrow{d} f$, then $f_n^{(m)} \xrightarrow{d} f^{(m)}$ for every order $m \in \mathbb{N}_o$.*

Proof. It suffices to observe that, if condition (3.5.1) is true, then also the condition

$$F_n^{(m+k)} = f_n^{(m)}, \quad F^{(m+k)} = f^{(m)}, \quad F_n \rightrightarrows F$$

is fulfilled. □

Remark 3.5.112 Theorem 3.5.129 *is very important in distributional calculus. It allows us to differentiate convergent sequences without any restrictions.*

Theorem 3.5.130 *If a sequence of distributions is convergent in (a,b), then it converges to a distribution in (a,b).*

Proof. Suppose that $(f_n)_{n\in\mathbb{N}}$ is convergent in (a,b). Let $(\delta_n)_{n\in\mathbb{N}}$ be any δ sequence and $\delta_n(x) = 0$ for $|x| > \alpha_n > 0, n \in \mathbb{N}$. We shall prove that the sequence $(\varphi_n)_{n\in\mathbb{N}}$, where $\varphi_n = f_n * \delta_n$, is fundamental in (a,b) and that the sequence $(\varphi_n)_{n\in\mathbb{N}}$ converges to $[\varphi_n]$.

In fact, let I be an arbitrary bounded open interval inside (a,b), and let I' be an bounded open interval inside (a,b) too, such that $I \subset I'_{-\alpha_n}$, where $I'_{-\alpha_n}$ denotes the set of all points $x \in I'$ whose distance from the boundary of I' is greater than α_n.

There exist an order $k \in \mathbb{N}_o$ and continuous functions F_n, $n \in \mathbb{N}$, F, such that

$$F_n^{(k)} = f_n \quad \text{and} \quad F_n \rightrightarrows F \quad \text{in} \quad I'.$$

By Property 3.4.9 of delta sequence, we have

$$F_n * \delta_n \rightrightarrows F \quad \text{in} \quad I. \tag{3.5.2}$$

Since

$$\left(F_n * \delta_n\right)^{(k)} = F_n^{(k)} * \delta_n = f_n * \delta_n = \varphi_n,$$

the sequence φ_n is fundamental in (a,b). It therefore represents a distribution f in (a,b). By (3.5.2), we can write

$$F_n * \delta_n \xrightarrow{d} F \quad \text{and} \quad [F_n * \delta_n] = F \quad \text{in} \quad I.$$

Differentiating k-times, we get

$$\varphi_n \xrightarrow{d} F^{(k)} \quad \text{and} \quad [\varphi_n] = F^{(k)} \quad \text{in} \quad I.$$

Consequently,

$$\varphi_n \xrightarrow{d} f \quad \text{in} \quad I, \quad \text{where} \quad f = F^{(k)}. \tag{3.5.3}$$

Since $F_n - F_n * \delta_n \rightrightarrows 0$ in I, thus differentiating k-time, we obtain

$$f_n - \varphi_n \xrightarrow{d} 0 \quad \text{in} \quad I.$$

Hence, and with (3.5.3), we get $f_n \xrightarrow{d} f$ in I. Since I is arbitrary, it follows that $f_n \xrightarrow{d} f$ in (a,b). □

Theorem 3.5.131 *If a sequence of distributions $(f_n)_{n\in\mathbb{N}}$ is convergent to f in every open bounded interval inside (a,b), then $f_n \xrightarrow{d} f$ in (a,b).*

Proof. For any bounded open interval I inside (a,b) there exists an bounded open interval I' inside (a,b) such that $I \subset I'$. Since $f_n \xrightarrow{d} f$ in I', there exist an order $k \in \mathbb{N}_o$ and continuous functions $F, F_n, n \in \mathbb{N}$, such that

$$F_n^{(k)} = f_n, \quad F^{(k)} = f \quad \text{and} \quad F_n \rightrightarrows F \quad \text{in} \quad I.$$

Thus, $f_n \xrightarrow{d} f$ in (a,b). □

Example 3.5.78 *The sequence $(f_n)_{n\in\mathbb{N}}$, where*

$$f_n(x) = \frac{1}{n}\sin(n^2 x)$$

converges to 0 for $n \to \infty$ both in a usual and distributional sense. It follows from Theorem 3.5.129 that we have distributional convergence of the sequence of distributions to 0, i.e.,

$$n\cos(n^2 x) \xrightarrow{d} 0.$$

However, the sequence is not convergent in the usual sense at any point.

Example 3.5.79 *We will show that the following sequences are distributionally convergent. Namely, we have*

$$f_n(x) = \frac{1}{\pi}\mathrm{arctg}\,(nx) + \frac{1}{2} \xrightarrow{d} 1_+(x);$$
$$g_n(x) = \frac{1}{\pi}\frac{n}{n^2 x^2 + 1} \xrightarrow{d} \delta(x)$$
$$h_n(x) = -\frac{2}{\pi}\frac{n^3 x}{(n^2 x^2 + 1)^2} \xrightarrow{d} \delta'(x),$$

where $1_+(x)$ is Heaviside's function and δ is the Dirac delta distribution.

In fact, we have the equality

$$f_n = F_n^{(1)}, \quad g_n = F_n^{(2)}, \quad h_n = F_n^{(3)},$$

where

$$F_n(x) = \frac{x}{2} + \frac{x}{\pi}\mathrm{arctg}\,(nx) - \frac{1}{2\pi n}\ln(1 + n^2 x^2)$$

and

$$F_n \overset{a.u.c.}{\rightrightarrows} F \quad \text{in} \quad \mathbb{R},$$

where

$$F(x) = \begin{cases} 0 & \text{for } x < 0, \\ x & \text{for } x \geq 0 \end{cases}$$

(see Example 3.3.76, *section 3.3.2). From* Theorem 3.5.124 *we have the following distribu-tional convergence*

$$F_n(x) \xrightarrow{d} F(x), \quad x \in \mathbb{R}.$$

Hence, by Theorem 3.5.129, *we obtain the desired distributional convergence:*

$$f_n(x) = F_n'(x) \xrightarrow{d} F'(x) = 1_+(x);$$
$$g_n(x) = F_n^{(2)}(x) \xrightarrow{d} F^{(2)}(x) = 1_+'(x) = \delta(x);$$
$$h_n(x) = F_n^{(3)}(x) \xrightarrow{d} F^{(3)}(x) = \delta'(x).$$

Example 3.5.80 *The sequence of Picard functions* $(f_n)_{n\in\mathbb{N}}$, *where*

$$f_n(x) = \sqrt{\frac{n}{2\pi}} e^{-\frac{nx^2}{2}},$$

converges to Dirac's delta distribution.

In fact, the conditions

$$f_n = F_n^{(2)}, \quad F_n(x) \overset{a.u.c.}{\rightrightarrows} F(x) \quad in \quad \mathbb{R},$$

are fulfilled, where

$$F_n(x) = \int_{-\infty}^{x} f_n(t)dt^2, \quad F(x) = \begin{cases} 0 & \text{for } x < 0, \\ x & \text{for } x \geq 0. \end{cases}$$

By Theorem 3.5.124, *it follows that* $F_n \xrightarrow{d} F$. *Hence, by* Theorem 3.5.129, *we have*

$$f_n(x) = F_n^{(2)}(x) \xrightarrow{d} F^{(2)}(x) = 1_+'(x) = \delta(x).$$

Remark 3.5.113 *In classical mathematical analysis one considers many sequences of func-tions which are convergent distributionally to the Dirac delta distribution, i.e.,*

(a) *the sequence of Picard functions (see Figure 16, section 3.3.2)*

$$f_n(x) = \sqrt{\frac{n}{2\pi}} e^{-\frac{nx^2}{2}}, \quad n \in \mathbb{N};$$

(b) *the sequence of Dirichlet functions (see Figure 19)*

$$f_n(x) = \begin{cases} \frac{\sin(nx)}{\pi x}, & \text{for } x \neq 0, \\ \frac{n}{\pi}, & \text{for } x = 0, \end{cases} \quad n \in \mathbb{N};$$

(c) *the sequence of Stieltjes functions (see Figure 20)*

$$f_n(x) = \frac{2}{\pi} \frac{n}{e^{nx} + e^{-nx}}, \quad n \in \mathbb{N};$$

(d) *the sequence of functions (see Figure 21)*

$$f_n(x) = \frac{2}{\pi}\frac{n}{(x^2 n^2 + 1)^2}, \quad n \in \mathbb{N};$$

(e) *the sequence of Cauchy functions (see Figure 17, section 3.3.2)*

$$f_n(x) = \frac{1}{\pi}\frac{n}{x^2 n^2 + 1}, \quad n \in \mathbb{N}.$$

For the proof see the examples in section 3.3.2.

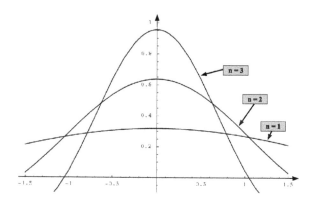

Figure 19. The graphs of Dirichlet's functions, $n = 1, 2, 3$

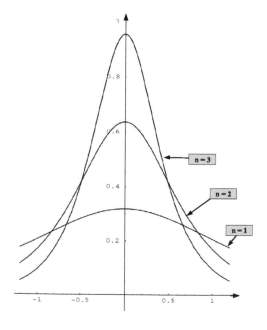

Figure 20. The graphs of Stieltjes functions, $n = 1, 2, 3$

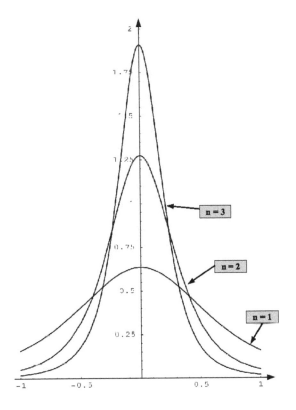

Figure 21. The graphs of functions $f_n(x) = \frac{2}{\pi} \frac{n}{(x^2 n^2 + 1)^2}$, $\quad n = 1, 2, 3$

3.5.2 Convergence and Regular Operations

The distributional limit commutes with all regular operators. The following formulas hold true

$$\lim_{n \to \infty} \lambda f_n(x) \stackrel{\mathrm{d}}{=} \lambda \lim_{n \to \infty} f_n(x), \quad \lambda \in \mathbb{R};$$

$$\lim_{n \to \infty} (f_n(x) + g_n(x)) \stackrel{\mathrm{d}}{=} \lim_{n \to \infty} f_n(x) + \lim_{n \to \infty} g_n(x);$$

$$\lim_{n \to \infty} f_n^{(m)}(x) \stackrel{\mathrm{d}}{=} (\lim_{n \to \infty} f_n(x))^{(m)}, \quad m \in \mathbb{N}_o;$$

$$\lim_{n \to \infty} \omega(x) f_n(x) \stackrel{\mathrm{d}}{=} \omega(x) \lim_{n \to \infty} f_n(x), \quad \omega \in \mathcal{D};$$

$$\lim_{n \to \infty} (f_n(x) * \omega(x)) \stackrel{\mathrm{d}}{=} \lim_{n \to \infty} f_n(x) * \omega(x), \quad \omega \in \mathcal{D}.$$

In the case of composite functions $(f_n \circ \sigma)(x) = f_n(\sigma(x))$ the limit

$$\lim_{n \to \infty} f_n(\sigma(x))$$

has two interpretations; as a limit of a sequence

$$\left(f_n(\sigma(x)) \right)_{n \in \mathbb{N}}$$

and as a substitution of a function $y = \sigma(x)$ in a distribution $\lim_{n \to \infty} f_n(y)$. The fact that passage to the limit commutes with substitution implies that both interpretations give the same result.

It is easy to check the commutativity of the convolution with a function of the class $\mathcal{D}$ and multiplying by a constant, addition, substraction, differentation. The commutativity of taking a limit with multiplication by a smooth function follows from Theorem 3.5.132 in this section. The commutativity of taking a limit with substitution follows from Theorem 3.5.133.

Theorem 3.5.132 *Let $(f_n)_{n \in \mathbb{N}}$ be a sequence of distributions, $(\omega_n)_{n \in \mathbb{N}}$ a sequence of functions in $\mathcal{D}$ and ω function in $\mathcal{D}$. If $f_n \xrightarrow{d} f$, $\omega_n^{(m)} \rightrightarrows \omega^{(m)}$ for each $m \in \mathbb{N}_o$, then $\omega_n f_n \xrightarrow{d} \omega f$.*

Proof. It follows from the convergence of the sequence of distributions $(f_n)_{n \in \mathbb{N}}$ in (a, b) that for every bounded open interval I inside (a, b) there exist continuous functions $F, F_n (n \in \mathbb{N})$, and the order $k \in \mathbb{N}_o$ for which

$$F_n^{(k)} = f_n, \quad F^{(k)} = f, \quad F_n(x) \rightrightarrows F(x) \quad \text{in} \quad I.$$

Therefore,

$$\omega_n F_n \rightrightarrows \omega F \quad \text{in} \quad I.$$

Hence, by Theorem 3.5.124, section 3.5.1, we have

$$\omega_n F_n \xrightarrow{d} \omega F \quad \text{in} \quad I.$$

Similarly,

$$\omega_n^{(k)} F_n \xrightarrow{d} \omega^{(k)} F \quad \text{in} \quad I. \tag{3.5.4}$$

One can check by induction that for smooth functions the following formula holds true:

$$\omega \varphi^{(k)} = \sum_{0 \le m \le k} (-1)^{(m)} \binom{k}{m} (\omega^{(m)} \varphi)^{(k-m)}. \tag{3.5.5}$$

For a fixed ω both sides are iterations of regular operations. Therefore, the formula (3.5.5) remains valid if φ is replaced by a distribution or a continuous function. In particular,

$$\omega_n F_n^{(k)} = \sum_{0 \le m \le k} (-1)^{(m)} \binom{k}{m} (\omega_n^{(m)} F_n)^{(k-m)}.$$

Hence,

$$\omega_n F_n^{(k)} \xrightarrow{d} \omega F^{(k)} \quad \text{in} \quad I,$$

thus,

$$\omega_n f_n \xrightarrow{d} \omega f \quad \text{in} \quad I.$$

Since I is arbitrary, it follows that

$$\omega_n f_n \xrightarrow{\mathrm{d}} \omega f \quad \text{in} \quad (a,b).$$

☐

Theorem 3.5.133 *Let $(f_n)_{n\in\mathbb{N}}$ be a sequence of distributions, $(\sigma_n)_{n\in\mathbb{N}}$ a sequence of functions of the class $\mathcal{D}$ and σ a function in $\mathcal{D}$. If $f_n \xrightarrow{\mathrm{d}} f$, $\sigma_n^{(m)} \overset{a.u.c.}{\rightrightarrows} \sigma^{(m)}$ for each $m \in \mathbb{N}_o$, $\sigma_n' \neq 0$ $(n \in \mathbb{N}_o)$ and $\sigma' \neq 0$, then*

$$(f_n \circ \sigma_n)(x) \xrightarrow{\mathrm{d}} (f \circ \sigma)(x).$$

The proof of this theorem will be omitted.

3.5.3 Distributionally Convergent Sequences of Smooth Functions

Theorem 3.5.134 *The sequence of constant functions is distributionally convergent iff it is convergent in an ordinary sense.*

Proof. a) Let $(c_n)_{n\in\mathbb{N}}$ be a sequence of constant functions convergent in an ordinary sense. This sequence is also uniformly convergent and hence by Theorem 3.5.124 (see section 3.5.1) distributionally convergent.

b) Let us assume now that the sequence $(c_n)_{n\in\mathbb{N}}$ of constant functions is distributionally convergent. Let us denote its distributional limit by c. The sequence $(c_n)_{n\in\mathbb{N}}$ is bounded. In fact, if it was not bounded there would exist a sequence $(c_{r_n})_{n\in\mathbb{N}}$, such that $\left(\frac{1}{c_{r_n}}\right)_{n\in\mathbb{N}}$ would be convergent to 0 in a ordinary sense, and hence convergent to 0 in a distributional sense. By Theorem 3.5.124 (see section 3.5.1) and the assumption $c_{r_n} \to c$ and hence,

$$1 = \frac{1}{c_{r_n}} \cdot c_{r_n} \to 0 \cdot c = 0, \quad \text{as} \quad n \to \infty,$$

which produces a contradiction.

Suppose that (c_n) does not converge in the ordinary sense. Then there exist two subsequences which converge to different limits. That contradicts Theorem 3.5.124, section 3.5.1.

☐

Lemma 3.5.48 *A sequence of smooth functions $(\varphi_n)_{n\in\mathbb{N}}$ is fundamental in (a,b) if for every bounded open interval I inside (a,b) there exist continuous functions $F_n, n \in \mathbb{N}$, and an order $k \in \mathbb{N}$ such that in I*

$$F_n^{(k)} = \varphi_n, \quad F_n(x) \rightrightarrows \quad \text{in} \quad I. \tag{3.5.6}$$

Proof. a) Let $(\varphi_n)_{n\in\mathbb{N}}$ be a fundamental sequence. By the definition it follows that for every bounded open interval I inside (a,b), there exist a sequence of smooth functions

$(\Phi_n)_{n\in\mathbb{N}}$ and an order $k \in \mathbb{N}_o$ such that in I

$$\Phi_n^{(k)} = \varphi_n \quad \text{and} \quad \Phi_n \rightrightarrows .$$

Since smooth functions are continuous, the condition (3.5.6) is satisfied.

b) Let us assume now that the condition (3.5.6) holds for every bounded open interval I inside (a, b). Let I' be an arbitrary bounded open interval inside (a, b) such that $I \subset I'$. Let $(\delta_n)_{n\in\mathbb{N}}$ be a delta sequence. We define a sequence $\left(\Phi_{nr}\right)_{n,r\mathbb{N}}$ of smooth functions in the following way

$$\Phi_{nr}(x) = \left(F_n(x) - \int_{x_o}^x \varphi_n(t)dt^k\right) * \delta_r(x) + \int_{x_o}^x \varphi_n(t)dt^k,$$

where $x_o \in I$.

We note that $\Phi_{nr}^{(k)} = \varphi_n$ in I for sufficiently large r, say $r > p_n$. By Property 3.4.8, section 3.4.1, for the delta sequence, we have

$$\Phi_{nr} \rightrightarrows F_n \quad \text{in} \quad I \quad \text{as} \quad r \to \infty.$$

Let us denote by F the limit of the sequence $(F_n)_{n\in\mathbb{N}}$. By (3.5.6) we have $F_n \rightrightarrows F$. Therefore, there exists a sequence $(r_n)_{n\in\mathbb{N}}, r_n > p_n, n \in \mathbb{N}$, such that

$$\Phi_n = \Phi_{nr_n} \rightrightarrows F \quad \text{in} \quad I.$$

Of course,

$$\Phi_n^{(k)} = \varphi_n \quad \text{in} \quad I.$$

Hence, the function $\Phi_n, n \in \mathbb{N}$, have the desired properties.   ▯

Theorem 3.5.135 *The sequence of smooth functions $(\varphi_n)_{n\in\mathbb{N}}$ is distributionally convergent to the distribution f if it is fundamental for f, i.e., $f = [\varphi_n]$.*

Proof. (a) Let $(\varphi_n)_{n\in\mathbb{N}}$ be a fundamental sequence for f, i.e., $f = [\varphi_n]$. By the definition of fundamental sequences it follows that for every bounded open interval I inside (a, b) there exist smooth functions $\Phi_n, \quad n \in \mathbb{N}$, a continuous function F and an order $k \in \mathbb{N}_o$ such that

$$\Phi_n \rightrightarrows F, \quad \Phi_n^{(k)} = \varphi_n \quad \text{in} \quad I. \tag{3.5.7}$$

By Theorem 3.3.101, section 3.3.2, the sequence $(\Phi_n)_{n\in\mathbb{N}}$ is fundamental, thus

$$[\Phi_n] = F.$$

Differentiating this equation k-times

$$F^{(k)} = [\Phi_n^{(k)}] = [\varphi_n]$$

we get the desired properties: $F^{(k)} = f$ in I. This and condition (3.5.7) means that $\varphi_n \xrightarrow{\text{d}} f$ in (a, b).

(b) Let us assume now that the sequence of smooth functions $(\varphi_n)_{n \in \mathbb{N}}$ is distributionally convergent to f. By the definition of the distributional convergence we conclude that for each bounded open interval I inside (a, b) there exist a sequence of smooth functions $(F_n)_{n \in \mathbb{N}}$, a continuous function F and an order $k \in \mathbb{N}_o$ such that

$$F_n^{(k)} = \varphi_n, \quad F^{(k)} = f \quad \text{and} \quad F_n \rightrightarrows F \quad \text{in} \quad I.$$

By Lemma 3.5.46, section 3.5.1, the sequence $(\varphi_n)_{n \in \mathbb{N}}$ is fundamental. In (a) we have proved that every fundamental sequence converges to the distribution that it represents. This implies that $f = [\varphi_n]$. □

Remark 3.5.114 *An analogy to* Theorem 3.5.135 *is true in the Cantor theory of real numbers.*

It is well known that a sequence of rational numbers is convergent to a real number α if it is fundamental for α.

Example 3.5.81 *By* Theorem 3.5.135, Examples 3.5.79 *and* 3.5.80 *(section 3.5.1), we may write:*

(a) $1_+(x) = \left[\frac{1}{\pi} \text{arctg}\,(nx) + \frac{1}{2}\right]$;

(b) $\delta(x) = \left[\frac{1}{\pi}\,\frac{n}{n^2x^2+1}\right] = \left[\sqrt{\frac{n}{2\pi}}\,e^{-\frac{nx^2}{2}}\right]$;

(c) $\delta'(x) = \left[-\frac{2}{\pi}\,\frac{n^3 x}{(n^2x^2+1)^2}\right]$;

(d) $\delta^{(2)}(x) = \left[\frac{2n^3}{\pi}\,\frac{3n^2x^2-1}{(n^2x^2+1)^3}\right]$.

Example 3.5.82 *For each $x \neq 0$, we have*

$$\lim_{n \to \infty} \frac{1}{\pi}\,\frac{n}{n^2x^2+1} \overset{d}{=} 0.$$

It is easy to check that in each interval I such that $x = 0 \notin I$ the convergence is almost uniform. By the definition of equivalent sequence we get

$$\left(\frac{1}{\pi}\,\frac{n}{n^2x^2+1}\right)_{n \in \mathbb{N}} \sim (0)_{n \in \mathbb{N}}$$

in the interval $-\infty < x < 0$ and interval $0 < x < +\infty$. By definition of equal distributions in the interval and by Example 3.5.81 (b), *we have $\delta(x) = 0$ in the interval $-\infty < x < 0$ and in the interval $0 < x < +\infty$.*

We can see now that the distribution δ cannot be identified with any function in the whole interval. It is, however, equal to a constant function in the interval $-\infty < x < 0$ and in the interval $0 < x < +\infty$.

3.5.4 Convolution of Distribution with a Smooth Function of Bounded Support

According to section 3.3.5, Remark 3.3.108, convolution of a distribution f with a given function $\omega \in \mathcal{D}$ is a regular operation. If $(f_n)_{n\in\mathbb{N}}$ is a fundamental sequence of distribution f then $(f_n * \omega)_{n\in\mathbb{N}}$ is a fundamental sequence for the convolution $f * \omega$. Hence, we have

$$f * \omega \overset{\mathrm{d}}{=} \lim_{n\to\infty} (f_n * \omega),$$

where $f = [f_n], \quad f * \omega = [f_n * \omega]$. By Theorem 3.5.135, section 3.5.3 and Theorem 3.4.115, section 3.4.2, we get the following:

Theorem 3.5.136 *If f is any distribution in (a, b) and $(\delta_n)_{n\in\mathbb{N}}$ is a delta sequence, then $(f * \delta_n)_{n\in\mathbb{N}}$ converges distributionally to f in (a, b), i.e.,*

$$\lim_{n\to\infty} (f * \delta_n) \overset{d}{=} f \quad in \quad (a, b).$$

Theorem 3.5.137 *If a sequence of distributions $(f_n)_{n\in\mathbb{N}}$ converges to f in (a, b) and $(\delta_n)_{n\in\mathbb{N}}$ is a delta sequence, then $(f_n * \delta_n)_{n\in\mathbb{N}}$ also converges to f in (a, b):*

$$\lim_{n\to\infty} (f_n * \delta_n) \overset{d}{=} f \quad in \quad (a, b),$$

i.e., $(f_n * \delta_n)_{n\in\mathbb{N}}$ is a fundamental sequence for f.

Proof. Let $(\delta_n)_{n\in\mathbb{N}}$ be a delta sequence. By definition it follows that there exists a sequence of positive numbers $(\alpha_N)_{n\in\mathbb{N}}$, converging to 0, such that

$$\delta_n(x) = 0 \quad \text{for} \quad |x| \geq \alpha_n, \, n \in \mathbb{N}.$$

Let I be a bounded open interval inside (a, b) and $I' = (\alpha, \beta)$ be a bounded interval inside (a, b), such that $I \subset I'_{-\alpha_n}$ where $I'_{-\alpha_n} = (\alpha + \alpha_n, \beta - \alpha_n)$. Since sequence $(f_n)_{n\in\mathbb{N}}$ converges distributionally to f, thus there exist an order $k \in \mathbb{N}_o$, a sequence of continuous functions $(F_n)_{n\in\mathbb{N}}$ and a continuous function F such that

$$F_n \rightrightarrows F, \quad F_n^{(k)} = f_n, \quad F^{(k)} = f \quad in \quad I'.$$

Hence, by Property 3.4.9, section 3.4.1, we get

$$(F_n * \delta_n)_{n\in\mathbb{N}} \overset{\text{a.u.c.}}{\rightrightarrows} F \quad in \quad I,$$

by Lemma 3.3.42, section 3.3.5, we obtain

$$(F_n * \delta_n)^{(k)} = f_n * \delta_n \quad in \quad I.$$

Hence, the sequence $(f_n * \delta_n)_{n\in\mathbb{N}}$ converges distributionally to f in I. Since I is arbitrary, we have

$$f_n * \delta_n \overset{\mathrm{d}}{\longrightarrow} f \quad in \quad (a, b),$$

and the theorem is proved. □

Theorem 3.5.138 *Let $f_n \in \mathcal{D}'$, $\varphi, \varphi_n \in \mathcal{D}$, $n \in \mathbb{N}$. If $\lim\limits_{n \to \infty} f_n \overset{d}{=} f$ in (a, b), $\varphi_n(x) = 0$ for $|x| > \alpha_o > 0, n \in \mathbb{N}$ and for every $m \in \mathbb{N}_o$*

$$\varphi_n^{(m)} \rightrightarrows \varphi^{(m)} \quad as \quad n \to \infty,$$

then for every $m \in \mathbb{N}_o$

$$(f_n * \varphi_n)^{(m)} \overset{a.u.c.}{\rightrightarrows} (f * \varphi)^{(m)} \quad in \quad (a, b), \quad as \quad n \to \infty.$$

Proof. Let I be an arbitrary given bounded open interval inside (a, b), and let $I' = (\alpha, \beta)$ be a bounded interval inside (a, b) such that

$$I \subset I'_{-\alpha_o}, \quad \text{where} \quad I'_{-\alpha_o} = (\alpha + \alpha_o, \beta - \alpha_o), \quad \alpha_o > 0.$$

By distributional convergence $(f_n)_{n \in \mathbb{N}}$ to f it follows that there exist an order $k \in \mathbb{N}_o$ and continuous functions F, F_n, $(n \in \mathbb{N})$, such that

$$F_n \rightrightarrows F, \quad F_n^{(k)} = f_n, \quad F^{(k)} = f \quad \text{in} \quad I'.$$

For any given $m \in \mathbb{N}_o$, we have

$$(f_n * \varphi_n)^{(m)} = F_n * \varphi_n^{(k+m)}, \quad (f * \varphi)^{(m)} = F * \varphi^{(m+k)} \quad \text{in} \quad I.$$

Moreover, we have in I,

$$|F_n * \varphi_n^{(m+k)} - F * \varphi^{(m+k)}| \leq |F_n - F| * |\varphi_n^{(m+k)}| +$$
$$|F| * |\varphi_n^{(m+k)} - \varphi^{(m+k)}| \leq \epsilon_n \int |\varphi_n^{(m+k)}| + \eta_n \int |F|,$$

where $\epsilon_n \to 0$ and $\eta_n \to 0$, as $n \to \infty$. Hence

$$(f_n * \varphi_n)^{(m)} \rightrightarrows (f * \varphi)^{(m)} \quad \text{in} \quad I.$$

Since I is arbitrary, we have

$$(f_n * \varphi)^{(m)} \overset{a.u.c.}{\rightrightarrows} (f * \varphi)^{(m)} \quad \text{in} \quad (a, b),$$

and the theorem is proved. □

Definition 3.5.79 *The sequence*

$$(f_n)_{n \in \mathbb{N}} = (f * \delta_n)_{n \in \mathbb{N}} \tag{3.5.8}$$

where f is a distribution in (a, b) and $(\delta_n)_{n \in \mathbb{N}}$ a delta sequence, is called a regular sequence for f.

Remark 3.5.115 *Since expressions f_n of (3.5.8) are smooth functions, it is a fundamental sequence for f, by Theorem 3.5.137. The class of regular sequences is, therefore, a special subclass of fundamental sequences.*

3.5.5 Applications

Problem 3.5.13 Prove that the sequence of Stieltjes functions (see Figure 20, section 3.5.1):

$$f(x) = \frac{2}{\pi} \frac{n}{e^{nx} + e^{-nx}}, \quad n \in \mathbb{N},$$

converges distributionally to Dirac's delta distribution $\delta(x)$.

Proof. Indeed, the following conditions are true:

$$f_n = F_n^{(2)}, \quad F_n \overset{\text{a.u.c.}}{\rightrightarrows} F \quad \text{in} \quad \mathbb{R},$$

where

$$F_n(x) = \int\limits_{-\infty}^{x} f_n(t)dt^2 = \int\limits_{-\infty}^{x} \frac{2}{\pi} \text{arctg } e^{nt} dt,$$

$$F(x) = \begin{cases} 0 & \text{for } x < 0, \\ x & \text{for } x \geq 0. \end{cases}$$

Hence, by Theorem 3.5.123, section 3.5.1, and Theorem 3.5.132, section 3.5.2, we get

$$f_n = F_n^{(2)} \overset{\text{d}}{\longrightarrow} F^{(2)} = 1'_+(x) = \delta.$$

$\square$

Problem 3.5.14 Prove that the sequence of Cauchy functions (see Figure 17, section 3.3.2)

$$f_n(x) = \frac{1}{\pi} \frac{\epsilon_n}{x^2 + \epsilon_n^2}, \quad \epsilon_n > 0, n \in \mathbb{N}$$

converges distributionally to $\delta(x)$, where $(\epsilon_n)_{n \in \mathbb{N}}$ is a sequence of positive numbers converging to 0.

Proof. Indeed, the sequence of functions

$$F_n(x) = \frac{1}{\pi} \text{arctg } \frac{x}{\epsilon_n} + \frac{1}{2}, \quad n \in \mathbb{N},$$

has the following properties:

$$F_n^{(1)}(x) = f_n(x), \quad F_n(x) \overset{\text{d}}{\longrightarrow} 1_+(x) \quad \text{in} \quad \mathbb{R}.$$

Hence, by Theorem 3.5.128, section 3.5.1, we obtain $f_n = F_n^{(1)} \overset{\text{d}}{\longrightarrow} 1'_+(x) = \delta(x) \quad \text{in} \quad \mathbb{R}.$ $\square$

Problem 3.5.15 Prove that the sequence of functions (see Figure 18, section 3.3.2)

$$f_n(x) = \frac{-2}{\pi} \frac{\epsilon_n x}{(x^2 + \epsilon_n^2)^2}, \quad n \in \mathbb{N}, \quad \epsilon_n > 0,$$

converges distributionally to a derivative of Dirac's delta distribution $\delta'(x)$, where $(\epsilon_n)_{n \in \mathbb{N}}$ is a sequence of positive numbers converging to 0.

Proof. Indeed, the following equation is true:

$$-\frac{2}{\pi}\frac{\epsilon_n x}{(x^2+\epsilon_n^2)^2} = \Big(\frac{1}{\pi}\frac{\epsilon_n}{x^2+\epsilon_n^2}\Big)',$$

thus, by Problem 3.5.14 and Theorem 3.5.128, section 3.5.1, we have

$$\lim_{n\to\infty}\Big(-\frac{2}{\pi}\frac{\epsilon_n x}{(x^2+\epsilon_n^2)^2}\Big) \xrightarrow{\mathrm{d}} \delta'(x).$$

⬜

Problem 3.5.16 Prove that the sequence of functions (see Figure 21, section 3.5.1)

$$f_n(x)=\frac{2}{\pi}\frac{\epsilon_n^3}{(x^2+\epsilon_n^2)^2},\quad n\in\mathbb{N},\quad \epsilon_n>0,$$

converges distributionally to Dirac's delta distribution, where $(\epsilon_n)_{n\in\mathbb{N}}$ is a sequence of positive numbers converging to 0.

Proof. Indeed, the sequence of functions

$$F_n(x)=\frac{2}{\pi}\Big(\frac{x\epsilon_n}{x^2+\epsilon_n^2}+\operatorname{arctg}\frac{x}{\epsilon_n}\Big),\quad n\in\mathbb{N},$$

has the following properties:

$$F_n'(x)=f_n(x),\quad F_n(x)\xrightarrow{\mathrm{d}}1_+(x).$$

Hence, by Theorem 3.5.129, section 3.5.1, we get

$$f_n(x)\xrightarrow{\mathrm{d}}1_+'(x)=\delta(x).$$

⬜

Problem 3.5.17 Prove that the sequence of Dirichlet functions (see Figure 19, section 3.5.1) converges distributionally to Dirac's delta distribution, i.e.,

$$\lim_{n\to\infty}\frac{\sin(nx)}{\pi x}\xrightarrow{\mathrm{d}}\delta(x).$$

3.6 Local Properties

3.6.1 Inner Product of Two Functions

First we recall the definition of the inner product of two functions.

Definition 3.6.80 *We define the inner product of two functions* $f, g : \mathbb{R} \to \mathbb{R}$ *by means of the formula*

$$(f, g) = \int_{\mathbb{R}} f(x)g(x)dx, \qquad (3.6.1)$$

provided that the integral exists.

As in the case of convolution, the following convention is adopted: the (ordinary) product fg takes the value 0 at a point, whenever one of its factors is 0 at the point, no matter whether the other factor is finite, infinite or even undetermined. This convention implies that, e.g., the inner product (f, g) exists, when f is defined in an open interval (a, b), g is defined in $\mathbb{R}$, the support of g is inside (a, b) and the product fg is locally integrable in (a, b).

Since the integral is taken in the sense of Lebesgue, the existence of the inner product (f, g) implies the existence of the inner product $(|f|, |g|)$. Conversely, if the inner product $(|f|, |g|)$ exists and, moreover, the product fg is measurable, then the inner product (f, g) exists.

It is easy to check that the following equalities hold:

$$(f, g) = (g, f), \qquad (3.6.2)$$

$$(\lambda f, g) = (f, \lambda g) = \lambda(f, g), \quad \lambda \in \mathbb{R}, \qquad (3.6.3)$$

$$(f + g, h) = (f, g) + (g, h), \qquad (3.6.4)$$

$$(f, g + h) = (f, g) + (f, h), \qquad (3.6.5)$$

$$(fg, h) = (f, gh). \qquad (3.6.6)$$

Using the notation $f_-(x) = f(-x)$, we can write

$$(f, g) = \int_{\mathbb{R}} f_-(0 - t)g(t)dt. \qquad (3.6.7)$$

Hence, the inner product (f, g) exists whenever the convolution $f_- * g$ exists at 0. We also have

$$(f, g) = (f_- * g)(0). \qquad (3.6.8)$$

From the above, (3.6.2) and the commutativity both of inner product and the convolution we have

$$(f, g) = (f * g_-)(0). \qquad (3.6.9)$$

In order to state Theorem 3.6.139 we recall the definition of the convolution of three functions.

Definition 3.6.81 *By the convolution* $f*g*h$ *of three functions we mean the double integral*

$$\int_{\mathbb{R}^2} f(x - t)g(t - u)h(u)dtdu, \qquad (3.6.10)$$

provided that the integral exists.

The convolution exists at a point $x \in \mathbb{R}$, whenever the product $f(x - t)g(t - u)h(u)$ is (Lebesgue) integrable over $\mathbb{R}^2$. As before, we understand that if one of the factors $f(x - t), g(t - u)$, or $h(u)$ is 0 for x, t and u, then the product is always taken to be 0, even if the remaining factors are not defined.

Since the integral (3.6.10) is meant in the sense of Lebesgue, the existence of $f * g * h$ implies the existence of $|f| * |g| * |h|$. The converse implication also holds, provided the product $f(x - t)g(t - u)h(u)$ is measurable. If we know that all the functions f, g, h are measurable, then $f * g * h$ exists, if $|f| * |h| * |h|$ exist.

Theorem 3.6.139 *If the convolution of three functions f, g, h is associativity at the origin, i.e., at 0, then*

$$(f_-, g * h) = (f * g, h_-). \tag{3.6.11}$$

Proof. By (3.6.8) the left-hand side of (3.6.11) is equal to $(f * (g * h))(0)$. By (3.6.9) the right-hand side of (3.6.11) is equal to $[(f * g) * h)](0)$. Hence, the equality of two sides follows by the associativity. ☐

Corollary 3.6.42 *If the convolution of functions f, g, h exists at 0, i.e., $(f * g * h)(0)$ exists, then equality (3.6.11) holds.*

Definition 3.6.82 *By the inner product of three functions f, g, h, we mean the value of the convolution $f * g * h$ at the origin, i.e., at 0,*

$$(f, g, h) = (f * g * h)(0).$$

By the definition of inner product and properties of convolution it follows:

Theorem 3.6.140 *The inner product (f, g, h) exists, if the integral*

$$\int_{\mathbb{R}^2} f(t)g(u)h(-t - u)dtdu$$

exists in the sense of Lebesgue.

Theorem 3.6.141 *If the inner product (f, g, h) exists, then*

$$(f, g, h) = (f * g, h_-) = (f_-, g * h) = (f * h, g_-).$$

Inner Product with a Function of Class $\mathcal{D}$.

Let $\psi \in \mathcal{D}$. The inner product

$$(\varphi, \psi) = \int_{\mathbb{R}} \varphi(x)\psi(x)dx$$

is defined for every smooth function $\varphi \subset \mathbb{R}$.

Lemma 3.6.49 *If $\psi \in \mathcal{D}$, $\varphi \in \mathcal{C}^\infty$, then the inner product (φ, ψ) is a regular operation, i.e., if $(\varphi)_{n\in\mathbb{N}}$ is a fundamental sequence, then the sequence (φ_n, ψ), $n \in \mathbb{N}$, is fundamental.*

Proof. Let $supp\psi \subset I \subset \mathbb{R}$. Let the sequence $(\varphi)_{n\in\mathbb{N}}$ be a fundamental sequence. Thus, there exist a smooth function F, a sequence of smooth functions $(F_n)_{n\in\mathbb{N}}$, and an order $k \in \mathbb{N}_o$ such that

$$F_n^{(k)} = f_n, \quad F_n \rightrightarrows \quad \text{in} \quad I.$$

Integrating by part, we get

$$(\varphi_n, \psi) = \int_I F_n^{(k)} \psi = (-1)^k \int_I F_n \psi^{(k)} \longrightarrow (-1)^k \int_I F\psi^{(k)},$$

i.e., the sequence of numbers $(\varphi_n, \psi), n \in \mathbb{N}$, converges, thus by the convention in section 3.2.3, this sequence is fundamental too.   □

Using Lemma 3.6.49, we can formulate the following:

Definition 3.6.83 *We defined the inner product of a distribution f in $\mathbb{R}$ with a function $\psi \in \mathcal{D}$ in the following way:*

$$(f, \psi) = [(\varphi_n, \psi)] =: \lim_{n\to\infty} (\varphi_n, \psi),$$

where $(\varphi_n)_{n\in\mathbb{N}}$ is a fundamental sequence of a distribution f.

3.6.2   Distributions of Finite Order

In section 3.3.3 we introduced a definition of the distribution of finite order. This notion can be defined in a different way:

Definition 3.6.84 *A distribution f in $\mathbb{R}$ is said to be of a finite order, if there exist a continuous function F in $\mathbb{R}$ and an order $k \in \mathbb{N}_o$ such that $F^{(k)} = f$ in $\mathbb{R}$.*

Theorem 3.6.142 *If G is a locally integrable function in $\mathbb{R}$ and $k \in \mathbb{N}_o$, then the distribution $f = G^{(k)}$ is of finite order.*

Proof. To prove this theorem it suffices to see that the function

$$F(x) = \int\limits_0^x G(t)dt$$

is continuous and the equality $F^{(k+1)} = f$ holds. ▯

Theorem 3.6.143 *The set of all distributions of finite order in $\mathbb{R}$ is a linear space.*

Proof. It is clear that if f is of finite order, so is λf for every number λ. Therefore, it suffices to show that if f and g are of finite order, then their sum $f+g$ is also of finite order.

Let $f = F^{(k)}$, $g = G^{(l)}$, where F and G are continuous functions and $k, l \in \mathbb{N}_o$. Let $p \in \mathbb{N}_o$, and $p \geq max(k,l)$, and let

$$\widetilde{F}(x) = \int\limits_0^x F(t)dt^{p-k}, \quad \widetilde{G}(x) = \int\limits_0^x G(t)dt^{p-l},$$

where the integral is defined in the following way:

$$\int\limits_0^x F(t)dt^k = \int\limits_0^x dt_k \int\limits_0^{t_k} dt_{k-1} \cdots \int\limits_0^{t_2} F(t_1)dt_1.$$

Then

$$f + g = \left(\widetilde{F} + \widetilde{G}\right)^{(p)}, \quad \text{where} \quad \widetilde{F} + \widetilde{G} \quad \text{is a continuous function.}$$
▯

Theorem 3.6.144 *If a distribution f vanishes outside a bounded open interval I, it is of finite order.*

Proof. By Theorem 3.4.114, section 3.4.2, it follows that there exist a continuous function F in $\mathbb{R}$ and an order $k \in \mathbb{N}_o$ such that $F^{(k)} = f$ in the neighborhood $I_{2\alpha}, \alpha \in \mathbb{R}_+$, where I_α denotes the set of all points $x \in \mathbb{R}$ whose distance from the set I is less than α. Let φ be a smooth function such that

$$\varphi(x) = \begin{cases} 1 & \text{for } x \in I_\alpha, \\ 0 & \text{for } x \notin I_{2\alpha}. \end{cases}$$

Since

$$f = F^{(k)}\varphi = \sum_{0 \leq m \leq k} (-1)^m \binom{k}{m} \left(F\varphi^{(m)}\right)^{(k-m)},$$

thus, by Theorem 3.6.143, the assertion follows. ▯

3.6.3   The Value of a Distribution at a Point

Definition 3.6.85 *We say that a distribution f takes the value c at a point x_o if for every regular sequence $(f_n)_{n\in\mathbb{N}} = (f * \delta_n)_{n\in\mathbb{N}}$, we have*

$$\lim_{n\to\infty} f_n(x_o) =: c. \tag{3.6.12}$$

It is easy to check that if a distribution f is a continuous function at a point x_o, then the above definition coincides with the value of a function in the ordinary sense.

If the value of the distribution at a point x_o exists, i.e., if the limit (3.6.12) exists, the point x_o is said to be *regular*. Otherwise it is said to be *singular*.

Theorem 3.6.145 ([AMS], [Lo.1]) *If a distribution f has a value at x_o, then*

$$\lim_{\alpha\to 0} f(\alpha x + x_o) =: c. \tag{3.6.13}$$

Remark 3.6.116 *The limit (3.6.13) is a distributional limit of a sequence of distributions $f(\alpha x + x_o)$ that depend on a continuous parameter (α). It can be proved that if the limit exists then it is also a distribution and that distribution is identical to a certain constant function. S. Łojasiewicz defines (see [Lo.1]) the value of a distribution f at a point x_o as a value of the above-mentioned constant function (at an arbitrary point).*

It follows from Theorem 3.6.145 that if the limit (3.6.12) exists, then also the limit (3.6.13) exists, i.e., Ł-value (the value of a distribution in sense of Łojasiewicz) exists. We cite below Łojasiewicz's theorem on the existence of a value of a distribution at a point. In particular, it follows from that theorem the existence of a limit (Ł-value) implies the existence of a limit (3.6.12). This means that the definitions (3.6.12) and (3.6.13) are equivalent.

Theorem 3.6.146 (see [Lo.1, Lo.2]) *A distribution f has a value c at a point x_o (Ł-value) iff there exist a continuous function F and an order $k \in \mathbb{N}_o$ such that $F^{(k)} = f$ and*

$$\lim_{x\to x_o} \frac{F(x)}{(x - x_o)^k} = \frac{c}{k!}.$$

Theorem 3.6.147 *If a distribution f' has a value at a point x_o, then a distribution f has a value at a point x_o, too.*

Proof.   In fact, by Theorem 3.6.146, there exist an order $k \in \mathbb{N}_o$, a continuous function F such that $F^{(k)} = f$ and the limit

$$\lim_{x\to x_o} \frac{F(x)}{(x - x_o)^k}.$$

If $k = 0$, then f' is a continuous function and so is f; therefore, the theorem is true.

If $k > 0$, then

$$\lim_{x\to x_o} \frac{F(x)}{(x - x_o)^{k-1}} = 0.$$

It follows, by Theorem 3.6.146, that the distribution $F^{(k-1)}$ has a value at a point x_o. Since the distributions $F^{(k-1)}$ and f differs only by a constant number, thus the distribution f also has a value at a point. ▯

Theorem 3.6.148 *If a distribution f is a locally integrable function continuous at x_o, then*

$$f(\alpha x + x_o) \stackrel{a.u.c.}{\rightrightarrows} f(x_o) \quad as \quad \alpha \to 0.$$

Consequently, the point x_o is regular and the value of f at x_o in the distributional sense is equal to $f(x_o)$.

Theorem 3.6.149 *If a locally integrable function F has an ordinary derivative F' at a point x_o, then this derivative is the value of the distribution F' at that point, i.e., $F'(x_o)$.*

Proof. By assumption there exists a limit

$$\lim_{\alpha \to 0} \frac{F(\alpha x + x_o) - F(x_o)}{\alpha x} = F'(x_o),$$

where all the symbols are interpreted in the classical sense. Consequently,

$$\frac{F(\alpha x + x_o) - F(x_o)}{\alpha x} \stackrel{a.u.c.}{\rightrightarrows} x F'(x_o), \quad x \in (-\infty; +\infty), \quad as \quad \alpha \to 0.$$

Differentiating this formula distributionally, we obtain

$$\lim_{\alpha \to 0} F'(\alpha x + x_o) \stackrel{d}{=} F'(x_o),$$

which proves the theorem. ▯

Remark 3.6.117 *The converse theorem to Theorem 3.6.149 is not true. It is possible for the ordinary derivative not to exist at some point although the distributional derivative has a value at this point.*

For instance, the function

$$F(x) = \begin{cases} 3x^2 \sin \frac{1}{x} - x \cos \frac{1}{x}, & for\ x \neq 0 \\ 0, & for\ x = 0 \end{cases}$$

does not have an ordinary derivative at the point 0. However, the distributional derivative $F'(x)$ has the value 0 at that point 0.

In fact, we have

$$(\alpha x)^3 \sin \frac{1}{\alpha x} \stackrel{a.u.c.}{\rightrightarrows} 0 \quad as \quad \alpha \to 0.$$

Hence, by successive differentiation,

$$\lim_{\alpha \to 0} \frac{F(\alpha x)}{\alpha} \stackrel{d}{=} 0, \quad \lim_{\alpha \to 0} F'(\alpha x) \stackrel{d}{=} 0.$$

The following theorem is a particular case of Theorem 3.6.149.

Theorem 3.6.150 *If f is a locally integrable function and the function*

$$F(x) = \int_0^x f(t)dt$$

has an ordinary derivative at x_o, then this derivative is the value of the function f at the point x_o.

Remark 3.6.118 *The value of a distribution f at a point x_o will be denoted by $f(x_o)$ as in the case of functions. This notation does not give rise to any misunderstanding. In fact, if the distribution f is a continuous function, both meanings of $f(x_o)$ coincide by Theorem 3.6.148. If f is only locally integrable then, by Theorem 3.6.150, the values of the distribution f exist almost everywhere. Both meanings of $f(x_o)$ coincide then almost everywhere but, in general, not everywhere. When the two values differ, we adopt the convention of denoting by $f(x_o)$ the value in the distributional sense.*

Example 3.6.83 *By Theorem 3.6.148 each point $x_o \neq 0$ is a regular point of the Heaviside function $1_+(x)$, and the value at x_o in the distributional sense is the same as the value in the usual sense. The point $x_o = 0$ is singular since the limit of*

$$1_+(\alpha x) = \frac{\alpha}{|\alpha|}\left(1_+(x) - \frac{1}{2}\right) + \frac{1}{2} \tag{3.6.14}$$

does not exist as $\alpha \to 0$.

Remark 3.6.119 *It can be proved that if the value of a distribution f is 0 everywhere, then f is the null function (see [Lo.2]). Thus, a distribution is uniquely determined by its values provided they exist everywhere.*

Theorem 3.6.151 *If a distribution f has a value $f(x_o)$ at a point x_o, and $\omega \in C^\infty$, then the distribution ωf also has a value $f(x_o)\omega(x_o)$ at a point x_o.*

Theorem 3.6.152 *Let $\varphi \in C^\infty$ and $\varphi'(x) \neq 0$ for every $x \in \mathbb{R}$. If a distribution f has a value at the point $\varphi(x_o)$, then the distribution $(f \circ \varphi)(x) = f(\varphi(x))$ also has this value at x_o.*

Example 3.6.84 *The Dirac delta distribution δ has the value 0 at each point $x_o \neq 0$ and has no value at the point $x_o = 0$. In fact, differentiating (3.6.14) we obtain*

$$|\alpha|\delta(\alpha x) = \delta(x).$$

The existence of the limit $\lim\limits_{\alpha \to 0} \delta(\alpha x)$ would imply that $\delta(x)$ is the function identically equal to 0, which is not true.

3.6.4   The Value of a Distribution at Infinity

Definition 3.6.86 *The value of the distributional limit*

$$\lim_{\beta \to \infty} f(x + \beta), \qquad (3.6.15)$$

if it exists, is said to be the value of the distribution f at ∞ and is denoted by $f(\infty)$.

The value $f(-\infty)$ of the function f at $-\infty$ is defined similarly. Obviously, the symbols $f(\infty)$, $f(-\infty)$ have a meaning iff the corresponding limits exist.

If the limit (1) exists, then it is a constant function (see [AMS], p. 44).

Theorem 3.6.153 *If a distribution f is a continuous function and has the ordinary limit c at ∞,   (or at $-\infty$), then*

$$f(x + \beta) \overset{a.u.c.}{\rightrightarrows} c \quad as \quad \beta \to \infty, \quad (or \ as \quad \beta \to -\infty).$$

Consequently, $f(\infty) = c$ (or $f(-\infty) = c$).

3.6.5   Support of a Distribution

The notion of a support of a locally integrable function is introduced, for example, by J. Mikusiński in [Mi.6], p. 196. In this section we introduce the notion of support of a distribution. L. Schwartz [S.2] defined the support of a distribution as the smallest closed set outside of which the distribution vanishes. We will explain here the notion of a support of a distribution introduced by J. Mikusiński.

Following S. Łojasiewicz, the value of a distribution at a point can be used to sharpen the concept of the support of a distribution. Let f be a distribution in $(a;b)$. By L_f we denote he set of all points in $(a;b)$ at which the value of a distribution f does not exist or is different from zero. The support of a distribution f is the closure of the set L_f, i.e.,

$$supp f = cl \ L_f.$$

It can be proved that the closure of L_f is the support in Schwartz's sense of f.

3.7   Irregular Operations

3.7.1   Definition

An advantage of the sequential approach to the theory of distributions is the ease of extending to distributions many operations, which are defined for smooth functions, i.e., regular operations (see section 3.3). An example of a regular operation is differentiation (of a given order $k \in \mathbb{N}_o$): $A(f) = f^{(k)}$, which can be performed for an arbitrary distribution f. It is well known (see section 3.6) that every distribution is locally (i.e., on an arbitrary bounded open interval in $\mathbb{R}$) a distributional derivative of a finite order of a continuous function.

It should be noted that in practice operations there are both regular and not regular ones. For instance, the two-argument operations of product $A(\varphi, \psi) = \varphi \cdot \psi$ and the convolution $A(\varphi, \psi) = \varphi * \psi$ are not regular operations and they cannot be defined for arbitrary distributions.

J. Mikusiński pointed out a general method of defining *irregular operations* on distributions by using delta sequences (see [Mi.4], [Mi.3], and [AMS], pp. 256–257).

Let us assume that an operation A is feasible for arbitrary smooth functions $\varphi_1, \varphi_2, \cdots, \varphi_k$ and let $f_1, f_2, \cdots, f_k$ be arbitrary distributions in $\mathbb{R}$.

Definition 3.7.87 *If $f_1, \cdots, f_k$ are arbitrary distributions in $\mathbb{R}$, we say that $A(f_1, \cdots, f_k)$ exists if for an arbitrary delta sequence $(\delta_n)_{n \in \mathbb{N}}$ the sequence*

$$\Big(A(f_1 * \delta_n, \cdots, f_k * \delta_n) \Big)_{n \in \mathbb{N}}$$

is fundamental; then the operation A on $f_1, \cdots, f_k$ is defined by the formula

$$A(f_1, \cdots, f_k) = [A(f_1 * \delta_n, \cdots, f_k * \delta_n)].$$

Remark 3.7.120 *If $A(f_1, \cdots, f_k)$ exists then the distribution does not depend on the choice of a delta sequence $(\delta_n)_{n \in \mathbb{N}}$.*

If A is a *regular* operation then, of course, A exists and coincides with the earlier defined result of the regular operation. If A is *irregular*, it need not exist for all distributions, but the definition embraces not only earlier known cases, but also new ones. For instance, for the operation of the product $A(f_1, f_2) = f_1 \cdot f_2$, the definition can be expressed in the form

$$f_1 \cdot f_2 \stackrel{\mathrm{d}}{=} \lim_{n \to \infty} (f_1 * \delta_n)(f_2 * \delta_n) \tag{3.7.1}$$

and it exists in this sense for a wide class of distributions but not in the classical sense of L. Schwartz [S.2].

Among other irregular operations an especially important role is played by the convolution of distributions (see section 3.7.3).

Historical Remarks

The operations of integration, convolution and a product of distributions can be performed only for certain classes of distributions. The operation of convolution of two distributions can be done if, e.g., their supports are compatible (see [AMS], p. 124). These difficulties were the impetus for the further search of new definitions of that operation. This problem is especially visible in Fourier transform theory. In the classical mathematical analysis the Fourier transformation transforms the convolution of integrable or square integrable functions into the product of their transforms. The question arises: what similarities can be found for distributions? An answer to that question can be found in the book by L. Schwartz [S.2] and the papers by Y. Hiraty and H. Ogaty, [HiO], Shiraishi and M. Itano, [ShI]. Those results, however, have not embraced all possibilities or they were too general. R. Shiraishi in [Sh] stated a hypothesis whether a convolution of tempered distributions is a tempered distribution. A negative answer to that problem was given by A. Kamiński [Ka.1], and then independently by P. Dierolf and J. Voigt in [DV]. But R. Shiraishi's question motivated A. Kamiński to modify the notion of compatibile supports (the so-called polynomial compatible) that guarantees that a convolution of tempered distributions is a tempered distribution [Ka.4]. For a definition of tempered distribution see, for example, [AMS], p. 165.

3.7.2 The Integral of Distributions

One of the most important operations on distributions is integration. Various definitions of the *definite integral* of a distribution can be found in the literature (see [Mi.4], [Si.3], [MiS.1], [Sk.2], [KSk]).

The *indefinite integral* or an *antiderivative* of a distribution f in $\mathbb{R}$ is a distribution h such that

$$h'(x) = f(x).$$

The existence of an indefinite integral follows from the fact that every distribution is locally a derivative of a certain order of a continuous function (see Theorem 3.4.114, section 3.4.2). The integral of a distribution is uniquely given up to a constant, i.e., the following theorem holds:

Theorem 3.7.154 *For every distribution f there exists a family of antiderivatives. Any two antiderivatives of f differ by a constant function.*

Let f be any distribution in $\mathbb{R}$, and h its indefinite integral.

If there are values $h(a)$ and $h(b)$ of distribution h at the points $x = a$ and $x = b$, then the number

$$\int_a^b f(x)\, dx = h(b) - h(a); \tag{3.7.2}$$

we call it the *definite integral* of f in the interval (a, b).

In the statement one can have $a = -\infty$ or $b = +\infty$ and then the formula (3.7.2) defines an integral of a distribution in an unbounded interval.

One can show that the integral given by (3.7.2) has properties similar to those of the integral of a function. Moreover, if a distribution f can be identified with a locally integrable function then the distributional integral reduces to the integral of a function.

Remark 3.7.121 *From now on we will consider integrals in the distributional sense.*

Theorem 3.7.155 *Let f, g be distributions and φ, ϕ are smooth functions ($\phi' \neq 0$), $\lambda, a, b, c \in \mathbb{R}$. The following formulas hold for integrals:*

$$\int_a^b f(x)dx = -\int_b^a f(x)dx; \tag{3.7.3}$$

$$\int_a^b \lambda f(x)dx = \lambda \int_a^b f(x)dx, \quad \lambda \in \mathbb{R}; \tag{3.7.4}$$

$$\int_a^b f(x)dx = \int_a^c f(x)dx + \int_c^b f(x)dx; \tag{3.7.5}$$

$$\int_a^b [f(x) + g(x)]dx = \int_a^b f(x)dx + \int_a^b g(x)dx; \tag{3.7.6}$$

$$\left(\int_a^b f(t)dt \right)' = f(x); \tag{3.7.7}$$

$$\int_a^b \varphi(x)f'(x)dx = \varphi(x)f(x) \Big|_{x=a}^{x=b} - \int_a^b \varphi'(x)f(x)dx; \tag{3.7.8}$$

$$\int_{\phi(a)}^{\phi(b)} f(x)dx = \int_a^b f(\phi(x))\phi'(x)dx, \tag{3.7.9}$$

provided the integrals on the right exist; in the case of the equality (3.7.8) we additionally assume the existence of the value of the distribution $\varphi(x)f(x)$ at the points $x = a$ and $x = b$.

Moreover, for an arbitrary $k \in \mathbb{N}_o$, if the distributions $\varphi^{(l)}f, (0 \leq l \leq k)$ are integrable in $\mathbb{R}$, and $\varphi \in \mathcal{D}$, then

$$\int_{\mathbb{R}} \varphi(x)f^{(k)}(x)dx = (-1)^k \int_{\mathbb{R}} \varphi^{(k)}(x)f(x)dx. \tag{3.7.10}$$

Remark 3.7.122 *If the function f is locally integrable on $\mathbb{R}$, then the integral $\int_{\mathbb{R}} f(x)\,dx$ understood in the distributional sense coincides with the usual integral on $\mathbb{R}$ if the integral exists.*

We will take the notation

$$\int\limits_{a}^{b} f(x+t)\,dt = h(x+b) - h(x+a), \quad \text{where} \quad h'(x) = f(x).$$

We will prove the following theorem:

Theorem 3.7.156 *If $f_n \in \mathcal{D}'$, $n \in \mathbb{N}$ and $f_n \xrightarrow{d} f$, then*

$$\int\limits_{a}^{b} f_n(x+t)\,dt \xrightarrow{d} \int\limits_{a}^{b} f(x+t)dt. \tag{3.7.11}$$

Proof. It follows from the definition of distributional convergence of a sequence $(f_n)_{n\in\mathbb{N}}$ to the distribution f (see Definition 3.5.77, section 3.5.1), that for any bounded open interval I, there exist an order $k \in \mathbb{N}_o$, a sequence of continuous functions $(F_n)_{n\in\mathbb{N}}$, a continuous function F, such that

$$F_n^{(k)} = f_n, \quad F^{(k)} = f, \quad F_n \rightrightarrows F \quad \text{in} \quad I.$$

Consequently,

$$F_n(x+b) - F_n(x+a) \rightrightarrows F(x+b) - F(x+a).$$

Hence, by Theorem 3.5.129, section 3.5.1, after differentiating $(k-1)$-times we obtain desired convergence. □

Theorem 3.7.157 *If φ is an arbitrary element of $\mathcal{D}$ and $(\delta_n)_{n\in\mathbb{N}}$ an arbitrary delta sequence, then*

$$\lim_{n\to\infty} \int\limits_{\mathbb{R}} \delta_n(x)\varphi(x)dx \stackrel{d}{=} \varphi(0); \tag{3.7.12}$$

$$\int\limits_{\mathbb{R}} \delta(x)\varphi(x)dx = \varphi(0), \qquad \varphi \in \mathcal{D}; \tag{3.7.13}$$

$$\int\limits_{\mathbb{R}} \delta(x - x_o)\varphi(x)dx = \varphi(x_o), \qquad \varphi \in \mathcal{D}. \tag{3.7.14}$$

Proof. Let $\varphi \in \mathcal{D}$. For an arbitrary delta sequence $(\delta_n)_{n\in\mathbb{N}}$ we have:

$$|(\delta_n, \varphi) - \varphi(0)| \leq \int\limits_{-\alpha_n}^{\alpha_n} |\delta_n(x)|\,|\varphi(x) - \varphi(0)|dx \to 0,$$

for $n \to \infty$, where $(\alpha_n)_{n\in\mathbb{N}}$ is a sequence of positive numbers convergent to zero and such that $\delta_n(x) = 0$ for $|x| \geq \alpha_n, n \in \mathbb{N}$.

The proof of the equality (3.7.14) can be obtained in a similar way. □

Theorem 3.7.158 *If for every function $\varphi \in \mathcal{D}$ the equality*

$$\int_{\mathbb{R}} f(x)\varphi(x)\,dx = \varphi(0),$$

holds, then $f = \delta$.

Proof. Let $f_n = f * \delta_n, n \in \mathbb{N}$, be a regular sequence for the distribution f. Then, by Definition 3.6.83, section 3.6.1, and the assumption for every function $\varphi \in \mathcal{D}$ we have

$$(f_n, \varphi) \xrightarrow{\mathrm{d}} \varphi(0).$$

Similarly, by Definition 3.6.83, section 3.6.1 and the equality (3.7.13), we get

$$(\delta_n, \varphi) \xrightarrow{\mathrm{d}} \varphi(0).$$

We note that also for every function $\varphi \in \mathcal{D}$ the interlaced sequence

$$(f_1, \varphi), (\delta_1, \varphi), (f_2, \varphi), (\delta_2, \varphi), \cdots$$

is distributionally convergent to $\varphi(0)$ for each $\varphi \in \mathcal{D}$. Hence, the sequence

$$f_1, \delta_1, f_2, \delta_2, \cdots$$

is fundamental, which implies $f = \delta$.   ⟦

Remark 3.7.123 *The distribution δ is not a locally integrable function, which follows from:*

Lemma 3.7.50 *There is no locally integrable function f such that*

$$\int_{\mathbb{R}} f(x)\varphi(x)\,dx = \varphi(0), \quad \varphi \in \mathcal{D}.$$

For a proof see [Szm], p. 40, and Example 3.2.64, section 3.2.1; and Example 3.2.66, section 3.2.2.

Theorem 3.7.159 *If $f_n \in \mathcal{D}'$, $n \in \mathbb{N}$ and $f_n \xrightarrow{\mathrm{d}} f$, then for every function $\varphi \in \mathcal{D}$*

$$\int_{\mathbb{R}} f_n \varphi \xrightarrow{\mathrm{d}} \int_{\mathbb{R}} f\varphi, \quad i.e., \quad (f_n, \varphi) \xrightarrow{\mathrm{d}} (f, \varphi).$$

The proof of the above theorem follows from Lemma 3.6.49, section 3.6.1.

The opposite theorem to Theorem 3.7.159 is also true. Namely, we have:

Theorem 3.7.160 *If for every fixed function $\varphi \in \mathcal{D}$ the sequence of numbers (f_n, φ) is convergent to (f, φ), then $f_n \xrightarrow{\mathrm{d}} f$.*

Using the above theorems and the definition of equality of distributions (section 3.3.3) one can easily show that the equality of two distributions f and g in $\mathbb{R}$ can be defined in an equivalent manner:

$$(f = g) \quad \text{iff} \quad [(f, \varphi) = (g, \varphi) \quad \text{for every function} \quad \varphi \in \mathcal{D}]. \qquad (3.7.15)$$

By the du Bois Reymond lemma (see section 3.2.2, Lemma 3.2.37), in case f and g are locally integrable in $\mathbb{R}$ the definition (3.7.16) coincides with the definition of equality of locally integrable functions.

Example 3.7.85 *For a translated by point x_o Heaviside function*

$$1_+(x - x_o) = \begin{cases} 1 & \text{for } x \geq x_o, \\ 0 & \text{for } x < x_o \end{cases}$$

we have

$$1'_+(x - x_o) = \delta(x - x_o).$$

This equation can be obtained in the same way as (3.4.11), section 3.4.3.

According to the notation (3.7.2), we can write:

$$\int_a^b \delta(x - x_o)dx = 1 \quad \text{for} \quad a < x_o < b; \qquad (3.7.16)$$

$$\int_a^b \delta(x - x_o)dx = 0 \quad \text{for} \quad a < b < x_o \quad \text{or} \quad x_o < a < b; \qquad (3.7.17)$$

$$\int_{-\infty}^{+\infty} \delta(x - x_o)dx = 1. \qquad (3.7.18)$$

Example 3.7.86 *For every function $\omega \in \mathcal{C}^\infty$ the equalities are true:*

$$\int_a^b \omega(x)\delta(x - x_o)dx = \omega(x_o), \quad \text{for} \quad a < x_o < b; \qquad (3.7.19)$$

$$\int_{-\infty}^{+\infty} \omega(x)\delta(x - x_o)dx = \omega(x_o), \quad \omega \in \mathcal{C}^\infty. \qquad (3.7.20)$$

By equality (3.4.3), section 3.4.1 and (3.7.16), we obtain in turn

$$\int_a^b \omega(x)\delta(x - x_o)dx = \int_a^b \omega(x_o)\delta(x - x_o)dx$$

$$= \omega(x_o)\int_a^b \delta(x - x_o)dx = \omega(x_o), \quad \text{if} \quad a < x_o < b.$$

In particular, if $a = -\infty$, $b = +\infty$, *we get equality (3.7.20).*

From (3.7.19) and (3.7.20) for $x_o = 0$, *we obtain*

$$\int_a^b \omega(x)\delta(x)dx = \omega(0), \quad \text{if} \quad a < 0 < b; \tag{3.7.21}$$

$$\int_{-\infty}^{+\infty} \omega(x)\delta(x)dx = \omega(0), \quad \omega \in C^\infty. \tag{3.7.22}$$

Example 3.7.87 *Let us consider a function* φ *given in the example, section 3.2.1, i.e., the function*

$$\varphi(x) = \begin{cases} \exp \frac{\alpha^2}{(x-\alpha)^2}, & |x| < \alpha, \\ 0, & |x| \geq \alpha. \end{cases}$$

Then, according to the equality (3.7.22) we have

$$\int_{-\infty}^{+\infty} \exp \frac{\alpha^2}{(x-\alpha)^2}\delta(x)dx = e.$$

Remark 3.7.124 *The relations (3.7.19) and (3.7.22) describe one of the basic properties of Dirac's delta distribution* $\delta(x)$. *In the functional approach to theory of distributions those properties are regarded as definitions.*

Example 3.7.88 *For every function* $\omega \in C^\infty$ *and* $k \in \mathbb{N}_o$ *the equality*

$$\int_{-\infty}^{+\infty} \omega(x)\delta^{(k)}(x - x_o)dx = (-1)^k\omega^{(k)}(x_o), \quad \omega \in C^\infty \tag{3.7.23}$$

is true.

In fact, by Property 3.4.5, section 3.4.1 and (3.7.10), (3.7.19), we obtain in turn:

$$\int_{-\infty}^{+\infty} \omega(x)\delta^{(k)}(x - x_o)\,dx = (-1)^k \int_{-\infty}^{+\infty} \omega^{(k)}(x)\delta(x - x_o)\,dx$$

$$= (-1)^k \int_{-\infty}^{+\infty} \omega^{(k)}(x_o)\delta(x - x_o)\,dx = (-1)^k\omega^{(k)}(x_o).$$

Example 3.7.89 *Let us consider the sequence of functions (see Figure 22):*

$$f_n(x) = \begin{cases} 0, & x < -\frac{1}{n}, \\ n(1 + nx), & -\frac{1}{n} \leq x \leq 0, \\ n(1 - nx), & 0 \leq x \leq \frac{1}{n}, \\ 0, & x > \frac{1}{n}, \quad n \in \mathbb{N}. \end{cases}$$

The sequence $(f_n)_{n \in \mathbb{N}}$ *is distributionally convergent to the distribution* $\delta(x)$.

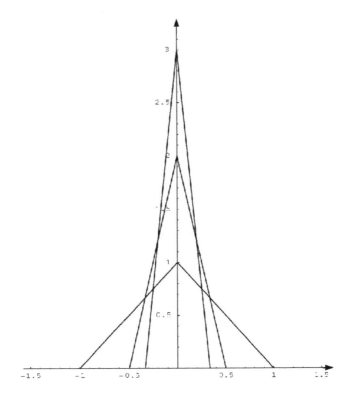

Figure 22. The graphs of functions $f_n(x)$, $n = 1, 2, 3$

In order to prove it we note that

$$\int\limits_{-\infty}^{+\infty} f_n(x)\, dx = 1 \quad \textit{for every} \quad n \in \mathbb{N},$$

and

$$\lim_{n \to \infty} \int\limits_{-1}^{x} f_n(t)\, dt = \begin{cases} 0, & \textit{if } x < 0, \\ 1, & \textit{if } x > 0. \end{cases}$$

Example 3.7.90 *Let us consider the functions (see Fig. 23)*

$$h_a(x) = \begin{cases} \frac{1}{2a}, & \textit{if } x \in [-a, a],\ a > 0, \\ 0, & \textit{if } x \notin [-a, a]. \end{cases}$$

We shall show that

$$\lim_{a \to 0+} h_a(x) \stackrel{d}{=} \delta(x).$$

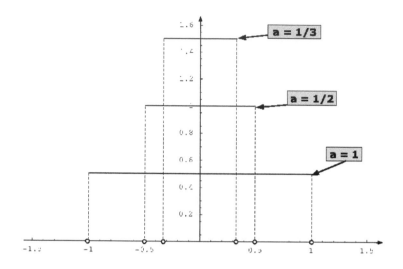

Figure 23. The graphs of functions $h_a(x)$, $a = 1, \frac{1}{2}, \frac{1}{3}$

In fact, for an arbitrary function $\varphi \in \mathcal{D}$, by Lagrange's theorem, we have

$$(h_a(x), \varphi(x)) = \int\limits_{-\infty}^{+\infty} h_a(x)\varphi(x)dx = \frac{1}{2a} \int\limits_{-a}^{a} \varphi(x)dx = \varphi(\xi),$$

where $-a < \xi < a$. Hence, for $a \to 0^+$, we obtain in turn

$$\lim_{a \to 0^+} (h_a(x), \varphi(x)) = \lim_{a \to 0^+} \varphi(\xi) = \varphi(0) = (\delta(x), \varphi(x)).$$

Thus, by Theorem 3.7.159, we get the required convergence.

Differentiation of a Piecewise Continuous Function

For piecewise continuous functions the following theorem holds true:

Theorem 3.7.161 *Let the function f have discontinuities of the first kind at points x_k, $k = 1, 2, \cdots, n$, with a jump equal to $\sigma_k = f(x_k + 0) - f(x_k - 0)$ and let f have the derivative $\frac{df}{dx}$ continuous everywhere without points x_k. Then the distributional derivative f' is given by the formula*

$$f'(x) = \frac{df(x)}{dx} + \sum_{k=1}^{n} \sigma_k \delta(x - x_k).$$

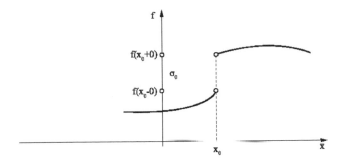

Figure 24. The graph of function with a jump σ_o at the point x_o

Proof. It is sufficient to consider the case when the function f has only one discontinuity of the first kind at a point x_o with a jump equal to

$$\sigma_o = f(x_o + 0) - f(x_o - 0), \quad \text{(see Figure 24)}.$$

It follows from (3.7.10) that for every function $\varphi \in \mathcal{D}$ there holds

$$\int_{-\infty}^{+\infty} f'(x)\varphi(x)\,dx = -\int_{-\infty}^{+\infty} f(x)\varphi'(x)\,dx.$$

By straightforward calculations we get

$$\int_{-\infty}^{+\infty} f(x)\varphi'(x)dx = \int_{-\infty}^{x_o} f(x)\varphi'(x)dx + \int_{x_o}^{+\infty} f(x)\varphi'(x)dx$$

$$= f(x)\varphi(x)\Big|_{-\infty}^{x_o} - \int_{-\infty}^{x_o} \frac{df(x)}{dx}\varphi(x)dx$$

$$+ f(x)\varphi(x)\Big|_{x_o}^{+\infty} - \int_{x_o}^{+\infty} \frac{df(x)}{dx}\varphi(x)dx$$

$$= f(x_o - 0)\varphi(x_o) - \int_{-\infty}^{+\infty} \frac{df(x)}{dx}\varphi(x)dx - f(x_o + 0)\varphi(x_o)$$

$$= -\varphi(x_o)\Big(f(x_o + 0) - f(x_o - 0)\Big) - \int_{-\infty}^{+\infty} \frac{df(x)}{dx}\varphi(x)dx$$

$$= -\sigma_o \varphi(x_o) - \int\limits_{-\infty}^{+\infty} \frac{df(x)}{dx} \varphi(x)dx$$

$$= -\int\limits_{-\infty}^{+\infty} \delta(x - x_o)\sigma_o\varphi(x)dx - \int\limits_{-\infty}^{+\infty} \frac{df(x)}{dx}\varphi(x)dx$$

$$= -\int\limits_{-\infty}^{+\infty} \left(\frac{df(x)}{dx} + \sigma_o\delta(x - x_o)\right)\varphi(x)dx,$$

where the equality

$$\int\limits_{-\infty}^{+\infty} \delta(x - x_o)\sigma_o\varphi(x)\, dx = \varphi(x_o)\sigma_o$$

follows from (3.7.20). Hence, for every function $f \in \mathcal{D}$ we have

$$\int\limits_{-\infty}^{+\infty} f'(x)\varphi(x)dx = \int\limits_{-\infty}^{+\infty} \left(\frac{df(x)}{dx} + \sigma_o\delta(x - x_o)\right)\varphi(x)dx.$$

By (3.7.15), we obtain

$$f'(x) = \frac{df(x)}{dx} + \sigma_o\delta(x - x_o). \tag{3.7.24}$$

$\square$

Conclusion 3.7.1 If the functions $f, f^{(1)}, f^{(2)}, \cdots, f^{(n-1)}$ have discontinuities of the first kind at a point x_o with jumps equal to $\sigma_o^{(o)}, \sigma_o^{(1)}, \sigma_o^{(2)}, \cdots, \sigma_o^{(n-1)}$, then

$$f'(x) = \frac{df(x)}{dx} + \sigma_o^{(o)}\delta(x - x_o);$$

$$f^{(2)}(x) = \frac{d^2 f(x)}{dx^2} + \sigma_o^{(o)}\delta'(x - x_o) + \sigma_o^{(1)}\delta(x - x_o);$$

$$f^{(3)}(x) = \frac{d^3 f(x)}{dx^3} + \sigma_o^{(o)}\delta^{(2)}(x - x_o) + \sigma_o^{(1)}\delta^{(1)}(x - x_o) + \sigma_o^{(2)}\delta(x - x_o);$$

$$\cdots\cdots\cdots\cdots\cdots\cdots\cdots\cdots\cdots \tag{3.7.25}$$

$$f^{(n)}(x) = \frac{d^n f(x)}{dx^n} + \sigma_o^{(o)}\delta^{(n-1)}(x - x_o) + \cdots + \sigma_o^{(n-1)}\delta(x - x_o)$$

$$= \frac{d^n f}{dx^n} + \sum_{l=0}^{n-1} \sigma_o^{(l)}\delta^{(k-l-1)}(x - x_o).$$

Proof. We will prove the above formulas by induction. For $n = 1$ the formula (3.7.25) takes the form (3.7.24).

Let us assume that formula (3.7.25) holds true for a certain $k \in \mathbb{N}$:

$$f^{(k)}(x) = \frac{d^k f(x)}{dx^k} + \sum_{l=0}^{k-1} \sigma_o^{(l)}\delta^{(k-l-1)}(x - x_o). \tag{3.7.26}$$

We will prove that the formula holds true for $k+1$. Indeed, differentiating the formula (3.7.26), we get

$$f^{(k+1)}(x) = \left(f^{(k)}(x)\right)' = \left(\frac{d^k f(x)}{dx^k} + \sum_{l=0}^{k-1} \sigma_o^{(l)} \delta^{(k-l-1)}(x-x_o)\right)'$$

$$= \frac{d^{k+1} f(x)}{dx^{k+1}} + \sigma_o^{(k)} \delta(x-x_o) + \sum_{l=0}^{k-1} \sigma_o^{(l)} \delta^{(k-1)}(x-x_o)$$

$$= \frac{d^{k+1} f(x)}{dx^{k+1}} + \sum_{l=0}^{k} \sigma_o^{(l)} \delta^{(k-l)}(x-x_o),$$

what was to be shown. ▯

It follows from Conclusion 3.7.1 that the n-th distributional derivative is equal to a sum of the n-th usual derivative and a sum of products of jumps of functions $f, f^{(1)}, f^{(2)}, \cdots,$ $f^{(n-1)}$ and relevant derivatives of δ – Dirac's distribution, translated to the point $x = x_o$. In particular, if f is a smooth function, then the distributional derivative is equal to derivative in the classical sense, i.e.,

$$f^{(k)}(x) = \frac{d^k f(x)}{dx^k}.$$

Example 3.7.91 *Let us consider a characteristic function of the interval $[-a, \ a]$*

$$h(x) = \begin{cases} 1, & \text{for } |x| \leq a, \\ 0, & \text{for } |x| > a > 0. \end{cases}$$

From formula (3.7.24) it follows that

$$h'(x) = \frac{dh}{dx} + \sigma_1 \delta(x+a) + \sigma_2 \delta(x-a) = 0 + \delta(x+a) - \delta(x-a).$$

Hence

$$h'(x) = \delta(x+a) - \delta(x-a).$$

Distributions with a One-Point Support

According to the definition of the support of the distribution (see section 3.6.3), a point $x_o \in \mathbb{R}$ is a support of the distribution f in $\mathbb{R}$ if and only if the distribution f is a zero distribution in the set $\mathbb{R} \setminus \{x_o\}$, i.e., in the space $\mathbb{R}$ without the point x_o, but f is not a zero distribution in the whole space $\mathbb{R}$.

We have the following theorem for distributions with one-point support.

Theorem 3.7.162 *If the support of a distribution f in $\mathbb{R}$ is the origin, then f is a linear combination of the Dirac delta distribution and its derivatives. In other words, there exist a number $k \in \mathbb{N}_o$ and real numbers α_m $(0 \leq m \leq k)$ such that*

$$f(x) = \sum_{0 \leq m \leq k} \alpha_m \, \delta^{(m)}(x).$$

In order to prove this theorem it is enough to consider Theorem 3.4.121, section 3.4.4, for $x_o = 0$.

Applications

Problem 3.7.13 Prove that a sequence of Dirichlet functions is distributionally convergent to Dirac's delta distribution $\delta(x)$, i.e., prove that the following takes place:

$$\lim_{n\to\infty} \frac{\sin nx}{\pi x} \overset{\mathrm{d}}{=} \delta(x). \tag{3.7.27}$$

Proof. Using the properties of the Fourier series (see [F]) we can write:

$$\lim_{n\to\infty} \int_{-\infty}^{+\infty} \frac{\sin nx}{x}\varphi(x)\,dx = \pi\varphi(0), \quad \varphi \in \mathcal{D}. \tag{3.7.28}$$

By (3.7.21) we have

$$\int_{-\infty}^{+\infty} \varphi(x)\delta(x)dx = \varphi(0).$$

Hence, and from (3.7.28) we have the convergence of the sequence of functions

$$\left(\frac{\sin nx}{\pi x}\right)_{n\in\mathbb{N}}$$

to Dirac's delta distribution, as $n \to \infty$. This implies the equality (3.7.27). ◻

Problem 3.7.14 Prove the equality

$$\frac{1}{2\pi} \int_{-\infty}^{+\infty} e^{i\omega x}d\omega \overset{\mathrm{d}}{=} \delta(x). \tag{3.7.29}$$

Proof. Let us consider the sequence

$$f_n(x) = \int_{-n}^{n} e^{i\omega x}d\omega, \quad n \in \mathbb{N}.$$

From

$$\int_{-n}^{n} e^{i\omega x}\,d\omega = \frac{e^{inx} - e^{-inx}}{ix} = 2\frac{\sin nx}{x}$$

and (3.7.27) we obtain

$$\lim_{n\to\infty} \int_{-n}^{n} e^{i\omega x}d\omega = \lim_{n\to\infty} 2\frac{\sin nx}{x} \overset{\mathrm{d}}{=} 2\pi\delta(x);$$

thus we get the required equality (3.7.29). ◻

3.7.3 Convolution of Distributions

Among other irregular operations an especially important role is played by the convolution of distributions. The sequential theory of convolution was developed in [AMS].

1. Convolution of Two Smooth Functions

J. Mikusiński has shown that if φ and ψ are smooth functions in $\mathbb{R}$, then the convolution $\varphi * \psi$ is not necessarily a smooth function, even if it exists at every point $x \in \mathbb{R}$. We give as an example, see [AMS], p. 130, the following:

Example 3.7.92 *Let ω be a smooth function in $\mathbb{R}$, such that*

$$\omega(x) = 0, \quad if \quad |x| \le \frac{1}{4} \quad or \quad |x| \ge \frac{1}{2}$$

and let

$$\varphi(x) = \psi(x) = \sum_{n=-\infty}^{+\infty} \omega\left(2^n(x-n)\right).$$

It is easy to see that this series converges almost uniformly. Thus,

$$\varphi'(x) = \psi'(x) = \sum_{n=-\infty}^{+\infty} 2^{|n|}\omega\left(2^n(x-n)\right).$$

If $x \ne 0$ the infinite integrals

$$\int_{-\infty}^{+\infty} \varphi(x-t)\psi(t)dt \quad and \quad \int_{-\infty}^{+\infty} \varphi(x-t)\psi'(t)dt$$

reduce to integrals over a bounded interval. Thus, the convolution

$$h = \varphi * \psi \quad and \quad h' = \varphi' * \psi$$

exist for every $x \ne 0$. We have

$$h(0) = \sum_{n=-\infty}^{+\infty} 2^{-|n|}p \quad where \quad p = \int_{-\infty}^{+\infty} \omega(-t)\omega(t)dt.$$

From this follows that the convolution $\varphi * \psi$ is also defined at $x = 0$, and is thus defined everywhere in $\mathbb{R}$. On the other hand, we have

$$h'(0) = \sum_{n=-\infty}^{+\infty} 2^{-|n|}2^{|n|}q = \sum_{n=-\infty}^{+\infty} q, \quad where \quad q = \int_{-\infty}^{+\infty} \omega(-t)\omega'(t)dt.$$

The series $\sum_{n=-\infty}^{+\infty} q$ is divergent if $q \ne 0$. Whether or not $h' = \varphi' * \psi$ is defined at $x = 0$ depends on q, and we can choose ω in such a way that $q \ne 0$. Then $h'(0) = \pm\infty$ and $h'(x)$ does not tend to any finite limit as $x \to 0$. Thus, although the convolution $\varphi * \psi$ of the smooth functions φ and ψ exists at $x = 0$, it is not a smooth function.

Definition 3.7.88 *We say that the convolution $\varphi * \psi$ of the smooth functions $\varphi, \psi \in \mathbb{R}$ exists smoothly if for any order $m, k \in \mathbb{N}_o$ the convolutions $\varphi^{(k)} * \psi^{(m)}$ exist in $\mathbb{R}$ and are continuous, and the convolutions $|\varphi^{(k)}| * |\psi^{(m)}|$ are locally integrable functions.*

If the convolution of smooth functions *exists smoothly* then it has the following properties:

Property 3.7.17 If the convolution of two smooth functions φ and ψ exists smoothly, then it is itself a smooth function. Moreover,

$$(\varphi * \psi)^{(k)} = \varphi^{(k)} * \psi = \varphi * \psi^{(k)}, \quad \text{for each} \quad k \in \mathbb{N}_o.$$

Property 3.7.18 If the convolution of two smooth functions φ and ψ exists smoothly and if λ is a real number, then the convolutions

$$(\lambda \varphi) * \psi, \quad \text{and} \quad \varphi * (\lambda \psi)$$

exist smoothly and the equalities

$$(\lambda \varphi) * \psi = \varphi * (\lambda \psi)$$

holds everywhere.

Property 3.7.19 If φ, ψ and χ are smooth functions and the convolutions $\varphi * \psi$ and $\varphi * \chi$ exist smoothly, then the convolution $\varphi * (\psi * \chi)$ also exists smoothly and

$$\varphi * (\psi + \chi) = \varphi * \psi + \varphi * \chi$$

holds everywhere.

Property 3.7.20 If f, g are locally integrable functions such that the convolution $|f| * |g|$ is a locally integrable function and h, u are smooth functions of bounded supports, then the convolution

$$(f * h) * (g * u)$$

exists smoothly.

2. Convolution of Two Distributions

Now we will introduce the operation of convolution of two distributions and will give its basic properties. We have already mentioned different possibilities of introducing the definition of convolution. Below we cite a definition of convolution in Mikusiński's sense, [AMS].

Let f and g be distributions in $\mathbb{R}$ and let $(f_n)_{n\in\mathbb{N}} = (f * \delta_n)_{n\in\mathbb{N}}$ and $(g_n)_{n\in\mathbb{N}} = (g * \delta_n)_{n\in\mathbb{N}}$ be their regular sequences with the same delta sequence.

Definition 3.7.89 *We say that the convolution of f and g exists if for every delta sequence* $(\delta_n)_{n\in\mathbb{N}}$ *the corresponding convolutions* $f_n * g_n$ $(n \in \mathbb{N})$ *exist smoothly and represent a fundamental sequence. The distribution determined by the fundamental sequence is, by definition, the convolution of f and g, i.e.,*

$$f * g : \overset{d}{=} \lim_{n\to\infty} [(f * \delta_n) * (g * \delta_n)]. \tag{3.7.30}$$

Remark 3.7.125 Definition 3.7.89 *does not depend on the choice of delta sequence* $(\delta_n)_{n\in\mathbb{N}}$.

Proof. Let $(\delta_{1n})_{n\in\mathbb{N}}$ and $(\delta_{2n})_{n\in\mathbb{N}}$ are two different delta sequences. If the convolution $f * g$ exists, then both sequences

$$\left((f * \delta_{1n}) * (g * \delta_{1n})\right)_{n\in\mathbb{N}} \quad \text{and} \quad \left((f * \delta_{2n}) * (g * \delta_{2n})\right)_{n\in\mathbb{N}} \tag{3.7.31}$$

are fundamental. We have to show that they represent the same distribution.

Let δ_n be the n-th element of the interlaced sequence

$$\delta_{11}, \delta_{21}, \delta_{12}, \delta_{22}, \delta_{13}, \delta_{23}, \cdots ;$$

then the sequence $(\delta_n)_{n\in\mathbb{N}}$ is also a delta sequence. This implies that the sequence

$$\left((f * \delta_n) * (g * \delta_n)\right)_{n\in\mathbb{N}}$$

is fundamental. Since sequences (3.7.31) are fundamental, they thus represent the same distribution. $\quad\Box$

Remark 3.7.126 *The convolution* (3.7.30) *is not a regular operation, since in this definition we have restricted ourselves to those particular fundamental sequences* $(f_n)_{n\in\mathbb{N}}$ *and* $(g_n)_{n\in\mathbb{N}}$ *which are of the form*

$$f_n = f * \delta_n, \quad g_n = g * \delta_n, \quad n \in \mathbb{N}.$$

Remark 3.7.127 *The convolution* $f * g$ *does not exist for any pairs of distributions* f, g.

Remark 3.7.128 Definition 3.7.89 *is compatible with the definition of the convolution of two locally integrable functions in* $\mathbb{R}$.

Proof. If f and g are locally integrable functions in $\mathbb{R}$, such that convolution of $|f| * |g|$ exists almost everywhere and represents a locally integrable function, then (by Theorem 1, p. 286 in [Sk.4]) convolution $f_n * g_n$ exists smoothly. Moreover, by Property 3.4.13, section 3.4, the sequence $(f_n * g_n)_{n\in\mathbb{N}}$ converges locally in mean, and thus distributionally to $f * g$. Therefore, convolution $f * g$ is compatible with Definition 3.7.89 of the convolution of distributions. $\quad\Box$

Remark 3.7.129 *The definition of convolution (3.7.30) is compatible with the definition of convolution, if the distribution g is a smooth function of bounded support (see Remark 3.3.108, section 3.3.5)*

$$f * g \stackrel{d}{=} \lim_{n \to \infty} (f_n * g) \quad g \in \mathcal{D}.$$

Proof. It is easy to check that the convolutions $f_n * g_n$ exist smoothly. It thus suffices to check that

$$f_n * g_n - f_n * g \stackrel{d}{\longrightarrow} 0, \quad \text{i.e.} \quad \lim_{n \to \infty} (f_n * (g_n - g)) \stackrel{d}{=} 0.$$

In fact, by Property 3.4.10, section 3.4.1, we have

$$(g_n - g)^{(k)} \stackrel{\text{a.u.c.}}{\Rightarrow} 0 \quad \text{for every} \quad k \in \mathbb{N}_o.$$

Since $f_n \stackrel{d}{\longrightarrow} f$, by Theorem 3.5.137, section 3.5.4, this implies that

$$f_n * (g_n - g) \stackrel{d}{\longrightarrow} f * 0 = 0,$$

which completes the proof.   ⬚

3. Properties of Convolution of Distributions

Property 3.7.21 If the convolution $f * g$ of the distributions f, g exists, then the convolutions $(\lambda f) * g$ and $f * (\lambda g)$ also exist for every real number λ and we have

$$(\lambda f) * g = f * (\lambda g) = \lambda (f * g), \quad \lambda \in \mathbb{R}. \tag{3.7.32}$$

Property 3.7.22 If the convolutions $f * g$ and $f * h$ of the distributions f, g, h exist, then the convolution $f * (g + h)$ also exists and we have

$$f * (g + h) = f * g + f * h. \tag{3.7.33}$$

Property 3.7.23 If the convolution $f * g$ of the distributions f, g exists, then $g * f$ exists and we have

$$f * g = g * f. \tag{3.7.34}$$

Property 3.7.24 If the convolution $f * g$ of the distributions f, g exists, then the convolutions $f^{(k)} * g$ and $f * g^{(k)}$ also exist for every $k \in \mathbb{N}_o$ and we have

$$(f * g)^{(k)} = f^{(k)} * g = f * g^{(k)}, \quad k \in \mathbb{N}_o. \tag{3.7.35}$$

Property 3.7.25 If the convolution $f * g$ of the distributions f, g exists, then $f^{(k)} * g^{(l)}$ also exists for any $k, l \in \mathbb{N}_o$ and we have

$$f^{(k)} * g^{(l)} = (f * g)^{(k+l)}, \quad k, l \in \mathbb{N}_o. \tag{3.7.36}$$

Proof. (of Properties 3.7.21 through 3.7.25) Let $(\delta_n)_{n \in \mathbb{N}_o}$ be a delta sequence and let

$$f_n = f * \delta_n, \quad g_n = g * \delta_n, \quad h_n = h * \delta_n.$$

If the convolution $f * g$ exists, then the convolutions $f_n * g_n$ exist smoothly and we have

$$(\lambda f_n) * g_n = f_n * (\lambda g_n) = \lambda(f_n * g_n),$$

$$f_n * g_n = g_n * f_n,$$

$$(f_n * g_n)^{(k)} = f_n^{(k)} * g_n = f_n * g_n^{(k)},$$

and all the convolutions involved exist smoothly. This proves equality (3.7.32), (3.7.34), (3.7.35), and (3.7.36). Moreover, if $f * h$ exists, then

$$f_n * (g_n + h_n) = f_n * g_n + f_n * h_n$$

and all the convolutions exist smoothly. This proves (3.7.33).

By (3.7.35), we have

$$(f * g)^{(k+l)} = [(f * g)^{(k)}]^{(m)} = (f^{(k)} * g)^{(m)} = f^{(k)} * g^{(m)};$$

this proves (3.7.36). ∎

For convolution of two distributions, when at least one of them is of bounded support, we have:

Property 3.7.26 If f, g are distributions in $\mathbb{R}$ and at least one of them is of bounded support, then the convolution $f * g$ exists in $\mathbb{R}$.

Property 3.7.27 If two of the distributions f, g, h are of bounded support, then all the convolutions involved in

$$(f * g) * h = f * (g * h) \tag{3.7.37}$$

exist and the equality holds.

Property 3.7.28 If one of the distributions f, g, h has a bounded support and the convolution of the remaining two exists, then all the convolutions involved in the equality (3.7.37) exist and the equality holds.

Property 3.7.29 If f, g, h are distributions such that the convolution $f * g$ exists and each of distributions h, u is of bounded support, then

$$(f * h) * (g * u) = (f * g) * (h * u) \tag{3.7.38}$$

and all the convolutions appearing in the equality exist.

Corollary 3.7.43 *The Dirac delta distribution is a unit for the convolution, i.e.,*

$$\delta * f = f * \delta = f \quad for \quad f \in \mathcal{D}',$$

(see Problem 3.7.13*).*

The operation of convolution is connected to the scalar product and to the value of distribution at a point. From Theorem 3.6.139, formula (3.6.8), (3.6.9), section 3.6.1, we have

Theorem 3.7.163 *Let f and g be distributions and $g_-(x) = g(-x)$. If there exist convolution $f * g$ and its value $(f * g)(0)$ at a point 0, then there exists scalar product (f, g_-) and the equality holds*

$$(f * g)(0) = (f, g_-). \tag{3.7.39}$$

Theorem 3.7.164 *If a convolution of distributions f, g, h is associative, i.e., convolutions $(f * g) * h$, $f * (g * h)$ exist and their values $\big((f * g) * h\big)(0)$, $\big(f * (g * h)\big)(0)$ at a point 0, then*

$$(f_-, g * h) = (f * g, h_-).$$

Problem 3.7.15 Prove the following formulas:

(a) $f * \delta = f$,

(b) $f * \delta' = f'$,

(c) $\delta(x - \alpha) * \delta(x - \beta) = \delta(x - \alpha - \beta)$,

(d) $(1_+ * \varphi)(x) = \int\limits_{-\infty}^{x} \varphi(t)dt, \quad \varphi \in \mathcal{D}.$

Proof. (a) By Property 3.7.26, section 3.7.3, there exists convolution $f * \delta$, i.e.,

$$f * \delta \overset{\mathrm{d}}{=} \lim_{n\to\infty} \big((f * \delta_n) * \delta_n\big),$$

where $(\delta_n)_{n\in\mathbb{N}}$ is a delta sequence. By Property 3.7.27, section 3.7.3, and Property 3.4.3, section 3.4.1, we have

$$(f * \delta_n) * \delta_n = f * (\delta_n * \delta_n) = f * \bar\delta_n,$$

where $(\bar\delta_n)_{n\in\mathbb{N}}$ is a delta sequaence, too. The sequence $(f * \bar\delta_n)_{n\in\mathbb{N}}$ is a regular sequence of a distribution f, thus

$$\lim_{n\to\infty} (f * \bar\delta_n) \overset{\mathrm{d}}{=} f.$$

Hence

$$f * \delta = f, \quad f \in \mathcal{D}'.$$

(b) By Property 3.7.26, section 3.7.3, there exists a convolution $f * \delta'$, i.e.,

$$f * \delta' \overset{\mathrm{d}}{=} \lim_{n \to \infty} ((f * \delta_n) * \delta_n').$$

Similarly, we have in (a)

$$(f * \delta_n) * \delta_n' = f * (\delta_n * \delta_n)' = f * \overline{\delta}_n{}' = f' * \overline{\delta}_n.$$

Since the sequence $(f' * \delta_n)_{n \in \mathbb{N}}$ is regular for a distribution f', thus we obtain

$$f * \delta' = f', \quad f \in \mathcal{D}'.$$

⬛

Problem 3.7.16 Let

$$f_k(x) = 1_+(x) \frac{x^{k-1}}{\Gamma(k)} e^{\lambda x}, \quad k > 0, \quad \Gamma(k) = (k-1)!$$

Prove that for $k = \alpha, \beta$, we have

$$(f_\alpha * f_\beta)(x) = 1_+(x) f_{\alpha+\beta}(x).$$

Remark. Put $\tau = xu$.

3.7.4 Multiplication of Distributions

The product of distributions is an irregular operation. It has caused problems to the creators of the theory of distributions from the very beginning. At first one could only multiply distributions by smooth functions. L. Schwartz's example (see Problem 3.7.20, section 3.7.5) shows that even then problems with associativity of this operation arise. The task of generalizing the product of distributions has been undertaken by many mathematicians (see [BoL], [BoP], [Br], [BrD]). H. König ([Kö.1], [Kö.2], [Kö.3]) has shown that one cannot define a general product of distributions without losing its "good" properties.

In the sequential approach of the theory of generalized functions one has J. Mikusiński's 1962 definition of the product (see [Mi.5]):

$$f \cdot g : \overset{\mathrm{d}}{=} \lim_{n \to \infty} (f * \delta_{1n})(g * \delta_{2n}) \tag{3.7.40}$$

if the limit exists for arbitrary delta sequences $(\delta_{1n})_{n \in \mathbb{N}}$, $(\delta_n)_{n \in \mathbb{N}}$. R. Shiraishi and M. Itano have shown in [ShI] that the definition (3.7.40) is equivalent to the definition given by Y. Hirata and H. Ogata in [HiO] also with the use of delta sequences. Their definition is not as general as (3.7.42) because the product $\frac{1}{x} \delta(x)$ does not exist in the sense (3.7.40) (see [It],

[Ka.3]). The product of distributions was also of interest to many other mathematicians, i.e., A. Kamiński ([Ka.1], [Ka.2], [Ka.3]), P. Antosik and J. Ligęza [AL].

We start below with the definition of a product of distributions given by Mikusiński in 1966 (see [Mi.8], [AMS], p. 242):

Definition 3.7.90 *We say that the product of distributions f and g exists in (a, b) if for every delta sequence $(\delta_n)_{n \in \mathbb{N}}$ the sequence*

$$\Big((f * \delta_n)(g * \delta_n)\Big)_{n \in \mathbb{N}} \tag{3.7.41}$$

is distributionally convergent in (a, b).

   Then we write

$$f \cdot g := \lim_{n \to \infty} (f * \delta_n)(g * \delta_n). \tag{3.7.42}$$

The definition (3.7.42) is very general and enables us to consider very interesting and important products of distributions that do not exist in the sense of Schwartz. For example, the significant in physics product $\frac{1}{x} \cdot \delta(x)$ exists and equals, according to expectations of physicists, $-\frac{1}{2} \cdot \delta(x)$ (see [Mi.8], [AMS], p. 249). The generalizations of this formula can be found in [Fs.2] and [Is].

Theorem 3.7.165 *If g is a smooth function in (a, b) then the limit (3.7.42) exists for every distribution f defined in (a, b) and is equal to the product $f \cdot g$, where the product is understood as a regular operation.*

Proof.   It follows from the property of delta sequences, Property 3.4.10, section 3.4.1, that

$$(g * \delta_n)^{(k)} \overset{\text{a.u.c.}}{\rightrightarrows} g^{(k)} \quad \text{for all} \quad k \in \mathbb{N}_o.$$

Since

$$f * \delta_n \overset{\text{d}}{\longrightarrow} f \quad \text{in} \quad (a, b),$$

by the Theorem 3.5.132, section 3.5.2, follows that the sequence (3.7.41) is convergent to the product $f \cdot g$. ☐

Corollary 3.7.44 *It follows from the Theorem 1 that we can use the symbol $f \cdot g$ also in the case when g is not a smooth function.*

   Properties of the Product *The following equalities hold:*

$$f \cdot g = g \cdot f; \tag{3.7.43}$$

$$(\alpha f) \cdot g = f \cdot (\alpha g) = \alpha (f \cdot g), \quad \alpha \in \mathbb{R}; \tag{3.7.44}$$

$$(f \pm g) \cdot h = f \cdot h \pm g \cdot h; \tag{3.7.45}$$

$$(f \cdot g)^{(k)} = \sum_{l=0}^{k} \binom{k}{l} f^{(l)} \cdot g^{(k-l)}, \quad k \in \mathbb{N}_o, \tag{3.7.46}$$

provided the products on the right exist.

Theorem 3.7.166 *If a distribution f takes the value $f(0)$ at the point $x = 0$, then*

$$f(x)\delta(x) = f(0)\delta(x), \quad f \in \mathcal{D}'. \tag{3.7.47}$$

Proof. Let $\varphi \in \mathcal{D}$ and $\varphi(0) \neq 0$. Since, by Definition 3.6.83, section 3.6.1 and equality (3.7.14),

$$(\delta_n, \varphi) \xrightarrow{\mathrm{d}} \varphi(0), \quad \text{as} \quad n \to \infty \quad i.e., \quad \lim_{n \to \infty} \int_{\mathbb{R}} \delta_n(x)\varphi(x)dx \stackrel{\mathrm{d}}{=} \varphi(0),$$

we have

$$\gamma_n = (\delta_n, \varphi) \neq 0 \quad \text{for} \quad n \geq n_o.$$

It is easily checked that the sequence

$$\alpha_n = \frac{1}{\gamma_n} \delta_n \, \varphi, \quad n \in \mathbb{N},$$

is a delta sequence for $n \geq n_o$. Using the notation

$$\overline{\alpha}_n(x) := \alpha_n(-x),$$

we can write for $n \geq n_o$,

$$\left((f * \delta_n)\delta_n, \varphi \right) = \gamma_n \left(f * \delta_n, \alpha_n \right)$$
$$= \gamma_n [(f * \delta_n) * \overline{\alpha}_n](0) = \gamma_n [f * (\delta_n * \overline{\alpha}_n)](0),$$

by (3.6.6), (3.6.10), section 3.6.1, Property 3.7.27, section 3.7.3. Hence, by Theorem 3.5.132, section 3.5.2, we obtain

$$\left((f * \delta_n)\delta_n, \varphi \right) \xrightarrow{\mathrm{d}} \varphi(0)f(0),$$

since the sequence $(\delta_n * \overline{\alpha}_n)_{n \in \mathbb{N}}$ is a delta sequence, too.

If $\varphi(0) = 0$, we can write $\varphi = \varphi_1 + \varphi_2$, where $\varphi_1, \varphi_2 \in \mathcal{D}$ and $\varphi_1(0) \neq 0$ and $\varphi_2(0) \neq 0$. Applying the preceding result to φ_1 and φ_2 we have

$$\left((f * \delta_n)\delta_n, \varphi_1 \right) \xrightarrow{\mathrm{d}} \varphi_1(0)f(0),$$
$$\left((f * \delta_n)\delta_n, \varphi_2 \right) \xrightarrow{\mathrm{d}} \varphi_2(0)f(0).$$

Hence,

$$\left((f * \delta_n)\delta_n, \varphi \right) = \left((f * \delta_n)\delta_n, \varphi_1 \right) + \left((f * \delta_n)\delta_n, \varphi_2 \right)$$
$$\xrightarrow{\mathrm{d}} \varphi_1(0)f(0) + \varphi_2(0)f(0) = \varphi(0)f(0).$$

Thus,

$$\left((f * \delta_n)\delta_n, \varphi\right) \xrightarrow{\mathrm{d}} f(0)\varphi(0), \quad \text{for every} \quad \varphi \in \mathcal{D}. \tag{3.7.48}$$

According to (3.6.6), section 3.6.1 and Theorem 3.7.157, section 3.7.2, we have

$$(f(0)\delta, \varphi) = f(0)(\delta, \varphi) = f(0)\,\varphi(0). \tag{3.7.49}$$

By Definition 3.7.90 it follows that

$$\lim_{n\to\infty} (f * \delta_n)\delta_n \stackrel{\mathrm{d}}{=} f(x)\delta(x). \tag{3.7.50}$$

Comparing (3.7.49) and (3.7.50) we get

$$\left((f * \delta_n)\delta_n, \varphi\right) \xrightarrow{\mathrm{d}} (f(0)\delta, \varphi), \quad \text{for} \quad \varphi \in \mathcal{D}. \tag{3.7.51}$$

By (3.7.51) we have

$$\left((f * \delta_n)\delta_n, \varphi\right) \xrightarrow{\mathrm{d}} \left(f(x)\delta(x), \varphi\right), \quad \text{for} \quad \varphi \in \mathcal{D}. \tag{3.7.52}$$

According to (3.7.52), (3.7.53), and Theorem 3.5.123, section 3.5.1, we obtain

$$f(0)\delta(x) = f(x)\delta(x).$$

which proves our assertion.   ⧠

On the Associativity of the Product

The product of three functions is always associative, i.e., $(f \cdot g) \cdot h = f \cdot (g \cdot h)$. Therefore, it may seem odd that an analogous property does not hold for distributions. The example below given by L. Schwartz in [S.2] (see Problem 3.7.20) shows the difficulties that arise when multiplying distributions. However, it should be noted that associativity does hold whenever at least two of the factors are smooth functions. Explicity, if f is any distribution and φ, ψ are smooth functions, then

$$(f\varphi)\psi = f(\varphi\psi).$$

This equality follows from the remark that the products with φ, with ψ and with $\varphi\psi$ are regular operations. Namely, we have:

Theorem 3.7.167 (L. Schwartz) *If φ, ψ are smooth functions and f is a distribution, then*

$$(\varphi\psi)f = \varphi(\psi f).$$

One can state certain criteria for the existence of the product of distributions and its associativity. In order to do that we give the definition of the order of a distribution with respect to a continuous function. Let $k^+ = \max(k, 0)$, $k^- = \min(k, 0)$ for $k \in \mathbb{Z}$, where $\mathbb{Z}$ is the set of integers.

Definition 3.7.91 *We say that a distribution f is of the order $k \in \mathbb{Z}$ in (a, b) with respect to a continuous function if there exists such a function F that*

$$F^{(k^+)} = f \quad in \quad (a, b) \quad and$$

$$F^{(j)} \quad for \quad 0 \le j \le -k^- \quad is \ a \ continuous \ function \ in \quad (a, b).$$

The order of the distribution defined above is not unique.

Theorem 3.7.168 (Mi.5) *If f and g are distributions in (a, b) of orders $k, m \in \mathbb{Z}$, respectively, with respect to a continuous function and $k + m \le 0$ then the product $f \cdot g$ exists in (a, b) and it is of the order $\max(m, k)$.*

Theorem 3.7.169 *If f, g and h are distributions in (a, b) of orders $k, l, m \in \mathbb{Z}$ with respect to a continuous function and*

$$k + l \le 0, \quad k + m \le 0, \quad m + l \le 0$$

then

$$(f \cdot g) \cdot h = f \cdot (g \cdot h).$$

The Schwartz theorem (Theorem 3.7.167) can be obtained as a conclusion from Theorem 3.7.169. Using the notation of an order of the distribution with respect to a measure, P. Antosik generalized Theorem 3.7.168 and Theorem 3.7.169 (see [A]).

3.7.5 Applications

Nonexistence of δ^2

By the square of the delta distribution, δ^2, we mean the product $\delta \cdot \delta$.

Problem 3.7.17 Prove that the product $\delta \cdot \delta$ does not exist, i.e., that the symbol δ^2 is meaningless.

Proof. According to the definition of the product (Definition 3.7.90, section 3.7.4), we can write

$$\delta \cdot \delta \overset{\mathrm{d}}{=} \lim_{n \to \infty} (\delta * \delta_n)(\delta * \delta_n), \quad \text{i.e.,} \quad \delta \cdot \delta \overset{\mathrm{d}}{=} \lim_{n \to \infty} \delta_n^2,$$

where $(\delta_n)_{n \in \mathbb{N}}$ is an arbitrary delta sequence. Note that there exists a smooth function φ of bounded support such that

$$\varphi(x) = 1 \quad \text{for} \quad -\frac{1}{4} \le x \le \frac{1}{4}$$

$$\int_{\mathbb{R}} \varphi(x)\, dx = 1.$$

The sequence

$$\delta_n(x) = n\varphi(nx), \quad n \in \mathbb{N}$$

is a delta sequence, and furthermore,

$$(\delta_n(x))^2 = n^2 \quad \text{for} \quad x \in I_n = \left\{ x : -\frac{1}{4n} \le x \le \frac{1}{4n} \right\}.$$

Hence,

$$(\delta_n^2, \varphi) = \int_{\mathbb{R}} \delta_n(x)\varphi(x) \, dx \ge \int_{I_n} \delta_n^2(x) \, dx = \int_{I_n} n^2 \, dx = \frac{n}{2} \to \infty.$$

This means that the sequence $(\delta_n^2)_{n \in \mathbb{N}}$ does not converge, i.e., that the square δ^2 does not exist.   □

The Product $x \cdot \frac{1}{x}$

The distribution $\frac{1}{x}$ is defined as the distributional derivative of $\ln |x|$:

$$\frac{1}{x} = (\ln |x|)'. \tag{$*$}$$

Problem 3.7.18 Prove that the product $x \cdot \frac{1}{x}$ exists and

$$x \cdot \frac{1}{x} = 1. \tag{3.7.53}$$

Proof. Note that the function $x \cdot \ln |x|$ is a primitive function, in the ordinary sense, of

$$1 + \ln |x|. \tag{3.7.54}$$

This function is a locally integrable function; thus it is equal to the distributional derivative of $x \cdot \ln |x|$, i.e.,

$$(x \cdot \ln |x|)' = 1 + \ln |x|. \tag{3.7.55}$$

Since x is a smooth function the product $x \cdot \ln |x|$ can be regarded as a regular operation. Differentiating, we get

$$(x \cdot \ln |x|)' = x(\ln |x|)' + \ln |x|.$$

Hence, by $(*)$ we obtain

$$(x \cdot \ln |x|)' = x \cdot \frac{1}{x} + \ln |x|. \tag{3.7.56}$$

According to (3.7.56) and (3.7.55), we obtain the required equation, i.e.,

$$x \cdot \frac{1}{x} = 1.$$

□

The Product $1_+ \cdot \delta$

Problem 3.7.19 Prove that the product $1_+ \cdot \delta$ exists and

$$1_+ \cdot \delta = \frac{1}{2}\delta. \tag{3.7.57}$$

Proof. Let $(\delta_n)_{n \in \mathbb{N}}$ be any delta sequence and let $H_n = 1_+ * \delta_n, n \in \mathbb{N}$. At that time

$$H_n' = (1_+ * \delta_n)' = 1_+' * \delta_n = \delta * \delta_n = \delta_n$$

and

$$1_+ \stackrel{\mathrm{d}}{=} \lim_{n \to \infty} (1_+ * \delta_n)^2.$$

Note that

$$(H_n^2)' = 2H_n H_n' = 2H_n \delta_n.$$

From here

$$\lim_{n \to \infty} H_n \delta_n \stackrel{\mathrm{d}}{=} \lim_{n \to \infty} \frac{1}{2}(H_n^2)' \stackrel{\mathrm{d}}{=} \frac{1}{2}\delta.$$

The product $1_+ \cdot \delta$ exists and the equality (3.7.57) holds. ▢

Problem 3.7.20

$$\text{Prove that} \quad \left(\frac{1}{x} \cdot x\right)\delta \neq \frac{1}{x}(x \cdot \delta).$$

Proof. According to Problem 3.7.18 and (3.4.3), section 3.4.1, it suffices to see that the equalities hold:

$$\left(\frac{1}{x} \cdot x\right)\delta = 1 \cdot \delta = \delta,$$
$$\frac{1}{x}\left(x \cdot \delta\right) = \frac{1}{x} \cdot 0 = 0.$$

▢

3.8   Hilbert Transform and Multiplication Forms

3.8.1   Definition of the Hilbert Transform

Let us define the Hilbert transform of the function $\varphi \in \mathcal{D}$.

Definition 3.8.92 *By the Hilbert transform $\mathcal{H}$ of $\varphi \in \mathcal{D}$ we mean the limit*

$$\mathcal{H}(x) :\overset{d}{=} \lim_{n\to\infty} \int_{A_n} \varphi(x-t)\frac{dt}{t}, \tag{3.8.1}$$

where $A_n = \{x \in \mathbb{R}: \quad |x| \geq \frac{1}{n}\}$.

Lemma 3.8.51 *For every function $\varphi \in \mathcal{D}$ there exists the limit (3.8.1); moreover, the equality*

$$\mathcal{H}(x) = \varphi(x) * \frac{1}{x} \tag{3.8.2}$$

holds.

Proof. Let $\varphi \in \mathcal{D}$. Let us consider the integral

$$\int_{A_n} \varphi(x-t)\frac{dt}{t},$$

where $A_n = \{x \in \mathbb{R}: \quad |x| \geq \frac{1}{n}\}$.

Integrating by parts, we obtain

$$\int_{A_n} \varphi(x-t)\frac{dt}{t} = \ln n\left[\varphi(x-\frac{1}{n}) - \varphi(x+\frac{1}{n})\right] + \int_{A_n} \varphi'(x-t)\ln|t|dt. \tag{3.8.3}$$

Note that, by Lagrange's theorem, we have the equality

$$\varphi(x+\frac{1}{n}) - \varphi(x-\frac{1}{n}) = \frac{2}{n}\varphi'(\xi_n), \quad \text{where} \quad |\xi_n| < \frac{1}{n}.$$

Hence, by (3.8.3), we get

$$\mathcal{H}(x) = \lim_{n\to\infty} \int_{A_n} \varphi(x-t)\frac{dt}{t} = \varphi'(x) * \ln|x|. \tag{3.8.4}$$

We now show that $\mathcal{H}$ is a square integrable function. In fact, there is a number x_o such that

$$\varphi(x) = 0 \quad \text{for} \quad |x| > x_o.$$

This implies that for $|x| > x_o$ we can write

$$\left|\int_{A_n} \varphi(x-t)\frac{dt}{t}\right| \leq \int_{-\infty}^{+\infty} \left|\frac{\varphi(x-t)}{t}\right|dt = \int_{-\infty}^{+\infty} \left|\frac{\varphi(x)}{x-t}\right|dt \leq \frac{M}{|x|-x_o}, \quad \text{with} \quad M = \int_{\mathbb{R}} |\varphi|.$$

Thus,

$$\left|\mathcal{H}(x)\right| \leq \frac{M}{|x|-x_o} \quad |x| > x_o. \tag{3.8.5}$$

Since $\mathcal{H} \in C^\infty$, therefore, $\mathcal{H} \in L^2(\mathbb{R})$.

We recall that the distribution $\frac{1}{x}$ is defined as the distributional derivative of $\ln|x|$ (see section 3.7.5), hus the formula (4) can be rewritten as

$$\mathcal{H}(x) = \varphi(x) * \left(\ln|x|\right)' = \varphi(x) * \frac{1}{x}, \tag{3.8.6}$$

which completes the proof. $\square$

Remark 3.8.130 *By the formula (3.8.6) we can alternatively define the Hilbert transform of the function $\varphi \in \mathcal{D}$ as the convolution:*

$$\mathcal{H} = \varphi * \frac{1}{x}.$$

This definition suggests a generalization of the Hilbert transform onto any distribution f for which the convolution

$$\mathcal{H} = f * \frac{1}{x}$$

exists. For instance, the Hilbert transform of Dirac's delta distribution exists and the equality

$$\mathcal{H} = \delta * \frac{1}{x} = \frac{1}{x}$$

holds.

3.8.2 Applications and Examples

Example 3.8.93 *Nonexistence of $\left(\frac{1}{x}\right)^2$:*

By $\left(\frac{1}{x}\right)^2$ we mean the product $\left(\frac{1}{x}\right)^2 = \frac{1}{x} \cdot \frac{1}{x}$.

Lemma 3.8.52 *The product $\frac{1}{x} \cdot \frac{1}{x}$ does not exist in the distributional sense.*

Proof. If $\frac{1}{x} \cdot \frac{1}{x}$ existed, then, by the definition of product of distributions (see section 3.7.4), for any delta sequence $(\delta_n)_{n \in \mathbb{N}}$ the sequence

$$\left(\left(\delta_n * \frac{1}{x}\right)^2\right)_{n \in \mathbb{N}}$$

would distributionally converge and

$$\frac{1}{x} \cdot \frac{1}{x} = \lim_{n \to \infty} \left(\delta_n * \frac{1}{x}\right)^2. \tag{3.8.7}$$

Taking a special delta sequence we show that this limit does not exist. Let $\psi \in \mathcal{D}$ and $\psi(x) \geq 0$ for $x \in \mathbb{R}$. Thus, there is a number x_o such that

$$\psi(x) = 0 \quad \text{for} \quad |x| > x_o.$$

This implies that for every x satisfying $|x| > x_o$ there is an index n_o such that

$$\left| \int_{A_n} \psi(x-t) \frac{dt}{t} \right|' = \int_{-\infty}^{+\infty} \frac{\psi(x-t)}{|x|} dt = \int_{-\infty}^{+\infty} \frac{\psi(t)}{|x-t|} dt \geq \frac{\int \psi(t) dt}{|x_o| + |x|} \quad \text{for} \quad n > n_o.$$

Hence,

$$\left| \psi(x) * \frac{1}{x} \right| \geq \frac{\int \psi(t) dt}{|x_o| + |x|} \quad \text{for} \quad |x| > x_o.$$

Let φ be a nonnegative function of class $\mathcal{D}$ such that

$$\varphi(x) : \begin{cases} \geq \frac{1}{2} & \text{for } |x| \leq \frac{1}{2}, \\ = 0 & \text{for } |x| \geq 1 \end{cases}$$

and $\int \varphi = 1$. Then the sequence

$$\Big(\delta_n(x) \Big)_{n \in \mathbb{N}} = \Big(n\varphi(nx) \Big)_{n \in \mathbb{N}}$$

is a delta sequence and we have

$$\left| \delta_n(x) * \frac{1}{x} \right| \geq \frac{1}{\frac{1}{n} * |x|} \quad \text{for} \quad |x| > \frac{1}{n}.$$

Hence,

$$\left(\left(\delta_n * \frac{1}{x} \right)^2, \varphi \right) \geq \frac{1}{2} \int_{\frac{1}{n}}^{\frac{1}{2}} \frac{dx}{\left(\frac{1}{n} * |x| \right)^2} = \frac{n(n-2)}{4(n+2)} \longrightarrow \infty.$$

This shows that the product (3.8.7) does not exist. $\qquad\Box$

Remark 3.8.131 *It should be noted that by definition*

$$\frac{1}{x^2} = \left(-\frac{1}{x} \right)' = \left(-\ln|x| \right)^{(2)}.$$

The symbol $\frac{1}{x^2}$ represents a distribution and should not be confused, in the theory of distribution, with the square $\left(\frac{1}{x} \right)^2$.

Example 3.8.94 *The formulas of Gonzalez-Dominguez and Scarfiello*

The formulas of Gonzalez-Dominguez and Scarfiello are of the form

$$\left(\varphi * \frac{1}{x}, \varphi \right) = 0 \quad \varphi \in \mathcal{D}, \tag{3.8.8}$$

$$\left(\varphi * \frac{1}{x}, x\varphi \right) = \frac{1}{2} \left(\int \varphi \right)^2, \quad \varphi \in \mathcal{D}. \tag{3.8.9}$$

We will prove the formula (3.8.10). It is easy to see that for $t \neq 0$, we have the identity

$$x \frac{\varphi(x-t)}{t} - \frac{(x-t)\varphi(x-t)}{t} = \varphi(x-t). \tag{3.8.10}$$

Let $\varphi \in \mathcal{D}$ and $A_n = \{x \in \mathbb{R} : \ |x| \geq \frac{1}{n}\}$. Then, by (3.8.10) it follows that

$$\int_{A_n} x\frac{\varphi(x-t)}{t}dt - \int_{A_n} \frac{(x-t)\varphi(x-t)}{t}dt = \int_{A_n} \varphi(x-t)dt.$$

Hence, by (3.8.1), (3.8.6) in section 3.8.1, as $n \to \infty$, we have

$$x\left(\varphi * \frac{1}{x}\right) - (\tau\varphi) * \frac{1}{x} = \int \varphi. \tag{3.8.11}$$

Note that from the properties of the inner product the following equality can be derived:

$$\left(x\left(\varphi * \frac{1}{x}\right), \varphi\right) = \left(\varphi * \frac{1}{x}, x\varphi\right),$$
$$\left((x\varphi) * \frac{1}{x}, \varphi\right) = \left(x\varphi, -\frac{1}{x} * \varphi\right) = -\left(\varphi * \frac{1}{x}, x\varphi\right). \tag{3.8.12}$$

By (3.8.11) for the inner product with any function $\tau \in \mathcal{D}$ in particular for $\varphi \in \mathcal{D}$, we have

$$\left(x\left(\varphi * \frac{1}{x}\right) - (x\varphi) * \frac{1}{x}, \varphi\right) = \left(\int \varphi, \varphi\right), \quad \varphi \in \mathcal{D}. \tag{3.8.13}$$

By the property of the inner product and (3.8.12) the left side of (3.8.13) is of the form

$$\left(x\left(\varphi * \frac{1}{x}\right), \varphi\right) - \left((x\varphi) * \frac{1}{x}, \varphi\right) = 2\left(\varphi * \frac{1}{x}, x\varphi\right).$$

Thus,

$$\left(\varphi * \frac{1}{x}, x\varphi\right) = \frac{1}{2}\left(\int \varphi, \varphi\right) = \frac{1}{2}\left(\int \varphi\right)^2, \quad \varphi \in \mathcal{D}.$$

and the formula (3.8.9) of Gonzalez–Dominguez and Scarfiello is proved.

Example 3.8.95 *The product $\frac{1}{x} \cdot \delta$*

The product $\frac{1}{x} \cdot \delta(x)$ appears in physics and its result $-\frac{1}{2}\delta'(x)$, in the distributional sense of J. Mikusiński, fulfills the expectations of physicists.

According to the definition (3.7.42), section 3.7.4, the product $\frac{1}{x} \cdot \delta$, is given by the formula

$$\frac{1}{x} \cdot \delta(x) \overset{\mathrm{d}}{=} \lim_{n \to \infty}\left(\delta_n * \frac{1}{x}\right)\delta_n.$$

We shall show that the limit exists and is equal to $-\frac{1}{2}\delta'(x)$, i.e.,

$$\frac{1}{x} \cdot \delta = -\frac{1}{2}\delta'. \tag{3.8.14}$$

Let $f_n = \left(\delta_n * \frac{1}{x}\right)\delta_n, \quad n \in \mathbb{N}$ and

$$F_n(x) = \frac{1}{2}\int_{-\infty}^{x} (x-t)^2 f_n(t)dt, \quad n \in \mathbb{N};$$

then

$$F_n^{(3)} = f_n, \quad n \in \mathbb{N}.$$

We show that $F_n \rightrightarrows$ in $(-\infty, +\infty)$.

Since

$$f_n(x) = 0 \quad \text{for} \quad x \leq -\alpha_n, \quad n \in \mathbb{N}, \tag{3.8.15}$$

thus, $F_n(x) = 0$ for $x \leq -\alpha_n$. Since $f_n(x) = 0$ for $x \geq \alpha_n, n \in \mathbb{N}$, thus

$$F_n'(x) = \int_{-\infty}^{x} (x-t)f_n(t)dt = x \int_{-\infty}^{x} f_n(t)dt - \int_{-\infty}^{+\infty} t f_n(t)dt$$

$$= x\left(\delta_n * \frac{1}{x}, \delta_n\right) - \left(\delta_n * \frac{1}{x}, x\delta_n\right) = -\frac{1}{2}\left(\int \delta_n\right)^2 = -\frac{1}{2},$$

by (3.8.8), (3.8.9). Hence,

$$F_n(x) = F_n(\alpha_n) - \frac{x}{2} + \frac{\alpha_n}{2} \quad \text{for} \quad x \geq \alpha_n, \quad n \in \mathbb{N}. \tag{3.8.16}$$

Finally, if $|x| \leq \alpha_n$, we write

$$F_n(x) = \frac{1}{2} \int_{-\alpha_n}^{x} (x-t)^2 \left(\int_{-\alpha_n}^{\alpha_n} \delta_n(u) \frac{1}{t-u} du \right) \delta_n(t) dt$$

$$= \frac{1}{2} \int_{-\alpha_n}^{x} (x-t)^2 \left(\int_{-\alpha_n}^{\alpha_n} \delta^{(2)}(u) L(t-u) du \right) \delta_n(t) dt,$$

where

$$L(x) = \begin{cases} x\ln|x| - x, & \text{for } x \neq 0, \\ 0, & \text{for } x = 0. \end{cases}$$

Hence,

$$|F_n(x)| \leq \frac{1}{2} \int_{-\alpha_n}^{\alpha_n} (2\alpha_n)^2 \left(\int_{-\alpha_n}^{\alpha_n} |\delta_n^{(2)}(u) L(2\alpha_n)| du \right) |\delta_n(t)| dt$$

$$= 2L(2\alpha_n)\alpha_n^2 \int_{-\alpha_n}^{\alpha_n} |\delta_n^{(2)}(u)| du \int_{-\alpha_n}^{\alpha_n} |\delta_n(t)| dt$$

$$= 2L(2\alpha_n) M_2 M_o = \epsilon_n,$$

where for $M_i (i = 0, 2)$; see section 3.4.1, Definition 3.4.74.

Since the function L is continuous and $L(0) = 0$, it follows that for $|x| \leq \alpha_n$

$$|F_n(x)| \leq \epsilon_n \to 0, \quad \text{as} \quad n \to \infty. \tag{3.8.17}$$

By (3.8.15), (3.8.16), (3.8.17) it follows that the sequence $(F_n)_{n \in \mathbb{N}}$ converges uniformly in $(-\infty, +\infty)$ to the function F given by

$$F(x) = \begin{cases} 0, & \text{for } x < 0, \\ -\frac{x}{2}, & \text{for } x \geq 0. \end{cases}$$

Hence,

$$\lim_{n \to \infty} f_n(x) \overset{\mathrm{d}}{=} F^{(3)}(x) = -\frac{1}{2}\delta'(x),$$

which proves (3.8.14).

Example 3.8.96 Some Other Formulas with Dirac's Delta Distribution

As an interesting application we have the following formulas:

$$\frac{1}{x} * \frac{1}{x} = -\pi^2 \delta,$$

$$\left(\delta + \frac{1}{\pi i}\frac{1}{x}\right)^2 = -\frac{1}{\pi i}\delta^2 - \frac{1}{\pi^2}\frac{1}{x^2},$$

$$\delta^2 - \frac{1}{\pi^2}\left(\frac{1}{x}\right)^2 = -\frac{1}{\pi^2}\frac{1}{x^2}.$$

The proofs for existence of the left-hand sides of the above equalities and proofs of the above equalities can be obtained with application of the properties of the Fourier transform. We omit the proofs and refer interested readers to the book [AMS].

References

[A]   P. Antosik, *Poriadok otnositielno miery i jego primienienije k issledowaniju proizwie-dienija obobszczennych funkcji*, Studia Math., **26** (1966), 247–262.

[AKV] M. Ju. Antimirov, A.A. Kolyshkin, R. Vaillancourt, *Applied Integral Transforms*, AMS–CRM Monograph Series, vol. 2, Providence, Rhode Island 1993.

[AL] P. Antosik, J. Ligęza, *Product of measures and functions of finite variation*, Proceedings of the Conference of Generalized Functions and Operational Calculi, Varna, (1975).

[AMS] P. Antosik, J. Mikusiński, R. Sikorski, *Theory of Distributions. The Sequential Approach*, Elsevier-PWN, Amsterdam and Warsaw 1973.

[Be.1] L. Berg, *Einführung in die Operatorenrechnung*, VEB Deutscher Verlag der Wissenschaften, Berlin 1962.

[Be.2] L. Berg, *Operatorenrechnung I. Algebraische Methoden*, VEB Deutscher Verlag der Wissenschaften, Berlin 1972.

[Be.3] L. Berg, *Operatorenrechnung II. Funktionentheoretische Methoden*, VEB Deutscher Verlag der Wissenschaften, Berlin 1974.

[BL] G. Birkhoff, S. MacLane, *A Survey of Modern Algebra*, The MacMillan Company, New York 1958.

[BM] T.K. Boehme, J. Mikusiński, *Operational Calculus*, vol. 2, PWN and Pergamon Press, Warsaw and Oxford 1987.

[BoL] N.N. Bogolubov, A.A. Łogunov, I.T. Todorov, *Osnowy aksiomaticzeskogo podchoda k kwantowej teorii pola* (in Russian), "Nauka," Moscow 1969.

[BoP] N.N. Bogolubov, O.S. Parasiuk, *Über die Multiplikation der Quantentheorie der Felder*, Acta Math. **97** (1957), 227–266.

[Br] H.J. Bremerman, *Distributions, Complex Analysis and Fourier Transforms*, Addison-Wesley, Reading, Massachusetts 1965.

[BrD] N.J. Bremerman, L. Durand, *On analytic continuations and Fourier transformations of Schwartz distributions*, J. Math. Phys., **2** (1961), 240–258.

[BGPV] Yu. A. Brychkov, H.-J. Glaeske, A.P. Prudnikov, Vu Kim Tuan, *Multidimensional Integal Transforms*, Gordon & Breach Science Publishers, Philadelphia 1992.

[BP] Yu. A. Brychkov, A. P. Prudnikov, *Integral Transforms of Generalized Functions*, Gordon & Breach Science Publishers, New York 1989.

[BuN] P.L. Butzer, R.J. Nessel, *Fourier Analysis and Approximation I, One-Dimensional Theory*, Birkhäuser, Basel 1971.

[BuS] P.L. Butzer, R.L. Stens, *The Operational Properties of the Chebyshev Transform I. General Properties*, Functiones et Approximatio **5** (1977), 129–160.

[Ch.1] R.V. Churchill, *The operational calculus of Legendre transforms*, J. Math. and Physics **33** (1954), 165–178.

[Ch.2] R.V. Churchill, *Operational Mathematics, 3rd Edition*, McGraw-Hill Book Company, New York 1972.

[ChD] R.V. Churchill, C.L. Dolph, *Inverse transforms of products of Legendre transforms*, Proc. Amer. Math. Soc. **5** (1954), 93–100.

[Co] S.D. Conte, *Gegenbauer transforms*, Quart. J. Math. Oxford (2), **6** (1955), 48–52.

[Da] B. Davis, *Integral Transforms and Their Applications, 2nd Edition*, Springer-Verlag, New York 1985.

[De.1] L. Debnath, *On Laguerre transform*, Bull. Calcutta Math. Soc. **52** (1960), 69–77.

[De.2] L. Debnath, *On Jacobi transforms*, Bull. Calcutta Math. Soc. **55** (1963), 113–120.

[De.3] L. Debnath, *On Hermite transforms*, Math. Vesnik **1** (1964), 285–292.

[De.4] L. Debnath, *Some operational properties of the Hermite transform*, Math. Vesnik **5** (1968), 29–36.

[De.5] L. Debnath, *On the Faltung Theorem of Laguerre transforms*, Studia Univ. Babes-Bolyai. Ser. Phys. **2** (1969), 41–45.

[De.6] L. Debnath, *Integral Transforms and Their Applications*, CRC Press, Boca Raton 1995.

[DV] P. Dierolf, J. Voigt, *Convolution and S'-convolution of distributions*, Collect. Math., **29** (3) (1978), 185–196.

[Di] I.H. Dimovski, *Convolutional Calculus, 2nd Edition*, Kluwer Acad. Publ., Dordrecht, Boston and London 1990.

[DiK] I.H. Dimovski, S.L. Kalla, *Convolution for Hermite transforms*, Math. Japonica **33** (1988), 345–351.

[Dir] P. Dirak, *The physical interpretation of quantum dynamics*, Proc. Roy. Soc., Sect. A **113** (1926–1927), 621–641.

[DP] V.A. Ditkin, A.P. Prudnikov, *Integral Transforms and Operational Calculus*, Pergamon Press, Oxford 1965.

[Doe.1] G. Doetsch, *Handbuch der Laplace Transformation*, Verlag Birkhäuser, Basel and Stuttgart, 1050 (vol. 1), 1955 (vol. 2), 1956 (vol. 3).

[Doe.2] G. Doetsch, *Anleitung zum praktischen Gebrauch der Laplace-Transformation*, R. Oldenbourg, München 1956.

[Doe.3] G. Doetsch, *Einführung in Theorie und Anwendung der Laplace-Transformation*, *2nd Edition*, Basel and Stuttgart 1970.

[E.1] A. Erdélyi, *Higher Transcesdental Functions*, vol. 1 and vol. 2, McGraw–Hill Book Comp. New York 1953.

[E.2] A. Erdélyi, *Operational Calculus and Generalized Functions*, Holt, Rinehart and Winston, New York 1962.

[EMOT] A. Erdélyi, W. Magnus, F. Oberhettinger, F. Tricomi, *Tables of Integral Transforms*, vol. 1 and vol. 2, McGraw-Hill Book Company, New York 1954.

[F] G.M. Fichtenholz, *Differential and Integral Calculus* (in Polish), Vol. I, II, III, PWN Warsaw 1964.

[Fi] J.M. Firth, *Discrete Transforms*, Chapman & Hall, London 1992.

[Fö] V.A. Fock, *On the representation of an arbitrary function by integrals involving the Legendre function with a complex index*, Dokl. AN SSSR **39** (1943), 253–256.

[Fs.1] B. Fisher, *Products of generalized functions*, Studia Math., 33 **(2)** (1969), 227–230.

[Fs.2] B. Fisher, *The product of distributions*, Quart. J. Math., Oxford Ser., **2** (1971), 291–298.

[Fo] O. Föllinger, *Laplace und Fourier–Transformation*, AEG-Telefunken (Aktiengesellschaft), Berlin, Frankfurt/Main 1982.

[Ga.1] G. Gasper, *Positivity and the convolution structure for Jacobi series*, Ann. of Math. (2) **93** (1971), 122–135.

[Ga.2] G. Gasper, *Banach algebras for Jacobi series and positivity of a kernel*, Ann. of Math. (2) **95** (1972), 261–280.

[GS] I.M. Gel'fand, G.E. Shilov, *Generalized Functions*, vol. 1, Academic Press, New York and London 1964.

[GV]   I.M. Gel'fand, N.Ja. Vilenkin, *Generalized Functions*, vol. 4, Academic Press, New York and London 1964.

[Gl.1] H.-J. Glaeske, *Convolution structure of (generalized) Hermite transforms*, Banach Center Publ. **53** (2000), 113–120.

[Gl.2] H.-J. Glaeske, *On the convolution structure of Hermite transforms — A Survey*, Proc. Int. Conf. on Geometry, Analysis and Applications, World Scientific Publishers, Varanasi, India, Singapore (2000), 217–225.

[Gl.3] H.-J. Glaeske, *On a Hermite transform in spaces of generalized functions on R^n*, Integral Transforms Spec. Funct. **13** (2002), 309–319.

[GlS]   H.-J. Glaeske, M. Saigo, *On a hybrid Laguerre–Fourier transform*, Integral Transforms Spec. Funct. **10** (2000), 227–238.

[GlV]   H.-J. Glaeske, Vu Kim Tuan, *Some applications of the convolutions theorem of the Hilbert transform*, Int. Transforms Spec. Funct. **3** (1995), 263–268.

[GöM]   E. Görlich, C. Markett, *A convolution structure for Laguerre series*, Indag. Math. (N.S.) **44** (1982), 161–171.

[G-DS] A. Gonzales-Dominguez, R. Scarfiello, *Nota sobre la formula v.p. $\frac{1}{x}\delta = -\frac{1}{2}\delta'$*, Rev. de la Union Matem. Argen., **1** (1956), 53-67.

[Ha]   I. Halperin, *Introduction to the Theory of Distributions*, University of Toronto Press, 1952.

[H.1]   O. Heaviside, *Operators in mathematical physics*, Proc. Roy. Soc. A. **52** (1893), p. 504.

[H.2]   O. Heaviside, *Operators in mathematical physics*, Proc. Roy. Soc. A. **54** (1894), p. 105.

[H.3]   O. Heaviside, *Electromagnetic theory*, London 1899.

[Hi]   I. I. Hirschmann, *Variation diminishing Hankel transforms*, J. Anal. Math. **8** (1960), 307–336.

[HiO]   Y. Hirata, N. Ogata, *On the exchange formula for distributions*, J. Sci. Hiroshima Univ., Ser. A-1, **22** (1958), 147–152.

[Is]   S. Ishikova, *Products of Wiener functions on an abstract Wiener space*, in: *Generalized Functions, Convergence Structures and Their Applications*, Plenum Press, New York and London 1988, 179–185.

[It]   M. Itano, *On the theory of multiplicative products of distributions*, J. Sci. Hiroshima Univ., Ser. A-1, **30** (1966), 151–181.

[Je]   A.J. Jerri, *Integral and Discrete Transforms with Applications and Error Analysis*, Marcel Dekker, New York 1992.

[Jo] D.S. Jones, *The convolution of generalized functions*, Quart. J. Math. Oxford Ser. (2) **24** (1973), 145–163.

[K] E. Krätzel, *Bemerkungen zur Meijer–transformation und Anwendungen*, Math. Nachr. **30** (1965), 328–334.

[Ka.1] A. Kamiński, *On convolutions, products and Fourier transforms of distributions*, Bull. Acad. Pol. Sci., Ser. Sci. Math., **25** (1977), 369-374.

[Ka.2] A. Kamiński, *On the product $\frac{1}{x^k}\delta^{(k-1)}(x)$*, Bull. Acad. Pol. Sci., Ser. Sci. Math., **25** (1977), 375-379.

[Ka.3] A. Kamiński, *Remarks on delta- and unit-sequences*, Bull. Acad. Pol. Sci., Ser. Sci. Math., **26** (1978), 25-30.

[Ka.4] A. Kamiński, *Convolution, product and Fourier transform of distributions*, Studia Math. 74, **1** (1982), 83–96.

[KSA] A.A. Kilbas, M. Saigo, *H Transforms—Theory and applications*, Chapman & Hall/CRC Press, Boca Raton 2004.

[KS.1] W. Kierat, K. Skórnik, *On generalized functions approach to the Cauchy problem for heat equation*, Integral Transforms & Special Functions, **2** (1994), 107–116.

[KS.2] W. Kierat, K. Skórnik, *A remark on solutions of the Laguerre differential equation*, Dissertationes Math. **340** (1995), 137–141.

[KSk] W. Kierat, U. Skórnik, *A remark on the integral of integrable distributions*, Integral Transforms & Special Functions, **11** (2001), 189–195.

[KSz] W. Kierat, U. Sztaba, *Distributions, Integral Transforms and Applications*, Analytic Methods and Special Functions, vol. 7, Taylor & Francis, London and New York 2003.

[Koe] J. Koekoek, *On the Fourier integral theorem*, Nieuw Arch. Wisk. IV, Ser. **5** (1987), 83–85.

[Kö.1] H. König, *Neue Begründung der Theorie der Distributionen von L. Schwartz*, Math. Nachr., **9** (1953), 129–148.

[Kö.2] K. König, *Multiplikation von Distributionen*, Math. Ann. **128** (1955), 420–452.

[Kö.3] H. König, *Multiplikation und Variablentransformation in der Theorie der Distributionen*, Arch. Math., **6** (1955), 391–396.

[Koo] T. Koornwinder, *The addition formula for Jacobi plynomials, I*, Summary of results, Indag. Math. **34** (1972), 181–191.

[Ko.1] J. Korevaar, *Distributions Defined from the Point of View of Applied Mathematics*, Proceedings Kon. Neder. Akad. Wetenschappen, Series A. Nr 2 and Indag. Math. **17.2**

(1955), 368–383; Preceedings Kon. Neder. Akad. Wetenschappen, Series A, **58.5** and **17.4** (1955), 463–503; Preceedings Kon. Neder. Akad. Wetenschappen, Series A, **58.5** and **17.5** (1955), 563–674.

[Ko.2]   J. Korevaar, *On the Fourier Integral Theorem*, Nieuw Arch. Wiskcd IV. Sci. **5** (1987), 83–85.

[La]   S.K. Lakshmanarao, *Gegenbauer Transforms*, Math. Student **22** (1954), 157–165.

[Le]   N.N. Lebedev, *Special Functions and Their Applications*, Prentice-Hall, Englewood Cliffs, New Jersey 1965.

[Li]   M. J. Lighthill, *An introduction to Fourier Analysis and Generalized Functions*, Cambridge University Press, New York 1958.

[Lo.1]   S. Łojasiewicz, *Sur la valeur d'une distribution dans un point*, Bull. Pol. Acad. Sci. Cl. III, **4** (1956), 239–242.

[Lo.2]   S. Łojasiewicz, *Sur la valeur et le limite distribution dans un point*, Studia Math., **16** (1957), 1–36.

[LWZ]   S. Łojasiewicz, J. Wloka, Z. Zieleźny, *Über eine Definition des Wertes einer Distribution*, Bull. Acad. Pol. Sci. Cl III, **3** (1955), 479–481.

[Mh]   H. Macinkowska, *Distributions, Soboleff Spaces, Differential Equations* (in Polish), Wydawnictwo Naukowe PWN, Wrocław 1993.

[M]   O.I. Marichev, *Handbook of Integral Transforms of Higher Transcendental Functions, Theory and Algorithmic Tables*, Ellis Horwood, Chichester 1982.

[Ma.1]   C. Markett, *Mean Cesáro summability of Laguerre expansions and norm estimates with shifted parameter*, Anal. Math. **8** (1982), 19–37.

[Ma.2]   C. Markett, *The Product Formula and Convolution Structure Associated with the Generalized Hermite Polynomials*, J. Approx. Theory **73** (1993), 199–217.

[MC]   J. McCully, *The Laguerre Transform*, SIAM Rev. **2** (1960), 185–191.

[Me]   E. Meister, *Integraltransformationen mit Anwendungen auf Probleme der Mathematischen Physik*, Verlag Peter Lang, Frankfurt/Main–Bern–New York 1983.

[Meh]   F.G. Mehler, *Über eine mit den Kugel- und Cylinderfunktionen verwandte Funktion und ihre Anwendung in der Theorie der Elektricitätsvertheilung*, Math. Ann. **18** (1881), 161–194.

[Mi.1]   J. Mikusiński, *Sur la méthode de généralisation de M. Laurent Schwartz et sur la convergence faible*, Fund. Math., **35** (1948), 235–239.

[Mi.2] J. Mikusiński, *Une definition de distribution*, Bull. Acad. Pol. Sci. Cl. III **3** (1955), 589–591.

[Mi.3] J. Mikusiński, *On the value of a distribution at a point*, Bull. Acad. Pol. Sci., Ser. Sci. Math., **8** (10) (1960), 681–683.

[Mi.4] J. Mikusiński, *Irregular operations on distributions*, Studia Math., **20** (1961), 163–169.

[Mi.5] J. Mikusiński, *Criteria of the existence and of the associativity of the product of distributions*, Studia Math., **21** (1962), 253–259.

[Mi.6] J. Mikusiński, *The Bochner Integral*, Birkhäuser, Basel and Stuttgart 1978.

[Mi.7] J. Mikusiński, *Operational Calculus*, vol. I, PWN and Pergamon Press, Warsaw and Oxford 1983.

[Mi.8] J. Mikusiński, *On the square of the Dirac delta-distribution*, Bull. Acad. Pol. Sci., Ser Sci. Math. 14(**9**) (1969), 511–513.

[Mi.9] J. Mikusiński, *Sequential theory of the convolutions of distributions*, Studia Math., (**29**) (1968), 151–160.

[MiS.1] J. Mikusiński, R. Sikorski, *The elementary theory of distributions. I*, Rozprawy Mat., **12** (1957).

[MiS.2] J. Mikusiński, R. Sikorski, *The elementary theory of distributions. II*, Rozprawy Mat., **25** (1961).

[MiR] J. Mikusiński, C. Ryll–Nardzewski, *Un théorèms sur le produit de composition des functions de plusieurs variables*, Studia Math. **13** (1953), 62–68.

[NU] A.F. Nikiforov, V.B. Uvarov, *Special Functions of Mathematical Physics*, Birkhäuser, Basel and Boston 1988.

[O.1] F. Oberhettinger, *Tabellen zur Fourier–Transformation*, Springer-Verlag, Berlin–Göttingen–Heidelberg 1957.

[O.2] F. Oberhettinger, *Tables of Bessel Transforms*, Springer-Verlag, New York 1972.

[O.3] F. Oberhettinger, *Tables of Mellin Transforms*, Springer-Verlag, Berlin 1974.

[OB] F. Oberhettinger, L. Baldii, *Tables of Laplace Transforms*, Springer-Verlag, Berlin 1973.

[Ob] E.I. Obolashvili, *The Fourier Transform and Its Application in the Theory of Elasticity* (in Russian), Metsniereba, Tbilissi 1979.

[OH] F. Oberhettinger, T.P. Higgins, *Tables of Lebedev, Mehler and generalized Mehler transforms*, Boeing Sci. Res. Lab. Math. Note **246**, Seattle 1961.

[Pa] A. Papoulis, *The Fourier Integral and Its Applications*, McGraw-Hill Book Company 1963.

[PB] B. van der Pool, H. Bremmer, *Operational Calculus Based on the Two-Sided Laplace Integral*, Springer, Berlin–Heidelberg–New York 1955.

[PBM] A.P. Prudnikov, Yu. A. Brychkov, O.J. Marichev, *Integral and Series*, Gordon & Breach Science Publishers, New York, vol. I, *Elementary Functions*, 1986; vol. II, *Special Functions*, 1986; vol. III, *More Special Functions*, 1989; vol. IV, *Direct Laplace Transforms*, 1992; vol. V, *Inverse Laplace Transforms*, 1992.

[Po] H. Pollard, *The mean convergence of orthogonal series I: Summary of results*, Trans. Amer. Math. Soc. **62** (1947), 387–403.

[R] Th.J. Rivlin, *The Chebyshev Polynomials, 2nd Edition*, John Wiley & Sons, New York 1990.

[Sa] G. Sansone, *Orthogonal Functions*, Dover Publications Inc., New York 1991.

[S.1] L. Schwartz, *Génééralisation de la notion de fonction, de dérivation, de transformation de Fourier, et applications mathématiques et physiques*, Annales Univ. Grenoble **21** (1945), 57–74.

[S.2] L. Schwartz, *Théorie des distributions*, Hermann, Paris, vol. I, 1950; vol. II, 1951.

[Sc] E.J. Scott, *Jacobi tranforms*, Quart. J. Math. Oxford **4** (1953), 36–40.

[Se] J. Sebastiao e Silva, *Sur une construction axiomatique de la théorie des distributions*, Univ. Lisboa, Revista Fac. Ci. (2), **4** (1955), 79–186.

[Sh] R. Shiraishi, *On the definition of convolutions for distributions*, J. Sci. Hiroshima Univ., Ser A, **23** (1959), 19–32.

[ShI] R. Shiraishi, M. Itano, *On the multiplicative products of distributions*, J. Sci. Hiroshima Univ., Ser. A-I, 28 **(3)** (1964), 223–236.

[Si.1] R. Sikorski, *A definition of the notion of distribution*, Bull. Acad. Pol. Sci. Cl.III, **2** (1954), 207–211.

[Si.2] R. Sikorski, *On substitution of the Dirac delta-distribution*, Bull. Acad. Pol. Sci. (1960), 685–689.

[Si.3] R. Sikorski, *Integrals of distributions*, Studia Math. **20** (1961), 119–139.

[Sk.1] K. Skórnik, *An estimation of Fourier coefficients of periodic distributions*, Bull. Acad. Pol. Sci., Ser. Sci. Math., **16** (7)(1968), 581–585.

[Za]   A.I. Zayed, *Handbook of Function and Generalized Function Transformations*, CRC Press, Boca Raton 1996.

[Ze.1] A.H. Zemanian, *Distribution Theory and Transform Analysis*, Dover Publications, New York 1987.

[Ze.2] A.H. Zemanian, *Generalized Integral Transformations*, Dover Publications, New York 1987.

[Zi]   Z. Zieleźny, *Súr la définition de Łojasiewicz de la valeúr d'unc distribution dans un point*, Bull. Pol. Acad. Sci. Cl. III, **3** (1955), 519–520.

Index

Milton Keynes UK
Ingram Content Group UK Ltd.
UKHW052022071024
449327UK00027B/2380

9 780367 390495